# GENERAL MOTORS FULL-SIZE TRUCKS 2007-13 REPAIR MANUAL

## CHILTON'S

Covers U.S. and Canadian models of Chevrolet Silverado, GMC Sierra and Sierra Denali Pick-ups, Chevrolet Suburban and Tahoe, GMC Yukon, Yukon XL, Yukon Denali and Chevrolet Avalanche
Two and four-wheel drive versions

*Does not include 2007 Silverado Classic, Sierra Classic or Sierra Denali Classic, or information specific to diesel engine models, CNG models, hybrids or models equipped with rear-wheel steering*

### by Mike Stubblefield

**CHILTON** Automotive Books
PUBLISHED BY HAYNES NORTH AMERICA, Inc.

Manufactured in USA
©2009, 2012, 2014 Haynes North America, Inc.
ISBN-13: 978-1-62092-127-2
ISBN-10: 1-62092-127-8
Library of Congress Control Number: 2014947613

**Haynes Publishing Group**
Sparkford Nr Yeovil
Somerset BA22 7JJ England

ABCDE
FGHIJ
K

**Haynes North America, Inc**
861 Lawrence Drive
Newbury Park
California 91320 USA

8S3

Chilton is a registered trademark of W.G. Nichols, Inc., and has been licensed to Haynes North America, Inc.

# Contents

**INTRODUCTORY PAGES**

About this manual – 0-5
Introduction – 0-5
Vehicle identification numbers – 0-6
Recall information - 0-7
Buying parts – 0-10
Maintenance techniques, tools and working facilities – 0-11
Jacking and towing – 0-19
Booster battery (jump) starting – 0-20
Automotive chemicals and lubricants – 0-21
Conversion factors – 0-22
Fraction/decimal/millimeter equivalents – 0-23
Safety first! – 0-24
Troubleshooting – 0-25

**1** TUNE-UP AND ROUTINE MAINTENANCE – 1-1

**2** V6 ENGINE – 2A-1
V8 ENGINE – 2B-1
GENERAL ENGINE OVERHAUL PROCEDURES – 2C-1

**3** COOLING, HEATING AND AIR CONDITIONING SYSTEMS – 3-1

**4** FUEL AND EXHAUST SYSTEMS – 4-1

**5** ENGINE ELECTRICAL SYSTEMS – 5-1

**6** EMISSIONS AND ENGINE CONTROL SYSTEMS – 6-1

| | |
|---|---|
| AUTOMATIC TRANSMISSION– 7A-1<br>TRANSFER CASE – 7B-1 | **7** |
| DRIVELINE – 8-1 | **8** |
| BRAKES – 9-1 | **9** |
| SUSPENSION AND STEERING SYSTEMS – 10-1 | **10** |
| BODY – 11-1 | **11** |
| CHASSIS ELECTRICAL SYSTEM – 12-1<br>WIRING DIAGRAMS – 12-27 | **12** |
| GLOSSARY – GL-1 | **GLOSSARY** |
| MASTER INDEX – IND-1 | **MASTER INDEX** |

**Mechanic and photographer with 2007 Chevrolet Silverado pick-up**

## ACKNOWLEDGEMENTS

Technical writers who contributed to this project include Joe Hamilton, Robert Maddox, Tim Imhoff and Jamie Sarte. Wiring diagrams originated exclusively for Haynes North America, Inc. by Valley Forge Technical Information Services.

All rights reserved. No part of this book may be reproduced or transmitted in any form or by any means, electronic or mechanical, including photocopying, recording or by any information storage or retrieval system, without permission in writing from the copyright holder.

While every attempt is made to ensure that the information in this manual is correct, no liability can be accepted by the authors or publishers for loss, damage or injury caused by any errors in, or omissions from, the information given.

# INTRODUCTION 0-5

## About this manual

### ITS PURPOSE

The purpose of this manual is to help you get the best value from your vehicle. It can do so in several ways. It can help you decide what work must be done, even if you choose to have it done by a dealer service department or a repair shop; it provides information and procedures for routine maintenance and servicing; and it offers diagnostic and repair procedures to follow when trouble occurs.

We hope you use the manual to tackle the work yourself. For many simpler jobs, doing it yourself may be quicker than arranging an appointment to get the vehicle into a shop and making the trips to leave it and pick it up. More importantly, a lot of money can be saved by avoiding the expense the shop must pass on to you to cover its labor and overhead costs. An added benefit is the sense of satisfaction and accomplishment that you feel after doing the job yourself.

### USING THE MANUAL

The manual is divided into Chapters. Each Chapter is divided into numbered Sections, which are headed in bold type between horizontal lines. Each Section consists of consecutively numbered paragraphs.

At the beginning of each numbered Section you will be referred to any illustrations which apply to the procedures in that Section. The reference numbers used in illustration captions pinpoint the pertinent Section and the Step within that Section. That is, illustration 3.2 means the illustration refers to Section 3 and Step (or paragraph) 2 within that Section.

Procedures, once described in the text, are not normally repeated. When it's necessary to refer to another Chapter, the reference will be given as Chapter and Section number. Cross references given without use of the word "Chapter" apply to Sections and/or paragraphs in the same Chapter. For example, "see Section 8" means in the same Chapter.

References to the left or right side of the vehicle assume you are sitting in the driver's seat, facing forward.

Even though we have prepared this manual with extreme care, neither the publisher nor the author can accept responsibility for any errors in, or omissions from, the information given.

### ➡ NOTE
A *Note* provides information necessary to properly complete a procedure or information which will make the procedure easier to understand.

### ✱✱ CAUTION
A *Caution* provides a special procedure or special steps which must be taken while completing the procedure where the Caution is found. Not heeding a Caution can result in damage to the assembly being worked on.

### ✱✱ WARNING
A *Warning* provides a special procedure or special steps which must be taken while completing the procedure where the Warning is found. Not heeding a Warning can result in personal injury.

## Introduction

The chassis layout on the models covered by this manual is conventional with the engine mounted at the front and the power being transmitted through either a four-speed or six-speed automatic transmission or driveshaft to the rear axle. On 4WD models, a transfer case transmits the power to the front differential and then to the front wheels through independent driveaxles.

Engines are either V6 or V8, equipped with electronic, sequential multi-port fuel injection.

These models feature independent front suspension with coil-over shock absorbers (1500 models) or torsion bars and shock absorbers (2500 and 3500 models). At the rear, all models have a solid rear axle supported by leaf springs and shock absorbers on pick-ups, and coil springs, upper and lower control arms and shock absorbers on SUV models.

The power-assisted steering is either rack-and-pinion or conventional recirculating ball type. The rack-and-pinion steering unit is mounted in front of the engine and the conventional steering box is located on the left-side frame rail.

Most models have a power assisted disc-type front and rear brake system with an Anti-lock Brake System (ABS) as standard equipment. Some models are equipped with rear drum brakes.

## Vehicle identification numbers

### VEHICLE IDENTIFICATION NUMBERS

Modifications are a continuing and unpublicized process in vehicle manufacturing. Since spare parts manuals and lists are compiled on a numerical basis, the individual vehicle numbers are essential to correctly identify the component required.

### VEHICLE IDENTIFICATION NUMBER (VIN)

The Vehicle Identification Number (VIN), which appears on the Vehicle Certificate of Title and Registration, is also embossed on a gray plate located on the upper left (driver's side) corner of the dashboard, near the windshield (see illustration). The VIN tells you when and where a vehicle was manufactured, its country of origin, make, type, passenger safety system, line, series, body style, engine and assembly plant.

### VIN ENGINE AND MODEL YEAR CODES

Two particularly important pieces of information found in the VIN are the engine code and the model year code. Counting from the left, the engine code letter designation is the 8th character and the model year code is the 10th character.

**On the model years covered by this manual the engine codes are:**

| | |
|---|---|
| 0 | 5.3L V8 (LMG) |
| 3 | 5.3L V8 (LC9) |
| 8 | 6.2L V8 (L92) |
| A | 4.8L V8 (L20) |
| C | 4.8L V8 (LY2) |
| G | 6.0L V8 (L96) |
| J | 5.3L V8 (LY5) |
| K | 6.0L V8 (LY6) |
| M | 5.3L V8 (LH6) |
| X | 4.3L V6 (LU3) |
| Y | 6.0L V8 (L76) |

**On the models covered by this manual the model year codes are:**

| | |
|---|---|
| 7 | 2007 |
| 8 | 2008 |
| 9 | 2009 |
| A | 2010 |
| B | 2011 |
| C | 2012 |
| D | 2013 |

### VEHICLE SAFETY CERTIFICATION LABEL

The Vehicle Safety Certification label is attached to the rear edge of the driver's door (see illustration). The label contains the name of the manufacturer, the month and year of production, the Gross Vehicle Weight Rating (GVWR), the Gross Axle Weight Rating (GAWR) and the certification statement. On most models, the label also includes the OEM tire sizes and pressures.

### ENGINE IDENTIFICATION NUMBER (EIN)

The Engine Identification Number (EIN) is stamped into the rear of the engine block just below the left cylinder head.

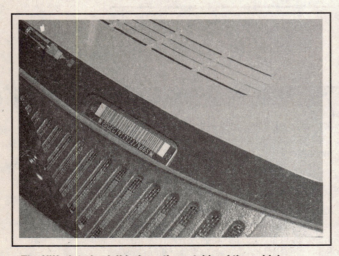

**The VIN plate is visible from the outside of the vehicle, through the driver's side of the windshield**

**The Vehicle Safety Certification label is affixed to the driver's side door end or post**

## Recall information

Vehicle recalls are carried out by the manufacturer in the rare event of a possible safety-related defect. The vehicle's registered owner is contacted at the address on file at the Department of Motor Vehicles and given the details of the recall. Remedial work is carried out free of charge at a dealer service department.

If you are the new owner of a used vehicle which was subject to a recall and you want to be sure that the work has been carried out, it's best to contact a dealer service department and ask about your individual vehicle - you'll need to furnish them your Vehicle Identification Number (VIN).

The table below is based on information provided by the National Highway Traffic Safety Administration (NHTSA), the body which oversees vehicle recalls in the United States. The recall database is updated constantly. For the latest information on vehicle recalls, check the NHTSA website at www.nhtsa.gov, www.safercar.gov, or call the NHTSA hotline at 1-888-327-4236.

| Recall date | Recall campaign number | Model(s) affected | Concern |
|---|---|---|---|
| Aug 09, 2006 | 06V307000 | 2007 Avalanche 1500 | Certain models fail to comply with the requirements of federal motor vehicle safety standard no. 110, "Tire selection and rims." These vehicles have an incomplete tire rim designation on the certification/tire label located on the driver's door edge. In addition, the label may also be missing the tire load rating on the label. Although this information is not required by the standard, if a tire of a lesser load rating is installed, the tire may not be able to sustain the loads encountered during use. If a customer replaces a wheel and only relies on the rim size designation that is indicated on the label, a wheel that is of a different rim contour designation may be installed. It may be difficult or impossible to mount the tire on a wheel with the wrong contour. If the tire is mounted on the wheel with the wrong contour, the wheel and tire may not perform as intended, which could increase the risk of a crash. |
| May 23, 2007 | 07V228000 | 2007 Chevrolet 2500HD and 3500HD | On certain vehicles equipped with series 0307, HR0307A, HR0307B, HR1000A or HR1500HB HY-RAIL guide wheel equipment, these steering stops may not be correctly installed. Damage occurs to the vehicle components around the rims on the front of the vehicle. The anti-lock brake sensor wires could be cut. |
| Dec 13, 2007 | 07E106000 | 2007 GMC Sierra 1500 | Certain Federal-Mogul replacement wheel hub assemblies with the brand names: National, Carquest p/nos. 515020, 515021, 515025, 515053, 515054, 515059, and 515060, shipped between January 23, 2006, and December 20, 2007, sold for light duty and medium duty trucks. The inboard retention nut used to maintain the hub bearing assembly can loosen, resulting in an ABS light indication, noise, and/or wheel separation. Wheel separation can result in a vehicle crash. |

| Recall date | Recall campaign number | Model(s) affected | Concern |
|---|---|---|---|
| Jul 18, 2008 | 08V344000 | 2007 and 2008 Chevrolet 3500 | Knapheide is recalling 108 2006-2008 Pro Series Truck Beds equipped with an ICC bumper which included a provision for a ball hitch. There may be an insufficient weld attaching the hitch channel to the supporting structure. This could result in bending of the hitch channel or the hitch could be pulled off the vehicle, increasing the risk of a crash. |
| Aug 04, 2008 | 08E050000 | 2008 Chevrolet Silverado | K2 Motor is recalling 1,921 aftermarket headlamps of various models sold for use on the above listed vehicles. These headlamps are missing the amber side reflex reflector which fails to conform with the requirements of Federal Motor Vehicle Safety Standard no. 108, "Lamps, Reflective Devices, and Associated Equipment." Without the amber side reflex reflectors, the lighting visibility may be affected, possibly resulting in a vehicle crash. |
| Aug 27, 2008 | 08V441000 | 2007 and 2008 Chevrolet Avalanche, Suburban, Tahoe, GMC Sierra, Yukon, Yukon XL | GM is recalling vehicles equipped with a heated wiper washer fluid system. A short circuit on the printed circuit board for the washer fluid heater may overheat the control-circuit ground wire. This may cause other electrical features to malfunction, create an odor, or cause smoke increasing the risk of a fire. |
| May 06, 2009 | 09V154000 | 2009 Chevrolet Avalanche, Tahoe, Suburban, GMC Yukon and Yukon XL | GM is recalling 27,188 vehicles. The fuel system control modules may have a condition in which an adhesive separation of the room temperature vulcanizing (RTV) seal between the seal and the housing may allow water to seep into the module. Water in the module could cause a short or open circuit, illumination of the service engine soon lamp, setting of diagnostic trouble codes or the engine may be hard to start, may not start or may stall, increasing the risk of a crash. |
| June 4, 2010 | 10V240000 | 2009 Avalanche, Silverado, Suburban, Tahoe, Sierra, Yukon, Yukon XL | GM is recalling vehicles equipped with a heated washer fluid system (HWFS). This recall applies to vehicles that have had a previous recall (08V441000) repair made. There have been reported failures of the device's thermal protection feature. The significance varies from minor distortion to considerable melting of the plastic around the HWFS fluid chamber. It is possible for the heated washer module to ignite and a fire may occur. |
| June 28, 2011 | 11V339000 | 2011 Chevrolet Silverado and GMC Sierra | GM is recalling certain vehicles because the intermediate steering shaft bolt may not have been tightened to the correct specification. This could allow the two shafts to separate, and the driver could experience loss of steering control, increasing the risk of a crash. |

| Recall date | Recall campaign number | Model(s) affected | Concern |
|---|---|---|---|
| Jan 02, 2013 | 13V001000 | 2013 Chevrolet Silverado, Suburban, Tahoe, Avalanche, GMC Sierra, Yukon, Yukon XL | Some models may have been built with a fractured park lock cable or a malformed steering column lock actuator gear in the lock module assembly. As a result, the vehicle may shift from Park with the ignition key removed or the ignition key in the OFF position. The vehicle may also shift out of Park without the application of the brake pedal while the key is off. Either of these scenarios may cause the vehicle to roll away after the driver has exited the vehicle, increasing the risk of a crash. |

## Buying parts

Replacement parts are available from many sources, which generally fall into one of two categories - authorized dealer parts departments and independent retail auto parts stores. Our advice concerning these parts is as follows:

**Retail auto parts stores:** Good auto parts stores will stock frequently needed components which wear out relatively fast, such as clutch components, exhaust systems, brake parts, tune-up parts, etc. These stores often supply new or reconditioned parts on an exchange basis, which can save a considerable amount of money. Discount auto parts stores are often very good places to buy materials and parts needed for general vehicle maintenance such as oil, grease, filters, spark plugs, belts, touch-up paint, bulbs, etc. They also usually sell tools and general accessories, have convenient hours, charge lower prices and can often be found not far from home.

**Authorized dealer parts department:** This is the best source for parts which are unique to the vehicle and not generally available elsewhere (such as major engine parts, transmission parts, trim pieces, etc.).

**Warranty information:** If the vehicle is still covered under warranty, be sure that any replacement parts purchased - regardless of the source - do not invalidate the warranty!

To be sure of obtaining the correct parts, have engine and chassis numbers available and, if possible, take the old parts along for positive identification.

## Maintenance techniques, tools and working facilities

## MAINTENANCE TECHNIQUES

There are a number of techniques involved in maintenance and repair that will be referred to throughout this manual. Application of these techniques will enable the home mechanic to be more efficient, better organized and capable of performing the various tasks properly, which will ensure that the repair job is thorough and complete.

### Fasteners

Fasteners are nuts, bolts, studs and screws used to hold two or more parts together. There are a few things to keep in mind when working with fasteners. Almost all of them use a locking device of some type, either a lockwasher, locknut, locking tab or thread adhesive. All threaded fasteners should be clean and straight, with undamaged threads and undamaged corners on the hex head where the wrench fits. Develop the habit of replacing all damaged nuts and bolts with new ones. Special locknuts with nylon or fiber inserts can only be used once. If they are removed, they lose their locking ability and must be replaced with new ones.

Rusted nuts and bolts should be treated with a penetrating fluid to ease removal and prevent breakage. Some mechanics use turpentine in a spout-type oil can, which works quite well. After applying the rust penetrant, let it work for a few minutes before trying to loosen the nut or bolt. Badly rusted fasteners may have to be chiseled or sawed off or removed with a special nut breaker, available at tool stores.

If a bolt or stud breaks off in an assembly, it can be drilled and removed with a special tool commonly available for this purpose. Most automotive machine shops can perform this task, as well as other repair procedures, such as the repair of threaded holes that have been stripped out.

Flat washers and lockwashers, when removed from an assembly, should always be replaced exactly as removed. Replace any damaged washers with new ones. Never use a lockwasher on any soft metal surface (such as aluminum), thin sheet metal or plastic.

### Fastener sizes

For a number of reasons, automobile manufacturers are making wider and wider use of metric fasteners. Therefore, it is important to be able to tell the difference between standard (sometimes called U.S. or SAE) and metric hardware, since they cannot be interchanged.

All bolts, whether standard or metric, are sized according to diameter, thread pitch and length. For example, a standard 1/2 - 13 x 1 bolt is 1/2 inch in diameter, has 13 threads per inch and is 1 inch long. An M12 - 1.75 x 25 metric bolt is 12 mm in diameter, has a thread pitch of 1.75 mm (the distance between threads) and is 25 mm long. The two bolts are nearly identical, and easily confused, but they are not interchangeable.

In addition to the differences in diameter, thread pitch and length, metric and standard bolts can also be distinguished by examining the bolt heads. To begin with, the distance across the flats on a standard bolt head is measured in inches, while the same dimension on a metric bolt is sized in millimeters (the same is true for nuts). As a result, a standard wrench should not be used on a metric bolt and a metric wrench should not be used on a standard bolt. Also, most standard bolts have slashes radiating out from the center of the head to denote the grade or strength of the bolt, which is an indication of the amount of torque that can be applied to it. The greater the number of slashes, the greater the strength of the bolt. Grades 0 through 5 are commonly used on automobiles. Metric bolts have a property class (grade) number, rather than a slash, molded into their heads to indicate bolt strength. In this case, the higher the number, the stronger the bolt. Property class numbers 8.8, 9.8 and 10.9 are commonly used on automobiles.

Strength markings can also be used to distinguish standard hex nuts from metric hex nuts. Many standard nuts have dots stamped into one side, while metric nuts are marked with a number. The greater the number of dots, or the higher the number, the greater the strength of the nut.

Metric studs are also marked on their ends according to property class (grade). Larger studs are numbered (the same as metric bolts), while smaller studs carry a geometric code to denote grade.

It should be noted that many fasteners, especially Grades 0 through 2, have no distinguishing marks on them. When such is the case, the only way to determine whether it is standard or metric is to measure the thread pitch or compare it to a known fastener of the same size.

Standard fasteners are often referred to as SAE, as opposed to metric. However, it should be noted that SAE technically refers to a non-metric fine thread fastener only. Coarse thread non-metric fasteners are referred to as USS sizes.

Since fasteners of the same size (both standard and metric) may have different strength ratings, be sure to reinstall any bolts, studs or nuts removed from your vehicle in their original locations. Also, when replacing a fastener with a new one, make sure that the new one has a strength rating equal to or greater than the original.

### Tightening sequences and procedures

Most threaded fasteners should be tightened to a specific torque value (torque is the twisting force applied to a threaded component such as a nut or bolt). Overtightening the fastener can weaken it and cause it to break, while undertightening can cause it to eventually come loose. Bolts, screws and studs, depending on the material they are made of and their thread diameters, have specific torque values, many of which are noted in the Specifications at the end of each Chapter. Be sure to follow the torque recommendations closely. For fasteners not assigned a specific torque, a general torque value chart is presented here as a guide. These torque values are for dry (unlubricated) fasteners threaded into steel or cast iron (not aluminum). As was previously mentioned, the size and grade of a fastener determine the amount of torque that can safely be applied to it. The figures listed here are approximate for Grade 2 and Grade 3 fasteners. Higher grades can tolerate higher torque values.

Fasteners laid out in a pattern, such as cylinder head bolts, oil pan bolts, differential cover bolts, etc., must be loosened or tightened in sequence to avoid warping the component. This sequence will normally be shown in the appropriate Chapter. If a specific pattern is not given, the following procedures can be used to prevent warping.

Initially, the bolts or nuts should be assembled finger-tight only. Next, they should be tightened one full turn each, in a criss-cross or diagonal pattern. After each one has been tightened one full turn, return to the first one and tighten them all one-half turn, following the same

## 0-12 MAINTENANCE TECHNIQUES, TOOLS AND WORKING FACILITIES

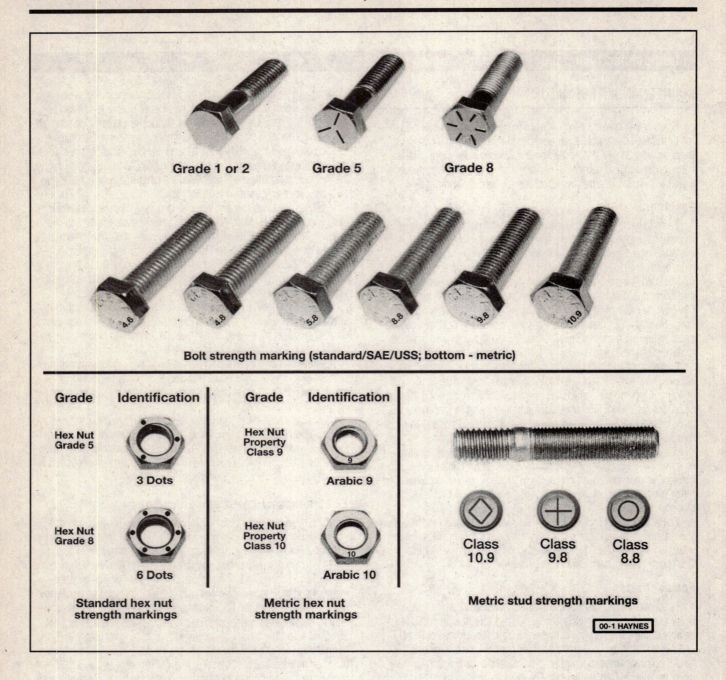

Bolt strength marking (standard/SAE/USS; bottom - metric)

Standard hex nut strength markings

Metric hex nut strength markings

Metric stud strength markings

pattern. Finally, tighten each of them one-quarter turn at a time until each fastener has been tightened to the proper torque. To loosen and remove the fasteners, the procedure would be reversed.

### Component disassembly

Component disassembly should be done with care and purpose to help ensure that the parts go back together properly. Always keep track of the sequence in which parts are removed. Make note of special characteristics or marks on parts that can be installed more than one way, such as a grooved thrust washer on a shaft. It is a good idea to lay the disassembled parts out on a clean surface in the order that they were removed. It may also be helpful to make sketches or take instant photos of components before removal.

When removing fasteners from a component, keep track of their locations. Sometimes threading a bolt back in a part, or putting the washers and nut back on a stud, can prevent mix-ups later. If nuts and bolts cannot be returned to their original locations, they should be kept in a compartmented box or a series of small boxes. A cupcake or muffin tin is ideal for this purpose, since each cavity can hold the bolts and nuts from a particular area (i.e. oil pan bolts, valve cover bolts, engine

# MAINTENANCE TECHNIQUES, TOOLS AND WORKING FACILITIES  0-13

| | Ft-lbs | Nm |
|---|---|---|
| **Metric thread sizes** | | |
| M-6 | 6 to 9 | 9 to 12 |
| M-8 | 14 to 21 | 19 to 28 |
| M-10 | 28 to 40 | 38 to 54 |
| M-12 | 50 to 71 | 68 to 96 |
| M-14 | 80 to 140 | 109 to 154 |
| **Pipe thread sizes** | | |
| 1/8 | 5 to 8 | 7 to 10 |
| 1/4 | 12 to 18 | 17 to 24 |
| 3/8 | 22 to 33 | 30 to 44 |
| 1/2 | 25 to 35 | 34 to 47 |
| **U.S. thread sizes** | | |
| 1/4 - 20 | 6 to 9 | 9 to 12 |
| 5/16 - 18 | 12 to 18 | 17 to 24 |
| 5/16 - 24 | 14 to 20 | 19 to 27 |
| 3/8 - 16 | 22 to 32 | 30 to 43 |
| 3/8 - 24 | 27 to 38 | 37 to 51 |
| 7/16 - 14 | 40 to 55 | 55 to 74 |
| 7/16 - 20 | 40 to 60 | 55 to 81 |
| 1/2 - 13 | 55 to 80 | 75 to 108 |

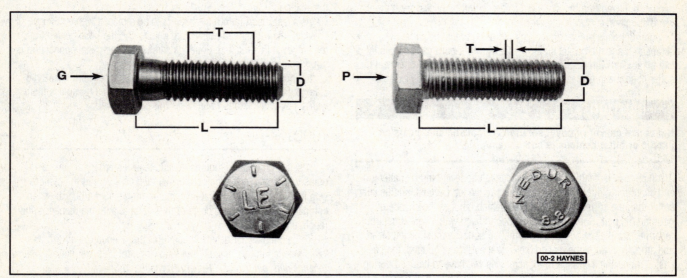

**Standard (SAE and USS) bolt dimensions/grade marks**
- G  Grade marks (bolt strength)
- L  Length (in inches)
- T  Thread pitch (number of threads per inch)
- D  Nominal diameter (in inches)

**Metric bolt dimensions/grade marks**
- P  Property class (bolt strength)
- L  Length (in millimeters)
- T  Thread pitch (distance between threads in millimeters)
- D  Diameter

mount bolts, etc.). A pan of this type is especially helpful when working on assemblies with very small parts, such as the carburetor, alternator, valve train or interior dash and trim pieces. The cavities can be marked with paint or tape to identify the contents.

Whenever wiring looms, harnesses or connectors are separated, it is a good idea to identify the two halves with numbered pieces of masking tape so they can be easily reconnected.

## Gasket sealing surfaces

Throughout any vehicle, gaskets are used to seal the mating surfaces between two parts and keep lubricants, fluids, vacuum or pressure contained in an assembly.

Many times these gaskets are coated with a liquid or paste-type gasket sealing compound before assembly. Age, heat and pressure can sometimes cause the two parts to stick together so tightly that they are very difficult to separate. Often, the assembly can be loosened by striking it with a soft-face hammer near the mating surfaces. A regular hammer can be used if a block of wood is placed between the hammer and the part. Do not hammer on cast parts or parts that could be easily damaged. With any particularly stubborn part, always recheck to make sure that every fastener has been removed.

Avoid using a screwdriver or bar to pry apart an assembly, as they

# 0-14 MAINTENANCE TECHNIQUES, TOOLS AND WORKING FACILITIES

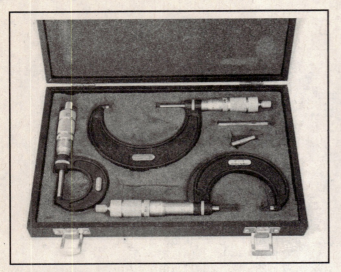

**Micrometer set**

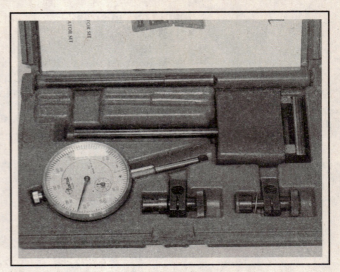

**Dial indicator set**

can easily mar the gasket sealing surfaces of the parts, which must remain smooth. If prying is absolutely necessary, use an old broom handle, but keep in mind that extra clean up will be necessary if the wood splinters.

After the parts are separated, the old gasket must be carefully scraped off and the gasket surfaces cleaned. Stubborn gasket material can be soaked with rust penetrant or treated with a special chemical to soften it so it can be easily scraped off.

### ❋❋ CAUTION:
**Never use gasket removal solutions or caustic chemicals on plastic or other composite components.**

A scraper can be fashioned from a piece of copper tubing by flattening and sharpening one end. Copper is recommended because it is usually softer than the surfaces to be scraped, which reduces the chance of gouging the part. Some gaskets can be removed with a wire brush, but regardless of the method used, the mating surfaces must be left clean and smooth. If for some reason the gasket surface is gouged, then a gasket sealer thick enough to fill scratches will have to be used during reassembly of the components. For most applications, a non-drying (or semi-drying) gasket sealer should be used.

## Hose removal tips

### ❋❋ WARNING:
**If the vehicle is equipped with air conditioning, do not disconnect any of the A/C hoses without first having the system depressurized by a dealer service department or a service station.**

Hose removal precautions closely parallel gasket removal precautions. Avoid scratching or gouging the surface that the hose mates against or the connection may leak. This is especially true for radiator hoses. Because of various chemical reactions, the rubber in hoses can bond itself to the metal spigot that the hose fits over. To remove a hose, first loosen the hose clamps that secure it to the spigot. Then, with slip-joint pliers, grab the hose at the clamp and rotate it around the spigot. Work it back and forth until it is completely free, then pull it off. Silicone or other lubricants will ease removal if they can be applied between the hose and the outside of the spigot. Apply the same lubricant to the inside of the hose and the outside of the spigot to simplify installation.

As a last resort (and if the hose is to be replaced with a new one anyway), the rubber can be slit with a knife and the hose peeled from the spigot. If this must be done, be careful that the metal connection is not damaged.

If a hose clamp is broken or damaged, do not reuse it. Wire-type clamps usually weaken with age, so it is a good idea to replace them with screw-type clamps whenever a hose is removed.

## TOOLS

A selection of good tools is a basic requirement for anyone who plans to maintain and repair his or her own vehicle. For the owner who has few tools, the initial investment might seem high, but when compared to the spiraling costs of professional auto maintenance and repair, it is a wise one.

To help the owner decide which tools are needed to perform the tasks detailed in this manual, the following tool lists are offered: *Maintenance and minor repair, Repair/overhaul and Special.*

The newcomer to practical mechanics should start off with the *maintenance and minor repair* tool kit, which is adequate for the simpler jobs performed on a vehicle. Then, as confidence and experience grow, the owner can tackle more difficult tasks, buying additional tools as they are needed. Eventually the basic kit will be expanded into the *repair and overhaul* tool set. Over a period of time, the experienced do-it-yourselfer will assemble a tool set complete enough for most repair and overhaul procedures and will add tools from the special category when it is felt that the expense is justified by the frequency of use.

### Maintenance and minor repair tool kit

The tools in this list should be considered the minimum required for performance of routine maintenance, servicing and minor repair work. We recommend the purchase of combination wrenches (box-end and open-end combined in one wrench). While more expensive than open end wrenches, they offer the advantages of both types of wrench.

*Combination wrench set (1/4-inch to 1 inch or 6 mm to 19 mm)*
*Adjustable wrench, 8 inch*
*Spark plug wrench with rubber insert*

# MAINTENANCE TECHNIQUES, TOOLS AND WORKING FACILITIES    0-15

Spark plug gap adjusting tool
Feeler gauge set
Brake bleeder wrench
Standard screwdriver (5/16-inch x 6 inch)
Phillips screwdriver (No. 2 x 6 inch)
Combination pliers - 6 inch
Hacksaw and assortment of blades
Tire pressure gauge
Grease gun

Oil can
Fine emery cloth
Wire brush
Battery post and cable cleaning tool
Oil filter wrench
Funnel (medium size)
Safety goggles
Jackstands (2)
Drain pan

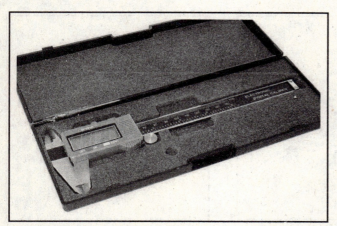

**Dial caliper**

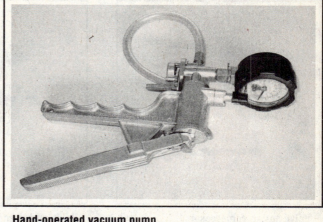

**Hand-operated vacuum pump**

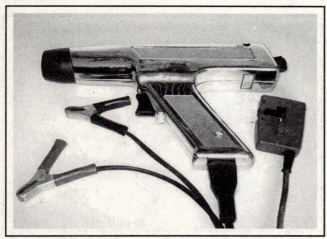

**Timing light**

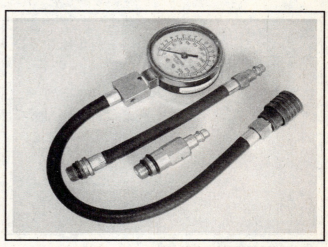

**Compression gauge with spark plug hole adapter**

**Damper/steering wheel puller**

**General purpose puller**

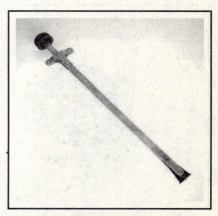

**Hydraulic lifter removal tool**

## 0-16 MAINTENANCE TECHNIQUES, TOOLS AND WORKING FACILITIES

➡ **Note:** If basic tune-ups are going to be part of routine maintenance, it will be necessary to purchase a good quality stroboscopic timing light and combination tachometer/dwell meter. Although they are included in the list of special tools, it is mentioned here because they are absolutely necessary for tuning most vehicles properly.

### Repair and overhaul tool set

These tools are essential for anyone who plans to perform major repairs and are in addition to those in the maintenance and minor repair tool kit. Included is a comprehensive set of sockets which, though expensive, are invaluable because of their versatility, especially when various extensions and drives are available. We recommend the 1/2-inch drive over the 3/8-inch drive. Although the larger drive is bulky and more expensive, it has the capacity of accepting a very wide range of large sockets. Ideally, however, the mechanic should have a 3/8-inch drive set and a 1/2-inch drive set.

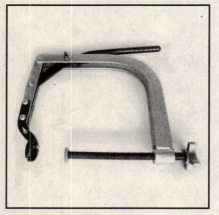
**Valve spring compressor**

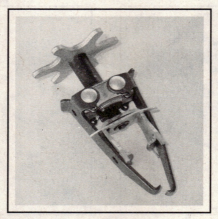
**Valve spring compressor**

**Ridge reamer**

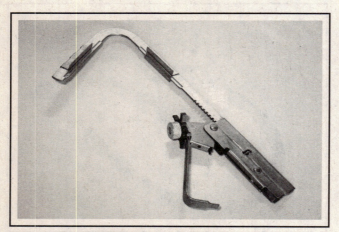
**Piston ring groove cleaning tool**

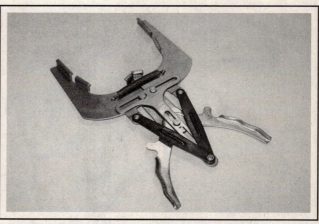
**Ring removal/installation tool**

**Ring compressor**

**Cylinder hone**

**Brake hold-down spring tool**

# MAINTENANCE TECHNIQUES, TOOLS AND WORKING FACILITIES    0-17

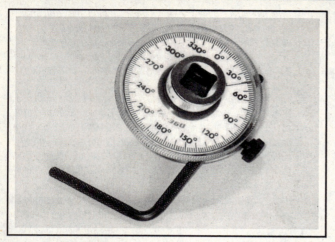

**Torque angle gauge**

**Clutch plate alignment tool**

Socket set(s)
Reversible ratchet
Extension - 10 inch
Universal joint
Torque wrench (same size drive as sockets)
Ball peen hammer - 8 ounce
Soft-face hammer (plastic/rubber)
Standard screwdriver (1/4-inch x 6 inch)
Standard screwdriver (stubby - 5/16-inch)
Phillips screwdriver (No. 3 x 8 inch)
Phillips screwdriver (stubby - No. 2)
Pliers - vise grip
Pliers - lineman's
Pliers - needle nose
Pliers - snap-ring (internal and external)
Cold chisel - 1/2-inch
Scribe
Scraper (made from flattened copper tubing)
Centerpunch
Pin punches (1/16, 1/8, 3/16-inch)
Steel rule/straightedge - 12 inch
Allen wrench set (1/8 to 3/8-inch or 4 mm to 10 mm)
A selection of files
Wire brush (large)
Jackstands (second set)
Jack (scissor or hydraulic type)

➡ **Note: Another tool which is often useful is an electric drill with a chuck capacity of 3/8-inch and a set of good quality drill bits.**

## Special tools

The tools in this list include those which are not used regularly, are expensive to buy, or which need to be used in accordance with their manufacturer's instructions. Unless these tools will be used frequently, it is not very economical to purchase many of them. A consideration would be to split the cost and use between yourself and a friend or friends. In addition, most of these tools can be obtained from a tool rental shop on a temporary basis.

This list primarily contains only those tools and instruments widely available to the public, and not those special tools produced by the vehicle manufacturer for distribution to dealer service departments. Occasionally, references to the manufacturer's special tools are included in the text of this manual. Generally, an alternative method of doing the job without the special tool is offered. However, sometimes there is no alternative to their use. Where this is the case, and the tool cannot be purchased or borrowed, the work should be turned over to the dealer service department or an automotive repair shop.

Valve spring compressor
Piston ring groove cleaning tool
Piston ring compressor
Piston ring installation tool
Cylinder compression gauge
Cylinder ridge reamer
Cylinder surfacing hone
Cylinder bore gauge
Micrometers and/or dial calipers
Hydraulic lifter removal tool
Balljoint separator
Universal-type puller
Impact screwdriver
Dial indicator set
Stroboscopic timing light (inductive pick-up)
Hand operated vacuum/pressure pump
Tachometer/dwell meter
Universal electrical multimeter
Cable hoist
Brake spring removal and installation tools
Floor jack

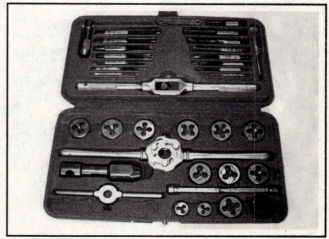

**Tap and die set**

# MAINTENANCE TECHNIQUES, TOOLS AND WORKING FACILITIES

## Buying tools

For the do-it-yourselfer who is just starting to get involved in vehicle maintenance and repair, there are a number of options available when purchasing tools. If maintenance and minor repair is the extent of the work to be done, the purchase of individual tools is satisfactory. If, on the other hand, extensive work is planned, it would be a good idea to purchase a modest tool set from one of the large retail chain stores. A set can usually be bought at a substantial savings over the individual tool prices, and they often come with a tool box. As additional tools are needed, add-on sets, individual tools and a larger tool box can be purchased to expand the tool selection. Building a tool set gradually allows the cost of the tools to be spread over a longer period of time and gives the mechanic the freedom to choose only those tools that will actually be used.

Tool stores will often be the only source of some of the special tools that are needed, but regardless of where tools are bought, try to avoid cheap ones, especially when buying screwdrivers and sockets, because they won't last very long. The expense involved in replacing cheap tools will eventually be greater than the initial cost of quality tools.

## Care and maintenance of tools

Good tools are expensive, so it makes sense to treat them with respect. Keep them clean and in usable condition and store them properly when not in use. Always wipe off any dirt, grease or metal chips before putting them away. Never leave tools lying around in the work area. Upon completion of a job, always check closely under the hood for tools that may have been left there so they won't get lost during a test drive.

Some tools, such as screwdrivers, pliers, wrenches and sockets, can be hung on a panel mounted on the garage or workshop wall, while others should be kept in a tool box or tray. Measuring instruments, gauges, meters, etc. must be carefully stored where they cannot be damaged by weather or impact from other tools.

When tools are used with care and stored properly, they will last a very long time. Even with the best of care, though, tools will wear out if used frequently. When a tool is damaged or worn out, replace it. Subsequent jobs will be safer and more enjoyable if you do.

## HOW TO REPAIR DAMAGED THREADS

Sometimes, the internal threads of a nut or bolt hole can become stripped, usually from overtightening. Stripping threads is an all-too-common occurrence, especially when working with aluminum parts, because aluminum is so soft that it easily strips out.

Usually, external or internal threads are only partially stripped. After they've been cleaned up with a tap or die, they'll still work. Sometimes, however, threads are badly damaged. When this happens, you've got three choices:

1) Drill and tap the hole to the next suitable oversize and install a larger diameter bolt, screw or stud.
2) Drill and tap the hole to accept a threaded plug, then drill and tap the plug to the original screw size. You can also buy a plug already threaded to the original size. Then you simply drill a hole to the specified size, then run the threaded plug into the hole with a bolt and jam nut. Once the plug is fully seated, remove the jam nut and bolt.
3) The third method uses a patented thread repair kit like Heli-Coil or Slimsert. These easy-to-use kits are designed to repair damaged threads in straight-through holes and blind holes. Both are available as kits which can handle a variety of sizes and thread patterns. Drill the hole, then tap it with the special included tap. Install the Heli-Coil and the hole is back to its original diameter and thread pitch.

Regardless of which method you use, be sure to proceed calmly and carefully. A little impatience or carelessness during one of these relatively simple procedures can ruin your whole day's work and cost you a bundle if you wreck an expensive part.

## WORKING FACILITIES

Not to be overlooked when discussing tools is the workshop. If anything more than routine maintenance is to be carried out, some sort of suitable work area is essential.

It is understood, and appreciated, that many home mechanics do not have a good workshop or garage available, and end up removing an engine or doing major repairs outside. It is recommended, however, that the overhaul or repair be completed under the cover of a roof.

A clean, flat workbench or table of comfortable working height is an absolute necessity. The workbench should be equipped with a vise that has a jaw opening of at least four inches.

As mentioned previously, some clean, dry storage space is also required for tools, as well as the lubricants, fluids, cleaning solvents, etc. which soon become necessary.

Sometimes waste oil and fluids, drained from the engine or cooling system during normal maintenance or repairs, present a disposal problem. To avoid pouring them on the ground or into a sewage system, pour the used fluids into large containers, seal them with caps and take them to an authorized disposal site or recycling center. Plastic jugs, such as old antifreeze containers, are ideal for this purpose.

Always keep a supply of old newspapers and clean rags available. Old towels are excellent for mopping up spills. Many mechanics use rolls of paper towels for most work because they are readily available and disposable. To help keep the area under the vehicle clean, a large cardboard box can be cut open and flattened to protect the garage or shop floor.

Whenever working over a painted surface, such as when leaning over a fender to service something under the hood, always cover it with an old blanket or bedspread to protect the finish. Vinyl covered pads, made especially for this purpose, are available at auto parts stores.

# JACKING AND TOWING  0-19

## Jacking and towing

### JACKING

The jack supplied with the vehicle should only be used for raising the vehicle when changing a tire or placing jackstands under the frame. NEVER work under the vehicle or start the engine when the vehicle is supported only by a jack.

The vehicle should be parked on level ground with the wheels blocked, the parking brake applied and the transmission in Park (automatic) or Reverse (manual). If the vehicle is parked alongside the roadway, or in any other hazardous situation, turn on the emergency hazard flashers. If a tire is to be changed, loosen the lug nuts one-half turn before raising off the ground.

Place the jack under the vehicle in the indicated positions (see illustrations). Operate the jack with a slow, smooth motion until the wheel is raised off the ground. Remove the lug nuts, pull off the wheel, install the spare and thread the lug nuts back on with the beveled side facing in. Tighten the lug nuts snugly, lower the vehicle until some weight is on the wheel, tighten them completely in a criss-cross pattern and remove the jack.

### TOWING

Equipment specifically designed for towing should be used and attached to the main structural members of the vehicle. Optional tow hooks may be attached to the frame at both ends of the vehicle; they are intended for emergency use only, for rescuing a stranded vehicle. Do not use the tow hooks for highway towing. Stand clear when using tow straps or chains (they may break, causing serious injury.)

Safety is a major consideration when towing and all applicable state and local laws must be obeyed. In addition to a tow bar, a safety chain must be used for all towing.

The manufacturer states that the best way to tow these vehicles is with wheel lift equipment or a flatbed car carrier.

Two-wheel drive vehicles with automatic transmission may be towed with the rear wheels on a towing dolly. They should NOT be towed with the rear wheels on the ground, as this could damage the automatic transmission.

Four-wheel drive models can be towed with the front wheels on or off the ground and the transfer case in Neutral.

If any vehicle is to be towed with the front wheels on the ground and the rear wheels raised, the ignition key must be turned to the OFF position to unlock the steering column and a steering wheel clamping device designed for towing must be used or damage to the steering column lock may occur.

**Rear jacking position - the axle must be secure in the jack head grooves**

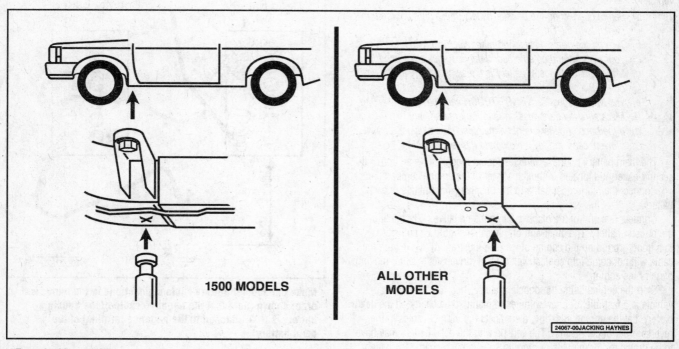

**Front jacking position**

## Booster battery (jump) starting

Observe these precautions when using a booster battery to start a vehicle:

- a) Before connecting the booster battery, make sure the ignition switch is in the Off position.
- b) Turn off the lights, heater and other electrical loads.
- c) Your eyes should be shielded. Safety goggles are a good idea.
- d) Make sure the booster battery is the same voltage as the dead one in the vehicle.
- e) The two vehicles MUST NOT TOUCH each other!
- f) Make sure the transmission is in Park (automatic).
- g) If the booster battery is not a maintenance-free type, remove the vent caps and lay a cloth over the vent holes.

The main battery on these vehicles (some models have an optional second battery) is located in the left corner of the engine compartment.

Connect the red jumper cable to the positive (+) terminals of each battery.

Connect one end of the black cable to the negative (-) terminal of the booster battery. The other end of this cable should be connected to a good ground on the engine block (see illustration). Make sure the cable will not come into contact with the fan, drivebelts or other moving parts of the engine.

Start the engine using the booster battery, then run the booster vehicle at a fast idle for a few minutes to instill some charge in the dead battery. Let the engine idle, then disconnect the jumper cables in the reverse order of connection. The vehicle with the dead battery may have to be driven for 20 minutes or more to sufficiently recharge the battery for independent starting.

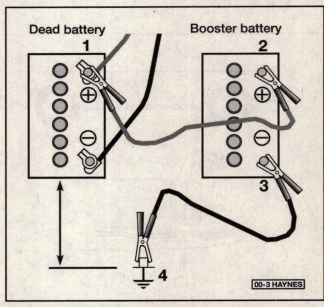

**Make the booster battery cable connections in the numerical order shown (note that the negative cable of the booster battery is NOT attached to the negative terminal of the dead battery)**

## Automotive chemicals and lubricants

A number of automotive chemicals and lubricants are available for use during vehicle maintenance and repair. They include a wide variety of products ranging from cleaning solvents and degreasers to lubricants and protective sprays for rubber, plastic and vinyl.

## CLEANERS

**Carburetor cleaner and choke cleaner** is a strong solvent for gum, varnish and carbon. Most carburetor cleaners leave a dry-type lubricant film which will not harden or gum up. Because of this film it is not recommended for use on electrical components.

**Brake system cleaner** is used to remove brake dust, grease and brake fluid from the brake system, where clean surfaces are absolutely necessary. It leaves no residue and often eliminates brake squeal caused by contaminants.

**Electrical cleaner** removes oxidation, corrosion and carbon deposits from electrical contacts, restoring full current flow. It can also be used to clean spark plugs, carburetor jets, voltage regulators and other parts where an oil-free surface is desired.

**Demoisturants** remove water and moisture from electrical components such as alternators, voltage regulators, electrical connectors and fuse blocks. They are non-conductive and non-corrosive.

**Degreasers** are heavy-duty solvents used to remove grease from the outside of the engine and from chassis components. They can be sprayed or brushed on and, depending on the type, are rinsed off either with water or solvent.

## LUBRICANTS

**Motor oil** is the lubricant formulated for use in engines. It normally contains a wide variety of additives to prevent corrosion and reduce foaming and wear. Motor oil comes in various weights (viscosity ratings) from 0 to 50. The recommended weight of the oil depends on the season, temperature and the demands on the engine. Light oil is used in cold climates and under light load conditions. Heavy oil is used in hot climates and where high loads are encountered. Multi-viscosity oils are designed to have characteristics of both light and heavy oils and are available in a number of weights from 0W-20 to 20W-50.

**Gear oil** is designed to be used in differentials, manual transmissions and other areas where high-temperature lubrication is required.

**Chassis and wheel bearing grease** is a heavy grease used where increased loads and friction are encountered, such as for wheel bearings, balljoints, tie-rod ends and universal joints.

**High-temperature wheel bearing grease** is designed to withstand the extreme temperatures encountered by wheel bearings in disc brake equipped vehicles. It usually contains molybdenum disulfide (moly), which is a dry-type lubricant.

**White grease** is a heavy grease for metal-to-metal applications where water is a problem. White grease stays soft under both low and high temperatures (usually from -100 to +190-degrees F), and will not wash off or dilute in the presence of water.

**Assembly lube** is a special extreme pressure lubricant, usually containing moly, used to lubricate high-load parts (such as main and rod bearings and cam lobes) for initial start-up of a new engine. The assembly lube lubricates the parts without being squeezed out or washed away until the engine oiling system begins to function.

**Silicone lubricants** are used to protect rubber, plastic, vinyl and nylon parts.

**Graphite lubricants** are used where oils cannot be used due to contamination problems, such as in locks. The dry graphite will lubricate metal parts while remaining uncontaminated by dirt, water, oil or acids. It is electrically conductive and will not foul electrical contacts in locks such as the ignition switch.

**Moly penetrants** loosen and lubricate frozen, rusted and corroded fasteners and prevent future rusting or freezing.

**Heat-sink grease** is a special electrically non-conductive grease that is used for mounting electronic ignition modules where it is essential that heat is transferred away from the module.

## SEALANTS

**RTV sealant** is one of the most widely used gasket compounds. Made from silicone, RTV is air curing, it seals, bonds, waterproofs, fills surface irregularities, remains flexible, doesn't shrink, is relatively easy to remove, and is used as a supplementary sealer with almost all low and medium temperature gaskets.

**Anaerobic sealant** is much like RTV in that it can be used either to seal gaskets or to form gaskets by itself. It remains flexible, is solvent resistant and fills surface imperfections. The difference between an anaerobic sealant and an RTV-type sealant is in the curing. RTV cures when exposed to air, while an anaerobic sealant cures only in the absence of air. This means that an anaerobic sealant cures only after the assembly of parts, sealing them together.

**Thread and pipe sealant** is used for sealing hydraulic and pneumatic fittings and vacuum lines. It is usually made from a Teflon compound, and comes in a spray, a paint-on liquid and as a wrap-around tape.

## CHEMICALS

**Anti-seize compound** prevents seizing, galling, cold welding, rust and corrosion in fasteners. High-temperature anti-seize, usually made with copper and graphite lubricants, is used for exhaust system and exhaust manifold bolts.

**Anaerobic locking compounds** are used to keep fasteners from vibrating or working loose and cure only after installation, in the absence of air. Medium strength locking compound is used for small nuts, bolts and screws that may be removed later. High-strength locking compound is for large nuts, bolts and studs which aren't removed on a regular basis.

**Oil additives** range from viscosity index improvers to chemical treatments that claim to reduce internal engine friction. It should be noted that most oil manufacturers caution against using additives with their oils.

**Gas additives** perform several functions, depending on their chemical makeup. They usually contain solvents that help dissolve gum and varnish that build up on carburetor, fuel injection and intake parts. They also serve to break down carbon deposits that form on the inside surfaces of the combustion chambers. Some additives contain upper cylinder lubricants for valves and piston rings, and others contain chemicals to remove condensation from the gas tank.

## MISCELLANEOUS

**Brake fluid** is specially formulated hydraulic fluid that can withstand the heat and pressure encountered in brake systems. Care must be taken so this fluid does not come in contact with painted surfaces or plastics. An opened container should always be resealed to prevent contamination by water or dirt.

**Weatherstrip adhesive** is used to bond weatherstripping around doors, windows and trunk lids. It is sometimes used to attach trim pieces.

**Undercoating** is a petroleum-based, tar-like substance that is designed to protect metal surfaces on the underside of the vehicle from corrosion. It also acts as a sound-deadening agent by insulating the bottom of the vehicle.

**Waxes and polishes** are used to help protect painted and plated surfaces from the weather. Different types of paint may require the use of different types of wax and polish. Some polishes utilize a chemical or abrasive cleaner to help remove the top layer of oxidized (dull) paint on older vehicles. In recent years many non-wax polishes that contain a wide variety of chemicals such as polymers and silicones have been introduced. These non-wax polishes are usually easier to apply and last longer than conventional waxes and polishes.

## CONVERSION FACTORS

### LENGTH (distance)
| | | | | | |
|---|---|---|---|---|---|
| Inches (in) | X | 25.4 | = Millimeters (mm) | X 0.0394 | = Inches (in) |
| Feet (ft) | X | 0.305 | = Meters (m) | X 3.281 | = Feet (ft) |
| Miles | X | 1.609 | = Kilometers (km) | X 0.621 | = Miles |

### VOLUME (capacity)
| | | | | | |
|---|---|---|---|---|---|
| Cubic inches (cu in; in$^3$) | X | 16.387 | = Cubic centimeters (cc; cm$^3$) | X 0.061 | = Cubic inches (cu in; in$^3$) |
| Imperial pints (Imp pt) | X | 0.568 | = Liters (l) | X 1.76 | = Imperial pints (Imp pt) |
| Imperial quarts (Imp qt) | X | 1.137 | = Liters (l) | X 0.88 | = Imperial quarts (Imp qt) |
| Imperial quarts (Imp qt) | X | 1.201 | = US quarts (US qt) | X 0.833 | = Imperial quarts (Imp qt) |
| US quarts (US qt) | X | 0.946 | = Liters (l) | X 1.057 | = US quarts (US qt) |
| Imperial gallons (Imp gal) | X | 4.546 | = Liters (l) | X 0.22 | = Imperial gallons (Imp gal) |
| Imperial gallons (Imp gal) | X | 1.201 | = US gallons (US gal) | X 0.833 | = Imperial gallons (Imp gal) |
| US gallons (US gal) | X | 3.785 | = Liters (l) | X 0.264 | = US gallons (US gal) |

### MASS (weight)
| | | | | | |
|---|---|---|---|---|---|
| Ounces (oz) | X | 28.35 | = Grams (g) | X 0.035 | = Ounces (oz) |
| Pounds (lb) | X | 0.454 | = Kilograms (kg) | X 2.205 | = Pounds (lb) |

### FORCE
| | | | | | |
|---|---|---|---|---|---|
| Ounces-force (ozf; oz) | X | 0.278 | = Newtons (N) | X 3.6 | = Ounces-force (ozf; oz) |
| Pounds-force (lbf; lb) | X | 4.448 | = Newtons (N) | X 0.225 | = Pounds-force (lbf; lb) |
| Newtons (N) | X | 0.1 | = Kilograms-force (kgf; kg) | X 9.81 | = Newtons (N) |

### PRESSURE
| | | | | | |
|---|---|---|---|---|---|
| Pounds-force per square inch (psi; lbf/in$^2$; lb/in$^2$) | X | 0.070 | = Kilograms-force per square centimeter (kgf/cm$^2$; kg/cm$^2$) | X 14.223 | = Pounds-force per square inch (psi; lbf/in$^2$; lb/in$^2$) |
| Pounds-force per square inch (psi; lbf/in$^2$; lb/in$^2$) | X | 0.068 | = Atmospheres (atm) | X 14.696 | = Pounds-force per square inch (psi; lbf/in$^2$; lb/in$^2$) |
| Pounds-force per square inch (psi; lbf/in$^2$; lb/in$^2$) | X | 0.069 | = Bars | X 14.5 | = Pounds-force per square inch (psi; lbf/in$^2$; lb/in$^2$) |
| Pounds-force per square inch (psi; lbf/in$^2$; lb/in$^2$) | X | 6.895 | = Kilopascals (kPa) | X 0.145 | = Pounds-force per square inch (psi; lbf/in$^2$; lb/in$^2$) |
| Kilopascals (kPa) | X | 0.01 | = Kilograms-force per square centimeter (kgf/cm$^2$; kg/cm$^2$) | X 98.1 | = Kilopascals (kPa) |

### TORQUE (moment of force)
| | | | | | |
|---|---|---|---|---|---|
| Pounds-force inches (lbf in; lb in) | X | 1.152 | = Kilograms-force centimeter (kgf cm; kg cm) | X 0.868 | = Pounds-force inches (lbf in; lb in) |
| Pounds-force inches (lbf in; lb in) | X | 0.113 | = Newton meters (Nm) | X 8.85 | = Pounds-force inches (lbf in; lb in) |
| Pounds-force inches (lbf in; lb in) | X | 0.083 | = Pounds-force feet (lbf ft; lb ft) | X 12 | = Pounds-force inches (lbf in; lb in) |
| Pounds-force feet (lbf ft; lb ft) | X | 0.138 | = Kilograms-force meters (kgf m; kg m) | X 7.233 | = Pounds-force feet (lbf ft; lb ft) |
| Pounds-force feet (lbf ft; lb ft) | X | 1.356 | = Newton meters (Nm) | X 0.738 | = Pounds-force feet (lbf ft; lb ft) |
| Newton meters (Nm) | X | 0.102 | = Kilograms-force meters (kgf m; kg m) | X 9.804 | = Newton meters (Nm) |

### VACUUM
| | | | | | |
|---|---|---|---|---|---|
| Inches mercury (in. Hg) | X | 3.377 | = Kilopascals (kPa) | X 0.2961 | = Inches mercury |
| Inches mercury (in. Hg) | X | 25.4 | = Millimeters mercury (mm Hg) | X 0.0394 | = Inches mercury |

### POWER
| | | | | | |
|---|---|---|---|---|---|
| Horsepower (hp) | X | 745.7 | = Watts (W) | X 0.0013 | = Horsepower (hp) |

### VELOCITY (speed)
| | | | | | |
|---|---|---|---|---|---|
| Miles per hour (miles/hr; mph) | X | 1.609 | = Kilometers per hour (km/hr; kph) | X 0.621 | = Miles per hour (miles/hr; mph) |

### FUEL CONSUMPTION*
| | | | | | |
|---|---|---|---|---|---|
| Miles per gallon, Imperial (mpg) | X | 0.354 | = Kilometers per liter (km/l) | X 2.825 | = Miles per gallon, Imperial (mpg) |
| Miles per gallon, US (mpg) | X | 0.425 | = Kilometers per liter (km/l) | X 2.352 | = Miles per gallon, US (mpg) |

### TEMPERATURE
Degrees Fahrenheit = (°C x 1.8) + 32     Degrees Celsius (Degrees Centigrade; °C) = (°F - 32) x 0.56

*It is common practice to convert from miles per gallon (mpg) to liters/100 kilometers (l/100km), where mpg (Imperial) x l/100 km = 282 and mpg (US) x l/100 km = 235

## FRACTION/DECIMAL/MILLIMETER EQUIVALENTS

### DECIMALS TO MILLIMETERS

| Decimal | mm | Decimal | mm |
|---|---|---|---|
| 0.001 | 0.0254 | 0.500 | 12.7000 |
| 0.002 | 0.0508 | 0.510 | 12.9540 |
| 0.003 | 0.0762 | 0.520 | 13.2080 |
| 0.004 | 0.1016 | 0.530 | 13.4620 |
| 0.005 | 0.1270 | 0.540 | 13.7160 |
| 0.006 | 0.1524 | 0.550 | 13.9700 |
| 0.007 | 0.1778 | 0.560 | 14.2240 |
| 0.008 | 0.2032 | 0.570 | 14.4780 |
| 0.009 | 0.2286 | 0.580 | 14.7320 |
|  |  | 0.590 | 14.9860 |
| 0.010 | 0.2540 |  |  |
| 0.020 | 0.5080 |  |  |
| 0.030 | 0.7620 |  |  |
| 0.040 | 1.0160 | 0.600 | 15.2400 |
| 0.050 | 1.2700 | 0.610 | 15.4940 |
| 0.060 | 1.5240 | 0.620 | 15.7480 |
| 0.070 | 1.7780 | 0.630 | 16.0020 |
| 0.080 | 2.0320 | 0.640 | 16.2560 |
| 0.090 | 2.2860 | 0.650 | 16.5100 |
|  |  | 0.660 | 16.7640 |
| 0.100 | 2.5400 | 0.670 | 17.0180 |
| 0.110 | 2.7940 | 0.680 | 17.2720 |
| 0.120 | 3.0480 | 0.690 | 17.5260 |
| 0.130 | 3.3020 |  |  |
| 0.140 | 3.5560 |  |  |
| 0.150 | 3.8100 |  |  |
| 0.160 | 4.0640 | 0.700 | 17.7800 |
| 0.170 | 4.3180 | 0.710 | 18.0340 |
| 0.180 | 4.5720 | 0.720 | 18.2880 |
| 0.190 | 4.8260 | 0.730 | 18.5420 |
|  |  | 0.740 | 18.7960 |
| 0.200 | 5.0800 | 0.750 | 19.0500 |
| 0.210 | 5.3340 | 0.760 | 19.3040 |
| 0.220 | 5.5880 | 0.770 | 19.5580 |
| 0.230 | 5.8420 | 0.780 | 19.8120 |
| 0.240 | 6.0960 | 0.790 | 20.0660 |
| 0.250 | 6.3500 |  |  |
| 0.260 | 6.6040 |  |  |
| 0.270 | 6.8580 | 0.800 | 20.3200 |
| 0.280 | 7.1120 | 0.810 | 20.5740 |
| 0.290 | 7.3660 | 0.820 | 21.8280 |
|  |  | 0.830 | 21.0820 |
| 0.300 | 7.6200 | 0.840 | 21.3360 |
| 0.310 | 7.8740 | 0.850 | 21.5900 |
| 0.320 | 8.1280 | 0.860 | 21.8440 |
| 0.330 | 8.3820 | 0.870 | 22.0980 |
| 0.340 | 8.6360 | 0.880 | 22.3520 |
| 0.350 | 8.8900 | 0.890 | 22.6060 |
| 0.360 | 9.1440 |  |  |
| 0.370 | 9.3980 |  |  |
| 0.380 | 9.6520 |  |  |
| 0.390 | 9.9060 |  |  |
|  |  | 0.900 | 22.8600 |
| 0.400 | 10.1600 | 0.910 | 23.1140 |
| 0.410 | 10.4140 | 0.920 | 23.3680 |
| 0.420 | 10.6680 | 0.930 | 23.6220 |
| 0.430 | 10.9220 | 0.940 | 23.8760 |
| 0.440 | 11.1760 | 0.950 | 24.1300 |
| 0.450 | 11.4300 | 0.960 | 24.3840 |
| 0.460 | 11.6840 | 0.970 | 24.6380 |
| 0.470 | 11.9380 | 0.980 | 24.8920 |
| 0.480 | 12.1920 | 0.990 | 25.1460 |
| 0.490 | 12.4460 | 1.000 | 25.4000 |

### FRACTIONS TO DECIMALS TO MILLIMETERS

| Fraction | Decimal | mm | Fraction | Decimal | mm |
|---|---|---|---|---|---|
| 1/64 | 0.0156 | 0.3969 | 33/64 | 0.5156 | 13.0969 |
| 1/32 | 0.0312 | 0.7938 | 17/32 | 0.5312 | 13.4938 |
| 3/64 | 0.0469 | 1.1906 | 35/64 | 0.5469 | 13.8906 |
| 1/16 | 0.0625 | 1.5875 | 9/16 | 0.5625 | 14.2875 |
| 5/64 | 0.0781 | 1.9844 | 37/64 | 0.5781 | 14.6844 |
| 3/32 | 0.0938 | 2.3812 | 19/32 | 0.5938 | 15.0812 |
| 7/64 | 0.1094 | 2.7781 | 39/64 | 0.6094 | 15.4781 |
| 1/8 | 0.1250 | 3.1750 | 5/8 | 0.6250 | 15.8750 |
| 9/64 | 0.1406 | 3.5719 | 41/64 | 0.6406 | 16.2719 |
| 5/32 | 0.1562 | 3.9688 | 21/32 | 0.6562 | 16.6688 |
| 11/64 | 0.1719 | 4.3656 | 43/64 | 0.6719 | 17.0656 |
| 3/16 | 0.1875 | 4.7625 | 11/16 | 0.6875 | 17.4625 |
| 13/64 | 0.2031 | 5.1594 | 45/64 | 0.7031 | 17.8594 |
| 7/32 | 0.2188 | 5.5562 | 23/32 | 0.7188 | 18.2562 |
| 15/64 | 0.2344 | 5.9531 | 47/64 | 0.7344 | 18.6531 |
| 1/4 | 0.2500 | 6.3500 | 3/4 | 0.7500 | 19.0500 |
| 17/64 | 0.2656 | 6.7469 | 49/64 | 0.7656 | 19.4469 |
| 9/32 | 0.2812 | 7.1438 | 25/32 | 0.7812 | 19.8438 |
| 19/64 | 0.2969 | 7.5406 | 51/64 | 0.7969 | 20.2406 |
| 5/16 | 0.3125 | 7.9375 | 13/16 | 0.8125 | 20.6375 |
| 21/64 | 0.3281 | 8.3344 | 53/64 | 0.8281 | 21.0344 |
| 11/32 | 0.3438 | 8.7312 | 27/32 | 0.8438 | 21.4312 |
| 23/64 | 0.3594 | 9.1281 | 55/64 | 0.8594 | 21.8281 |
| 3/8 | 0.3750 | 9.5250 | 7/8 | 0.8750 | 22.2250 |
| 25/64 | 0.3906 | 9.9219 | 57/64 | 0.8906 | 22.6219 |
| 13/32 | 0.4062 | 10.3188 | 29/32 | 0.9062 | 23.0188 |
| 27/64 | 0.4219 | 10.7156 | 59/64 | 0.9219 | 23.4156 |
| 7/16 | 0.4375 | 11.1125 | 15/16 | 0.9375 | 23.8125 |
| 29/64 | 0.4531 | 11.5094 | 61/64 | 0.9531 | 24.2094 |
| 15/32 | 0.4688 | 11.9062 | 31/32 | 0.9688 | 24.6062 |
| 31/64 | 0.4844 | 12.3031 | 63/64 | 0.9844 | 25.0031 |
| 1/2 | 0.5000 | 12.7000 | 1 | 1.0000 | 25.4000 |

## Safety first!

Regardless of how enthusiastic you may be about getting on with the job at hand, take the time to ensure that your safety is not jeopardized. A moment's lack of attention can result in an accident, as can failure to observe certain simple safety precautions. The possibility of an accident will always exist, and the following points should not be considered a comprehensive list of all dangers. Rather, they are intended to make you aware of the risks and to encourage a safety conscious approach to all work you carry out on your vehicle.

### ESSENTIAL DOS AND DON'TS

**DON'T** rely on a jack when working under the vehicle. Always use approved jackstands to support the weight of the vehicle and place them under the recommended lift or support points.
**DON'T** attempt to loosen extremely tight fasteners (i.e. wheel lug nuts) while the vehicle is on a jack - it may fall.
**DON'T** start the engine without first making sure that the transmission is in Neutral (or Park where applicable) and the parking brake is set.
**DON'T** remove the radiator cap from a hot cooling system - let it cool or cover it with a cloth and release the pressure gradually.
**DON'T** attempt to drain the engine oil until you are sure it has cooled to the point that it will not burn you.
**DON'T** touch any part of the engine or exhaust system until it has cooled sufficiently to avoid burns.
**DON'T** siphon toxic liquids such as gasoline, antifreeze and brake fluid by mouth, or allow them to remain on your skin.
**DON'T** inhale brake lining dust - it is potentially hazardous (see Asbestos below).
**DON'T** allow spilled oil or grease to remain on the floor - wipe it up before someone slips on it.
**DON'T** use loose fitting wrenches or other tools which may slip and cause injury.
**DON'T** push on wrenches when loosening or tightening nuts or bolts. Always try to pull the wrench toward you. If the situation calls for pushing the wrench away, push with an open hand to avoid scraped knuckles if the wrench should slip.
**DON'T** attempt to lift a heavy component alone - get someone to help you.
**DON'T** rush or take unsafe shortcuts to finish a job.
**DON'T** allow children or animals in or around the vehicle while you are working on it.
**DO** wear eye protection when using power tools such as a drill, sander, bench grinder, etc. and when working under a vehicle.
**DO** keep loose clothing and long hair well out of the way of moving parts.
**DO** make sure that any hoist used has a safe working load rating adequate for the job.
**DO** get someone to check on you periodically when working alone on a vehicle.
**DO** carry out work in a logical sequence and make sure that everything is correctly assembled and tightened.
**DO** keep chemicals and fluids tightly capped and out of the reach of children and pets.
**DO** remember that your vehicle's safety affects that of yourself and others. If in doubt on any point, get professional advice.

### STEERING, SUSPENSION AND BRAKES

These systems are essential to driving safety, so make sure you have a qualified shop or individual check your work. Also, compressed suspension springs can cause injury if released suddenly - be sure to use a spring compressor.

### AIRBAGS

Airbags are explosive devices that can CAUSE injury if they deploy while you're working on the vehicle. Follow the manufacturer's instructions to disable the airbag whenever you're working in the vicinity of airbag components.

### ASBESTOS

Certain friction, insulating, sealing, and other products - such as brake linings, brake bands, clutch linings, torque converters, gaskets, etc. - may contain asbestos or other hazardous friction material. Extreme care must be taken to avoid inhalation of dust from such products, since it is hazardous to health. If in doubt, assume that they do contain asbestos.

### FIRE

Remember at all times that gasoline is highly flammable. Never smoke or have any kind of open flame around when working on a vehicle. But the risk does not end there. A spark caused by an electrical short circuit, by two metal surfaces contacting each other, or even by static electricity built up in your body under certain conditions, can ignite gasoline vapors, which in a confined space are highly explosive. Do not, under any circumstances, use gasoline for cleaning parts. Use an approved safety solvent.

Always disconnect the battery ground (-) cable at the battery before working on any part of the fuel system or electrical system. Never risk spilling fuel on a hot engine or exhaust component. It is strongly recommended that a fire extinguisher suitable for use on fuel and electrical fires be kept handy in the garage or workshop at all times. Never try to extinguish a fuel or electrical fire with water.

### FUMES

Certain fumes are highly toxic and can quickly cause unconsciousness and even death if inhaled to any extent. Gasoline vapor falls into this category, as do the vapors from some cleaning solvents. Any draining or pouring of such volatile fluids should be done in a well ventilated area.

When using cleaning fluids and solvents, read the instructions on the container carefully. Never use materials from unmarked containers.

Never run the engine in an enclosed space, such as a garage. Exhaust fumes contain carbon monoxide, which is extremely poisonous. If you need to run the engine, always do so in the open air, or at least have the rear of the vehicle outside the work area.

### THE BATTERY

Never create a spark or allow a bare light bulb near a battery. They normally give off a certain amount of hydrogen gas, which is highly explosive.

Always disconnect the battery ground (-) cable at the battery before working on the fuel or electrical systems.

If possible, loosen the filler caps or cover when charging the battery from an external source (this does not apply to sealed or maintenance-free batteries). Do not charge at an excessive rate or the battery may burst.

Take care when adding water to a non maintenance-free battery and when carrying a battery. The electrolyte, even when diluted, is very corrosive and should not be allowed to contact clothing or skin.

Always wear eye protection when cleaning the battery to prevent the caustic deposits from entering your eyes.

### HOUSEHOLD CURRENT

When using an electric power tool, inspection light, etc., which operates on household current, always make sure that the tool is correctly connected to its plug and that, where necessary, it is properly grounded. Do not use such items in damp conditions and, again, do not create a spark or apply excessive heat in the vicinity of fuel or fuel vapor.

### SECONDARY IGNITION SYSTEM VOLTAGE

A severe electric shock can result from touching certain parts of the ignition system (such as the spark plug wires) when the engine is running or being cranked, particularly if components are damp or the insulation is defective. In the case of an electronic ignition system, the secondary system voltage is much higher and could prove fatal.

### HYDROFLUORIC ACID

This extremely corrosive acid is formed when certain types of synthetic rubber, found in some O-rings, oil seals, fuel hoses, etc. are exposed to temperatures above 750-degrees F (400-degrees C). The rubber changes into a charred or sticky substance containing the acid. *Once formed, the acid remains dangerous for years. If it gets onto the skin, it may be necessary to amputate the limb concerned.*

When dealing with a vehicle which has suffered a fire, or with components salvaged from such a vehicle, wear protective gloves and discard them after use.

# TROUBLESHOOTING 0-25

## Troubleshooting

### CONTENTS

Section   Symptom

### Engine
1. Engine will not rotate when attempting to start
2. Engine rotates but will not start
3. Starter motor operates without rotating engine
4. Engine hard to start when cold
5. Engine hard to start when hot
6. Starter motor noisy or excessively rough in engagement
7. Engine starts but stops immediately
8. Engine lopes while idling or idles erratically
9. Engine misses at idle speed
10. Engine misses throughout driving speed range
11. Engine stalls
12. Engine lacks power
13. Engine backfires
14. Pinging or knocking engine sounds during acceleration or uphill
15. Engine continues to run after switching off

### Engine electrical system
16. Battery will not hold a charge
17. Alternator light fails to go out.
18. Alternator light fails to come on when key is turned on

### Fuel system
19. Excessive fuel consumption
20. Fuel leakage and/or fuel odor

### Cooling system
21. Overheating
22. Overcooling
23. External coolant leakage
24. Internal coolant leakage
25. Coolant loss
26. Poor coolant circulation

### Automatic transmission
27. General shift mechanism problems
28. Transmission will not downshift with accelerator pedal pressed to the floor
29. Transmission slips, shifts rough, is noisy or has no drive in forward or reverse gears
30. Fluid leakage

Section   Symptom

### Transfer case
31. Transfer case is difficult to shift into the desired range
32. Transfer case noisy in all gears
33. Noisy or jumps out of four-wheel drive Low range
34. Lubricant leaks from the vent or output shaft seals

### Driveshaft
35. Oil leak at front of driveshaft
36. Knock or clunk when the transmission is under initial load (just after transmission is put into gear)
37. Metallic grinding sound consistent with vehicle speed
38. Vibration

### Axles
39. Noise
40. Vibration
41. Oil leakage

### Driveaxles (4WD)
42. Clicking
43. Shudder on acceleration
44. Vibration on highway

### Brakes
45. Vehicle pulls to one side during braking
46. Noise (grinding or high-pitched squeal with the brakes applied)
47. Excessive brake pedal travel
48. Brake pedal feels spongy when depressed
49. Excessive effort required to stop vehicle
50. Pedal travels to the floor with little resistance
51. Brake pedal pulsates during brake application

### Suspension and steering systems
52. Vehicle pulls to one side
53. Shimmy, shake or vibration
54. Excessive pitching and/or rolling around corners or during braking
55. Excessively stiff steering
56. Excessive play in steering
57. Lack of power assistance
58. Excessive tire wear (not specific to one area)
59. Excessive tire wear on outside edge
60. Excessive tire wear on inside edge
61. Tire tread worn in one place

# TROUBLESHOOTING

This section provides an easy reference guide to the more common problems that may occur during the operation of your vehicle. These problems and possible causes are grouped under various components or systems; i.e. Engine, Cooling System, etc., and also refer to the Chapter and/or Section that deals with the problem.

Remember that successful troubleshooting is not a mysterious black art practiced only by professional mechanics. It's simply the result of a bit of knowledge combined with an intelligent, systematic approach to the problem. Always work by a process of elimination, starting with the simplest solution and working through to the most complex - and never overlook the obvious. Anyone can forget to fill the gas tank or leave the lights on overnight, so don't assume that you are above such oversights.

Finally, always get clear in your mind why a problem has occurred and take steps to ensure that it doesn't happen again. If the electrical system fails because of a poor connection, check all other connections in the system to make sure that they don't fail as well. If a particular fuse continues to blow, find out why - don't just go on replacing fuses. Remember, failure of a small component can often be indicative of potential failure or incorrect functioning of a more important component or system.

## ENGINE

### 1  Engine will not rotate when attempting to start

1  Battery terminal connections loose or corroded. Check the cable terminals at the battery. Tighten the cable or remove corrosion as necessary.
2  Battery discharged or faulty. If the cable connections are clean and tight on the battery posts, turn the key to the On position and switch on the headlights and/or windshield wipers. If they fail to function, the battery is discharged.
3  Automatic transmission not completely engaged in Park.
4  Broken, loose or disconnected wiring in the starting circuit. Inspect all wiring and connectors at the battery, starter solenoid and ignition switch.
5  Starter motor pinion jammed in driveplate ring gear. Remove starter and inspect pinion and driveplate (Chapter 5).
6  Starter solenoid faulty (Chapter 5).
7  Starter motor faulty (Chapter 5).
8  Ignition switch faulty (Chapter 12).
9  Starter relay faulty.
10  Body Control Module (BCM) or Powertrain Control Module (PCM) faulty.

### 2  Engine rotates but will not start

1  Fuel tank empty, fuel filter plugged or fuel line restricted.
2  Fault in the fuel injection system (Chapter 4).
3  Battery discharged (engine rotates slowly). Check the operation of electrical components as described in the previous Section.
4  Battery terminal connections loose or corroded (see previous Section).
5  Fuel pump faulty (Chapter 4).
6  Ignition system faulty (see Chapter 5).
7  Worn, faulty or incorrectly gapped spark plugs (Chapter 1).
8  Broken, loose or disconnected wires at the ignition coils (Chapter 5).

### 3  Starter motor operates without rotating engine

1  Starter pinion sticking. Remove the starter (Chapter 5) and inspect.
2  Starter pinion or driveplate teeth worn or broken. Remove the driveplate access cover and inspect.

### 4  Engine hard to start when cold

1  Discharged or low battery. Check as described in Section 1.
2  Fault in the fuel or ignition systems (Chapters 4 and 5).
3  Injector(s) leaking (Chapter 4).

### 5  Engine hard to start when hot

1  Air filter clogged (Chapter 1).
2  Fault in the fuel or ignition systems (Chapters 4 and 5).
3  Fuel not reaching the injectors (see Chapter 4).
4  Low cylinder compression (Chapter 2).

### 6  Starter motor noisy or excessively rough in engagement

1  Pinion or flywheel gear teeth worn or broken. Remove the cover at the rear of the engine (if equipped) and inspect.
2  Starter motor mounting bolts loose or missing.

### 7  Engine starts but stops immediately

1  Fault in the fuel or ignition systems (Chapters 4 and 5).
2  Vacuum leak at the gasket surfaces of the intake manifold or throttle body. Make sure all mounting bolts/nuts are tightened securely and all vacuum hoses connected to the manifold are positioned properly and in good condition.
3  Restricted intake or exhaust systems (Chapter 4).

### 8  Engine lopes while idling or idles erratically

1  Vacuum leakage. Check the mounting bolts/nuts at the throttle body and intake manifold for tightness. Make sure all vacuum hoses are connected and in good condition. Use a stethoscope or a length of fuel hose held against your ear to listen for vacuum leaks while the engine is running. A hissing sound will be heard. A soapy water solution will also detect leaks.
2  Fault in the fuel or ignition systems (Chapters 4 and 5).
3  Plugged PCV valve or hose (see Chapter 1).
4  Air filter clogged (Chapter 1).
5  Fuel pump not delivering sufficient fuel to the fuel injectors (see Chapter 4).
6  Leaking head gasket. Perform a compression check (Chapter 2C).
7  Camshaft lobes worn (Chapter 2).
8  Faulty valve lifter (Chapter 2).

## 9  Engine misses at idle speed

1  Spark plugs worn, fouled or not gapped properly (Chapter 1).
2  Fault in the fuel or ignition systems (Chapters 4 and 5).
3  Vacuum leaks at intake manifold or hose connections. Check as described in Section 8.
4  Uneven or low cylinder compression. Check compression as described in Chapter 2C.

## 10  Engine misses throughout driving speed range

1  Fuel filter clogged and/or impurities in the fuel system (Chapter 1).
2  Faulty or incorrectly gapped spark plugs (Chapter 1).
3  Fault in the fuel or ignition systems (Chapters 4 and 5).
4  Faulty emissions system components (Chapter 6).
5  Low or uneven cylinder compression pressures. Remove the spark plugs and test the compression with a gauge (Chapter 2C).
6  Vacuum leaks at the throttle body, intake manifold or vacuum hoses (see Section 8).

## 11  Engine stalls

1  Fuel filter clogged and/or water and impurities in the fuel system (Chapter 1).
2  Fault in the fuel system or sensors (Chapters 4 and 6).
3  Faulty emissions system components (Chapter 6).
4  Faulty or incorrectly gapped spark plugs (Chapter 1).
5  Vacuum leak at the throttle body, intake manifold or vacuum hoses. Check as described in Section 8.

## 12  Engine lacks power

1  Fault in the fuel or ignition systems (Chapters 4 and 5).
2  Faulty or incorrectly gapped spark plugs (Chapter 1).
3  Faulty coils (Chapter 5).
4  Brakes binding (Chapter 1).
5  Automatic transmission fluid level incorrect (Chapter 1).
6  Fuel filter clogged and/or impurities in the fuel system (Chapter 1).
7  Emissions control system not functioning properly (Chapter 6).
8  Use of substandard fuel. Fill the tank with the proper fuel.
9  Low or uneven cylinder compression pressures. Test with a compression tester, which will detect leaking valves and/or a blown head gasket (Chapter 2).
10  Restriction in the intake or exhaust system (Chapter 4).

## 13  Engine backfires

1  Emissions system not functioning properly (Chapter 6).
2  Fault in the fuel or ignition systems (Chapters 4 and 5).
3  Vacuum leak at the throttle body, intake manifold or vacuum hoses. Check as described in Section 8.
4  Valves sticking (Chapter 2).

## 14  Pinging or knocking engine sounds during acceleration or uphill

1  Incorrect grade of fuel. Fill the tank with fuel of the proper octane rating.

2  Fault in the fuel or ignition systems (Chapters 4 and 5).
3  Improper spark plugs. Check the plug type against the VECI label located in the engine compartment. Also check the plugs for damage (Chapter 1).
4  Faulty emissions system (Chapter 6).
5  Vacuum leak. Check as described in Section 9.

## 15  Engine continues to run after switching off

Faulty ignition switch (Chapter 12).

# ENGINE ELECTRICAL SYSTEM

## 16  Battery will not hold a charge

1  Drivebelt or tensioner defective (Chapter 1).
2  Electrolyte level low or battery discharged (Chapter 1).
3  Battery terminals loose or corroded (Chapter 1).
4  Alternator not charging properly (Chapter 5).
5  Loose, broken or faulty wiring in the charging circuit (Chapter 5).
6  Short in the vehicle wiring causing a continuous drain on the battery (refer to Chapter 12 and the Wiring Diagrams).
7  Battery defective internally.

## 17  Alternator light fails to go out

1  Fault in the alternator or charging circuit (Chapter 5).
2  Drivebelt or tensioner defective (Chapter 1).

## 18  Alternator light fails to come on when key is turned on

1  Instrument cluster warning light bulb defective (Chapter 12).
2  Alternator faulty (Chapter 5).
3  Fault in the instrument cluster printed circuit, dashboard wiring or bulb holder (Chapter 12).

# FUEL SYSTEM

## 19  Excessive fuel consumption

1  Dirty or clogged air filter element (Chapter 1).
2  Emissions system not functioning properly (Chapter 6).
3  Fault in the fuel or ignition systems (Chapters 4 and 5).
4  Low tire pressure or incorrect tire size (Chapter 1).
5  Restricted exhaust system (Chapter 4).
6  Brakes binding (Chapter 9).

## 20  Fuel leakage and/or fuel odor

1  Leak in a fuel feed or vent line (Chapter 4).
2  Tank overfilled. Fill only to automatic shut-off.
3  Evaporative emissions system canister clogged (Chapter 6).
4  Vapor leaks from system lines or injectors (Chapter 4).

# 0-28 TROUBLESHOOTING

## COOLING SYSTEM

### 21 Overheating

1. Insufficient coolant in the system (Chapter 1).
2. Drivebelt or tensioner defective (Chapter 1).
3. Radiator core blocked or radiator grille dirty and restricted (see Chapter 3).
4. Thermostat faulty (Chapter 3).
5. Fan blades broken or cracked (Chapter 3).
6. Expansion tank cap not maintaining proper pressure. Have the cap pressure tested by a gas station or repair shop.
7. Fault in electrical circuit for cooling fans (Chapter 3).

### 22 Overcooling

1. Thermostat faulty (Chapter 3).
2. Inaccurate temperature gauge (Chapter 12).
3. Fault in electrical circuit for cooling fans (Chapter 3).

### 23 External coolant leakage

1. Deteriorated or damaged hoses or loose clamps. Replace hoses and/or tighten the clamps at the hose connections (Chapter 1).
2. Water pump seals defective. If this is the case, water will drip from the weep hole in the water pump body (Chapter 3).
3. Leakage from the radiator core or side tank(s). This will require the radiator to be professionally repaired (see Chapter 3 for removal procedures).
4. Engine drain plug(s) leaking (Chapter 1) or water jacket core plugs leaking.
5. Leakage at the heater core. Signs of leakage should show up on interior carpeting (Chapter 3).

### 24 Internal coolant leakage

➡ **Note: Internal coolant leaks can usually be detected by examining the oil. Check the dipstick and inside of the valve cover for water deposits and an oil consistency like that of a milkshake.**

1. Leaking cylinder head gasket. Have the cooling system pressure tested.
2. Cracked cylinder bore or cylinder head. Remove the head(s) and inspect (Chapter 2).
3. Leaking intake manifold gasket (Chapter 2).

### 25 Coolant loss

1. Too much coolant in the system (Chapter 1).
2. Coolant boiling away due to overheating (see Section 15).
3. External or internal leakage (see Sections 23 and 24).
4. Faulty expansion tank cap. Have the cap pressure tested.

### 26 Poor coolant circulation

1. Inoperative water pump. A quick test is to pinch the top radiator hose closed with your hand while the engine is idling, then let it loose. You should feel the surge of coolant if the pump is working properly (see Chapter 1).
2. Restriction in the cooling system. Drain, flush and refill the system (Chapter 1). If necessary, remove the radiator (Chapter 3) and have it reverse flushed.
3. Drivebelt or tensioner defective (Chapter 1).
4. Thermostat sticking (Chapter 3).
5. Drivebelt incorrectly routed, causing the pump to turn backward (Chapter 1).

## AUTOMATIC TRANSMISSION

➡ **Note: Due to the complexity of the automatic transmission, it's difficult for the home mechanic to properly diagnose and service this component. For problems other than the following, the vehicle should be taken to a dealer service department or a transmission shop.**

### 27 General shift mechanism problems

1. Chapter 7A deals with checking and adjusting the shift cable on automatic transmissions. Common problems that may be attributed to poorly adjusted cable are:
    a) *Engine starting in gears other than Park or Neutral.*
    b) *Indicator on shifter pointing to a gear other than the one actually being selected.*
    c) *Vehicle moves when in Park.*
2. Refer to Chapter 7A to adjust the linkage.
3. Problem with the electronic shift solenoid. Check for Diagnostic Trouble Codes (Chapter 6).

### 28 Transmission will not downshift with accelerator pedal pressed to the floor

Transmission pressure control solenoid valve faulty. Check for Diagnostic Trouble Codes (Chapter 6).

### 29 Transmission slips, shifts rough, is noisy or has no drive in forward or reverse gears

1. Of the many probable causes for the above problems, the home mechanic should be concerned with only one possibility - fluid level.
2. Before taking the vehicle to a repair shop, check the level and condition of the fluid as described in Chapter 1. Correct fluid level as necessary or change the fluid and filter if needed. If the problem persists, have a professional diagnose the probable cause.
3. If the transmission shifts late and the shifts are harsh, suspect a faulty transmission pressure control solenoid valve. Check for Diagnostic Trouble Codes (Chapter 6).

### 30 Fluid leakage

1. Automatic transmission fluid is a deep red color. Fluid leaks should not be confused with engine oil, which can easily be blown by airflow to the transmission.
2. To pinpoint a leak, first remove all built-up dirt and grime from around the transmission. Degreasing agents and/or steam cleaning will achieve this. With the underside clean, drive the vehicle at low speeds so airflow will not blow the leak far from its source. Raise the vehicle and determine where the leak is coming from. Common areas of leakage are:

a) *Pan:* Tighten the mounting bolts and/or replace the pan gasket as necessary (see Chapter 1).
b) *Filler pipe:* Replace the rubber seal where the pipe enters the transmission case.
c) *Transmission oil lines:* Tighten the connectors where the lines enter the transmission case and/or replace the lines.
d) *Vent pipe:* Transmission overfilled and/or water in fluid (see checking procedures, Chapter 1).
e) *Speed sensor connector:* Replace the O-ring where the vehicle speed sensor enters the transmission case (Chapter 6).

## TRANSFER CASE

### 31 Transfer case is difficult to shift into the desired range

1 Speed may be too great to permit engagement. Stop the vehicle and shift into the desired range.
2 Shift linkage loose, bent or binding. Check the linkage for damage or wear and replace or lubricate as necessary (Chapter 7B).
3 If the vehicle has been driven on a paved surface for some time, the driveline torque can make shifting difficult. Stop and shift into two-wheel drive on paved or hard surfaces.
4 Insufficient or incorrect grade of lubricant. Drain and refill the transfer case with the specified lubricant (Chapter 1).
5 Worn or damaged internal components. Disassembly and overhaul of the transfer case, by a qualified shop, may be necessary.
6 Fault in the electrical system of the front axle or automatic transfer case. Check for Diagnostic Trouble Codes (Chapter 6).

### 32 Transfer case noisy in all gears

Insufficient or incorrect grade of lubricant. Drain and refill (Chapter 1).

### 33 Noisy or jumps out of four-wheel drive Low range

1 Transfer case not fully engaged. Stop the vehicle, shift into Neutral and then engage 4L.
2 Shift linkage loose, worn or binding. Tighten, repair or lubricate linkage as necessary.
3 Shift fork cracked, inserts worn or fork binding on the rail. Disassemble and repair as necessary (Chapter 7B).
4 Fault in the electrical system of the front axle or automatic transfer case. Check for Diagnostic Trouble Codes (Chapter 6).

### 34 Lubricant leaks from the vent or output shaft seals

1 Transfer case is overfilled. Drain to the proper level (Chapter 1).
2 Vent is clogged or jammed closed. Clear or replace the vent.
3 Output shaft seal incorrectly installed or damaged. Replace the seal and check contact surfaces for nicks and scoring.

## DRIVESHAFT

### 35 Oil leak at seal end of driveshaft

Defective transmission or transfer case oil seal. See Chapter 7 for replacement procedures. While this is done, check the splined yoke for burrs or a rough condition that may be damaging the seal. Burrs can be removed with crocus cloth or a fine whetstone.

### 36 Knock or clunk when the transmission is under initial load (just after transmission is put into gear)

1 Loose or disconnected rear suspension components. Check all mounting bolts, nuts and bushings (see Chapter 10).
2 Loose driveshaft bolts. Inspect all bolts and nuts and tighten them to the specified torque.
3 Worn or damaged universal joint bearings. Check for wear (see Chapter 8).

### 37 Metallic grinding sound consistent with vehicle speed

Pronounced wear in the universal joint bearings. Check as described in Chapter 8.

### 38 Vibration

➥ **Note:** *Before assuming that the driveshaft is at fault, make sure the tires are perfectly balanced and perform the following test.*

1 Install a tachometer inside the vehicle to monitor engine speed as the vehicle is driven. Drive the vehicle and note the engine speed at which the vibration (roughness) is most pronounced. Now shift the transmission to a different gear and bring the engine speed to the same point.
2 If the vibration occurs at the same engine speed (rpm) regardless of which gear the transmission is in, the driveshaft is NOT at fault since the driveshaft speed varies.
3 If the vibration decreases or is eliminated when the transmission is in a different gear at the same engine speed, refer to the following probable causes.
4 Bent or dented driveshaft. Inspect and replace as necessary (see Chapter 8).
5 Undercoating or built-up dirt, etc. on the driveshaft. Clean the shaft thoroughly and recheck.
6 Worn universal joint bearings (see Chapter 8).
7 Driveshaft and/or companion flange out of balance. Check for missing weights on the shaft. Remove the driveshaft (see Chapter 8) and reinstall 180-degrees from original position, then retest. Have the driveshaft professionally balanced if the problem persists.

## AXLES

### 39 Noise

1 Road noise. No corrective procedures available.
2 Tire noise. Inspect tires and check tire pressures (Chapter 1).
3 Rear wheel bearings worn or damaged (Chapter 8).

### 40 Vibration

See probable causes under *Driveshaft*. Proceed under the guidelines listed for the driveshaft. If the problem persists, check the rear wheel

# 0-30 TROUBLESHOOTING

bearings by raising the rear of the vehicle and spinning the rear wheels by hand. Listen for evidence of rough (noisy) bearings. Remove and inspect (see Chapter 8).

### 41  Oil leakage

1  Pinion seal damaged (see Chapter 8).
2  Axleshaft oil seals damaged (see Chapter 8).
3  Differential inspection cover leaking. Tighten the bolts or replace the gasket as required (see Chapter 8).

## DRIVEAXLES (4WD)

### 42  Clicking noise on turns

Worn or damaged outboard CV joints (Chapter 8).

### 43  Shudder or vibration during acceleration

1  Excessive toe-in. Have alignment checked.
2  Incorrect spring heights (Chapter 10).
3  Worn or damaged inboard or outboard CV joints (Chapter 8).
4  Sticking inboard CV joint assembly (Chapter 8).

### 44  Vibration at highway speeds

1  Out-of-balance front wheels and/or tires (Chapters 1 and 10).
2  Out-of-round front tires (Chapters 1 and 10).
3  Worn CV joints (Chapter 8).

## BRAKES

➡Note: *Before assuming that a brake problem exists, make sure that the tires are in good condition and inflated properly (see Chapter 1), that the front-end alignment is correct and that the vehicle is not loaded with weight in an unequal manner.*

### 45  Vehicle pulls to one side during braking

1  Defective, damaged or oil contaminated disc brake pads on one side. Inspect as described in Chapter 9.
2  Excessive wear of brake pad material or disc on one side. Inspect and correct as necessary.
3  Loose or disconnected front suspension components. Inspect and tighten all bolts to the specified torque (Chapter 10).
4  Defective brake caliper assembly. Remove the caliper and inspect for a stuck piston or other damage (Chapter 9).
5  Inadequate lubrication of front brake caliper slide pins. Remove caliper and lubricate slide pins (Chapter 9).

### 46  Noise (grinding or high-pitched squeal with the brakes applied)

1  Brake pads or shoes worn out. Replace the pads or shoes with new ones immediately (Chapter 9).
2  Linings contaminated with dirt or grease. Replace pads.

3  Incorrect linings. Replace with correct linings.
4  Scored disc or drum (Chapter 9).

### 47  Excessive brake pedal travel

1  Partial brake system failure. Inspect the entire system (Chapter 9) and correct as required.
2  Insufficient fluid in the master cylinder. Check (Chapter 1), add fluid and bleed the system if necessary (Chapter 9).

### 48  Brake pedal feels spongy when depressed

1  Air in the hydraulic lines. Bleed the brake system (Chapter 9).
2  Faulty flexible hoses. Inspect all system hoses and lines. Replace parts as necessary.
3  Master cylinder mounting bolts/nuts loose.
4  Master cylinder defective (Chapter 9).

### 49  Excessive effort required to stop vehicle

1  Power brake booster not operating properly (see check in Chapter 1, repairs in Chapter 9).
2  Excessively worn pads or shoes. Inspect and replace if necessary (Chapter 9).
3  One or more caliper or wheel cylinder pistons seized or sticking. Inspect and replace as required (Chapter 9).
4  Brake pads contaminated with oil or grease. Inspect and replace as required (Chapter 9).
5  New pads installed and not yet seated. It will take a while for the new material to seat against the disc.

### 50  Pedal travels to the floor with little resistance

1  Little or no fluid in the master cylinder reservoir caused by leaking caliper piston(s), loose, damaged or disconnected brake lines. Inspect the entire system and correct as necessary.
2  Faulty master cylinder (Chapter 9).

### 51  Brake pedal pulsates during brake application

1  Disc(s) warped. Check for excessive lateral runout and parallelism. Have the discs resurfaced or replace them with new ones (Chapter 9).
2  Drums out-of-round. Have the drums resurfaced or replace them with new ones (Chapter 9).

## SUSPENSION AND STEERING SYSTEMS

### 52  Vehicle pulls to one side

1  Tire pressures uneven or tires mismatched (Chapter 1).
2  Defective tire (Chapter 1).
3  Excessive wear in suspension or steering components (Chapter 10).
4  Front end in need of alignment.
5  Front brakes dragging. Inspect the brakes as described in Chapter 9.

## 53 Shimmy, shake or vibration

1. Tire or wheel out-of-balance or out-of-round. Have professionally balanced.
2. Loose, worn or out-of-adjustment front wheel bearings (Chapter 1).
3. Shock absorbers and/or suspension components worn or damaged (Chapter 10).

## 54 Excessive pitching and/or rolling around corners or during braking

1. Defective shock absorbers. Replace as a set (Chapter 10).
2. Broken or weak springs and/or suspension components. Inspect as described in Chapters 1 and 10.

## 55 Excessively stiff steering

1. Lack of fluid in power steering fluid reservoir (Chapter 1).
2. Incorrect tire pressures (Chapter 1).
3. Lack of lubrication at steering joints (see Chapter 1).
4. Front end out of alignment.
5. Lack of power assistance (see Section 57).

## 56 Excessive play in steering

1. Worn front wheel bearings (Chapter 10).
2. Excessive wear in suspension or steering components (Chapter 10).
3. Steering gear damaged or out of adjustment (Chapter 10).

## 57 Lack of power assistance

1. Drivebelt or tensioner faulty (Chapter 1).
2. Fluid level low (Chapter 1).
3. Hoses or lines restricted. Inspect and replace parts as necessary.
4. Air in power steering system. Bleed the system (Chapter 10).

## 58 Excessive tire wear (not specific to one area)

1. Incorrect tire pressures (Chapter 1).
2. Tires out-of-balance. Have professionally balanced.
3. Wheels damaged. Inspect and replace as necessary.
4. Suspension or steering components excessively worn (Chapter 10).

## 59 Excessive tire wear on outside edge

1. Inflation pressures incorrect (Chapter 1).
2. Excessive speed in turns.
3. Front-end alignment incorrect. Have the front end professionally aligned.
4. Suspension arm bent or twisted (Chapter 10).

## 60 Excessive tire wear on inside edge

1. Inflation pressures incorrect (Chapter 1).
2. Front-end alignment incorrect. Have the front end professionally aligned.
3. Loose or damaged steering components (Chapter 10).

## 61 Tire tread worn in one place

1. Tires out-of-balance.
2. Damaged or buckled wheel. Inspect and replace if necessary.
3. Defective tire (Chapter 1).

## Notes

**Section**

1   Maintenance schedule
2   Introduction
3   Tune-up general information
4   Fluid level checks
5   Tire and tire pressure checks
6   Power steering fluid level check
7   Automatic transmission fluid level check
8   Engine oil and filter change
9   Seat belt check
10  Wiper blade inspection and replacement
11  Battery check, maintenance and charging
12  Drivebelt check, adjustment and replacement
13  Underhood hose check and replacement
14  Cooling system check
15  Tire rotation
16  Differential lubricant level check
17  Chassis lubrication
18  Fuel system check
19  Brake system check
20  Exhaust system check
21  Transfer case lubricant level check (4WD models)
22  Brake fluid change
23  Air filter replacement
24  Spark plug replacement
25  Cooling system servicing (draining, flushing and refilling)
26  Suspension, steering and driveaxle boot check
27  Automatic transmission fluid and filter change
28  Transfer case lubricant change (4WD models)
29  Differential lubricant change
30  Positive Crankcase Ventilation (PCV) valve replacement (V6 engine)
31  Spark plug wire check and replacement

**Reference to other Chapters**
CHECK ENGINE light on - See Chapter 6

# 1
# TUNE-UP AND ROUTINE MAINTENANCE

# 1-2 TUNE-UP AND ROUTINE MAINTENANCE

## 1 Maintenance schedule

The following maintenance intervals are based on the assumption that the vehicle owner will be doing the maintenance or service work, as opposed to having a dealer service department do the work. These are the minimum maintenance intervals recommended by the factory for vehicles that are driven daily. If you wish to keep your vehicle in peak condition at all times, you may wish to perform some of these procedures even more often. Because frequent maintenance enhances the efficiency, performance and resale value of your car, we encourage you to do so. If you drive in dusty areas, tow a trailer, idle or drive at low speeds for extended periods or drive for short distances (less than four miles) in below freezing temperatures, shorter intervals are also recommended.

When the vehicle is new, follow the maintenance schedule to the letter, record the maintenance performed in your owners manual and keep all receipts to protect the new vehicle warranty. In many cases the initial maintenance check is done at no cost to the owner (check with your dealer service department for more information).

### EVERY 250 MILES OR WEEKLY, WHICHEVER COMES FIRST

Check the engine oil level (Section 4)
Check the coolant level (Section 4)
Check the windshield washer fluid level (Section 4)
Check the brake and clutch fluid levels (Section 4)
Check the tires and tire pressures (Section 5)

### EVERY 3000 MILES OR 3 MONTHS, WHICHEVER COMES FIRST

*All items listed above, plus . . .*
Check the power steering fluid level (Section 6)
Check the automatic transmission fluid level (Section 7)
Change the engine oil and filter (Section 8)

### EVERY 6000 MILES OR 6 MONTHS, WHICHEVER COMES FIRST

*All items listed above, plus . . .*
Check the seat belts (Section 9)
Inspect the windshield wiper blades (Section 10)
Check and service the battery (Section 11)
Check the engine drivebelt (Section 12)
Inspect underhood hoses (Section 13)
Check the cooling system (Section 14)
Rotate the tires (Section 15)

### EVERY 15,000 MILES OR 12 MONTHS, WHICHEVER COMES FIRST

*All items listed above, plus . . .*
Check the differential lubricant level in the front (4x4) and rear axles (Section 16)
Lubricate the chassis (Section 17)
Check the fuel system (Section 18)
Check the brake system (Section 19)*
Check the exhaust system (Section 20)
Check the transfer case lubricant level (4WD models) (Section 21)
Check the air filter (Section 23)*

### EVERY 30,000 MILES OR 30 MONTHS, WHICHEVER COMES FIRST

*All items listed above, plus . . .*
Change the brake fluid (Section 22)
Replace the air filter (Section 23)*
Replace the spark plugs (conventional, non-platinum or non-iridium type) (Section 24)**
Service the cooling system (drain, flush and refill) (green-colored ethylene glycol antifreeze only) (Section 25)
Check the steering, suspension and driveaxle boots (Section 26)
Change the automatic transmission fluid and filter (Section 27)**

### EVERY 60,000 MILES OR 48 MONTHS, WHICHEVER COMES FIRST

Change the transfer case lubricant (Section 28)
Change the differential lubricant (Section 29)**
Replace the Positive Crankcase Ventilation (PCV) valve (V6 engine) (Section 30)

### EVERY 100,000 MILES OR 60 MONTHS, WHICHEVER COMES FIRST

Replace the spark plugs (platinum or iridium type) (Section 24)
Inspect/replace the spark plug wires (Section 31)
Service the cooling system (drain, flush and refill) (orange-colored DEX-COOL antifreeze only) (Section 25)

\* This item is affected by "severe" operating conditions, as described below. If the vehicle is operated under severe conditions, perform all maintenance indicated with an asterisk (*) at half the indicated intervals. Severe conditions exist if you mainly operate the vehicle . . .

in dusty areas
towing a trailer
idling for extended periods
driving at low speeds when outside temperatures remain below freezing and most trips are less than four miles long

\*\* Perform this procedure at half the recommended interval if operated under one or more of the following conditions:

in heavy city traffic where the outside temperature regularly reaches 90-degrees F or higher in hilly or mountainous terrain
frequent trailer towing
if the vehicle has been driven through deep water

# TUNE-UP AND ROUTINE MAINTENANCE

**Engine compartment components (V8 engine shown, V6 similar)**

1. Brake fluid reservoir
2. Underhood fuse/relay box
3. Windshield washer fluid reservoir
4. Power steering fluid reservoir
5. Drivebelt
6. Upper radiator hose
7. Air filter housing
8. Coolant expansion tank
9. Battery
10. Engine oil dipstick
11. Automatic transmission fluid dipstick
12. Engine oil filler cap

## 1-4 TUNE-UP AND ROUTINE MAINTENANCE

**Typical engine compartment underside components (V8, 4WD 1500 pick-up)**

1. Steering knuckle
2. Shock absorber lower mount
3. Steering gear boot
4. Lower balljoint
5. Driveaxle boot
6. Front brake caliper
7. Engine oil filter
8. Engine oil drain plug
9. Stabilizer bar
10. Automatic transmission fluid pan
11. Exhaust pipe/catalytic converter

# TUNE-UP AND ROUTINE MAINTENANCE   1-5

**Typical rear underside components**

1. Leaf spring
2. Shock absorber
3. Drum brake assembly
4. Parking brake cables
5. Fuel tank
6. Muffler
7. Differential cover

# 1-6 TUNE-UP AND ROUTINE MAINTENANCE

## 2 Introduction

This Chapter is designed to help the home mechanic maintain the Chevrolet/GMC pickup or Avalanche/Tahoe/Yukon/Suburban with the goals of maximum performance, economy, safety and reliability in mind.

Included is a master maintenance schedule, followed by procedures dealing specifically with each item on the schedule. Visual checks, adjustments, component replacement and other helpful items are included. Refer to the accompanying illustrations of the engine compartment and the underside of the vehicle for the locations of various components.

Servicing your vehicle in accordance with the mileage/time maintenance schedule and the step-by-step procedures will result in a planned maintenance program that should produce a long and reliable service life. Keep in mind that it's a comprehensive plan, so maintaining some items but not others at the specified intervals will produce the same results.

As you service your vehicle, you will discover that many of the procedures can - and should - be grouped together because of the nature of the particular procedure you're performing or because of the close proximity of two otherwise unrelated components to one another.

For example, if the vehicle is raised for chassis lubrication, you should inspect the exhaust, suspension, steering and fuel systems while you're under the vehicle. When you're rotating the tires, it makes good sense to check the brakes since the wheels are already removed. Finally, let's suppose you have to borrow or rent a torque wrench. Even if you only need it to tighten the spark plugs, you might as well check the torque of as many critical fasteners as time allows.

The first step in this maintenance program is to prepare yourself before the actual work begins. Read through all the procedures you're planning to do, then gather up all the parts and tools needed. If it looks like you might run into problems during a particular job, seek advice from a mechanic or an experienced do-it-yourselfer.

### OWNER'S MANUAL AND VECI LABEL INFORMATION

Your vehicle owner's manual was written for your year and model and contains very specific information on component locations, specifications, fuse ratings, part numbers, etc. The Owner's Manual is an important resource for the do-it-yourselfer to have; if one was not supplied with your vehicle, it can generally be ordered from a dealer parts department.

Among other important information, the Vehicle Emissions Control Information (VECI) label contains specifications and procedures for tune-up adjustments (if applicable) and spark plugs (see Chapter 6 for more information on the VECI label). The information on this label is the exact maintenance data recommended by the manufacturer. This data often varies by intended operating altitude, local emissions regulations, month of manufacture, etc.

This Chapter contains procedural details, safety information and more ambitious maintenance intervals than you might find in the manufacturer's literature. However, you may also find procedures and specifications in your Owner's Manual or VECI label that differ with what's printed here. In these cases, the Owner's Manual or VECI label can be considered correct, since it is specific to your particular vehicle.

## 3 Tune-up general information

The term tune-up is used in this manual to represent a combination of individual operations rather than one specific procedure that will maintain a gasoline engine in proper tune.

If, from the time the vehicle is new, the routine maintenance schedule is followed closely and frequent checks are made of fluid levels and high wear items, as suggested throughout this manual, the engine will be kept in relatively good running condition and the need for additional work will be minimized.

More likely than not, however, there may be times when the engine is running poorly due to lack of regular maintenance. This is even more likely if a used vehicle, which has not received regular and frequent maintenance checks, is purchased. In such cases, an engine tune-up will be needed outside of the regular routine maintenance intervals.

The first step in any tune-up or diagnostic procedure to help correct a poor running engine is a cylinder compression check. A compression check (see Chapter 2C) will help determine the condition of internal engine components and should be used as a guide for tune-up and repair procedures. If, for instance, the compression check indicates serious internal engine wear, a conventional tune-up won't improve the performance of the engine and would be a waste of time and money. Because of its importance, the compression check should be done by someone with the proper equipment and the knowledge to use it properly.

The following procedures are those most often needed to bring a generally poor running engine back into a proper state of tune.

### MINOR TUNE-UP

*Check all engine related fluids (Section 4)*
*Clean, inspect and test the battery (Section 11)*
*Check the drivebelt (Section 12)*
*Check all underhood hoses (Section 13)*
*Check the cooling system (Section 14)*
*Check the air filter (Section 23)*
*Inspect the spark plug wires (Section 30)*

### MAJOR TUNE-UP

*All items listed under Minor tune-up, plus . . .*
*Replace the air filter (Section 23)*
*Replace the spark plugs (Section 24)*
*Replace the PCV valve (V6 engine) (Section 30)*
*Replace the spark plug wires (Section 31)*
*Check the ignition system (Chapter 5)*
*Check the charging system (Chapter 5)*

# TUNE-UP AND ROUTINE MAINTENANCE

## 4  Fluid level checks (every 250 miles or weekly)

➡ **Note:** *The following are fluid level checks to be done on a 250 mile or weekly basis. Additional fluid level checks can be found in specific maintenance procedures that follow. Regardless of intervals, be alert to fluid leaks under the vehicle, which would indicate a fault to be corrected immediately.*

1  Fluids are an essential part of the lubrication, cooling, brake, clutch and windshield washer systems. Because the fluids gradually become depleted and/or contaminated during normal operation of the vehicle, they must be periodically replenished. See *Recommended lubricants and fluids* in this Chapter's Specifications before adding fluid to any of the following components.

➡ **Note:** *The vehicle must be on level ground when fluid levels are checked.*

### ENGINE OIL

▸ **Refer to illustrations 4.2, 4.4 and 4.6**

2  The engine oil level is checked with a dipstick that extends through a tube and into the oil pan at the bottom of the engine (see illustration).

3  The oil level should be checked before the vehicle has been driven, or about 5 minutes after the engine has been shut off. If the oil is checked immediately after driving the vehicle, some of the oil will remain in the upper engine components, resulting in an inaccurate reading on the dipstick.

4  Pull the dipstick out of the tube and wipe all the oil from the end with a clean rag or paper towel. Insert the clean dipstick all the way back into the tube, then pull it out again. Note the oil at the end of the dipstick. Add oil as necessary to keep the level between the cross-hatched area on the dipstick (see illustration).

5  Do not overfill the engine by adding too much oil since this may result in oil-fouled spark plugs, oil leaks or oil seal failures.

6  Oil is added to the engine after unscrewing a cap from the valve cover (see illustration).

7  Checking the oil level is an important preventive maintenance step. A consistently low oil level indicates oil leakage through damaged seals, defective gaskets or past worn rings or valve guides. If the oil looks milky or has water droplets in it, the cylinder head gasket(s) may be blown or the head(s) or block may be cracked. The engine should be checked immediately. The condition of the oil should also be checked. Whenever you check the oil level, slide your thumb and index finger up the dipstick before wiping off the oil. If you see small dirt or metal particles clinging to the dipstick, the oil should be changed (see Section 8).

**4.2  The engine oil dipstick is located on the right side of the engine (V8 engine shown, V6 similar)**

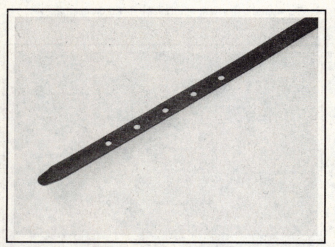

**4.4  The oil level must be maintained in the cross-hatched area, but not above it**

**4.6  Location of the oil filler cap - always make sure the area around the opening is clean before removing the cap to prevent dirt from contaminating the engine**

# 1-8 TUNE-UP AND ROUTINE MAINTENANCE

## ENGINE COOLANT

▶ Refer to illustration 4.9

### ※※ WARNING:
Do not allow antifreeze to come in contact with your skin or painted surfaces of the vehicle. Rinse off spills immediately with plenty of water. Antifreeze is highly toxic if ingested. Never leave antifreeze lying around in an open container or in puddles on the floor; children and pets are attracted by its sweet smell and may drink it. Check with local authorities on disposing of used antifreeze. Many communities have collection centers that will see that antifreeze is disposed of safely.

### ※※ CAUTION:
Never mix green-colored ethylene glycol antifreeze and orange-colored DEX-COOL silicate-free coolant because doing so will destroy the efficiency of the DEX-COOL coolant which is designed to last for 100,000 miles or five years.

➡ Note: Non-toxic antifreeze is now manufactured and available at local auto parts stores, but even this type should be disposed of properly.

8  All vehicles covered by this manual are equipped with a pressurized coolant expansion tank, located at the right side of the engine compartment, and connected by hoses to the radiator and cooling system.
9  The coolant level in the expansion tank should be checked regularly.

### ※※ WARNING:
Do not remove the pressure cap to check the coolant level when the engine is warm.

The level of coolant in the expansion tank varies with the temperature of the engine. When the engine is cold, the coolant level should be at or slightly above the COLD mark on the expansion tank (see illustration). To add coolant, slowly unscrew the cap, then add coolant until it is up to the COLD mark (see the **Caution** at the beginning of this Section).

### ※※ WARNING:
If coolant or steam escapes as you unscrew the cap, let the engine cool down longer, then remove the cap.

10  Install the cap, start the engine and run it at approximately 2000 rpm until the coolant temperature gauge reads approximately 195-degrees F. If the coolant level drops, let the engine cool, then add more until it is up to the COLD mark. In order to maintain the proper ratio of antifreeze and water, always top up the coolant level with the correct mixture. Do not use rust inhibitors or additives.
11  If the coolant level drops consistently, there may be a leak in the system. Inspect the radiator, hoses, pressure cap, drain plugs and water pump (see Section 14).
12  If no leaks are noted, have the pressure cap tested by a service station.
13  Check the condition of the coolant as well. It should be relatively clear. If it is brown or rust colored, the system should be drained, flushed and refilled. Even if the coolant appears to be normal, the corrosion inhibitors wear out, so it must be replaced at the specified intervals. If the system is filled with standard green coolant/water, it must be flushed and replaced more frequently than if the original DEX-COOL coolant is retained.

## WINDSHIELD WASHER FLUID

▶ Refer to illustration 4.14

14  Fluid for the windshield washer system is located in a plastic reservoir in the left side of the engine compartment (see illustration).
15  In milder climates, plain water can be used in the reservoir, but it should be kept no more than 2/3 full to allow for expansion if the water freezes. In colder climates, use windshield washer system antifreeze, available at any auto parts store, to lower the freezing point of the fluid. Mix the antifreeze with water in accordance with the manufacturer's directions on the container.

### ※※ CAUTION:
Don't use cooling system antifreeze - it will damage the vehicle's paint.

**4.9** The coolant expansion tank is located on the right side, near the air filter - make sure the level is up to the COLD mark when the engine is cold

**4.14** Location of the windshield washer fluid reservoir

# TUNE-UP AND ROUTINE MAINTENANCE 1-9

16  To help prevent icing in cold weather, warm the windshield with the defroster before using the washer.

## BATTERY ELECTROLYTE

17  These vehicles are equipped with a battery which is permanently sealed (except for vent holes) and has no filler caps. Water doesn't have to be added to these batteries at any time. If a maintenance-type battery is installed, the caps on the top of the battery should be removed periodically to check for a low electrolyte level. This check is most critical during the warm summer months. Add only distilled water to any battery.

## BRAKE FLUID

♦ Refer to illustration 4.19

18  The brake master cylinder is mounted on the upper left of the engine compartment firewall.

19  The translucent plastic reservoir allows the fluid inside to be checked without removing the cap (see illustration). Be sure to wipe the top of reservoir cap with a clean rag to prevent contamination of the brake system before removing the cover.

20  When adding fluid, pour it carefully into the reservoir to avoid spilling it on surrounding painted surfaces. Be sure the specified fluid is used, since mixing different types of brake fluid can cause damage to the system. See *Recommended lubricants and fluids* in this Chapter's Specifications or your owner's manual.

> ❋❋ **WARNING:**
>
> **Brake fluid can harm your eyes and damage painted surfaces, so use extreme caution when handling or pouring it. Do not use brake fluid that has been standing open or is more than one year old. Brake fluid absorbs moisture from the air. Moisture in the system can cause a dangerous loss of brake performance.**

21  At this time, the fluid and master cylinder can be inspected for contamination. The system should be drained and refilled if deposits, dirt particles or water droplets are seen in the fluid.

22  After filling the reservoir to the proper level, make sure the cover or cap is on tight to prevent fluid leakage.

23  The brake fluid level in the master cylinder will drop slightly as the pads at the front wheels wear down during normal operation. If the master cylinder requires repeated additions to keep it at the proper level, it's an indication of leakage in the brake system, which should be corrected immediately. Check all brake lines and connections (see Section 19 for more information).

24  If, upon checking the master cylinder fluid level, you discover the reservoir is empty or nearly empty, the brake system should be bled and thoroughly inspected (see Chapter 9).

4.19  Never let the brake fluid level drop below the MIN mark

## 5  Tire and tire pressure checks (every 250 miles or weekly)

♦ Refer to illustrations 5.2, 5.3, 5.4a, 5.4b and 5.8

1  Periodic inspection of the tires may spare you the inconvenience of being stranded with a flat tire. It can also provide you with vital information regarding possible problems in the steering and suspension systems before major damage occurs.

2  The original tires on this vehicle are equipped with 1/2-inch wide wear bands that will appear when tread depth reaches 1/16-inch, at which point the tires can be considered worn out. Tread wear can be monitored with a simple, inexpensive device known as a tread depth indicator (see illustration).

3  Note any abnormal tread wear (see illustration). Tread pattern irregularities such as cupping, flat spots and more wear on one side than the other are indications of front end alignment and/or balance problems. If any of these conditions are noted, take the vehicle to a tire shop or service station to correct the problem.

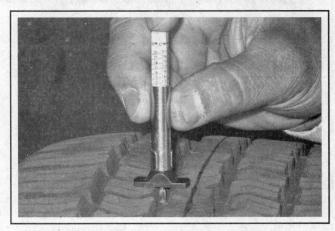

5.2  Use a tire tread depth indicator to monitor tire wear

# 1-10 TUNE-UP AND ROUTINE MAINTENANCE

UNDERINFLATION

CUPPING

Cupping may be caused by:
- Underinflation and/or mechanical irregularities such as out-of-balance condition of wheel and/or tire, and bent or damaged wheel.
- Loose or worn steering tie-rod or steering idler arm.
- Loose, damaged or worn front suspension parts.

OVERINFLATION

INCORRECT TOE-IN OR EXTREME CAMBER

FEATHERING DUE TO MISALIGNMENT

5.3 This chart will help you determine the condition of the tires and the probable cause(s) of abnormal wear

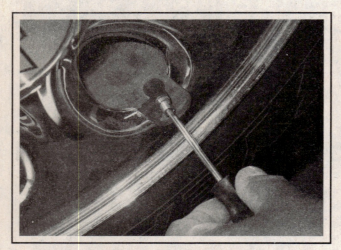

5.4a If a tire loses air on a steady basis, check the valve stem core first to make sure it's snug (valve core drivers are commonly available at auto parts stores)

5.4b If the valve stem core is tight, raise the corner of the vehicle with the low tire and spray a soapy water solution onto the tread as the tire is turned slowly - leaks will cause small bubbles to appear

4  Look closely for cuts, punctures and embedded nails or tacks. Sometimes a tire will hold air pressure for a short time or leak down very slowly after a nail has embedded itself in the tread. If a slow leak persists, check the valve stem core to make sure it's tight (see illustration). Examine the tread for an object that may have embedded itself in the tire or for a plug that may have begun to leak (radial tire punctures are repaired with a plug that's installed in a puncture). If a puncture is suspected, it can be easily verified by spraying a solution of soapy water onto the puncture area (see illustration). The soapy solution will bubble if there's a leak. Unless the puncture is unusually large, a tire shop or service station can usually repair the tire.

5  Carefully inspect the inner sidewall of each tire for evidence of brake fluid leakage. If you see any, inspect the brakes immediately.

6  Correct air pressure adds miles to the lifespan of the tires,

improves mileage and enhances overall ride quality. Tire pressure cannot be accurately estimated by looking at a tire, especially if it's a radial. A tire pressure gauge is essential. Keep an accurate gauge in the vehicle. The pressure gauges attached to the nozzles of air hoses at gas stations are often inaccurate.

7  Always check tire pressure when the tires are cold. Cold, in this case, means the vehicle has not been driven over a mile in the three hours preceding a tire pressure check. A pressure rise of four to eight pounds is not uncommon once the tires are warm.

8  Unscrew the valve cap protruding from the wheel or hubcap and push the gauge firmly onto the valve stem (see illustration). Note the reading on the gauge and compare the figure to the recommended tire pressure shown on the placard on the driver's side door pillar. Be sure to reinstall the valve cap to keep dirt and moisture out of the valve stem mechanism. Check all four tires and, if necessary, add enough air to bring them up to the recommended pressure.

9  Don't forget to keep the spare tire inflated to the specified pressure (see your owner's manual or the tire sidewall).

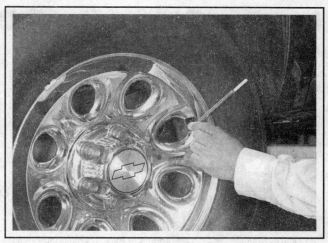

5.8  To extend the life of the tires, check the air pressure at least once a week with an accurate gauge (don't forget the spare!)

## 6  Power steering fluid level check (every 3000 miles or 3 months)

▸ Refer to illustrations 6.2 and 6.6

1  The power steering system relies on fluid which may, over a period of time, require replenishing.

2  On all models, the fluid reservoir for the power steering pump is located on the pump body at the front of the engine (see illustration).

3  For the check, the front wheels should be pointed straight ahead and the engine should be off.

4  Use a clean rag to wipe off the reservoir cap and the area around the cap. This will help prevent any foreign matter from entering the reservoir during the check.

5  Twist off the cap and check the temperature of the fluid at the end of the dipstick with your finger.

6  Wipe off the fluid with a clean rag, reinsert the dipstick, then withdraw it and read the fluid level. The fluid should be at the proper level, depending on whether it was checked hot or cold (see illustration). Never allow the fluid level to drop below the lower mark on the dipstick.

7  If additional fluid is required, pour the specified type into the reservoir, a little at a time to avoid overfilling.

8  If the reservoir requires frequent fluid additions, all power steering hoses, hose connections, steering gear and the power steering pump should be carefully checked for leaks.

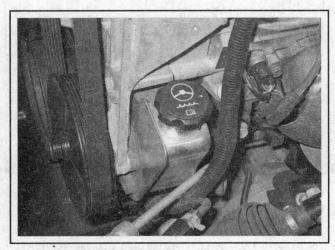

6.2  The power steering fluid dipstick is located in the power steering pump reservoir - turn the cap counterclockwise to remove it

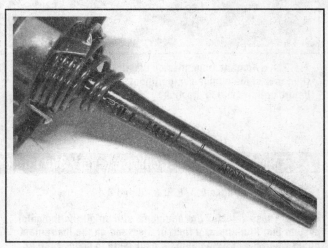

6.6  The power steering fluid dipstick has marks on it so the fluid can be checked hot or cold

# 1-12 TUNE-UP AND ROUTINE MAINTENANCE

## 7 Automatic transmission fluid level check (every 3000 miles or 3 months)

▶ Refer to illustrations 7.4 and 7.7

➡ **Note:** On some models, the Driver's Information Center (DIC) on the instrument panel is capable of displaying the transmission fluid temperature. Before starting this procedure, check to see if the DIC in your vehicle can do this. With the engine running and in Park, press the Trip/Fuel button or the trip odometer reset button and see if TRANS TEMP can be displayed. If so, you'll do this in Step 3.

1  The automatic transmission fluid level should be carefully maintained. Low fluid level can lead to slipping or loss of drive, while overfilling can cause foaming. Either condition can damage the transmission.

2  With the parking brake set, start the engine, then move the shift lever through all the gear ranges, ending in Park. The fluid level must be checked with the vehicle level and the engine running at idle.

➡ **Note:** Incorrect fluid level readings will result if the vehicle has just been driven at high speeds for an extended period, in hot weather in city traffic, or if it has been pulling a trailer. If any of these conditions apply, wait until the fluid has cooled (about 30 minutes).

3  If the DIC in your vehicle is capable of reading the temperature of the transmission fluid, display that value on the DIC now. If the fluid temperature is between 80 and 90-degrees F, you'll be using the COLD markings on the dipstick. If it's between 160 and 200-degrees F, you'll be using the HOT markings on the dipstick. If it's above 200-degrees F, wait until the fluid has cooled back down and falls into the "hot" temperature range. If it's between 90 and 160-degrees F, operate the vehicle until warms up to the "hot" range.

➡ **Note:** If the DIC in your vehicle can't display the transmission fluid temperature, proceed to the next step. You'll have to estimate the temperature of the fluid based on recent operation.

4  With the transmission at normal operating temperature, remove the dipstick from the filler tube. The dipstick is located at the rear of the engine compartment on the passenger's side (see illustration).

5  Wipe the fluid from the dipstick with a clean rag and push it back into the filler tube until the cap seats.

6  Pull the dipstick out again and note the fluid level.

7  If the fluid is cold or slightly warm (80 to 90-degrees F), the level should be in the COLD range on the dipstick (see illustration). If it's hot (160 to 200-degrees F), the level should be in the crosshatched HOT range. If additional fluid is required, add it into the dipstick tube using a funnel. Add the fluid a little at a time and keep checking the level until it's correct. It is important to not overfill the transmission.

8  The condition of the fluid should also be checked along with the level. If the fluid at the end of the dipstick is a dark reddish-brown color, or if it smells burned, it should be changed. If you are in doubt about the condition of the fluid, purchase some new fluid and compare the two for color and smell.

7.4 The automatic transmission dipstick is located at the right rear of the engine compartment - flip up the handle before pulling out the dipstick

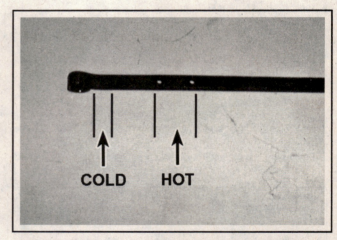

7.7 The automatic transmission fluid must be kept in the regions marked, depending on the fluid temperature

## 8 Engine oil and filter change (every 3000 miles or 3 months)

▶ Refer to illustrations 8.3, 8.8, 8.10 and 8.14

➡ **Note:** These vehicles are equipped with an oil life indicator system that illuminates a light or message on the instrument panel when the system deems it necessary to change the oil. A number of factors are taken into consideration to determine when the oil should be considered worn out. Generally, this system will allow the vehicle to accumulate more miles between oil changes than the traditional 3000-mile interval, but we believe that frequent oil changes are cheap insurance and will prolong engine life. If you do decide not to change your oil every 3000 miles and rely on the oil life indicator instead, make sure you don't exceed 7,500 miles before the oil is changed, regardless of what the oil life indicator shows.

1  Frequent oil changes are the most important preventive maintenance procedures that can be done by the home mechanic. As engine

# TUNE-UP AND ROUTINE MAINTENANCE

oil ages, it becomes diluted and contaminated, which leads to premature engine wear.

2  Although some sources recommend oil filter changes every other oil change, we feel that the minimal cost of an oil filter and the relative ease with which it is installed dictate that a new filter be installed every time the oil is changed.

3  Gather together all necessary tools and materials before beginning this procedure (see illustration). You should also have plenty of clean rags and newspapers handy to mop up any spills.

4  Warm the engine to normal operating temperature. If the new oil or any tools are needed, use this warm-up time to gather everything necessary for the job. The correct type of oil for your application can be found in *Recommended lubricants and fluids* in this Chapter's Specifications.

5  Raise the vehicle and support it securely on jackstands.

### ✳✳ WARNING:
**Do not work under a vehicle which is supported only by a jack.**

6  If this is your first oil change, familiarize yourself with the locations of the oil drain plug and the oil filter.

7  Set the drain pan under the drain plug. Keep in mind that the oil will initially flow from the pan with some force; position the pan accordingly.

8  Being careful not to touch any of the hot exhaust components, use a wrench to remove the drain plug near the bottom of the oil pan (see illustration). Depending on how hot the oil is, you may want to wear gloves while unscrewing the plug the final few turns.

9  After all the oil has drained, wipe off the drain plug with a clean rag. Clean the area around the drain plug opening and reinstall the plug. Tighten the plug securely with the wrench. If a torque wrench is available, use it to tighten the plug to the torque listed in this Chapter's Specifications.

10  Move the drain pan into position under the oil filter, then use the oil filter wrench to loosen the oil filter (see illustration).

11  Completely unscrew the old filter. Be careful: it's full of oil. Empty the oil inside the filter into the drain pan.

12  Compare the old filter with the new one to make sure they're the same type.

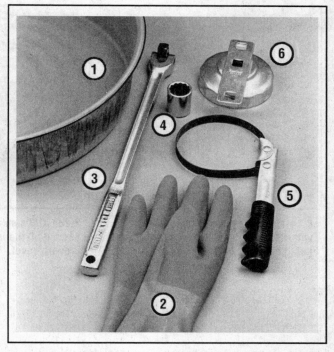

**8.3  These tools are required when changing the engine oil and filter**

1  **Drain pan** - It should be fairly shallow in depth, but wide to prevent spills
2  **Rubber gloves** - When removing the drain plug and filter, you will get oil on your hands (the gloves will prevent burns)
3  **Breaker bar** - Sometimes the oil drain plug is tight, and a long breaker bar is needed to loosen it
4  **Socket** - To be used with the breaker bar or a ratchet (must be the correct size to fit the drain plug - six-point preferred)
5  **Filter wrench** - This is a metal band-type wrench, which requires clearance around the filter to be effective
6  **Filter wrench** - This type fits on the bottom of the filter and can be turned with a ratchet or breaker bar (different-size wrenches are available for different types of filters)

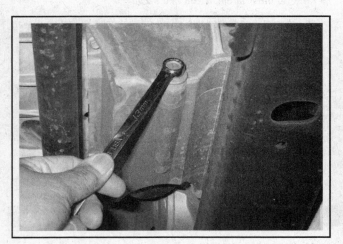

**8.8  Use a proper size box-end wrench or socket to remove the oil drain plug and avoid rounding it off**

**8.10  Since the oil filter is on very tight, you'll need a special wrench for removal - DO NOT use the wrench to tighten the new filter**

## 1-14 TUNE-UP AND ROUTINE MAINTENANCE

8.14 Lubricate the oil filter gasket with clean engine oil before installing the filter on the engine

8.25a DIC display should read 99 percent oil life left when properly reset

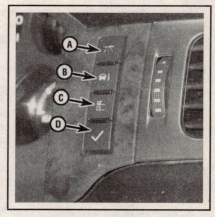

8.25b DIC buttons (on models so equipped) are trip/fuel (A), vehicle information (B), customization (C), and set/reset (D)

13  Clean the area where the oil filter mounts to the engine. Check the old filter to make sure the rubber gasket isn't stuck to the engine. If the gasket is stuck to the engine (use a flashlight if necessary), remove it.

14  Apply a light coat of clean oil to the rubber gasket on the new oil filter (see illustration).

15  Attach the new filter to the engine, following the tightening directions printed on the filter canister or packing box. Most filter manufacturers recommend against using a filter wrench due to the possibility of overtightening and damage to the seal.

16  Lower the vehicle, open the hood and locate the oil filler cap.

17  Refer to the engine oil capacity in this Chapter's Specifications and add the proper amount of fresh oil into the engine. Wait a few minutes to allow the oil to drain into the pan, then check the level on the oil dipstick (see Section 4 if necessary). If the oil level is above the hatched area, start the engine and allow the new oil to circulate.

18  Run the engine for only about a minute and then shut it off. Immediately look under the vehicle and check for leaks at the oil pan drain plug and around the oil filter.

19  Wait about five minutes, then recheck the level on the dipstick. Add more oil as necessary.

20  During the first few trips after an oil change, make it a point to check frequently for leaks and proper oil level.

21  The old oil drained from the engine cannot be reused in its present state and should be disposed of. Check with your local auto parts store, disposal facility or environmental agency to see if they will accept the oil for recycling. After the oil has cooled it can be drained into a container (capped plastic jugs, topped bottles, milk cartons, etc.) for transport to one of these disposal sites. Don't dispose of the oil by pouring it on the ground or down a drain!

### OIL LIFE MONITOR

22  The Oil Life Monitor is a function of the PCM that tracks engine operating temperature and rpm. If the PCM determines that your engine's oil has been used long enough, an indicator that shows "CHANGE ENGINE OIL SOON" will light on the instrument panel.

23  When you change your engine oil and filter, whether you change it at the interval recommended in this Chapter or only when the light comes on, you will have to reset the system to make the indicator go out and allow the system to accurately calculate when the next oil change is due.

**Vehicles without a Driver Information Center (DIC)**

24  To reset, turn the ignition key to the Run position (not Start) with the engine off, then fully depress and let up the accelerator pedal three times within a five-second period. Turn the key to Off, then start the engine. If the "CHANGE ENGINE OIL SOON" message still appears, repeat the resetting procedure.

**Vehicles equipped with a DIC**

▶ Refer to illustrations 8.25a and 8.25b

25  To reset, turn the ignition key to the Run position (not Start) with the engine off, then press the DIC (Driver Information Center) vehicle information button (see illustration) to display the "OIL LIFE LEFT/HOLD SET TO RESET" or "OIL LIFE REMAINING" message on the DIC, depending on model. Press the SET/RESET button (or the odometer trip reset button if your DIC doesn't have buttons) until 99 or 100 percent (depending on model) is displayed on the DIC (see illustration). The system will chime three times then the "CHANGE ENGINE OIL SOON" message will go out. Turn the key off, after a minute start the engine - if the CHANGE ENGINE OIL SOON message comes back on, repeat the procedure.

## 9  Seat belt check (every 6000 miles or 6 months)

1  Check seat belts, buckles, latch plates and guide loops for obvious damage and signs of wear.

2  Where the seat belt receptacle bolts to the floor of the vehicle, check that the bolts are secure.

3  See if the seat belt reminder light comes on when the key is turned to the Run or Start position. A chime should also sound.

# TUNE-UP AND ROUTINE MAINTENANCE   1-15

## 10  Wiper blade inspection and replacement (every 6000 miles or 6 months)

♦ **Refer to illustrations 10.3 and 10.6**

1  The windshield wiper blade elements should be checked periodically for cracks and deterioration.
2  Lift the wiper blade assembly away from the glass.
3  Squeeze both wiper blade release tabs together, then rotate away and unhook the blade from the arm (see illustration)
4  Compare the new blade with the old for length, design, etc.
5  Install the new blade by reversing the removal procedure.
6  Check to be sure the wiper blade is properly secured (see illustration), then wet the windshield and test for proper operation

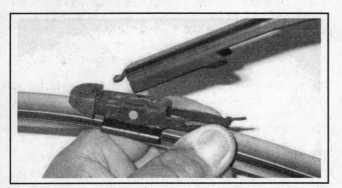

**10.3  Squeeze the release tabs and rotate the wiper blade away from the arm, then unhook it from the arm**

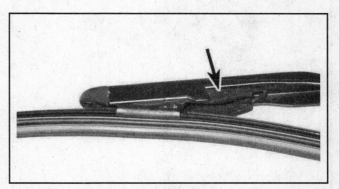

**10.6  Make sure both retaining tabs are properly seated into the wiper arm**

## 11  Battery check, maintenance and charging (every 6000 miles or 6 months)

♦ **Refer to illustrations 11.1, 11.5, 11.7a, 11.7b and 11.7c**

### ※※ WARNING:

Certain precautions must be followed when checking and servicing the battery. Hydrogen gas, which is highly flammable, is always present in the battery cells, so keep lighted tobacco and all other open flames and sparks away from the battery. The electrolyte inside the battery is actually dilute sulfuric acid, which will cause injury if splashed on your skin or in your eyes. It will also ruin clothes and painted surfaces. When removing the battery cables, always detach the negative cable first and hook it up last!

1  A routine preventive maintenance program for the battery in your vehicle is the only way to ensure quick and reliable starts. But before performing any battery maintenance, make sure that you have the proper equipment necessary to work safely around the battery (see illustration).

→**Note:** Some models have an auxiliary battery in addition to the standard battery. All of the following care and maintenance should be applied to both batteries.

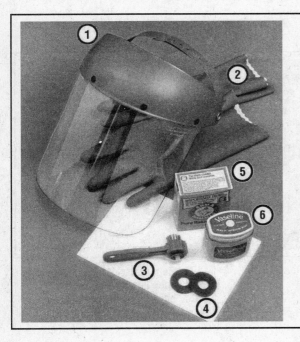

**11.1  Tools and materials required for battery maintenance**

1  **Face shield/safety goggles** - When removing corrosion with a brush, the acidic particles can easily fly up into your eyes
2  **Rubber gloves** - Another safety item to consider when servicing the battery - remember that's acid inside the battery!
3  **Battery terminal/cable cleaner** - This wire brush cleaning tool will remove all traces of corrosion from the battery posts and cable clamps
4  **Treated felt washers** - Placing one of these on each post, directly under the cable clamps, will help prevent corrosion
5  **Baking soda** - A solution of baking soda and water can be used to neutralize corrosion
6  **Petroleum jelly** - A layer of this on the battery posts will help prevent corrosion

# 1-16 TUNE-UP AND ROUTINE MAINTENANCE

2  There are also several precautions that should be taken whenever battery maintenance is performed. Before servicing the battery, always turn the engine and all accessories off and disconnect the cable from the negative terminal of the battery.

3  The battery produces hydrogen gas, which is both flammable and explosive. Never create a spark, smoke or light a match around the battery. Always charge the battery in a ventilated area.

4  Electrolyte contains poisonous and corrosive sulfuric acid. Do not allow it to get in your eyes, on your skin or on your clothes. Never ingest it. Wear protective safety glasses when working near the battery. Keep children away from the battery.

5  Note the external condition of the battery. If the positive terminal and cable clamp on your vehicle's battery is equipped with a rubber protector, make sure that it's not torn or damaged. It should completely cover the terminal. Look for any corroded or loose connections, cracks in the case or cover or loose hold-down clamps. Also check the entire length of each cable for cracks and frayed conductors (see illustration).

6  If corrosion, which looks like white, fluffy deposits is evident, particularly around the terminals, the battery should be removed for cleaning. Loosen the cable bolts with a wrench, being careful to remove the ground cable first, and slide them off the terminals. Then disconnect the hold-down clamp bolt and nut, remove the clamp and lift the battery from the engine compartment.

7  Clean the cable ends thoroughly with a battery brush or a terminal cleaner and a solution of warm water and baking soda. Wash the terminals and the side of the battery case with the same solution but make sure that the solution doesn't get into the battery. When cleaning the cables, terminals and battery case, wear safety goggles and rubber gloves to prevent any solution from coming in contact with your eyes or hands. Wear old clothes too - even diluted, sulfuric acid splashed onto clothes will burn holes in them. If the terminals have been corroded, clean them up with a terminal cleaner (see illustrations). Thoroughly wash all cleaned areas with plain water.

8  Make sure that the battery tray is in good condition and the hold-down clamp bolts are tight. If the battery is removed from the tray, make sure no parts remain in the bottom of the tray when the battery is reinstalled. When reinstalling the hold-down clamp bolts, do not over-tighten them.

9  Any metal parts of the vehicle damaged by corrosion should be covered with a zinc-based primer, then painted.

10  Information on removing and installing the battery can be found in Chapter 5. Information on jump-starting can be found at the front of this manual.

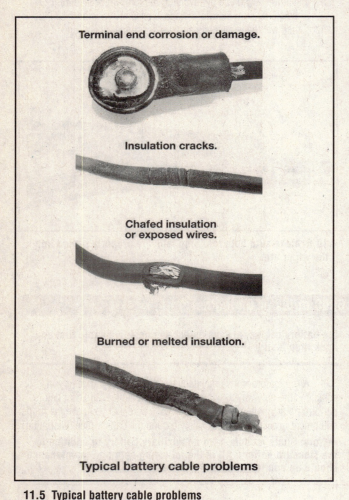

**11.5  Typical battery cable problems**

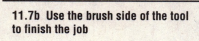

**11.7a**  A tool like this one (available at auto parts stores) is used to clean the side-terminal type battery-cable contact area

**11.7b**  Use the brush side of the tool to finish the job

**11.7c**  Regardless of the type of tool used on the battery and cables, a clean, shiny surface should be the result

# TUNE-UP AND ROUTINE MAINTENANCE

## CHARGING

> ⚠️ **WARNING:**
> When batteries are being charged, hydrogen gas, which is very explosive and flammable, is produced. Do not smoke or allow open flames near a charging or a recently charged battery. Wear eye protection when near the battery during charging. Also, make sure the charger is unplugged before connecting or disconnecting the battery from the charger.

➡️ **Note:** The manufacturer recommends the battery be removed from the vehicle for charging because the gas that escapes during this procedure can damage the paint. Fast charging with the battery cables connected can result in damage to the electrical system.

11  Slow-rate charging is the best way to restore a battery that's discharged to the point where it will not start the engine. It's also a good way to maintain the battery charge in a vehicle that's only driven a few miles between starts. Maintaining the battery charge is particularly important in the winter when the battery must work harder to start the engine and electrical accessories that drain the battery are in greater use.

12  It's best to use a one or two-amp battery charger (sometimes called a "trickle" charger). They are the safest and put the least strain on the battery. They are also the least expensive. For a faster charge, you can use a higher amperage charger, but don't use one rated more than 1/10th the amp/hour rating of the battery. Rapid boost charges that claim to restore the power of the battery in one to two hours are hardest on the battery and can damage batteries not in good condition. This type of charging should only be used in emergency situations.

13  The average time necessary to charge a battery should be listed in the instructions that come with the charger. As a general rule, a trickle charger will charge a battery in 12 to 16 hours.

14  Remove all the cell caps (if equipped) and cover the holes with a clean cloth to prevent spattering electrolyte. Disconnect the negative battery cable and hook the battery charger cable clamps up to the battery posts (positive to positive, negative to negative), then plug in the charger. Make sure it is set at 12-volts if it has a selector switch.

15  If you're using a charger with a rate higher than two amps, check the battery regularly during charging to make sure it doesn't overheat. If you're using a trickle charger, you can safely let the battery charge overnight after you've checked it regularly for the first couple of hours.

16  If the battery has removable cell caps, measure the specific gravity with a hydrometer every hour during the last few hours of the charging cycle. Hydrometers are available inexpensively from auto parts stores - follow the instructions that come with the hydrometer. Consider the battery charged when there's no change in the specific gravity reading for two hours and the electrolyte in the cells is gassing (bubbling) freely. The specific gravity reading from each cell should be very close to the others. If not, the battery probably has a bad cell(s).

17  Some batteries with sealed tops have built-in hydrometers on the top that indicate the state of charge by the color displayed in the hydrometer window. Normally, a bright-colored hydrometer indicates a full charge and a dark hydrometer indicates the battery still needs charging.

18  If the battery has a sealed top and no built-in hydrometer, you can hook up a digital voltmeter across the battery terminals to check the charge. A fully charged battery should read 12.5 volts or higher.

19  Further information on the battery and jump-starting can be found in Chapter 5 and at the front of this manual.

## 12  Drivebelt check, adjustment and replacement (every 6000 miles or 6 months)

▶ **Refer to illustrations 12.2, 12.4a, 12.4b, 12.5a and 12.5b**

1  A serpentine belt is located at the front of the engine and plays an important role in the overall operation of the engine and its components. Due to its function and material make up, the belt is prone to wear and should be periodically inspected. The serpentine belt drives the alternator, power steering pump, water pump and air conditioning compressor.

➡️ **Note:** V6 models have one belt. V8 models have one main belt and a second, inner belt just for the air conditioning compressor (if equipped).

2  With the engine off, open the hood and use your fingers (and a flashlight, if necessary), to move along the belt checking for cracks and separation of the belt plies. Also check for fraying and glazing, which gives the belt a shiny appearance (see illustration). Both sides of the belt should be inspected, which means you will have to twist the belt to check the underside.

3  Check the underside of the belt. They should be all the same depth, with none of the surfaces uneven.

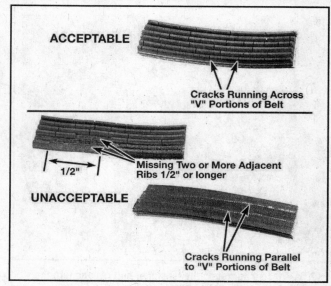

12.2  Check the belts for signs of wear like these - if it looks worn, replace it

## 1-18 TUNE-UP AND ROUTINE MAINTENANCE

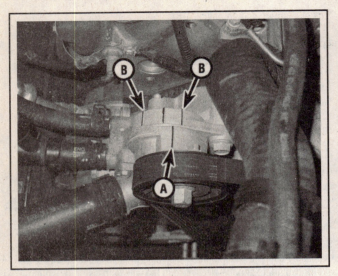

**12.4a** The mark (A) on the main belt's tensioner must remain between the marks (B) on the tensioner mount (V8 shown, main belt)

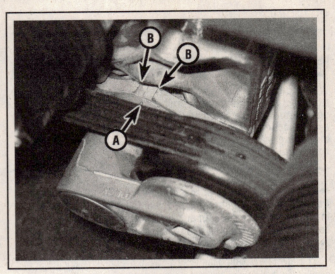

**12.4b** The tensioner for the air conditioning compressor drivebelt is harder to see. The mark (A) on the moveable part of the tensioner must be within the acceptable range marks (B)

4  The tension of the belt is maintained by a spring-loaded tensioner assembly and isn't adjustable. The belt should be replaced when the mark on the moveable part of the tensioner moves out of the acceptable range on the on the stationary part of the tensioner (see illustrations).

5  To replace the belt, rotate the tensioner to release belt tension (see illustrations).

➡ **Note:** *Before removing the belt, make a sketch of how it is routed around the pulleys, if you don't think you can remember.*

6  Remove the belt from the tensioner and auxiliary components and slowly release the tensioner.

7  Route the new belt over the various pulleys, again rotating the tensioner to allow the belt to be installed, then release the belt tensioner.

### TENSIONER REPLACEMENT

▶ **Refer to illustrations 12.8a and 12.8b**

8  To replace a tensioner that has lost its spring tension, exhibits binding or has a worn-out pulley/bearing, remove the mounting bolts (see illustrations).

9  Installation is the reverse of the removal procedure.

**12.5a** Place a socket on the tensioner bolt (A) and turn the tensioner clockwise (B) for belt removal

**12.5b** On V8 models with air conditioning, the inner belt tensioner is best accessed from below - use a 3/8-inch-drive tool in the tensioner's square hole (A) to rotate it (B) for belt removal

# TUNE-UP AND ROUTINE MAINTENANCE

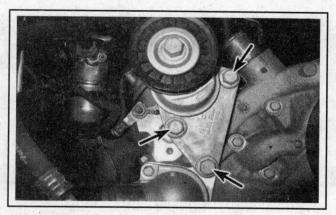

12.8a Main belt tensioner mounting bolts (V8 engine shown)

12.8b Mounting bolts for V8 engine air conditioning belt tensioner

## 13 Underhood hose check and replacement (every 6000 miles or 6 months)

### GENERAL

**※※ CAUTION:**

Replacement of air conditioning hoses must be left to a dealer service department or air conditioning shop that has the equipment to depressurize the system safely and recover the refrigerant. Never remove air conditioning components or hoses until the system has been depressurized.

1  High temperatures in the engine compartment can cause the deterioration of the rubber and plastic hoses used for engine, accessory and emission systems operation. Periodic inspection should be made for cracks, loose clamps, material hardening and leaks. Information specific to the cooling system hoses can be found in Section 14.

2  Some, but not all, hoses are secured to their fittings with clamps. Where clamps are used, check to be sure they haven't lost their tension, allowing the hose to leak. If clamps aren't used, make sure the hose has not expanded and/or hardened where it slips over the fitting, allowing it to leak.

### VACUUM HOSES

3  It's quite common for vacuum hoses, especially those in the emissions system, to be color-coded or identified by colored stripes molded into them. Various systems require hoses with different wall thickness, collapse resistance and temperature resistance. When replacing hoses, be sure the new ones are made of the same material.

4  Often the only effective way to check a hose is to remove it completely from the vehicle. If more than one hose is removed, be sure to label the hoses and fittings to ensure correct installation.

5  When checking vacuum hoses, be sure to include any plastic T-fittings in the check. Inspect the fittings for cracks and the hose where it fits over the fitting for distortion, which could cause leakage.

6  A small piece of vacuum hose (1/4-inch inside diameter) can be used as a stethoscope to detect vacuum leaks. Hold one end of the hose to your ear and probe around vacuum hoses and fittings, listening for the hissing sound characteristic of a vacuum leak.

**※※ WARNING:**

When probing with the vacuum hose stethoscope, be very careful not to come into contact with moving engine components such as the drivebelt, cooling fan, etc.

### FUEL HOSE

**※※ WARNING:**

Gasoline is extremely flammable, so take extra precautions when you work on any part of the fuel system. Don't smoke or allow open flames or bare light bulbs near the work area, and don't work in a garage where a gas-type appliance (such as a water heater or clothes dryer) is present. Since gasoline is carcinogenic, wear latex gloves when there's a possibility of being exposed to fuel, and, if you spill any fuel on your skin, rinse it off immediately with soap and water. Mop up any spills immediately and do not store fuel-soaked rags where they could ignite. When you perform any kind of work on the fuel system, wear safety glasses and have a Class B type fire extinguisher on hand. The fuel system is under pressure, so if any lines must be disconnected, the pressure in the system must be relieved first (see Chapter 4 for more information).

7  Check all rubber fuel lines for deterioration and chafing. Check especially for cracks in areas where the hose bends and just before fittings, such as where a hose attaches to the fuel filter and fuel injection unit.

8  High quality fuel line, specifically designed for high-pressure fuel injection applications, must be used for fuel line replacement. Never, under any circumstances, use regular fuel line, unreinforced vacuum line, clear plastic tubing or water hose for fuel lines.

9  Spring-type (pinch) clamps are commonly used on fuel lines. These clamps often lose their tension over a period of time, and can be sprung during removal. Replace all spring-type clamps with screw clamps whenever a hose is replaced.

## METAL LINES

10 Sections of metal line are routed along the frame, between the fuel tank and the engine. Check carefully to be sure the line has not been bent or crimped and no cracks have started in the line.

11 If a section of metal fuel line must be replaced, only seamless steel tubing should be used, since copper and aluminum tubing don't have the strength necessary to withstand normal engine vibration.

12 Check the metal brake lines where they enter the master cylinder and brake proportioning unit for cracks in the lines or loose fittings. Any sign of brake fluid leakage calls for an immediate and thorough inspection of the brake system.

## 14 Cooling system check (every 6000 miles or 6 months)

◆ Refer to illustration 14.4

**WARNING:**
Wait until the engine is completely cool before beginning this procedure.

**CAUTION:**
Never mix green-colored ethylene glycol antifreeze and orange-colored DEX-COOL silicate-free coolant because doing so will destroy the efficiency of the DEX-COOL coolant, which is designed to last for 100,000 miles or five years.

1 Many major engine failures can be attributed to a faulty cooling system. The cooling system also cools the transmission fluid and thus plays an important role in prolonging transmission life.

2 The cooling system should be checked with the engine cold. Do this before the vehicle is driven for the day or after it has been shut off for at least three hours.

3 Remove the coolant pressure cap on the expansion tank by slowly unscrewing it. If you hear any hissing sounds (indicating there is still pressure in the system), wait until it stops. Thoroughly clean the cap, inside and out, with clean water. Also clean the filler neck on the expansion tank. All traces of corrosion should be removed. The coolant inside the expansion tank should be relatively transparent. If it is rust colored, the system should be drained and refilled (see Section 25). If the coolant level is not up to the Cold mark, add additional antifreeze/coolant mixture (see Section 4).

4 Carefully check the large upper and lower radiator hoses along with any smaller diameter heater hoses that run from the engine to the firewall. Inspect each hose along its entire length, replacing any hose that is cracked, swollen or shows signs of deterioration. Cracks may become more apparent if the hose is squeezed (see illustration).

5 Make sure all hose connections are tight. A leak in the cooling system will usually show up as white or rust-colored deposits on the areas adjoining the leak. If wire-type clamps are used at the ends of the hoses, it may be wise to replace them with more secure, screw-type clamps.

6 Use compressed air or a soft brush to remove bugs, leaves, etc. from the front of the radiator or air conditioning condenser. Be careful not to damage the delicate cooling fins or cut yourself on them.

7 Every other inspection, or at the first indication of cooling system problems, have the cap and system pressure tested. If you don't have a pressure tester, most gas stations and repair shops will do this for a minimal charge.

### OVERHEAT PROTECTION OPERATING MODE

8 Models with V8 engines have a system to protect the engine from damage caused by severe overheating. When the computer senses an

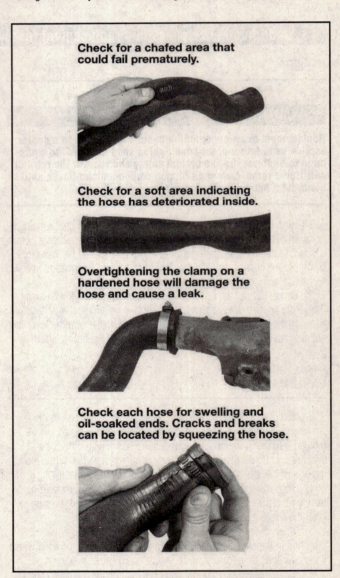

**14.4 Hoses, like drivebelts, have a habit of failing at the worst possible time - to prevent the inconvenience of a blown radiator or heater hose, inspect them carefully as shown here**

# TUNE-UP AND ROUTINE MAINTENANCE 1-21

overheat condition, an instrument panel warning light comes on that says "reduced power." In this mode, the computer switches the firing of the individual coils on and off at each cylinder to allow cooling cycles between the firing cycles. The engine will have a dramatic loss of power, but will allow vehicle operation in an emergency.

9  If this light is on, find a safe place to get off the road as soon as possible, and allow the engine to cool thoroughly.

10  Check the coolant level and inspect for a split hose or other obvious signs of coolant leakage.

11  The engine oil will be ruined after this mode has operated, since unburned fuel will get into the oil. After fixing the overheating problem, change the oil and filter right away and reset the Oil Life Monitor (see Section 8).

## 15  Tire rotation (every 6000 miles or 6 months)

♦ Refer to illustration 15.2

1  The tires should be rotated at the specified intervals and whenever uneven wear is noticed.

2  Tires must be rotated in the recommended pattern (see illustration).

3  Refer to the information in *Jacking and towing* at the front of this manual for the proper procedures to follow when raising the vehicle and changing a tire. If the brakes are to be checked, don't apply the parking brake as stated. Make sure the tires are blocked to prevent the vehicle from rolling as it's raised.

4  Preferably, the entire vehicle should be raised at the same time. This can be done on a hoist or by jacking up each corner, then lowering the vehicle onto jackstands placed under the frame rails. Always use four jackstands and make sure the vehicle is safely supported.

5  After rotation, check and adjust the tire pressures as necessary. Tighten the lug nuts to the torque listed in this Chapter's Specifications.

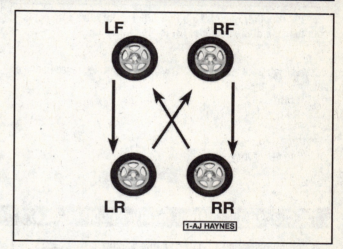

15.2  The recommended four-tire rotation pattern for non-directional radial tires

## 16  Differential lubricant level check (every 15,000 miles or 12 months)

♦ Refer to illustrations 16.2a, 16.2b and 16.2c

➡ Note: *4WD vehicles have two differentials - one in the center of each axle. 2WD vehicles have one differential in the center of the rear axle. On 4WD vehicles, be sure to check the lubricant level in both differentials.*

1  If the vehicle is raised to gain access to the plug, be sure to support it safely on jackstands - DO NOT crawl under the vehicle when only the jack supports it. Be sure the vehicle is level or the check may not be accurate. If you're checking the front differential lubricant, remove the splash shield or rock guard, if equipped.

2  Remove the plug from the filler hole in the differential housing or cover (see illustrations).

16.2a  On some models, the rear differential filler plug is on the differential cover. On this particular model, the cover must be removed to drain the lubricant

# 1-22 TUNE-UP AND ROUTINE MAINTENANCE

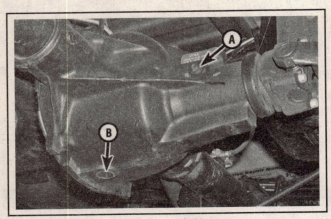

**16.2b On other models, the rear differential filler plug (A) is on the side of the housing. (B) is the drain plug**

**16.2c Front differential filler plug (A) and drain plug (B)**

3  The lubricant level should be within the specified range from the bottom of the filler hole:
**Front axle** - *0 to 1/4-inch*
**Rear axle**
  *8.6-inch axle = 3/64 to 3/4-inch*
  *9.5-inch axle = 5/8 to 1-9/16 inch*
  *10.5-inch axle = 0 to 3/8-inch*
  *11.5-inch axle = 5/8 to 3/4-inch*

If the level is low, use a pump or squeeze bottle to add the recommended lubricant until it is up to the proper level.

4  Install the plug into the filler hole and tighten it to the torque listed in this Chapter's Specifications.

## 17 Chassis lubrication (every 15,000 miles or 12 months)

♦ **Refer to illustration 17.1, 17.2a and 17.2b**

1  Refer to *Recommended lubricants and fluids* in this Chapter's Specifications to obtain the necessary grease, etc. You'll also need a grease gun (see illustration). If a suspension component has no grease fitting in place, this indicates the part is sealed and doesn't require periodic lubrication. Some components on 4WD models have fittings that aren't on 2WD versions, and vice versa.

2  Look under the vehicle and locate the grease fittings (see illustrations).

3  For easier access under the vehicle, raise it with a jack and place jackstands under the frame. Make sure it's safely supported by the stands. If the wheels are to be removed at this interval for tire rotation or brake inspection, loosen the lug nuts slightly while the vehicle is still on the ground.

4  Before beginning, force a little grease out of the nozzle to remove any dirt from the end of the gun. Wipe the nozzle clean with a rag.

5  With the grease gun and plenty of clean rags, crawl under the vehicle and begin lubricating the components.

6  Wipe one of the grease fittings clean and push the nozzle firmly over it. Pump the gun until the component is completely lubricated. On balljoints, stop pumping when the rubber seal is firm to the touch. Do not pump too much grease into the fitting as it could rupture the seal. For all other suspension and steering components, continue pumping grease into the fitting until it oozes out of the joint between the two components. If it escapes around the grease gun nozzle, the fitting is clogged or the nozzle is not completely seated on the fitting. Resecure the gun nozzle to the fitting and try again. If necessary, replace the fitting with a new one.

7  Wipe the excess grease from the components and the grease fitting. Repeat the procedure for the remaining fittings.

**17.1 Materials required for chassis and body lubrication**

1  **Engine oil** - *Light engine oil in a can like this can be used for door and hood hinges*
2  **Graphite spray** - *Used to lubricate lock cylinders*
3  **Grease** - *Grease, in a variety of types and weights, is available for use in a grease gun. Check the Specifications for your requirements*
4  **Grease gun** - *A common grease gun, shown here with a detachable hose and nozzle, is needed for chassis lubrication. After use, clean it thoroughly!*

# TUNE-UP AND ROUTINE MAINTENANCE

**17.2a  Wipe the dirt from the grease fittings before pushing the grease gun nozzle onto the fitting - lube the tie-rod ends**

**17.2b  On driveshafts with a sliding joint, lube the fitting (arrow) during chassis lubrication**

8  Clean the fitting and pump grease into the driveline universal joints until the grease can be seen coming out of the contact points. The other U-joints are sealed and do not require lubrication.

➡ Note: *Most replacement driveshaft U-joints aren't permanently sealed, and are sold with grease fittings. If your U-joints have been replaced, make sure you include them in your routine chassis lubrication.*

9  Also clean and lubricate the parking brake cable guides and levers.

### ✳✳ CAUTION:
**Do not use chassis lubrication on the brake cables themselves. The grease will cause the cable housings to deteriorate.**

## 18  Fuel system check (every 15,000 miles or 12 months)

▸ Refer to illustration 18.9

### ✳✳ WARNING:
**Gasoline is extremely flammable, so take extra precautions when you work on any part of the fuel system. Don't smoke or allow open flames or bare light bulbs near the work area, and don't work in a garage where a gas-type appliance (such as a water heater or clothes dryer) is present. Since gasoline is carcinogenic, wear fuel-resistant gloves when there's a possibility of being exposed to fuel, and, if you spill any fuel on your skin, rinse it off immediately with soap and water. Mop up any spills immediately and do not store fuel-soaked rags where they could ignite. When you perform any kind of work on the fuel system, wear safety glasses and have a Class B type fire extinguisher on hand. The fuel system is under constant pressure, so, before any lines are disconnected, the fuel system pressure must be relieved (see Chapter 4).**

➡ Note: *These vehicles are not equipped with fuel filters. A filter is integral with the fuel pump/fuel level sensor module and requires no maintenance.*

1  If you smell gasoline while driving or after the vehicle has been sitting in the sun, inspect the fuel system immediately.

2  Remove the gas filler cap and inspect if for damage and corrosion. The gasket should have an unbroken sealing imprint. If the gasket is damaged or corroded, install a new cap.

3  Inspect the fuel feed and return lines for cracks. Make sure that the connections between the fuel lines and the fuel injection system are tight.

### ✳✳ WARNING:
**Your vehicle is fuel injected, so you must relieve the fuel system pressure before servicing fuel system components. The fuel system pressure relief procedure is outlined in Chapter 4.**

4  If the fuel injectors are visible, look for signs of fuel leakage (wet spots) around any of the injectors, they may need new O-rings (see Chapter 4).

5  Since some components of the fuel system - the fuel tank and part of the fuel feed and return lines, for example - are underneath the vehicle, they can be inspected more easily with the vehicle raised on a hoist. If that's not possible, raise the vehicle and support it on jackstands.

6  With the vehicle raised and safely supported, inspect the gas tank and filler neck for punctures, cracks and other damage. The connection between the filler neck and the tank is particularly critical. Sometimes a rubber filler neck will leak because of loose clamps or deteriorated rubber. Inspect all fuel tank mounting brackets and straps to be sure that the tank is securely attached to the vehicle.

### ✳✳ WARNING:
**Do not, under any circumstances, try to repair a fuel tank (except rubber components). A welding torch or any open flame can easily cause fuel vapors inside the tank to explode.**

# 1-24 TUNE-UP AND ROUTINE MAINTENANCE

7  Carefully check all rubber hoses and metal lines leading away from the fuel tank. Check for loose connections, deteriorated hoses, crimped lines and other damage. Repair or replace damaged sections as necessary (see Chapter 4).

8  The evaporative emissions control system can also be a source of fuel odors. The function of the system is to store fuel vapors from the fuel tank in a charcoal canister until they can be routed to the intake manifold where they mix with incoming air before being burned in the combustion chambers.

9  The most common symptom of a faulty evaporative emissions system is a strong odor of fuel. If a fuel odor has been detected, and you have already checked the areas described above, check the charcoal canister, located under the rear of the vehicle, and the hoses connected to it (see illustration).

**18.9 Check the evaporative emissions system canister and its lines for damage**

## 19  Brake system check (every 15,000 miles or 12 months)

> **WARNING:**
> The dust created by the brake system is harmful to your health. Never blow it out with compressed air and don't inhale any of it. An approved filtering mask should be worn when working on the brakes. Do not, under any circumstances, use petroleum-based solvents to clean brake parts. Use brake system cleaner only! Try to use non-asbestos replacement parts whenever possible.

➡ **Note: For detailed photographs of the brake system, refer to Chapter 9.**

1  In addition to the specified intervals, the brakes should be inspected every time the wheels are removed or whenever a defect is suspected.

2  Any of the following symptoms could indicate a potential brake system defect: The vehicle pulls to one side when the brake pedal is depressed; the brakes make squealing or dragging noises when applied; brake pedal travel is excessive; the pedal pulsates; or brake fluid leaks, usually onto the inside of the tire or wheel.

3  Loosen the wheel lug nuts.

4  Raise the vehicle and place it securely on jackstands.

5  Remove the wheels (see *Jacking and towing* at the front of this book, or your owner's manual, if necessary).

### DISC BRAKES

▶ **Refer to illustrations 19.7a, 19.7b, 19.9 and 19.11**

6  There are two pads (an outer and an inner) in each caliper. The pads are visible with the wheels removed. Most vehicles covered by this manual have disc brakes front and rear, with a mechanical, drum-type parking brake mechanism inside the rear discs.

7  Check the pad thickness by looking at each end of the caliper and through the inspection window in the caliper body (see illustrations). If the lining material is less than the thickness listed in this Chapter's Specifications, replace the pads.

➡ **Note: Keep in mind that the lining material is riveted or bonded to a metal backing plate and the metal portion is not included in this measurement.**

8  If it is difficult to determine the exact thickness of the remaining pad material by the above method, or if you are at all concerned about the condition of the pads, remove the caliper(s), then remove the pads from the calipers for further inspection (see Chapter 9).

9  Once the pads are removed from the calipers, clean them with brake cleaner and re-measure them with a ruler or a vernier caliper (see illustration).

10  Measure the disc thickness with a micrometer to make sure that it still has service life remaining. If any disc is thinner than the specified minimum thickness, replace it (see Chapter 9). Even if the disc has service life remaining, check its condition. Look for scoring, gouging and burned spots. If these conditions exist, remove the disc and have it resurfaced (see Chapter 9).

11  Before installing the wheels, check all brake lines and hoses for damage, wear, deformation, cracks, corrosion, leakage, bends and twists, particularly in the vicinity of the rubber hoses at the calipers (see illustration). Check the clamps for tightness and the connections for leakage. Make sure that all hoses and lines are clear of sharp edges, moving parts and the exhaust system. If any of the above conditions are noted, repair, reroute or replace the lines and/or fittings as necessary (see Chapter 9).

### DRUM BRAKES

▶ **Refer to illustrations 19.15 and 19.17**

12  With the rear wheels removed, make sure the parking brake is off and tap on the outside of the brake drums with a mallet to loosen them.

13  Remove the brake drums. If they won't come off, refer to Chapter 9.

14  Carefully clean the brake assemblies with brake system cleaner.

> **WARNING:**
> Don't blow the dust out with compressed air and don't inhale any of it - it's harmful to your health.

15  Note the thickness of the lining material on both front and rear shoes (see illustration). Compare the measurement with the limit given in this Chapter's Specifications; if any lining is thinner than specified, then all of the brake shoes must be replaced (see Chapter 9). They

# TUNE-UP AND ROUTINE MAINTENANCE

19.7a With the wheel off, check the thickness of the inner pad through the inspection hole (front caliper)

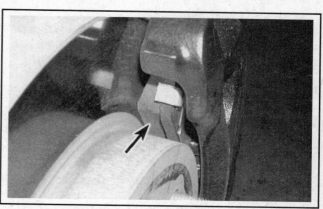

19.7b The outer pad on front calipers (and both pads on rear calipers) is checked at the edge of the caliper

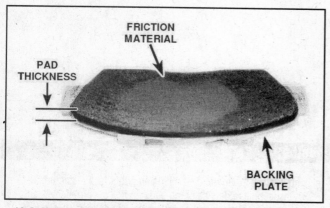

19.9 If a more precise measurement of pad thickness is necessary, remove the pads and measure the remaining friction material

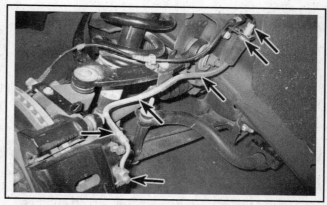

19.11 Check along the brake hoses and at each fitting for deterioration, cracks and leakage

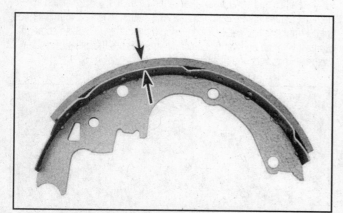

19.15 If the lining is bonded to the brake shoe, measure it's thickness from the outer surface to the steel shoe, as shown here; if the lining is riveted to the shoe, measure from the lining outer surface to the rivet head

19.17 Carefully peel back the wheel cylinder boot and check for leaking fluid, indicating that the cylinder must be replaced

should also be replaced if they're cracked, glazed with shiny areas or soaked with brake fluid.

16 Make sure that all of the brake springs are connected and in good condition.

17 Check the wheel cylinders for fluid leakage. Carefully pry back the wheel cylinder rubber cups with your fingers or a small screwdriver (see illustration). Any leakage is an indication that the wheel cylinders should be replaced at once (see Chapter 9). Also check all hoses and connections for signs of leakage.

18 Wipe the inside of the brake drums with a clean rag and brake cleaner or alcohol. Again, be careful no to breathe the dangerous brake dust.

# 1-26 TUNE-UP AND ROUTINE MAINTENANCE

19 Check the inside of the drum for score marks, deep scratches and hard spots that appear as small discolored areas. If imperfections can't be removed with fine emery cloth, the drum must be taken to an automotive machine shop for machining.

20 If your inspection reveals that all parts are in good condition, install the drums, install the wheels and lower the vehicle.

## BRAKE BOOSTER CHECK

21 Sit in the driver's seat and perform the following sequence of tests.

22 With the brake fully depressed, start the engine - the pedal should move down a little when the engine starts.

23 With the engine running, depress the brake pedal several times - the travel distance should not change.

24 Depress the brake, stop the engine and hold the pedal in for about 30 seconds - the pedal should neither sink nor rise.

25 Restart the engine, run it for about a minute and turn it off. Then firmly depress the brake several times - the pedal travel should decrease with each application.

26 If your brakes do not operate as described, the brake booster has failed. Refer to Chapter 9 for the replacement procedure.

## PARKING BRAKE

27 One method of checking the parking brake is to park the vehicle on a steep hill with the parking brake set and the transmission in Neutral (be sure to stay in the vehicle for this check). If the parking brake cannot prevent the vehicle from rolling, it's in need of attention (see Chapter 9).

## 20 Exhaust system check (every 15,000 miles or 12 months)

♦ Refer to illustrations 20.2a and 20.2b

1 With the engine cold (at least three hours after the vehicle has been driven), check the complete exhaust system from the manifold to the end of the tailpipe. Be careful around the catalytic converter, which may be hot even after three hours. The inspection should be done with the vehicle on a hoist to permit unrestricted access. If a hoist isn't available, raise the vehicle and support it securely on jackstands.

2 Check the exhaust pipes and connections for signs of leakage and/or corrosion indicating a potential failure. Make sure that all brackets and hangers are in good condition and tight (see illustrations).

3 Inspect the underside of the body for holes, corrosion, open seams, etc. which may allow exhaust gasses to enter the passenger compartment. Seal all body openings with silicone sealant or body putty.

4 Rattles and other noises can often be traced to the exhaust system, especially the hangers, mounts and heat shields. Try to move the pipes, mufflers and catalytic converter. If the components can come in contact with the body or suspension parts, secure the exhaust system with new brackets and hangers.

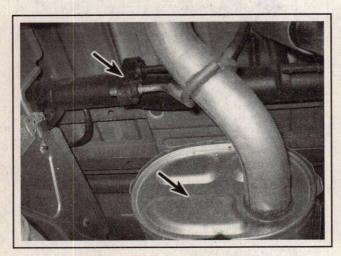

**20.2a Inspect the muffler and all hangers for signs of deterioration**

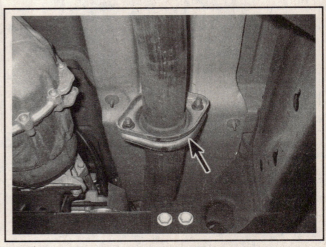

**20.2b Inspect all flanged joints for signs of exhaust gas leakage**

## 21 Transfer case lubricant level check (4WD models) (every 15,000 miles or 12 months)

♦ Refer to illustration 21.1

1  The transfer case lubricant level is checked by removing the upper plug located at the rear of the case (see illustration).
2  After removing the plug, reach inside the hole. The lubricant level should be just at the bottom of the hole. If not, add the appropriate lubricant through the opening.

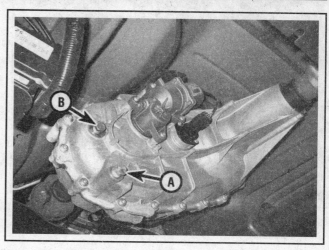

21.1  The drain plug (A) and fill plug (B) are on the rear of the transfer case

## 22 Brake fluid change (every 30,000 miles or 24 months)

**※ WARNING:**

**Brake fluid can harm your eyes and damage painted surfaces, so use extreme caution when handling or pouring it. Do not use brake fluid that has been standing open or is more than one year old. Brake fluid absorbs moisture from the air. Excess moisture can cause a dangerous loss of braking effectiveness.**

1  At the specified intervals, the brake fluid should be drained and replaced. Since the brake fluid may drip or splash when pouring it, place plenty of rags around the master cylinder to protect any surrounding painted surfaces.
2  Before beginning work, purchase the specified brake fluid (see *Recommended lubricants and fluids* in this Chapter's Specifications).
3  Remove the cap from the master cylinder reservoir.
4  Using a hand suction pump or similar device, withdraw the fluid from the master cylinder reservoir.
5  Add new fluid to the master cylinder until it rises to the line indicated on the reservoir.
6  Bleed the brake system as described in Chapter 9 at all four brakes until new and uncontaminated fluid is expelled from the bleeder screw. Be sure to maintain the fluid level in the master cylinder as you perform the bleeding process. If you allow the master cylinder to run dry, air will enter the system.
7  Refill the master cylinder with fluid and check the operation of the brakes. The pedal should feel solid when depressed, with no sponginess.

**※ WARNING:**

**Do not operate the vehicle if you are in doubt about the effectiveness of the brake system.**

## 23 Air filter replacement (every 30,000 miles or 24 months)

♦ Refer to illustrations 23.3a and 23.3b

1  At the specified intervals, the air filter element should be replaced with a new one.
2  On all models, the air filter is housed in a black plastic box mounted on the inner fenderwell on the right side of the engine compartment. If you drive in conditions that are particularly dusty, the need for a filter change may be necessary before the normally recommended mileage interval.
3  Loosen the captive screws and pull the housing cover up, then lift the air filter element out of the housing (see illustrations). Wipe out the inside of the air filter housing with a clean rag.
4  Place the new filter element in the air filter housing. Make sure it seats properly in the groove of the housing.
5  Installation is the reverse of removal. After installing the new filter, push in on the top of the filter indicator to reset it.

# 1-28 TUNE-UP AND ROUTINE MAINTENANCE

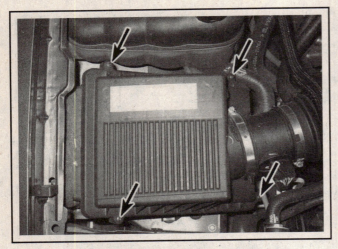

23.3a Loosen the screws and lift up the air filter housing cover . . .

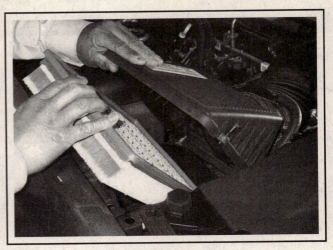

23.3b . . . then remove the filter element

## 24 Spark plug replacement (see maintenance schedule for service intervals)

▶ Refer to illustrations 24.2, 24.5a, 24.5b, 24.6, 24.8, 24.9 24.10 and 24.11

1  The spark plugs are threaded into the sides of the cylinder heads, adjacent to the exhaust ports.

2  In most cases, the tools necessary for spark plug replacement include a spark plug socket, which fits onto a ratchet (spark plug sockets are padded inside to prevent damage to the porcelain insulators on the new plugs), various extensions and a gap gauge to check and adjust the gaps on the new plugs (see illustration). A special plug wire removal tool is available for separating the wire boots from the spark plugs, but it isn't absolutely necessary. A torque wrench should be used to tighten the new plugs.

3  The best approach when replacing the spark plugs is to purchase the new ones in advance, adjust them to the proper gap and replace them one at a time. When buying the new spark plugs, be sure to obtain the correct plug type for your particular engine. This information can be found in the factory owner's manual and this Chapter's Specifications.

4  Allow the engine to cool completely before attempting to remove any of the plugs. While you're waiting for the engine to cool, check the new plugs for defects and adjust the gaps.

5  The gap is checked by inserting the proper-thickness gauge between the electrodes at the tip of the plug (see illustration). The gap between the electrodes should be the same as the one specified on the Emissions Control Information label or in this Chapter's Specifications. The wire should just slide between the electrodes with a slight amount of drag. If the gap is incorrect, use the adjuster on the gauge body to bend the curved side electrode slightly until the proper gap is obtained (see illustration). If the side electrode is not exactly over the center

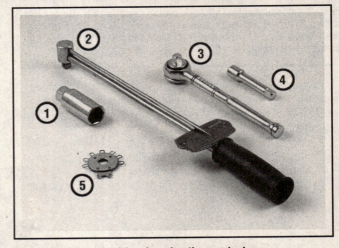

24.2 Tools required for changing the spark plugs

1  **Spark plug socket** - This will have special padding inside to protect the spark plug's porcelain insulator
2  **Torque wrench** - Although not mandatory, using this tool is the best way to ensure the plugs are tightened properly
3  **Ratchet** - Standard hand tool to fit the spark plug socket
4  **Extension** - Depending on model and accessories, you may need special extensions and universal joints to reach one or more of the plugs
5  **Spark plug gap gauge** - This gauge for checking the gap comes in a variety of styles. Make sure the gap for your engine is included

# TUNE-UP AND ROUTINE MAINTENANCE  1-29

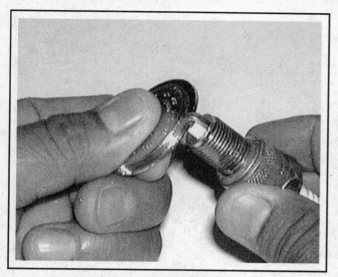

24.5a  Spark plug manufacturers recommend using a tapered thickness gauge when checking the gap - slide the thin side into the gap and turn until the gauge just fills the gap, then read the thickness on the gauge - do not force the tool into the gap or use the tapered portion to widen a gap

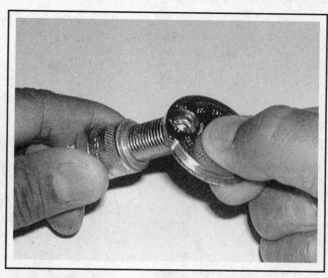

24.5b  To change the gap, bend the side electrode only, using the adjuster hole in the tool, and be very careful not to crack or chip the porcelain insulator surrounding the center electrode

24.6  A tool like this one makes the job of removing the spark plug boot easier - twist it back-and-forth and pull only on the boot

24.8  Use a socket and extension to unscrew the spark plugs - various length extensions and perhaps a flex-joint may be required to reach some plugs

electrode, bend it with the adjuster until it is. Check for cracks in the porcelain insulator (if any are found, the plug should not be used).

➡**Note:** Manufacturers recommend using a tapered thickness gauge when checking platinum or iridium-type spark plugs. Other types of gauges may scrape the thin coating from the electrodes, thus dramatically shortening the life of the plugs.

6  With the engine cool, remove the spark plug wire from one spark plug. Pull only on the rubber boot on the end of the wire - do not pull on the wire. A plug wire removal tool should be used if available (see illustration).

7  If compressed air is available, use it to blow any dirt or foreign material away from the spark plug hole. The idea here is to eliminate the possibility of debris falling into the cylinder as the spark plug is removed.

8  Place the spark plug socket over the plug and remove it from the engine by turning it in a counterclockwise direction (see illustration).

# 1-30 TUNE-UP AND ROUTINE MAINTENANCE

A normally worn spark plug should have light tan or gray deposits on the firing tip.

A carbon fouled plug, identified by soft, sooty, black deposits, may indicate an improperly tuned vehicle. Check the air cleaner, ignition components and engine control system.

An oil fouled spark plug indicates an engine with worn piston rings and/or bad valve seals allowing excessive oil to enter the chamber.

This spark plug has been left in the engine too long, as evidenced by the extreme gap- Plugs with such an extreme gap can cause misfiring and stumbling accompanied by a noticeable lack of power.

A physically damaged spark plug may be evidence of severe detonation in that cylinder. Watch that cylinder carefully between services, as a continued detonation will not only damage the plug, but could also damage the engine.

A bridged or almost bridged spark plug, identified by a build-up between the electrodes caused by excessive carbon or oil build-up on the plug.

**24.9 Inspect the spark plug to determine engine running conditions**

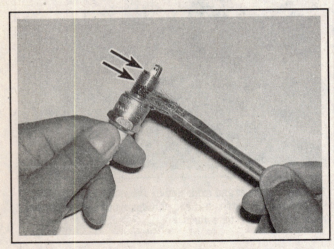

**24.10 Apply a thin coat of anti-seize compound to the spark plug threads, being careful not to get any near the lower threads**

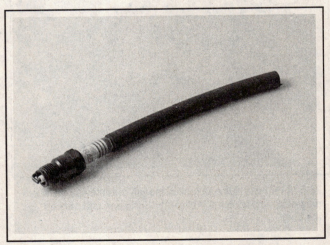

**24.11 A length of snug-fitting rubber hose will save time and prevent damaged threads when installing the spark plugs**

9  Compare the spark plug with the chart shown (see illustration) to get an indication of the general running condition of the engine.

10  Before installing the new plugs, it is a good idea to apply a thin coat of anti-seize compound to the threads (see illustration).

11  Thread one of the new plugs into the hole until you can no longer turn it with your fingers, then tighten it with a torque wrench (if available) or the ratchet. Where plugs are at the rear of the engine and harder to reach, it might be a good idea to slip a short length of rubber hose over the end of the plug to use as a tool to thread it into place (see illustration). The hose will grip the plug well enough to turn it, but will start to slip if the plug begins to cross-thread in the hole - this will prevent damaged threads and the accompanying repair costs.

12  Attach the plug wire to the new spark plug, again using a twisting motion on the boot until it's seated on the spark plug.

13  Repeat the procedure for the remaining spark plugs, replacing them one at a time to prevent mixing up the spark plug wires.

# TUNE-UP AND ROUTINE MAINTENANCE

## 25 Cooling system servicing (draining, flushing and refilling) (see maintenance schedule for interval)

**WARNING:**
Do not allow antifreeze to come in contact with your skin or painted surfaces of the vehicle. Rinse off spills immediately with plenty of water. Antifreeze is highly toxic if ingested. Never leave antifreeze lying around in an open container or in puddles on the floor; children and pets are attracted by its sweet smell and may drink it. Check with local authorities on disposing of used antifreeze. Many communities have collection centers that will see that antifreeze is disposed of safely. Antifreeze is flammable under certain conditions - be sure to read the precautions on the container.

➡ **Note:** Non-toxic coolant is available at local auto parts stores. Although the coolant is non-toxic when fresh, proper disposal is still required.

**CAUTION:**
Never mix green-colored ethylene glycol antifreeze and orange-colored DEX-COOL silicate-free coolant because doing so will destroy the efficiency of the DEX-COOL coolant, which is designed to last for 100,000 miles or five years.

### DRAINING

▸ **Refer to illustrations 25.3 and 25.4**

1  Periodically, the cooling system should be drained, flushed and refilled to replenish the antifreeze mixture and prevent formation of rust and corrosion, which can impair the performance of the cooling system and cause engine damage. When the cooling system is serviced, all hoses and the expansion tank cap should be checked and replaced if necessary.

2  Apply the parking brake and block the wheels.

**WARNING:**
If the vehicle has just been driven, wait several hours to allow the engine to cool down before beginning this procedure.

3  Move a large container under the lower radiator hose to catch the coolant. The lower radiator hose is located on the lower right side of the radiator (see illustration). Release the lower radiator hose spring clamp, detach the radiator hose from the radiator and direct it into the container. Remove the expansion tank cap.

4  After coolant stops flowing out of the radiator and hose, move the container under the engine block drain plugs - there's one on each side of the block (see illustration).

➡ **Note:** On some models, a block heater is installed in place of a plug.

Remove the plugs and allow the coolant in the block to drain.

➡ **Note:** Frequently, the coolant will not drain from the block after the plug is removed. This is due to a rust layer that has built up behind the plug. Insert a Phillips screwdriver into the hole to break the rust barrier.

5  While the coolant is draining, check the condition of the radiator hoses, heater hoses and clamps.

6  Replace any damaged clamps or hoses. Reinstall the drain plugs and tighten them securely, using Permatex #2 sealant on the threads of the plugs.

### FLUSHING

▸ **Refer to illustration 25.9**

7  Once the system is completely drained, remove the thermostat from the engine (see Chapter 3). Then reinstall the thermostat housing without the thermostat. This will allow the system to be thoroughly flushed.

8  Reinstall the lower radiator hose and tighten the radiator drain

**25.3 The lower radiator hose is located at the lower right corner of the radiator - release the spring clamp, remove the radiator hose from the radiator and aim it down into your drain pan**

**25.4 Cylinder block drain plug - there is one on each side of the block**

# 1-32 TUNE-UP AND ROUTINE MAINTENANCE

plug. Turn the heating controls to Hot, so that the heater core will be flushed at the same time as the rest of the cooling system.

9  Disconnect the upper radiator hose, then place a garden hose in the upper radiator inlet and flush the system until the water runs clear at the upper radiator hose (see illustration).

10  In severe cases of contamination or clogging of the radiator, remove the radiator (see Chapter 3) and have a radiator repair facility clean and repair it if necessary.

11  Many deposits can be removed by the chemical action of a cleaner available at auto parts stores. Follow the procedure outlined in the manufacturer's instructions.

➡ **Note:** When the coolant is regularly drained and the system refilled with the correct antifreeze/water mixture, there should be no need to use chemical cleaners or descalers.

## REFILLING

12  To refill the system, install the thermostat and reconnect any radiator hoses.

13  Place the heater temperature control in the maximum heat position.

14  Be sure to use the proper coolant listed in this Chapter's Specifications. Slowly fill the expansion tank with the recommended mixture of antifreeze and water to the COLD mark.

15  Install the cap on the expansion tank. Start the engine and run it at approximately 2000 to 2500 rpm until the engine reaches normal operating temperature and the thermostat opens.

16  Let the engine idle for three minutes.

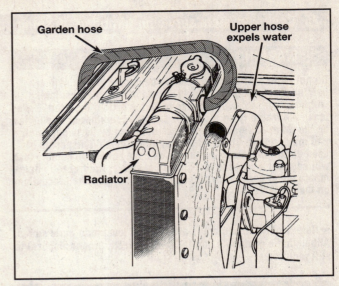

**25.9  With the thermostat removed, disconnect the upper radiator hose and flush the radiator and engine block with a garden hose**

17  Turn the engine off and let it cool. Add coolant to the expansion tank until the level is up to the COLD mark, repeat Steps 15 and 16, then turn the engine off.

18  Check the cooling system for leaks.

## 26  Suspension, steering and driveaxle boot check (every 30,000 miles or 30 months)

➡ **Note:** The steering linkage and suspension components should be checked periodically. Worn or damaged suspension and steering linkage components can result in excessive and abnormal tire wear, poor ride quality and vehicle handling and reduced fuel economy. For detailed illustrations of the steering and suspension components, refer to Chapter 10.

### SHOCK ABSORBER CHECK

▸ **Refer to illustration 26.6**

1  Park the vehicle on level ground, turn the engine off and set the parking brake. Check the tire pressures.

2  Push down at one corner of the vehicle, then release it while noting the movement of the body. It should stop moving and come to rest in a level position within one or two bounces.

3  If the vehicle continues to move up-and-down or if it fails to return to its original position, a worn or weak shock absorber is probably the reason.

4  Repeat the above check at each of the three remaining corners of the vehicle.

5  Raise the vehicle and support it securely on jackstands.

6  Check the shock absorbers for evidence of fluid leakage (see illustration). A light film of fluid is no cause for concern. Make sure that any fluid noted is from the shocks and not from some other source. If leakage is noted, replace the shocks as a set.

7  Check the shocks to be sure that they are securely mounted and undamaged. Check the upper mounts for damage and wear. If damage or wear is noted, replace the shocks as a set (front or rear).

8  If the shocks must be replaced, refer to Chapter 10 for the procedure.

**26.6  Check for signs of fluid leakage at this point on shock absorbers (front shock shown)**

# TUNE-UP AND ROUTINE MAINTENANCE 1-33

26.9a Examine the bushings for the upper . . .

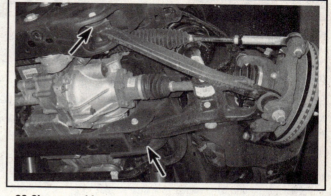

26.9b . . . and lower control arms on the front suspension

## STEERING AND SUSPENSION CHECK

▶ Refer to illustrations 26.9a, 26.9b, 26.9c and 26.11

9  Visually inspect the steering and suspension components (front and rear) for damage and distortion. Look for damaged seals, boots and bushings and leaks of any kind. Examine the bushings where the control arms meet the chassis (see illustrations).

10  Clean the lower end of the steering knuckle. Have an assistant grasp the lower edge of the tire and move the wheel in-and-out while you look for movement at the steering knuckle-to-control arm balljoint. If there is any movement the suspension balljoint(s) must be replaced.

11  Grasp each front tire at the front and rear edges, push in at the front, pull out at the rear and feel for play in the steering system components. If any freeplay is noted, check the idler arm and the tie-rod ends for looseness (see illustration).

12  Additional steering and suspension system information and illustrations can be found in Chapter 10.

## DRIVEAXLE BOOT CHECK (4WD MODELS)

▶ Refer to illustration 26.14

13  The driveaxle boots are very important because they prevent dirt, water and foreign material from entering and damaging the constant velocity (CV) joints. Oil and grease can cause the boot material to dete-

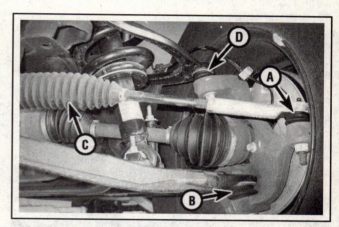

26.9c Inspect the tie-rod ends (A), the lower balljoints (B), the steering gear boots (C, models with rack-and-pinion steering) and the upper balljoints (D)

riorate prematurely, so it's a good idea to wash the boots with soap and water. Because it constantly pivots back and forth following the steering action of the front hub, the outer CV boot wears out sooner and should be inspected regularly.

14  Inspect the boots for tears and cracks as well as loose clamps (see illustration). If there is any evidence of cracks or leaking lubricant, they must be replaced as described in Chapter 8.

26.11 With the steering wheel in the locked position and the vehicle raised, grasp the front tire as shown and try to move it back-and-forth - if any play is noted, check the steering gear mounts and tie-rod ends for looseness

26.14 Inspect the inner and outer driveaxle boots on 4WD models for loose clamps, cracks or signs of leaking lubricant (inner boot shown)

## 1-34 TUNE-UP AND ROUTINE MAINTENANCE

### 27  Automatic transmission fluid and filter change (every 30,000 miles or 30 months)

▶ Refer to illustrations 27.5, 27.6, 27.7, 27.11, 27.12 and 27.13

1  At the specified intervals, the transmission fluid should be drained and replaced. Since the fluid will remain hot long after driving, perform this procedure only after the engine has cooled down completely.

2  Before beginning work, purchase the specified transmission fluid (see *Recommended lubricants and fluids* in this Chapter's Specifications) and a new filter and pan gasket.

3  Other tools necessary for this job include a floor jack, jackstands to support the vehicle in a raised position, a drain pan capable of holding at least eight quarts, newspapers and clean rags.

4  Raise the vehicle and support it securely on jackstands. Remove the front portion of the exhaust system if it will interfere with removal of the transmission fluid pan.

5  Place the drain pan underneath the transmission pan. Remove the drain plug, if equipped, and allow the fluid to drain until it barely comes out, then reinsert the drain plug (see illustration). If the pan is not equipped with a drain plug, remove the bolts from three sides of the pan, then loosen the remaining bolts and allow the pan to drop down to drain the fluid. Once most of the fluid has been drained, remove the bolts and carefully lower the pan and drain the rest of the fluid.

6  To access the pan bolts on the right side of the vehicle, remove the heat shield next to the catalytic converter (see illustration).

7  On the driver's side of the transmission, the shift linkage must be removed to access the pan bolts (see illustration).

8  Remove the transmission pan mounting bolts, then carefully pry the transmission pan loose with a screwdriver.

**※※ WARNING:**

**There will still be some transmission fluid in the pan.**

9  Carefully clean the gasket surface of the transmission to remove all traces of the old gasket and sealant.

10  Clean the pan with solvent and dry it with compressed air, if available.

➡Note: Some models are equipped with magnets in the transmission pan to catch metal debris. Clean the magnet thoroughly. A small amount of metal material is normal at the magnet. If there is considerable debris, consult a dealer or transmission specialist.

11  Remove the filter from the valve body inside the transmission (see illustration).

➡Note: Be very careful not to gouge the delicate aluminum gasket surface on the valve body.

12  Install a new seal and filter. On many replacement filters, the seal is attached to the filter to simplify installation (see illustration).

13  Make sure the gasket surface on the transmission pan is clean,

**27.5  Drain the transmission fluid pan by removing the drain plug (if equipped)**

**27.6  Remove the two bolts and the heat shield at the right side of the transmission**

**27.7  Disconnect the shift cable end from the ball-stud (A), then remove the two bolts securing the shift cable bracket to the transmission (B) (this allows room to access the oil pan bolts)**

# TUNE-UP AND ROUTINE MAINTENANCE

then install a new gasket on the pan (see illustration). Put the pan in place against the transmission and install all of the bolts. Working around the pan, tighten each bolt a little at a time to the torque listed in this Chapter's Specifications.

14  Reinstall the components removed for access to the pan bolts.

15  Lower the vehicle and add approximately 4 quarts of the specified type of automatic transmission fluid through the filler tube (see Section 7).

16  With the transmission in Park and the parking brake set, run the engine at a fast idle, but don't race it.

17  Move the gear selector through each range and back to Park, then let the engine idle. Check the fluid level. It will probably be low. Add enough fluid to bring the level to the proper mark on the dipstick. Be careful not to overfill.

18  Check under the vehicle for leaks during the first few trips. Check the fluid level again when the transmission is hot (see Section 7).

27.11 Remove the filter from the transmission by pulling it straight down

27.12 Use a seal removal tool to remove the transmission filter seal from the valve body, then replace it with a new seal - be careful not to scratch the aluminum cavity

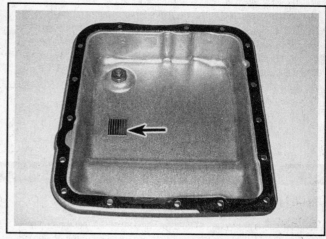

27.13 Clean the transmission pan, position the magnet back in place, and install the new pan gasket

## 28  Transfer case lubricant change (4WD models) (every 60,000 miles or 48 months)

1  This procedure should be performed after the vehicle has been driven so the lubricant will be warm and therefore will flow out of the transfer case more easily.

2  Raise the vehicle and support it securely on jackstands.

3  Remove the filler plug from the case (see illustration 21.1).

4  Remove the drain plug from the lower part of the case and allow the lubricant to drain completely.

5  After the case is completely drained, carefully clean and install the drain plug. Tighten the plug securely.

6  Fill the case with the specified lubricant until it is level with the lower edge of the filler hole.

7  Install the filler plug and tighten it securely.

8  Drive the vehicle for a short distance and recheck the lubricant level. In some instances a small amount of additional lubricant will have to be added.

# 1-36 TUNE-UP AND ROUTINE MAINTENANCE

## 29 Differential lubricant change (every 60,000 miles or 48 months)

1  This procedure should be performed after the vehicle has been driven, so the lubricant will be warm and therefore will flow out of the differential more easily.

2  Raise the vehicle and support it securely on jackstands. You'll be draining the lubricant by removing the drain plug, so move a drain pan, rags, newspapers and wrenches under the vehicle.

3  Remove the filler plug and the drain plug and allow the lubricant to drain into the pan, then clean and reinstall the drain plug (see Section 16).

➥ **Note: On differentials not equipped with a drain plug, the differential cover must be removed to drain the lubricant (see Chapter 8).**

4  Use a hand pump, syringe or squeeze bottle to fill the differential housing with the specified lubricant until it's at the level specified in Section 16. Reinstall the drain plug.

## 30 Positive Crankcase Ventilation (PCV) valve replacement (V6 engine) (every 60,000 miles or 48 months)

1  The PCV valve is located in the left valve cover.

2  With the engine idling at normal operating temperature, pull the valve (with hose attached) from the rubber grommet in the cover.

3  Place your finger over the valve opening. If there's no vacuum at the valve, check for a plugged hose, manifold port, or the valve itself. Replace any plugged or deteriorated hoses.

4  Turn off the engine and shake the PCV valve, listening for a rattle. If the valve doesn't rattle, replace it with a new one.

5  To replace the valve, pull it from the end of the hose, noting its installed position.

6  When purchasing a replacement PCV valve, make sure it's for your particular vehicle and engine size. Compare the old valve with the new one to make sure they're the same.

7  Push the valve into the end of the hose until it's seated.

8  Inspect the rubber grommet for damage and hardening. Replace it with a new one if necessary.

9  Push the PCV valve and hose securely into position.

## 31 Spark plug wire check and replacement (every 100,000 miles or 60 months)

▶ **Refer to illustration 31.6**

1  The spark plug wires should be checked at the recommended intervals and whenever new spark plugs are installed in the engine. V6 engines have spark plug wires that go from the spark plugs all the way to an ignition coil assembly. V8 engines have individual coils for each cylinder and short plug wires from each coil to the corresponding spark plug.

2  Begin this procedure by making a visual check of the spark plug wires while the engine is running. In a darkened garage (make sure there is adequate ventilation) start the engine and observe each plug wire. Be careful not to come into contact with any moving engine parts. If there is a break in the wire, you will see arcing or a small spark at the damaged area. If arcing is noticed, make a note to obtain new wires.

3  Disconnect the plug wire from one spark plug (with the engine off). To do this, grab the rubber boot, twist slightly and pull the wire free. Do not pull on the wire itself, only on the rubber boot. A boot-pulling tool is helpful (see illustration 24.6).

4  Check inside the boot for corrosion, which will look like a white crusty powder. Push the wire and boot back onto the end of the spark plug. It should be a tight fit on the plug. If it isn't, remove the wire and use a pair of pliers to carefully crimp the metal connector inside the boot until it fits securely on the end of the spark plug.

5  Using a clean rag, wipe the entire length of the wire to remove any built-up dirt and grease. Once the wire is clean, check for holes, burned areas, cracks and other damage. Don't bend the wire excessively or the conductor inside might break.

6  Disconnect the wire from the ignition coil assembly (V6 engines) or individual coil (V8 engines). Pull the wire straight out of the coil. Pull only on the rubber boot during removal (see illustration). Check for corrosion and a tight fit in the same manner as the spark plug end. Reattach the wire to the ignition coil.

7  Check the remaining spark plug wires one at a time, making sure they are securely fastened at both ends when the check is complete.

8  If new spark plug wires are required, purchase a new set for your specific engine model. Wire sets are available pre-cut, with the rubber boots already installed. Remove and replace the wires one at a time to avoid mix-ups in the firing order. The wire routing is extremely important, so be sure to note exactly how each wire is situated before removing it. On V6 engines, release the ignition wire loom clamps to exchange the wires, then snap the clamps back in place on the new wires.

**31.6 Use a spark plug boot pulling tool to remove each end of a spark plug wire (V8 shown) - never pull on the wire itself**

# TUNE-UP AND ROUTINE MAINTENANCE

## Specifications

### Recommended lubricants and fluids

→Note: Listed here are manufacturer recommendations at the time this manual was written. Manufacturers occasionally upgrade their fluid and lubricant specifications, so check with your local auto parts store for current recommendations.

| | |
|---|---|
| Engine oil | API "certified for gasoline engines" |
| Viscosity | 5W-30 |
| Fuel | Unleaded gasoline, 87 octane minimum |
| Automatic transmission fluid | DEXRON VI automatic transmission fluid |
| Transfer case | |
|     NVG 261-NP2 (manual shift) | Manual transmission fluid |
|     NVG 246-NP8 | Auto Trac II fluid |
|     All others | DEXRON VI automatic transmission fluid |
| Differential | |
|     Front | |
|         1500 series | SAE 80W90 GL-5 gear oil |
|         1500HD, 2500HD and 3500 series | SAE 75W90 synthetic axle lubricant |
|     Rear | SAE 75W90 synthetic axle lubricant |
| Power steering fluid | GM power steering fluid |
| Brake fluid | DOT 3 brake fluid |
| Engine coolant | 50/50 mixture of DEX-COOL coolant and demineralized water |
| Parking brake mechanism grease | White lithium-based grease NLGI no. 2 |
| Chassis lubrication grease | NLGI Grade 2 LB or GC-LB chassis grease |
| Hood, door and trunk hinge lubricant | Lubriplate, lubricant aerosol spray |
| Door hinge and check spring grease | NLGI no. 2 multi-purpose grease or equivalent |
| Key lock cylinder lubricant | Graphite spray |
| Hood latch assembly lubricant | NLGI no. 2 multi-purpose grease or equivalent |
| Door latch lubricant | NLGI no. 2 multi-purpose grease or equivalent |

### Capacities*

| | |
|---|---|
| Engine oil (including filter) | |
|     V6 engine | 4.5 quarts |
|     V8 engines | 6.0 quarts |
| Automatic transmission (fluid and filter change) | |
|     4L60-E/4L65-E/4L70-E | 5.0 quarts |
|     6L50/6L80 | 6.0 quarts |
|     6L90 | 6.3 quarts |
| Differential | |
|     Front (4WD models) | Up to 2.0 quarts |
|     Rear | |
|         8.6-inch axle | 2.2 quarts |
|         9.5-inch and 10.5-inch axles | 2.75 quarts |
|         11.5-inch axle | 3.2 quarts |
| Transfer case | |
|     NVG 261-NP2/246-NP8 | 2.0 quarts |
|     MP1222, MP12225, MP1226-NQG, MP1625, MP1626-NQF | |
|         and MP3023-NQH | 1.6 quarts |
|     BW 4485-NR3 | 1.5 quarts |

*All capacities approximate. Add as necessary to bring to appropriate level.

## 1-38 TUNE-UP AND ROUTINE MAINTENANCE

**Capacities* (continued)**

| | |
|---|---|
| Cooling system** | |
|   V6 engine | 16.5 quarts |
|   V8 engines | |
|     Without rear A/C | 16.9 quarts |
|     With rear A/C | |
|       4.8L | 17.8 quarts |
|       5.3L | 18.3 quarts |
|       6.0L | 17.4 quarts |

*All capacities approximate. Add as necessary to bring to appropriate level.
**Cooling system capacities vary depending on engine/transmission package, radiator and A/C system type. Add coolant as necessary to bring to appropriate level.

**Brakes**

| | |
|---|---|
| Disc brake pad wear limit | 3/32 inch |
| Drum brake shoe wear limit | 1/16 inch |
| Parking brake shoe wear limit | 1/16 inch |

**Ignition system**

| | |
|---|---|
| Spark plug type | |
|   V6 engine | AC Delco 41-993 or equivalent |
|   V8 engines | AC Delco 41-985 or equivalent |
| Spark plug gap | |
|   V6 engine | 0.060 inch |
|   V8 engines | 0.040 inch |
| Firing order | |
|   V6 engine | 1-6-5-4-3-2 |
|   V8 engines | 1-8-7-2-6-5-4-3 |

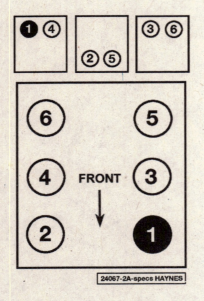

**V6 engine cylinder and coil terminal numbering**

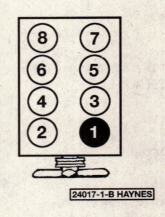

**V8 engine cylinder numbering diagram**

# TUNE-UP AND ROUTINE MAINTENANCE

## Torque specifications — Ft-lbs (unless otherwise indicated)

➡ Note: One foot-pound (ft-lb) of torque is equivalent to 12 inch-pounds (in-lbs) of torque. Torque values below approximately 15 ft-lbs are expressed in inch-pounds, since most foot-pound torque wrenches are not accurate at these smaller values.

| | |
|---|---|
| Spark plugs | 132 in-lbs |
| Wheel lug nuts | 140 |
| Automatic transmission | |
|     4L60-E/4L65-E/4L70-E | |
|         Pan bolts | 97 in-lbs |
|         Drain plug | 156 in-lbs |
|     6L50/6L80/6L90 | |
|         Pan bolts | 80 in-lbs |
| Engine oil drain plug | 18 |
| Transfer case drain/fill plug | |
|     MP manufacture | 156 in-lbs |
|     NVG manufacture | 15 |
|     BW manufacture | 18 |
| Differential drain and fill plugs | |
|     Front differential | 24 |
|     Rear differential | |
|         Drain plug (if equipped) | 24 |
|         Fill plug | 24 |
|         Differential cover bolts | 30 |

## Notes

**Section**

1. General information
2. Repair operations possible with the engine in the vehicle
3. Top Dead Center (TDC) for number one cylinder - locating
4. Valve covers - removal and installation
5. Rocker arms and pushrods - removal, inspection and installation
6. Valve springs, retainers and seals - replacement
7. Intake manifold - removal and installation
8. Exhaust manifolds - removal and installation
9. Cylinder heads - removal and installation
10. Vibration damper and pulley - removal and installation
11. Crankshaft front oil seal - replacement
12. Timing chain and sprockets - removal and installation
13. Camshaft and lifters - removal, inspection and installation
14. Oil pan - removal and installation
15. Oil pump - removal and installation
16. Driveplate - removal and installation
17. Rear main oil seal - replacement
18. Engine mounts - check and replacement

**Reference to other Chapters**

Balance shaft - removal and installation - See Chapter 2C
CHECK ENGINE light on - See Chapter 6
Cylinder compression check - See Chapter 2C
Drivebelt check, adjustment and replacement - See Chapter 1
Engine - removal and installation - See Chapter 2C
Engine oil and filter change - See Chapter 1
Engine overhaul - general information - See Chapter 2C
Spark plug replacement - See Chapter 1
Valves - servicing - See Chapter 2C
Water pump - removal and installation - See Chapter 3

# 2A
## 4.3L V6 ENGINE

# 2A-2  4.3L V6 ENGINE

## 1  General information

This Part of Chapter 2 is devoted to in-vehicle repair procedures for the 4.3L V6 engine. These engines utilize cast-iron blocks with six cylinders arranged in a "V" shape at a 90-degree angle between the two banks. The overhead valve cast iron cylinder heads are equipped with integral valve guides and seats. Hydraulic roller lifters actuate the valves through tubular pushrods and rocker arms. A balance shaft has been incorporated into this engine to smooth vibration. The balance shaft is located directly above the camshaft in the engine block and is driven off the camshaft. The oil pump is mounted to the rear main cap and is driven by the oil pump driveshaft.

To positively identify this engine, locate the Vehicle Identification Number (VIN) on the left front corner of the instrument panel. The VIN is visible from the outside of the vehicle through the windshield. The eighth character in the sequence is the engine designation:

**X = 4.3 liter V6 engine (LU3)**

Information concerning engine removal and installation and engine overhaul can be found in Part C of this Chapter. The following repair procedures are based on the assumption that the engine is installed in the vehicle. If the engine has been removed from the vehicle and mounted on a stand, many of the steps outlined in this Part of Chapter 2 will not apply.

## 2  Repair operations possible with the engine in the vehicle

Many major repair operations can be accomplished without removing the engine from the vehicle.

Clean the engine compartment and the exterior of the engine with some type of pressure washer before any work is done. It will make the job easier and help keep dirt out of the internal areas of the engine.

Remove the hood, if necessary, to improve access to the engine as repairs are performed (see Chapter 11 if necessary).

If vacuum, exhaust, oil or coolant leaks develop, indicating a need for gasket or seal replacement, the repairs can generally be made with the engine in the vehicle. The intake and exhaust manifold gaskets, timing chain cover gasket, oil pan gasket, crankshaft oil seals and cylinder head gaskets are all accessible with the engine in place.

Exterior engine components, such as the intake and exhaust manifolds, the oil pan and oil pump, the water pump, the starter motor, the alternator and the fuel system components can be removed for repair with the engine in place.

Since the cylinder heads can be removed without pulling the engine, valve component servicing can also be accomplished with the engine in the vehicle. Replacement of the timing chain and sprockets is also possible with the engine in the vehicle.

## 3  Top Dead Center (TDC) for number one cylinder- locating

▶ **Refer to illustration 3.6**

1  Top Dead Center (TDC) is the highest point in the cylinder that each piston reaches as it travels up the cylinder bore. Each piston reaches TDC on the compression stroke and again on the exhaust stroke, but TDC generally refers to piston position on the compression stroke.

2  Positioning the piston(s) at TDC is an essential part of certain procedures such as timing chain/sprocket removal.

3  Before beginning this procedure, be sure to place the transmission in Neutral and apply the parking brake or block the rear wheels. Disable the ignition system by disconnecting the primary (low voltage) wires from coil pack (see Chapter 5), then remove the spark plugs (see Chapter 1). If the engine will be rotated using methods b) or c) in the next Step, also disable the fuel system by disconnecting the electrical connector from the fuel meter body (see Chapter 4, Section 14).

4  In order to bring any piston to TDC, the crankshaft must be turned using one of the methods outlined below. When looking at the front of the engine, normal crankshaft rotation is clockwise.

a) *The preferred method is to turn the crankshaft with a socket and ratchet attached to the bolt threaded into the front of the crankshaft. Turn the bolt in a clockwise direction.*
b) *A remote starter switch, which may save some time, can also be used. Follow the instructions included with the switch. Once the piston is close to TDC, use a socket and ratchet as described in the previous paragraph.*
c) *If an assistant is available to turn the ignition switch to the Start position in short bursts, you can get the piston close to TDC without a remote starter switch. Make sure your assistant is out of the vehicle, away from the ignition switch, then use a socket and ratchet as described in Paragraph (a) to complete the procedure.*

5  Place your finger partially over the number one spark plug hole and rotate the crankshaft using one of the methods described above until air pressure is felt at the spark plug hole. Air pressure at the spark plug hole indicates that the cylinder has started the compression stroke. Once the compression stroke has begun, TDC for the number one cylinder is obtained when the piston reaches the top of the cylinder on the compression stroke.

## 4.3L V6 ENGINE  2A-3

6  To bring the piston to the top of the cylinder, continue to turn the crankshaft until the timing marks on the vibration damper align with the timing marks on the front cover (see illustration).

7  If you go past TDC, rotate the crankshaft counterclockwise 1/8 turn, then slowly rotate the crankshaft clockwise again until TDC is reached.

8  After the number one piston has been positioned at TDC on the compression stroke, TDC for any of the remaining pistons can be located by turning the crankshaft 120-degrees (1/3 turn) at a time and following the firing order.

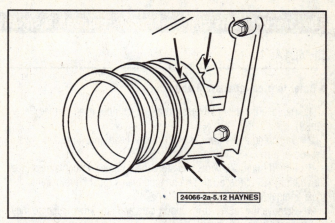

**3.6  After compression is felt at the number one cylinder, align the timing marks on the vibration damper with the timing marks on the front cover**

### 4  Valve covers - removal and installation

### REMOVAL

♦ **Refer to illustration 4.3**

1  Disconnect the cable from the negative terminal of the battery (see Chapter 5, Section 1).

**Right side**

2  Remove the duct from between the air filter housing and the throttle body.

3  Remove the three valve cover bolts and their grommets, then detach the cover from the cylinder head (see illustration).

➡ **Note: If the cover is stuck to the cylinder head, bump one end with a block of wood and a hammer to jar it loose. If that doesn't work, try to slip a flexible putty knife between the cylinder head and cover to break the gasket seal. Don't pry at the cover-to-head joint or damage to the sealing surfaces may occur (leading to oil leaks in the future).**

4  Remove the valve cover gasket.

**Left side**

5  Remove the bolts and clip securing the engine wiring harness. Disconnect the connector from the engine coolant temperature sensor and position the engine wiring harness aside.

6  Remove the vacuum hose from the power brake booster.

7  Disconnect the PCV hose.

8  Remove the three valve cover bolts, then detach the cover from the cylinder head.

➡ **Note: If the cover is stuck to the cylinder head, bump one end with a block of wood and a hammer to jar it loose. If that doesn't work, try to slip a flexible putty knife between the cylinder head and cover to break the gasket seal. Don't pry at the cover-to-head joint or damage to the sealing surfaces may occur (leading to oil leaks in the future).**

### INSTALLATION

9  The mating surfaces of each cylinder head and valve cover must be perfectly clean when the covers are installed. Use a gasket scraper to remove all traces of sealant and old gasket material, then clean the mating surfaces with brake system cleaner. If there's sealant or oil on the mating surfaces when the cover is installed, oil leaks may develop.

10  Clean the mounting bolt threads with a die to remove any corrosion and restore damaged threads. Make sure the threaded holes in the cylinder head are clean - run a tap into them to remove corrosion and restore damaged threads.

11  The gaskets should be mated to the covers before the covers are installed. Apply a thin coat of RTV sealant to the cover flange, then position the gasket inside the cover lip and allow the sealant to set up so the gasket adheres to the cover.

12  Install new valve cover bolt grommets to the valve cover and carefully position the cover(s) on the cylinder head and install the bolts.

13  Tighten the bolts in three or four steps to the torque listed in this Chapter's Specifications.

14  The remainder of installation is the reverse of removal.

15  Start the engine and check carefully for oil leaks as the engine warms up.

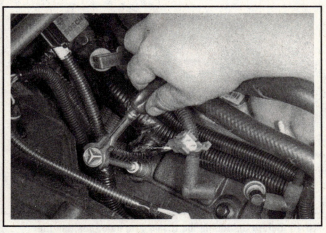

**4.3  Remove the three retaining bolts from the center of the valve cover**

## 5 Rocker arms and pushrods - removal, inspection and installation

### REMOVAL

▶ **Refer to illustrations 5.3 and 5.5**

1  Remove the valve cover(s) from the cylinder head(s) (see Section 4).
2  Beginning at the front of one cylinder head, loosen and remove the rocker arms.
3  Lift off the rocker arms and store them in marked containers (see illustration) (all of these valve train components must be reinstalled in their original locations).
4  Remove the rocker arm support bars and store them also so they can be installed as they were.
5  Remove the pushrods and store them separately to make sure they don't get mixed up during installation (see illustration).

### INSPECTION

6  Check each rocker arm for wear, cracks and other damage, especially where the pushrods and valve stems contact the rocker arm faces.
7  Make sure the hole at the pushrod end of each rocker arm is open.
8  Check each rocker arm pivot area for wear, cracks and galling. If the rocker arms are worn or damaged, replace them with new ones. Check the roller bearings in the rocker arms for free rotation.
9  Inspect the pushrods for cracks and excessive wear at the ends. Roll each pushrod across a piece of plate glass to see if it's bent (if it wobbles, it's bent).

### INSTALLATION

▶ **Refer to illustration 5.13**

10  Lubricate the lower end of each pushrod with clean engine oil or moly-base grease and install them in their original locations. Make sure each pushrod seats completely in the lifter.
11  If the rocker arm supports (below the rocker arms) were removed, make sure to reinstall them with the cast-in arrows pointing UP (away from the cylinder head) before installing the rocker arms. Apply moly-base grease to the ends of the valve stems and the upper ends of the pushrods.
12  Set the rocker arms in place. Lubricate the roller bearings in the rocker arms with clean engine oil.
13  Rotate the crankshaft so that the second timing mark on the vibration damper is 60-degrees before TDC (see illustration). This is a neutral position in the engine rotation and will allow minimum valve spring tension as the rocker arms are tightened. Evenly tighten the rocker arm bolts in several steps to the torque listed in this Chapter's Specifications.
14  The remainder of installation is the reverse of removal.
15  Start the engine and check for valve cover leaks and valvetrain noise.

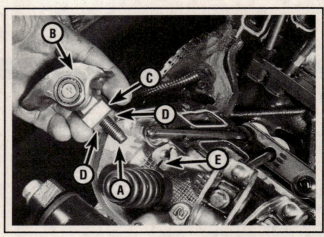

**5.3  Rocker arm details - the rocker arms are kept as an assembly by a small sleeve between the bolt and the pedestal. Note the projections on the pedestal; they fit into grooves in the head**

A  Rocker arm bolt
B  Rocker arm
C  Rocker arm pedestal
D  Pedestal projection
E  Grooves in the head

**5.5  A perforated cardboard box can be used to store the pushrods to ensure that they're installed in their original positions - note the label indicating the front of the engine**

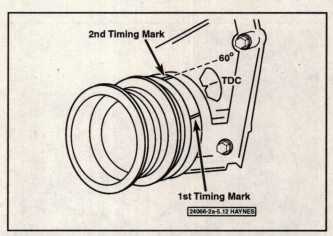

**5.13  Before tightening the rocker arm nuts, position the crankshaft so that the 2nd timing mark is 60 degrees before TDC**

# 4.3L V6 ENGINE

## 6  Valve springs, retainers and seals - replacement

♦ Refer to illustrations 6.5, 6.7a, 6.7b, 6.13a, 6.13b and 6.15

➡ **Note:** *Broken valve springs and defective valve stem seals can be replaced without removing the cylinder heads. Two special tools and a compressed air source are normally required to perform this operation, so read through this Section carefully and rent or buy the tools before beginning the job.*

1  Remove the valve cover from the cylinder head (see Section 4). If all of the valve stem seals are being replaced, remove both valve covers.

2  Remove the spark plug from the cylinder which has the defective component. If all of the valve stem seals are being replaced, all of the spark plugs should be removed.

3  Turn the crankshaft until the piston in the affected cylinder is at Top Dead Center on the compression stroke (see Section 3). If you are replacing all of the valve stem seals, begin with cylinder number 1 and work on the valves for one cylinder at a time. Move from cylinder-to-cylinder following the firing order sequence (1-6-5-4-3-2). Each cylinder in the firing order is 120-degrees of crankshaft rotation (clockwise) from the previous one.

4  Remove all of the rocker arms, pushrods and support bar from the cylinder head on which you're working.

5  Thread an adapter into the spark plug hole (see illustration) and connect an air hose from a compressed air source to it. Most auto parts stores can supply the air hose adapter.

➡ **Note:** *Many cylinder compression gauges utilize a screw-in fitting that may work with your air hose quick-disconnect fitting.*

6  Apply 90 to 100 psi of compressed air to the cylinder. The valves should be held in place by the air pressure.

### ✳✳ WARNING:

**If the cylinder isn't exactly at TDC, air pressure may cause the engine to rotate. Do not leave a socket or wrench on the vibration damper bolt; damage or personal injury may result.**

7  Stuff shop rags into the cylinder head holes to prevent parts and tools from falling into the engine, then use a valve spring compressor to compress the valve spring and the valve spring retainer. Remove the keepers with small needle-nose pliers or a magnet (see illustration).

➡ **Note:** *A couple of different types of tools are available for compressing the valve springs with the cylinder head in place. One type grips the lower spring coils and presses on the retainer as the knob is turned, while the other type utilizes a rocker arm bolt for leverage (see illustration). Both types work very well, although the lever type is usually less expensive.*

8  Remove the spring retainer or rotator and valve spring assembly (on some models there is both an inner and outer valve spring for each valve - the inner is called a spring damper), then remove the valve stem seal from the valve guide.

➡ **Note:** *If air pressure fails to hold the valve in the closed position during this operation, the valve face and/or seat is probably damaged. If so, the cylinder head will have to be removed for additional repair operations.*

**6.5** Use compressed air to hold the valve closed when the springs are removed - the air hose adapter (arrow) threads into the spark plug hole and accepts the hose from the compressor

**6.7a** Once the spring is depressed, the keepers can be removed with a small magnet or needle-nose pliers (a magnet is preferred to prevent dropping the keepers)

**6.7b** Lever-type valve spring compressors use a bolt as a pivot point to apply leverage to the valve spring - it is usually less expensive than the type that grips the spring coils

## 2A-6  4.3L V6 ENGINE

**6.13a  Using a deep socket and hammer, gently tap the new seals onto the valve guide to the specified depth**

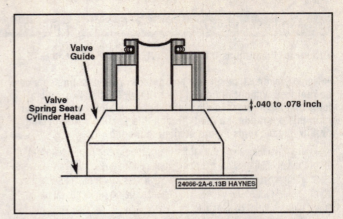

**6.13b  Valve seal installation depth - don't bottom the seal against the valve guide**

9  Wrap a rubber band or tape around the top of the valve stem so the valve won't fall into the combustion chamber, then release the air pressure.

10  Inspect the valve stem for damage. Rotate the valve in the guide and check the end for eccentric movement, which would indicate that the valve is bent.

11  Move the valve up-and-down in the guide and make sure it does not bind. If the valve stem binds, either the valve is bent or the guide is damaged. In either case, the cylinder head will have to be removed for repair.

12  Reapply air pressure to the cylinder to retain the valve in the closed position, then remove the tape or rubber band from the valve stem.

13  Lubricate the valve stems with engine oil and install the new valve stem seals. Using the stem of the valves as a guide, slide the seals down to the top of each valve guide, then use a hammer and a deep socket or seal installation tool to gently tap each seal into place until it's positioned to the specified depth (see illustrations).

### ❋❋ CAUTION:

**Intake and exhaust seals are color coded - do not mix them up. Intake seals are typically white or off-white in color, while exhaust valve stem seals are brown with a white stripe.**

Don't twist or cock the seals during installation or they won't seal properly on the valve stems. Make sure the garter spring is still in place around the top of the seal.

14  Install the valve spring and damper (if equipped) over the valve, with the more closely-wound spring coils toward the cylinder head.

15  Install the valve spring retainer or rotator.

➥ Note: Rotators are used on the exhaust valves only. Compress the valve springs and carefully position the valve stem keepers in the groove. Apply a small dab of grease to the inside of each keeper to hold it in place (see illustration).

16  Disconnect the air hose and remove the adapter from the spark plug hole.

17  Repeat the above procedure on the remaining cylinders, following the firing order sequence (see the Specifications). Bring each piston to Top Dead Center on the compression stroke before applying air pressure.

18  Install the rocker arms, support bars and pushrods (see Section 5).
19  Install the valve cover(s) (see Section 4).
20  Install the spark plug(s) and hook up the wire(s).
21  Start and run the engine, then check for oil leaks and unusual sounds coming from the valve cover area.

**6.15  Apply a small dab of grease to each keeper as shown here before installation - it will hold them in place on the valve stem as the spring is released**

### 7  Intake manifold - removal and installation

### ❋❋ WARNING:

**Wait until the engine is completely cool before beginning this procedure.**

➥ Note: The upper and lower intake manifolds can be removed as a unit, by removing only the lower intake manifold bolts. For repairs or inspection of the fuel meter body or injectors, refer to Chapter 4 for removal of the upper intake plenum.

### REMOVAL

1  Disconnect the cable from the negative terminal of the battery (see Chapter 5, Section 1).

# 4.3L V6 ENGINE 2A-7

7.17 The bolt hole threads must be clean and dry to ensure accurate torque readings when the manifold mounting bolts are installed

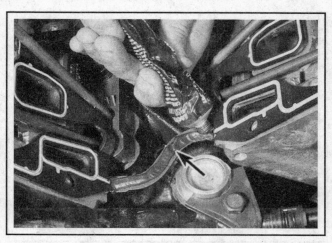

7.18 Apply a 3/16-inch bead of RTV sealant to the front and rear manifold mating surfaces of the engine block - be sure the beads extend up the cylinder heads 1/2-inch on each side

2  Refer to Chapter 1 and drain the coolant.
3  Remove the ignition coil (see Chapter 5).
4  Remove the EVAP canister tube.
5  Disconnect the heater hoses from the intake manifold and the water pump, then lay them aside.
6  Disconnect the wiring from the throttle body. Also disconnect the wiring harness from the air conditioning clutch, the purge control solenoid valve, the MAP sensor and the oil pressure sensor. Disconnect interfering wiring harness brackets. Disconnect the ground wire from the rear of the right cylinder head.
7  Disconnect the electrical connector from the fuel meter body.
8  Remove the PCV valve hose and the power brake booster hose. Disconnect the radiator hose from the thermostat housing.
9  Remove the water pump intake hose.
10  Refer to Chapter 1 and remove the drivebelt.
11  Loosen the nut at the power steering pump rear bracket. Remove the front nut from the power steering pump rear bracket.
12  Remove the power steering pump bracket nut and bolts.
13  Slide the power steering pump bracket forward enough to access the bolt at the front of the intake manifold. You don't need to remove the air conditioning compressor or the power steering pump from the bracket.
14  Loosen the intake lower manifold mounting bolts in 1/4-turn increments in the reverse order of the tightening sequence until they can be removed by hand (see illustration 7.23). The manifold will probably be stuck to the cylinder heads and force may be required to break the gasket seal. A pry bar can be positioned to pry up a casting projection at the front of the manifold to break the bond made by the gasket.

### ❋❋ CAUTION:

**Do not pry between the block and manifold or the heads and manifold, or damage to the gasket sealing surfaces may result and vacuum leaks could develop.**

15  Remove the intake manifold. As the manifold is lifted from the engine, be sure to check for and disconnect anything still attached to the manifold.

## INSTALLATION

▶ Refer to illustrations 7.17, 7.18, 7.20 and 7.23

➡ Note: The mating surfaces of the cylinder heads, block and manifold must be perfectly clean when the manifold is installed. Gasket removal solvents in aerosol cans are available at most auto parts stores and may be helpful when removing old gasket material that is stuck to the heads and manifold. Be sure to follow the directions printed on the container.

16  Use a gasket scraper to remove all traces of sealant and old gasket material, then clean the mating surfaces with brake system cleaner. If there's old sealant or oil on the mating surfaces when the manifold is installed, oil or vacuum leaks may develop. When working on the cylinder heads and block, cover the lifter valley with shop rags to keep debris out of the engine. Use a vacuum cleaner to remove any gasket material that falls into the intake ports in the cylinder heads.
17  Use a tap of the correct size to chase the threads in the bolt holes, then use compressed air (if available) to remove the debris from the holes (see illustration).

### ❋❋ WARNING:

**Wear safety glasses or a face shield to protect your eyes when using compressed air! Remove excessive carbon deposits and corrosion from the exhaust, EGR and coolant passages in the cylinder heads and manifold.**

18  Apply a 3/16-inch wide bead of RTV sealant to the front and rear manifold mating surfaces of the block (see illustration). Make sure the beads extend up the cylinder heads 1/2-inch on each side. Also apply a dab of sealant on top of the gaskets where they touch the longer beads of sealant.
19  If the new manifold gaskets do not come equipped with a rubber sealing ring around the coolant passages, apply a thin coat of RTV sealant around the coolant passage holes on the cylinder head side of the new intake manifold gaskets.

➡ Note: GM replacement gaskets come equipped with a rubber sealant ring around the coolant passages and do not require extra RTV sealant around the coolant passages.

## 2A-8  4.3L V6 ENGINE

7.20  Be sure the gaskets are installed with the marks UP

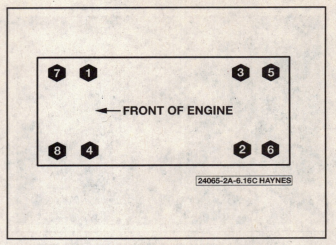

7.23  Lower intake manifold bolt tightening sequence

20  Position the gaskets on the cylinder heads, with the ears at each end overlapping the bead of RTV sealant on the cylinder head. The upper side of each gasket should have a THIS SIDE UP label stamped into it to ensure correct installation (see illustration).

21  Make sure all intake port openings, coolant passage holes and bolt holes are aligned correctly. Some gaskets may have small tabs which must be bent over until they're flush with the rear surface of each cylinder head.

22  Carefully set the manifold in place while the sealant is still wet.

### ✳✳ CAUTION:

**Don't disturb the gaskets and don't move the manifold fore-and-aft after it contacts the sealant on the block.**

23  Following the recommended sequence, install the bolts (with thread-locking compound on the bolts) and tighten them to the torque listed in this Chapter's Specifications (see illustration). Work up to the final torque in three stages.

24  The remainder of installation is the reverse of removal. If the upper intake manifold was removed, refer to Chapter 4 for installation.

25  Refill the cooling system (see Chapter 1).

26  Start the engine and check carefully for oil and coolant leaks at the intake manifold joints.

## 8  Exhaust manifolds - removal and installation

### REMOVAL

▶ Refer to illustrations 8.4 and 8.6

### ✳✳ WARNING:

**Use caution when working around the exhaust manifolds, the sheetmetal heat shields can be sharp on the edges. Also, the engine should be cold when this procedure is followed.**

1  Disconnect the cable from the negative terminal of the battery (see Chapter 5, Section 1).

2  Raise the vehicle and support it securely on jackstands.

3  Working under the vehicle, apply penetrating oil to the exhaust pipe-to-manifold studs and nuts (they're usually rusty).

4  Remove the nuts retaining the exhaust pipe(s) to the manifold(s) (see illustration).

5  Refer to Chapter 1 and disconnect the spark plug wires. Remove them from their retainers and lay them out of the way.

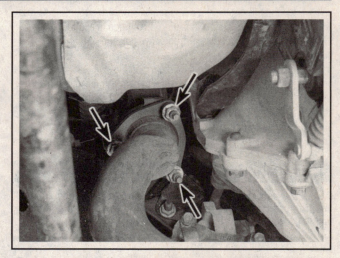

8.4  Access the exhaust pipe bolts/nuts from underneath the vehicle - on some models it may be easier to remove the wheel and work through the fenderwell opening

# 4.3L V6 ENGINE    2A-9

6  Remove the exhaust manifold fasteners (see illustration).
7  Remove the manifolds along with the heat shields. Throw the gaskets away. The heat shields can be removed at this time.

## INSTALLATION

8  Check the manifold for cracks and make sure the bolt threads are clean and undamaged. The manifold and cylinder head mating surfaces must be clean before the manifolds are reinstalled - use a gasket scraper to remove all carbon deposits.
9  Position the manifold on a bench and install the heat shields, bolts and gaskets onto the manifold. Retaining tabs surrounding the gasket bolt holes will hold the assembly together as the manifold is installed. Place the manifold on the cylinder head and install the mounting bolts finger tight.
10  When tightening the mounting bolts, work from the center to the ends and be sure to use a torque wrench. Tighten the bolts in two steps to the torque listed in this Chapter's Specifications. If equipped, bend the locking tabs back against the bolt heads.
11  The remainder of installation is the reverse of removal.
12  Start the engine and check for exhaust leaks.

**8.6  Typical right exhaust manifold mounting bolts with dipstick mounting bolt (A)**

## 9  Cylinder heads - removal and installation

### ※ WARNING:
**Wait until the engine is completely cool before beginning this procedure.**

## REMOVAL

1  Drain the cooling system (see Chapter 1).
2  Remove the intake manifold (see Section 7).
3  Remove the valve covers (see Section 4).
4  Remove the pushrods (see Section 5).
5  Refer to Section 8 and disconnect the exhaust manifold(s) from the cylinder head(s).

➡**Note: The manifolds can be left attached to the cylinder heads if they are disconnected from the exhaust pipes.**

### Right cylinder head

6  Remove the alternator bracket and its mounting stud.

### Left cylinder head

7  Remove the electrical junction block bracket bolt and move the bracket and wiring harness to the side.
8  Loosen the power steering pump rear bracket nut. Remove the front nut from the rear bracket.
9  Remove the bolts and nut from the power steering pump bracket.
10  Slide the bracket assembly off of the mounting stud, leaving the air conditioning compressor and the power steering pump attached.

11  Remove the bracket stud from the cylinder head.
12  Disconnect the ground wire from the rear of the cylinder head, then move the harness aside.
13  Remove the coolant temperature sensor (see Chapter 6).

### Both cylinder heads

14  Remove the spark plugs, then remove the spark plug wire support brackets.
15  Loosen the cylinder head bolts in 1/4-turn increments until they can be removed by hand. Work from bolt-to-bolt in a pattern that's the reverse of the tightening sequence shown in illustration 9.25.

➡**Note: Don't overlook the row of bolts on the lower edge of each cylinder head, near the spark plug holes. Discard the bolts.**

16  Lift the cylinder head(s) off the engine. If resistance is felt, DO NOT pry between the cylinder head and block as damage to the mating surfaces will result. To dislodge the cylinder head, place a block of wood against the end of it and strike the wood block with a hammer. Store the cylinder heads on wood blocks to prevent damage to the gasket sealing surfaces.
17  Cylinder head disassembly and inspection procedures are covered in detail in Chapter 2, Part C.

## INSTALLATION

▶ **Refer to illustrations 9.19, 9.22, 9.24 and 9.25**

18  The mating surfaces of the cylinder heads and block must be perfectly clean when the cylinder heads are installed.

# 2A-10  4.3L V6 ENGINE

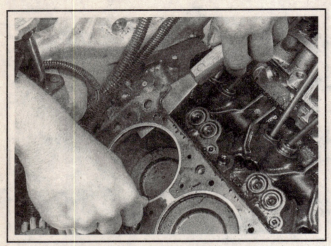

9.19  Thoroughly scrape the sealing surfaces, then clean them to remove all contaminants

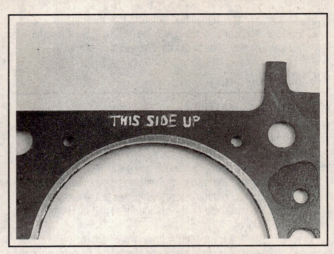

9.22  Be sure the gaskets are installed with the marks UP

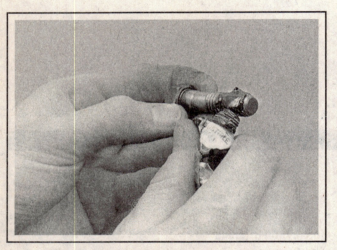

9.24  The cylinder head bolts MUST be coated with a non-hardening sealant (such as Permatex no. 2) before they're installed - coolant will leak past the bolts if this isn't done

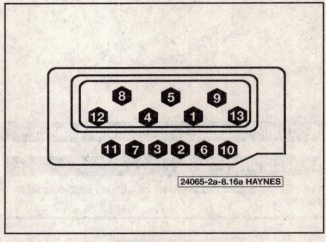

9.25  Cylinder head bolt TIGHTENING sequence

19  Use a gasket scraper to remove all traces of carbon and old gasket material, then clean the mating surfaces with brake system cleaner (see illustration). If there's oil on the mating surfaces when the cylinder heads are installed, the gaskets may not seal correctly and leaks may develop. When working on the block, cover the lifter valley with shop rags to keep debris out of the engine. Use a vacuum cleaner to remove any debris that falls into the cylinders.

20  Check the block and cylinder head mating surfaces for nicks, deep scratches and other damage. If damage is slight, it can be removed with a file - if it's excessive, machining may be the only alternative.

21  Use a tap of the correct size to chase the threads in the cylinder head bolt holes. Dirt, corrosion, sealant and damaged threads will affect torque readings.

22  Position the new gaskets over the dowel pins in the block (see illustration).

➡ Note: Composition-type gaskets are used on most engines, with a thin, sheetmetal core. Be very careful when handling because the edges may be very sharp. Composition gaskets do not require sealant.

23  Carefully position the cylinder heads on the block without disturbing the gaskets.

24  Purchase NEW cylinder head bolts. Before installing the cylinder head bolts, coat the threads with a non-hardening sealant such as Permatex no. 2 (see illustration).

25  Install the bolts in their correct locations according to their lengths and tighten them finger tight. Follow the recommended sequence and tighten the bolts in several steps to the torque and angle of rotation listed in this Chapter's Specifications (see illustration).

26  The remainder of installation is the reverse of removal.

27  Change the engine oil and filter (see Chapter 1), then start the engine and check carefully for oil and coolant leaks.

# 4.3L V6 ENGINE 2A-11

## 10  Vibration damper and pulley - removal and installation

♦ **Refer to illustrations 10.3, 10.5 and 10.6**

1  Disconnect the cable from the negative terminal of the battery (see Chapter 5, Section 1).

2  Refer to Chapter 3 and remove the engine cooling fan, then refer to Chapter 1 and remove the engine drivebelt.

3  Remove the bolts and separate the crankshaft pulley from the vibration damper (see illustration).

4  Remove the large vibration damper-to-crankshaft bolt. To keep the crankshaft from turning, remove the starter (see Chapter 5) and have an assistant wedge a large screwdriver against the ring gear teeth. A chain or strap wrench can also be used to secure the damper.

5  Using the proper puller (commonly available from auto parts stores), detach the vibration damper from the crankshaft (see illustration).

### ✱✱ CAUTION:

Do not use a puller with jaws that grip the outer edge of the damper. The puller must be the type that utilizes bolts to apply force to the center of the damper hub only.

Be careful not to lose the Woodruff key.

### ✱✱ CAUTION:

Insert a short bolt somewhat smaller than the damper bolt into the crankshaft for the tip of the tool to push against, to avoid damage to the threads in the crankshaft.

Make written note of any weights used in the damper, their locations and lengths. These weights must be installed in exactly the same way when the damper is installed. If a new vibration damper is being installed, transfer the weight(s) to the new damper in the proper location(s).

6  Make sure the woodruff key is in place, then position the vibration damper on the crankshaft and slide it on as far as it will go. Use a small dab of RTV sealant on the keyway of the damper before installation. Note that the slot (keyway) in the hub must be aligned with the Woodruff key in the end of the crankshaft (see illustration).

7  Using a vibration damper installation tool, press the damper onto the crankshaft. Note that the crankshaft bolt can also be used to press the crankshaft balancer into position, but when doing so, use a liberal amount of clean engine oil on the bolt threads to prevent galling. Make sure the raised crown of the damper bolt washer is away from the crankshaft.

8  Tighten the crankshaft bolt to the torque listed in this Chapter's Specifications.

9  The remainder of installation is the reverse of removal.

**10.3  Remove the crankshaft pulley bolts and separate the pulley from the vibration damper**

**10.5  Use a bolt-on-type puller to remove the vibration damper**

**10.6  The pulley keyway must be aligned with the Woodruff key (arrow) in the crankshaft nose**

## 2A-12  4.3L V6 ENGINE

### 11 Crankshaft front oil seal - replacement

**Refer to illustrations 11.2, 11.3, 11.5 and 11.6**

1  Remove the crankshaft pulley and vibration damper (see Section 10).

2  Note how the seal is installed - the new one must be installed to the same depth and facing the same way. Carefully pry the oil seal out of the cover with a seal puller or a large screwdriver (see illustration).

**CAUTION:**

**Be careful not to scratch, gouge or distort the area that the seal fits into or an oil leak will develop. Wrap electrician's tape around the tip of the screwdriver to avoid damage to the crankshaft.**

3  If the seal is being replaced with the timing chain cover removed, support the cover on top of two blocks of wood and drive the seal out from the backside with a hammer and punch (see illustration).

**CAUTION:**

**The manufacturer states that the front cover must be replaced with a new one any time it's removed. If you don't replace it, at least be careful not to scratch, gouge or distort the area that the seal fits into or a leak will develop.**

4  Clean the seal bore to remove any old seal material and corrosion. Position the new seal in the bore with the seal lip (usually the side with the spring) facing IN (toward the engine). A small amount of oil applied to the outer edge of the new seal will make installation easier.

5  Drive the seal into the bore with a seal driver or a large socket and hammer until it's completely seated (see illustration). Select a socket that's the same outside diameter as the seal and make sure the new seal is pressed into place until it bottoms against the cover flange.

6  Check the surface of the damper that the oil seal rides on. If the surface has been grooved from long-time contact with the seal, a press-on sleeve may be available to renew the sealing surface (see illustration). This sleeve is pressed into place with a hammer and a block of wood and is commonly available from auto parts stores.

7  Lubricate the seal lips with engine oil and reinstall the vibration damper. Use a vibration damper installation tool to press the damper onto the crankshaft.

8  Install the vibration damper-to-crankshaft bolt and tighten it to the torque listed in this Chapter's Specifications. Install the crankshaft pulley and tighten the bolts to the torque listed in this Chapter's Specifications.

9  The remainder of installation is the reverse of removal.

**11.2  If you're replacing the seal with the timing chain cover installed, pry it out with a seal removal tool or a large screwdriver**

**11.3  If you're replacing the seal with the timing chain cover removed, drive the old seal out from the inside with a hammer and punch or a screwdriver while supporting the cover near the seal bore; it's best to replace the timing cover to avoid leaks**

**11.5  Use a seal driver or large socket to drive the new seal into the cover**

**11.6  If the sealing surface of the damper hub has a wear groove from contact with the seal, repair sleeves are available at most auto parts stores**

# 4.3L V6 ENGINE  2A-13

## 12  Timing chain and sprockets - removal and installation

### ⚠️ WARNING:

**Wait until the engine is completely cool before beginning this procedure.**

### REMOVAL

◆ **Refer to illustrations 12.5, 12.10, 12.11 and 12.13**

1  Disconnect the cable from the negative terminal of the battery (see Chapter 5, Section 1). Drain the cooling system (see Chapter 1).

2  Refer to Chapter 3 and remove the cooling fan and water pump.

3  Position the number four piston at TDC on the compression stroke (see Section 3) to align the sprocket timing marks. It may be easiest to wait until you remove the front cover to do this.

### ⚠️ CAUTION:

**Once this has been done, DO NOT turn the crankshaft until the timing chain and sprockets have been reinstalled!**

4  Refer to Section 10 and remove the drivebelt pulley and the vibration damper.

5  Remove the crankshaft position sensor and the nut retaining the wiring harness to the front cover (see illustration).

6  Remove the oil pan (see Section 14).

7  Disconnect the wiring harness clips from the front cover. Refer to Chapter 6 and remove the crankshaft position sensor.

8  Remove the camshaft position sensor and its jumper wiring harness (see Chapter 6). Remove the sensor from the jumper.

9  Remove the cover bolts and separate the timing chain cover from the block. It may be stuck - if so, use a putty knife or screwdriver to break the gasket seal.

➡**Note:** This engine uses a timing chain cover that is made of a composite material, which is not reusable. It is recommended to replace the front cover and the oil seal rather than try to reseal it and have it leak later.

10  Remove the crankshaft position sensor reluctor ring just inside the front cover (see illustration). Measure the timing chain freeplay. If it is more than 5/8 inch, the chain and both sprockets should be replaced.

11  Make sure that the sprocket timing marks are aligned (see illustration).

12  Release the timing chain tensioner from its lower pin in order to release the force from the chain.

13  Remove the three bolts from the end of the camshaft, then detach the camshaft sprocket and chain as an assembly (see illustration).

➡**Note:** The balance shaft drive gear will stay attached to the camshaft and the driven gear will stay attached to the balance shaft.

If replacement of the timing chain is necessary, remove the sprocket on the crankshaft with a two- or three-jaw puller, but be careful not to damage the threads in the end of the crankshaft.

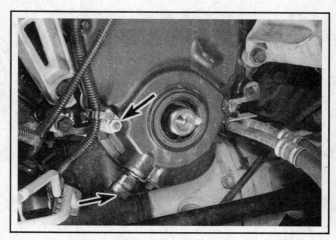

**12.5  Remove the crankshaft position sensor (lower arrow) and remove the nut (upper arrow) retaining the wiring harness to the cover stud**

**12.10  Before removing the sprockets or chain, remove the reluctor ring from the crankshaft**

**12.11  Remove the three bolts from the end of the camshaft and remove the camshaft sprocket and chain as an assembly**

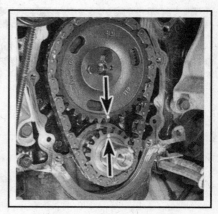

**12.13  Position the timing marks on the camshaft and crankshaft sprockets in the 6 and 12 o'clock positions, respectively. When aligned as shown, the number four piston is at TDC on the compression stroke**

# 2A-14  4.3L V6 ENGINE

## INSTALLATION

♦ **Refer to illustration 12.16**

14  Use a gasket scraper to remove all traces of old gasket material and sealant from the engine block. Clean the block sealing surfaces with brake system cleaner.

15  If a new timing chain is being installed, be sure to align the keyway in the crankshaft sprocket with the Woodruff key in the end of the crankshaft. Press the sprocket onto the crankshaft with the vibration damper bolt, a large socket and some washers, or tap it gently into place until it's completely seated.

**※ CAUTION:**

**If resistance is encountered, DO NOT hammer the sprocket onto the crankshaft. It may eventually move onto the shaft, but it may be cracked in the process and fail later, causing extensive engine damage.**

16  Before installing the timing chain and camshaft sprocket, align the balance shaft gears (see illustration). The camshaft should be positioned with the balance shaft drive gear timing mark at 12 o'clock and the driven gear mark at 6 o'clock.

17  Position the crankshaft so the sprocket timing mark is in the 12 o'clock position. Loop the chain over the camshaft sprocket, mesh the chain with the crankshaft sprocket and position the camshaft sprocket on the camshaft with the timing mark in the 6 o'clock position. When correctly installed, the marks on the sprockets will be aligned as shown (see illustration 12.11).

➡ **Note: The number four piston will be at TDC on the compression stroke with the sprockets aligned as shown.**

18  Apply a non-hardening thread locking compound to the camshaft sprocket bolt threads, then install and tighten them to the torque listed in this Chapter's Specifications. Lubricate the chain with clean engine oil.

**12.16  Before installing the camshaft sprocket, make sure the timing marks on the balance shaft gears are properly aligned (arrows)**

19  Install the crankshaft position sensor reluctor ring. Be sure to install the reluctor with the dished side facing OUT!

20  Apply a thin layer of RTV sealant to the engine block sealing surface, then position a new front cover and oil seal assembly on the engine block (the dowel pins and sealant will hold it in place).

➡ **Note: Composite covers do not have a gasket, they use sealant only. Always purchase a new front cover and oil seal assembly to avoid sealing problems caused by the distortion of prying the old cover off.**

21  Install the cover retaining bolts and tighten them to the torque listed in this Chapter's Specifications.

22  Refer to the appropriate Sections and install the oil pan, vibration damper and the crankshaft position sensor. Be sure to use a new O-ring on the crankshaft position sensor.

23  The remainder of installation is the reverse of removal.

## 13  Camshaft and lifters - removal, inspection and installation

➡ **Note: The camshaft should always be thoroughly inspected before installation and camshaft endplay should always be checked prior to camshaft removal. Refer to Chapter 2C for the camshaft inspection procedures.**

### REMOVAL

♦ **Refer to illustrations 13.2, 13.4 and 13.5**

1  Refer to the appropriate Sections and remove the intake manifold, the rocker arms, the pushrods, the timing chain, camshaft sprocket and the balance shaft drive gear. The balance shaft drive gear can easily be removed after the camshaft sprocket and timing chain are removed. The fan, radiator and condenser should be removed as well (see Chapter 3).

2  Before removing the lifters, arrange to store them in a clearly labeled box to ensure that they're reinstalled in their original locations.

Remove the lifter retainer (see illustration). Remove the lifters and store them where they won't get dirty (see illustration). DO NOT attempt to withdraw the camshaft with the lifters in place.

3  There are several ways to extract the lifters from the bores. A special tool designed to grip and remove lifters is manufactured by many tool companies and is widely available, but it may not be required in every case. On newer engines, without a lot of varnish buildup, the lifters can often be removed with a small magnet or even with your fingers. A machinist's scribe with a bent end can be used to pull the lifters out by positioning the point under the retainer ring in the top of each lifter.

**※ CAUTION:**

**Don't use pliers to remove the lifters unless you intend to replace them with new ones (along with the camshaft). The pliers could damage the precision machined and hardened lifters, rendering them useless.**

## 4.3L V6 ENGINE 2A-15

13.2 The roller lifters are held in place by a retainer - remove the four retainer bolts and remove the retainer

13.4 Remove the Torx bolts and take off the camshaft retainer plate, noting which side faces the block

13.5 Thread three long bolts into the camshaft to use as a handle - pull the camshaft straight out, without nicking the bearings

13.8 The roller on the lifters must spin freely - check for wear and excessive play

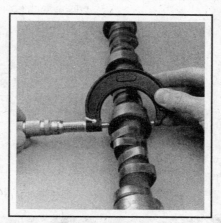

13.10a Measure the camshaft bearing journals with a micrometer

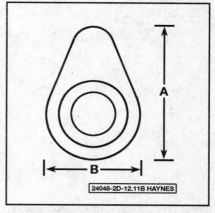

13.10b Measure the cam lobe maximum diameter (A) and the minimum (B) - subtract B from A to get the lobe lift

4   Remove the two bolts and the camshaft retainer plate, noting which direction faces the block (see illustration).

5   Thread three 6-inch long, 5/16-18 bolts into the camshaft sprocket bolt holes to use as a handle when removing the camshaft from the block (see illustration).

6   Carefully pull the camshaft out. Support the cam near the block so the lobes don't nick or gouge the bearings as it's withdrawn.

### INSPECTION

▸ Refer to illustrations 13.8, 13.10a and 13.10b

7   Clean the lifter with solvent and dry them thoroughly without mixing them up.

8   Check the rollers carefully for wear and make sure they turn smoothly, freely and without excessive play (see illustration). Inspect the pushrod ends if the lifter pushrod seats are worn.

9   Clean the camshaft with solvent and dry it. Inspect the bearing journals for uneven wear and other damage. If there is damage, then the camshaft bearings are probably damaged as well. Both the camshaft and bearings will have to be replaced.

10  Measure the journals with a micrometer (see illustration) to determine if they're worn or out-of-round. Measure each cam lobe to check for wear. Measure them at their highest points and at their smallest. The difference is the lobe lift (see illustration). Refer to the Specifications in this Chapter.

11  Inspect the cam lobes for scoring, chipped areas, pitting and uneven wear. If they're in good condition and the lift is as specified, you can reuse the camshaft.

12  Check the camshaft bearings in the block for wear and damage.

13  Camshaft bearing replacement requires special tools and expertise that place it outside the scope of the home mechanic. Take the block to an automotive machine shop to have the job done correctly.

# 2A-16    4.3L V6 ENGINE

## INSTALLATION

▸ Refer to illustration 13.14

14  Lubricate the camshaft bearing journals and cam lobes with camshaft and lifter assembly lube (see illustration).

15  Slide the camshaft into the engine. Support the cam near the block and be careful not to scrape or nick the bearings.

16  Turn the camshaft until the dowel pin is in the 3 o'clock position, install the camshaft thrust plate, then tighten the bolts to the torque listed in this Chapter's Specifications.

17  Install the balance shaft drive gear over the camshaft, aligning the dowel pin. Make sure the balance shaft timing marks are properly aligned (see illustration 12.16).

18  Install the timing chain and sprockets (see Section 12).

19  Lubricate the lifters with clean engine oil and install them in the block. If the original lifters are being reinstalled, be sure to return them to their original locations. If a new camshaft is being installed, install new lifters as well.

20  The remainder of installation is the reverse of removal.

21  Refill the cooling system, and change the engine oil and filter (see Chapter 1).

22  Start the engine and check for leaks.

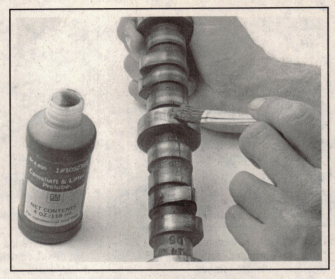

**13.14  Lubricate the camshaft journals and lobes with camshaft and lifter assembly lube before installation to provide initial lubrication**

## 14  Oil pan - removal and installation

▸ Refer to illustrations 14.15, 14.16a and 14.16b

### REMOVAL

1  Disconnect the cable from the negative terminal of the battery (see Chapter 5, Section 1).

2  Raise the vehicle and support it securely on jackstands, then refer to Chapter 1 and drain the engine oil and remove the oil filter.

3  Remove the oil pan skid plate if equipped.

4  Remove the lower crossmember from below the oil pan.

5  On 4WD models, remove the front differential carrier (see Chapter 8).

6  There is an engine wiring harness at the front of the oil pan. Disconnect the harness bracket.

7  Disconnect the front exhaust Y pipe from the engine and the exhaust system and remove it from the vehicle. This step is not absolutely necessary, but it will help facilitate removal of the oil pan.

8  Remove the starter motor (see Chapter 5). Also remove the plastic bellhousing side covers (see Chapter 7A). Disconnect the transmission cooler line bracket from the side of the oil pan.

9  Disconnect the remaining wiring harness brackets.

10  Remove the transmission-to-engine mounting bolts.

11  Disconnect the electrical connector from the oil level sensor, if equipped.

12  Remove the transmission-to-oil pan bolts (see Chapter 7).

13  Remove the oil pan bolts, then lower the pan from the engine. The pan will probably stick to the engine, so strike the pan with a rubber mallet until it breaks the gasket seal.

### ✱✱ CAUTION:

**Before using force on the oil pan, be sure all the bolts have been removed. Carefully slide the oil pan out, to the rear.**

### INSTALLATION

14  Wash out the oil pan with solvent. Thoroughly clean the mounting surfaces of the oil pan and engine block of old gasket material and sealer. Wipe the gasket surfaces clean with a rag soaked in brake system cleaner.

➡ Note: On models with a low-oil-level sensor, remove the sensor and install a new sensor upon assembly.

15  Apply a 3/16-inch wide, one inch long bead of RTV sealant to the corners where the front cover meets the block and at the rear where the rear main cap meets the block (see illustration). Then attach the new gasket to the pan, install the pan and tighten the bolts/studs finger-tight.

16  The alignment of the rear face of the aluminum pan to the rear of the block is important. Measure between the rear face of the pan and the front face of the transmission bellhousing with feeler gauges. Clearance should ideally be flush, but up to 0.011-inch is allowable. If the clearance is OK, tighten the pan bolts/studs in sequence to the torque listed in this Chapter's Specifications (see illustrations).

17  The remainder of installation is the reverse of removal.

18  Add the proper type and quantity of oil (see Chapter 1), start the engine and check for leaks before placing the vehicle back in service.

14.15  Apply a bead of RTV sealant to each corner of the oil pan rail where the gasket will meet it

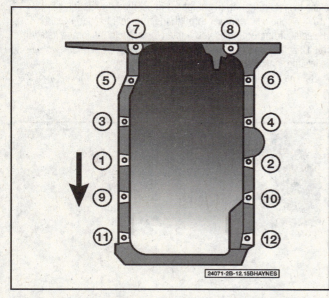

14.16a  Bolt tightening sequence for the cast aluminum oil pan

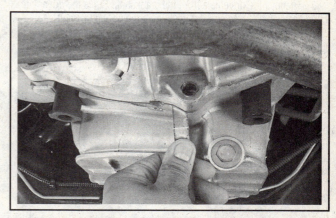

14.16b  Before tightening the oil pan bolts, measure the gap between the bellhousing and the oil pan in three places - if the gap is greater than 0.011-inch the oil pan will have to move towards the bellhousing - ideally the oil pan should be flush with the bellhousing

## 15  Oil pump - removal and installation

▶ Refer to illustration 15.2

1  Remove the oil pan (see Section 14).

2  While supporting the oil pump, remove the pump-to-rear main bearing cap bolt (see illustration).

3  Lower the pump and remove it along with the pump driveshaft.

4  If a new oil pump is installed, make sure the pump driveshaft is mated with the shaft inside the pump and a new driveshaft retainer is used.

5  Position the pump over the dowel pins on the rear main cap and make sure the slot in the upper end of the driveshaft is aligned with the tang on the lower end of the oil pump driveshaft. The oil pump driveshaft drives the oil pump, so it is absolutely essential that the components mate properly. Also note that no gasket is used between the oil pump and the rear main cap.

6  Install the mounting bolt and tighten it to the torque listed in this Chapter's Specifications.

7  Install the oil pan and refill the engine with fresh oil. The remainder of assembly is the reverse of the disassembly procedures.

15.2  Oil pump mounting bolt

# 2A-18  4.3L V6 ENGINE

## 16  Driveplate - removal and installation

### REMOVAL

◆ **Refer to illustrations 16.2a and 16.2b**

1  Raise the vehicle and support it securely on jackstands, then refer to Chapter 7A and remove the transmission.

2  Mark the relationship between the driveplate and the crankshaft with a marker or similar device, then remove the bolts that secure the driveplate to the crankshaft (see illustration). If the crankshaft turns, wedge a screwdriver in the ring gear teeth to jam the driveplate (see illustration).

➡**Note: If there is a retaining ring between the bolts and the driveplate, note which side faces the driveplate when removing it.**

3  Remove the driveplate from the crankshaft. Since the driveplate is fairly heavy, be sure to support it while removing the last bolt.

4  Clean the driveplate to remove grease and oil. Inspect the surface for cracks, and check for cracked and broken ring gear teeth. Lay the driveplate on a flat surface to check for warpage.

5  Clean and inspect the mating surfaces of the driveplate and the crankshaft. If the rear main oil seal is leaking, replace it before reinstalling the driveplate (see Section 17).

### INSTALLATION

6  Position the driveplate against the crankshaft. Be sure to align the marks made during removal. Note that some engines have an alignment dowel or staggered bolt holes to ensure correct installation. Before installing the bolts, apply thread locking compound to the threads and place the retaining ring (if equipped) in position on the driveplate.

7  Wedge a screwdriver through the ring gear teeth to keep the driveplate from turning as you tighten the bolts to the torque listed in this Chapter's Specifications. If the transmission front pump seal/O-ring is leaking, now would be a very good time to replace it.

8  The remainder of installation is the reverse of removal.

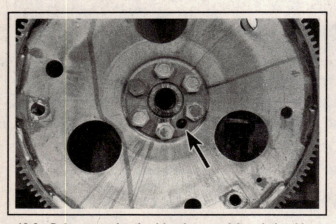

**16.2a  Before removing the driveplate, mark its relationship to the crankshaft**

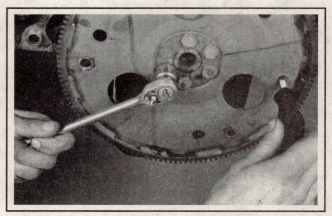

**16.2b  A large screwdriver wedged in one of the driveplate holes can be used to keep it from turning**

## 17  Rear main oil seal - replacement

◆ **Refer to illustration 17.4**

1  Remove the transmission (see Chapter 7).

2  Remove the driveplate (see Section 16).

3  Inspect the oil seal, as well as the oil pan and engine block surface for signs of leakage. Sometimes an oil pan gasket leak can appear to be a rear oil seal leak.

4  Pry the oil seal from the block with a screwdriver (see illustration).

➡**Note: There are notches at the two o'clock and ten o'clock positions into which you can insert a screwdriver for prying.**

Be careful not to nick or scratch the crankshaft or the seal bore. Thoroughly clean the seal bore in the block with a shop towel. Remove all traces of oil and dirt.

# 4.3L V6 ENGINE  2A-19

5  Lubricate all surfaces with a very small amount of engine oil. Install the seal over the end of the crankshaft and carefully tap it into place. Most seals are marked "This Side Out" to eliminate confusion as they have a different design than the old style seals previously used. A special aftermarket tool may be available at your local auto parts store. The tool tightly fits the diameter of the seal and, used with a hammer, drives the seal in.

➥**Note: Do not drive it in any farther than the original seal was installed. The seal should be installed flush with the block surface and not angled.**

6  Install the driveplate (see Section 16).
7  Install the transmission (see Chapter 7).

**17.4  Carefully pry out the old seal**

## 18  Engine mounts - check and replacement

The engine mount replacement for V6 engines is identical to the engine mount replacement procedure for the V8 engines. Refer to Chapter 2 Part B for the procedure.

## Specifications

### General

| | |
|---|---|
| Displacement | 262 cubic inches (4.3 liters) |
| Bore and stroke | 4.012 x 3.480 inches |
| Cylinder numbers | |
|     Left bank | 1-3-5 |
|     Right bank | 2-4-6 |
| Firing order | 1-6-5-4-3-2 |

### Camshaft

| | |
|---|---|
| Journal diameter | 1.8677 to 1.8696 inches |
| Endplay | 0.001 to 0.009 inch |
| Lobe lift | |
|     Intake | 0.270 inch |
|     Exhaust | 0.279 inch |
| Runout, maximum | 0.004 inch |

**Cylinder and coil terminal location diagram**

## 2A-20  4.3L V6 ENGINE

### Torque specifications — Ft-lbs (unless otherwise indicated)

➡ **Note:** One foot-pound (ft-lb) of torque is equivalent to 12 inch-pounds (in-lbs) of torque. Torque values below approximately 15 foot-pounds are expressed in inch-pounds, because most foot-pound torque wrenches are not accurate at these smaller values.

| Fastener | Torque |
|---|---|
| Balance shaft driven gear bolt | |
|     Step 1 | 15 |
|     Step 2 | Tighten an additional 35 degrees |
| Balance shaft retainer bolts | 106 in-lbs |
| Camshaft retainer bolts | 106 in-lbs |
| Camshaft sprocket bolts | 18 |
| Crankshaft pulley bolts | 43 |
| Crankshaft vibration damper bolt | 70 |
| Cylinder head bolts* (in sequence - see illustration 9.25) | |
|     Step 1 (all bolts) | 22 |
|     Step 2 | |
|         Long bolts | Tighten an additional 75 degrees |
|         Medium length bolts | Tighten an additional 65 degrees |
|         Short bolts | Tighten an additional 55 degrees |
| Driveplate-to-crankshaft bolts | 74 |
| Exhaust manifold bolt/studs | |
|     Step 1 | 132 in-lbs |
|     Step 2 | 22 |
| Upper intake manifold studs | |
|     Step 1 | 44 in-lbs |
|     Step 2 | 80 in-lbs |
| Lifter retainer bolts | 144 in-lbs |
| Lower intake manifold bolts | |
|     Step 1 | 27 in-lbs |
|     Step 2 | 106 in-lbs |
|     Step 3 | 132 in-lbs |
| Oil pan mounting bolt/nut | 18 |
| Oil pan-to-transmission bolts | 35 |
| Oil pan baffle bolts | 106 in-lbs |
| Oil pump mounting bolts | 66 |
| Rocker arm bolts | 22 |
| Valve cover-to-cylinder head bolts | 106 in-lbs |
| Timing chain cover-to-block bolts | 106 in-lbs |

*Use NEW cylinder head bolts. Used ones should be discarded.

# 2B

## V8 ENGINES

**Section**

1. General information
2. Repair operations possible with the engine in the vehicle
3. Top Dead Center (TDC) for number one piston - locating
4. Valve covers - removal and installation
5. Rocker arms and pushrods - removal, inspection and installation
6. Valve springs, retainers and seals - replacement
7. Intake manifold - removal and installation
8. Exhaust manifolds - removal and installation
9. Cylinder heads - removal and installation
10. Crankshaft balancer - removal and installation
11. Crankshaft front oil seal - removal and installation
12. Timing chain - removal, inspection and installation
13. Camshaft and lifters - removal, inspection and installation
14. Oil pan - removal and installation
15. Oil pump - removal, inspection and installation
16. Driveplate - removal and installation
17. Rear main oil seal - replacement
18. Engine mounts - check and replacement

**Reference to other Chapters**

CHECK ENGINE light on - See Chapter 6
Cylinder compression check - See Chapter 2C
Drivebelt check, adjustment and replacement - See Chapter 1
Engine - removal and installation - See Chapter 2C
Engine overhaul - See Chapter 2C
Water pump - removal and installation - See Chapter 3

# 2B-2  V8 ENGINES

## 1  General information

This Part of Chapter 2 is devoted to in-vehicle repair procedures for the 4.8L, 5.3L, 6.0L and 6.2L V8 engines. These engines utilize either aluminum or cast-iron blocks with eight cylinders arranged in a "V" shape at a 90-degree angle between the two banks. The cylinder heads utilize an overhead valve arrangement. Engines with aluminum cylinder heads use pressed-in valve guides and valve seats, while engines with cast iron cylinder heads have integral valve guides and hardened valve seats. Hydraulic roller lifters actuate the valves through tubular pushrods and rocker arms. The oil pump is mounted at the front of the engine behind the timing chain cover and is driven by the crankshaft.

To positively identify these engines, locate the Vehicle Identification Number (VIN) on the left front corner of the instrument panel. The VIN is visible from the outside of the vehicle through the windshield. The eighth character in the sequence is the engine designation:

    C = 4.8L V8 engine
    M, 0, 3, J = 5.3L V8 engine
    K, Y = 6.0L V8 engine
    8 = 6.2L V8 engine

Information concerning engine removal and installation and engine overhaul can be found in Part C of this Chapter. The following repair procedures are based on the assumption that the engine is installed in the vehicle. If the engine has been removed from the vehicle and mounted on a stand, many of the steps outlined in this Part of Chapter 2 will not apply.

## 2  Repair operations possible with the engine in the vehicle

Many major repair operations can be accomplished without removing the engine from the vehicle.

Clean the engine compartment and the exterior of the engine with some type of pressure washer before any work is done. A clean engine will make the job easier and will help keep dirt out of the internal areas of the engine.

Depending on the components involved, it may be a good idea to remove the hood to improve access to the engine as repairs are performed (refer to Chapter 11 if necessary).

If oil or coolant leaks develop, indicating a need for gasket or seal replacement, the repairs can generally be made with the engine in the vehicle. The oil pan gasket, the cylinder head gaskets, intake and exhaust manifold gaskets, timing chain cover gaskets and the crankshaft oil seals are all accessible with the engine in place.

Exterior engine components, such as the water pump, the starter motor, the alternator and the fuel injection components, as well as the intake and exhaust manifolds, can be removed for repair with the engine in place.

Since the cylinder heads can be removed without removing the engine, valve component servicing can also be accomplished with the engine in the vehicle.

Replacement of, repairs to or inspection of the timing chain and sprockets and the oil pump are all possible with the engine in place.

In extreme cases caused by a lack of necessary equipment, repair or replacement of piston rings, pistons, connecting rods and rod bearings is possible with the engine in the vehicle. However, this practice is not recommended because of the cleaning and preparation work that must be done to the components involved.

## 3  Top Dead Center (TDC) for number one piston - locating

1  Top Dead Center (TDC) is the highest point in the cylinder that each piston reaches as it travels up the cylinder bore. Each piston reaches TDC on the compression stroke and again on the exhaust stroke, but TDC generally refers to piston position on the compression stroke.

2  Positioning the piston(s) at TDC is an essential part of many procedures such as timing chain/sprocket removal.

3  Before beginning this procedure, be sure to place the transmission in Neutral and apply the parking brake or block the rear wheels. If method b) or c) will be used to rotate the engine in the next Step, disable the ignition system by disconnecting the primary electrical connectors at the ignition coil packs. Also, if method b) or c) will be used to rotate the engine in the next Step, disable the fuel system by removing the INJ A and INJ B fuses from the underhood fuse and relay box. Remove the spark plugs (see Chapter 1).

4  In order to bring any piston to TDC, the crankshaft must be turned using one of the methods outlined below. When looking at the front of the engine, normal crankshaft rotation is clockwise.

   *a) The preferred method is to turn the crankshaft with a socket and ratchet attached to the bolt threaded into the front of the crankshaft. Turn the bolt in a clockwise direction only. Never turn the bolt counterclockwise.*

   *b) A remote starter switch, which may save some time, can also be used. Follow the instructions included with the switch. Once the piston is close to TDC, use a socket and ratchet as described in the previous paragraph.*

   *c) If an assistant is available to turn the ignition switch to the Start position in short bursts, you can get the piston close to TDC without a remote starter switch. Make sure your assistant is out of the vehicle, away from the ignition switch, then use a socket and ratchet as described in Paragraph (a) to complete the procedure.*

5  Install a compression gauge in the No. 1 spark plug hole. It should be a gauge with a screw-in type fitting and a hose at least six inches long. Rotate the crankshaft using one of the methods described

# V8 ENGINES 2B-3

above while observing for pressure on the compression gauge. The moment the gauge shows pressure indicates that the No. 1 cylinder has begun the compression stroke.

6  To bring the piston to the top of the cylinder, insert a long wooden dowel or a piece of plastic coat hanger into the number one spark plug hole until it touches the top of the piston. Use the dowel (as a feeler gauge) to tell where the top of the piston is located in the cylinder while slowly rotating the crankshaft. As the piston rises the dowel will be pushed out. The point at which the dowel stops moving outward is TDC.

7  If you go past TDC, rotate the crank-shaft counterclockwise until the piston is approximately one inch below TDC, then slowly rotate the crankshaft clockwise again until TDC is reached.

8  After the number one piston has been positioned at TDC on the compression stroke, TDC for any of the remaining pistons can be located by turning the crankshaft 90-degrees (1/4 turn) at a time and following the firing order.

## 4  Valve covers - removal and installation

### REMOVAL

▶ **Refer to illustrations 4.4a, 4.4b and 4.8**

1  Disconnect the cable from the negative terminal of the battery (see Chapter 5, Section 1).
2  Remove the engine cover by pulling it upward.
3  Disconnect the interfering engine wiring harnesses and move them aside.
4  If you're working on the right valve cover, detach the heater hoses from the hose bracket on the valve cover (see illustrations). Remove the heater hose bracket.
5  Disconnect the spark plug wires from the coils by first twisting them a half turn to loosen them. Pull only on the boot - not on the wire.
6  Remove the ignition coil bracket along with the coils.
7  Remove the PCV tube or hose from the valve cover.
8  Remove the valve cover bolts, then detach the cover from the cylinder head (see illustration).

➡**Note: If the cover is stuck to the cylinder head, bump one end with a block of wood and a hammer to jar it loose. If that doesn't work, try to slip a flexible putty knife between the cylinder head and cover to break the gasket seal. Don't pry at the cover-to-head joint or damage to the sealing surfaces may occur (leading to oil leaks in the future).**

4.4a  The main wiring harness restricts access to the left valve cover - it must be released from its retainers and moved aside

4.4b  The heater hoses can be unclipped from their retainer to remove the right valve cover

4.8  Valve cover mounting bolts

## 2B-4 V8 ENGINES

### INSTALLATION

▶ **Refer to illustration 4.11**

9  The mating surfaces of each cylinder head and valve cover must be perfectly clean when the covers are installed. Use a gasket scraper to remove all traces of sealant and old gasket material, then clean the mating surfaces with brake system cleaner. If there's sealant or oil on the mating surfaces when the cover is installed, oil leaks may develop.

10  Clean the mounting bolt threads with a die to remove any corrosion and restore damaged threads. Make sure the threaded holes in the cylinder head are clean - run a tap into them to remove corrosion and restore damaged threads.

11  Position the gasket inside the cover lip (see illustration). If the gasket will not stay in place in the cover lip, apply a thin coat of RTV sealant to the cover flange, then allow the sealant to set up so the gasket adheres to the cover.

12  Tighten the bolts in three or four steps to the torque listed in this Chapter's Specifications.

13  The remainder of installation is the reverse of removal. Use thread locking material on the ignition coil bracket studs and nuts.

14  Start the engine and check carefully for oil leaks as the engine warms up.

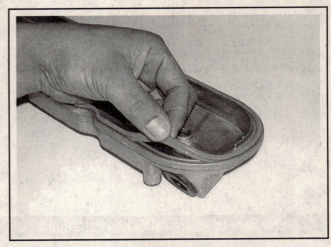

**4.11 Position the new gasket in the valve cover lip**

### 5  Rocker arms and pushrods - removal, inspection and installation

### REMOVAL

▶ **Refer to illustrations 5.2 and 5.3**

1  Refer to Section 4 and remove the valve covers from the cylinder heads.

2  Loosen the rocker arm pivot bolts one at a time and remove the rocker arms and bolts, then remove the pivot support pedestal (see illustration). Keep track of the component positions, since they must be returned to the same locations. Store each set of components separately in a marked container to ensure that they're reinstalled in their original locations. This includes the rocker arms and the pivot support

3  Remove the pushrods and store them in order to make sure they don't get mixed up during installation (see illustration).

**5.2 Remove the mounting bolts (A) and rocker arms, then remove the pivot support pedestal (B)**

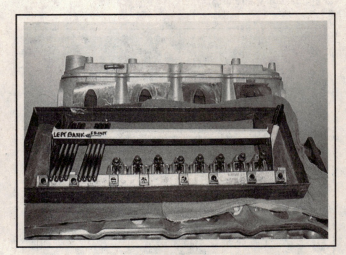

**5.3 Store the pushrods and rocker arms in order to ensure they are reinstalled in their original locations**

# V8 ENGINES  2B-5

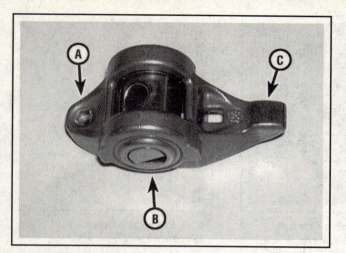

**5.4 Rocker arm wear points**

- A  Pushrod socket
- B  Pivot bearing
- C  Valve stem contact point

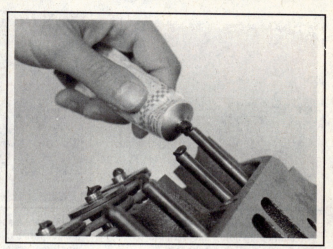

**5.9 Lubricate the pushrod ends and the valve stems with engine assembly lube before installing the rocker arms**

## INSPECTION

**Refer to illustration 5.4**

4  Check each rocker arm for wear, cracks and other damage, especially where the pushrods and valve stems contact the rocker arm (see illustration).

5  Check the pivot bearings for binding and roughness. If the bearings are worn or damaged, replacement of the entire rocker arm will be necessary.

→**Note:** *Keep in mind that there is no valve adjustment on these engines, so excessive wear or damage in the valve train can easily result in excessive valve clearance, which in turn will cause valve noise when the engine is running.*

Also check the rocker arm pivot support pedestal for cracks and other obvious damage.

6  Make sure the hole at the pushrod end of each rocker arm is open.

7  Inspect the pushrods for cracks and excessive wear at the ends, also check that the oil hole running through each pushrod is not clogged. Roll each pushrod across a piece of plate glass to see if it's bent (if it wobbles, it's bent).

## INSTALLATION

♦ **Refer to illustration 5.9**

8  Lubricate the lower end of each pushrod with clean engine oil or engine assembly lube and install them in their original locations. Make sure each pushrod seats completely in the lifter socket.

9  Apply engine assembly lube to the ends of the valve stems and to the upper ends of the pushrods to prevent damage to the mating surfaces on initial start-up (see illustration). Also apply clean engine oil to the pivot shaft and bearing of each rocker arm and install the rocker arms loosely in their original locations. DO NOT tighten the bolts at this time!

10  Rotate the crankshaft until the number one piston is at TDC on the compression stroke (see Section 3). Both pushrods for the number one cylinder should be in the lowered position. When the number one piston is at TDC, tighten the intake valve rocker arms for the Number 1, 3, 4, and 5 cylinders and the exhaust rocker arms for the Number 1, 2, 7, and 8 cylinders. Tighten each of the specified rocker arm bolts to the torque listed in this Chapter's Specifications.

11  Rotate the crankshaft 360 degrees. Tighten the intake valve rocker arms for the Number 2, 6, 7, and 8 cylinders and the exhaust rocker arms for the Number 3, 4, 5, and 6 cylinders. Tighten each of the rocker arm bolts to the torque listed in this Chapter's Specifications.

12  Refer to Section 4 and install the valve covers. Start the engine, listen for unusual valve train noses and check for oil leaks at the valve cover gaskets.

## 6  Valve springs, retainers and seals - replacement

♦ **Refer to illustrations 6.5, 6.8, 6.10, 6.15a, 6.15b and 6.19**

→**Note:** *Broken valve springs and defective valve stem seals can be replaced without removing the cylinder head. Two special tools and a compressed air source are normally required to perform this operation, so read through this Section carefully and rent or buy the tools before beginning the job.*

1  Remove the spark plugs (see Chapter 1).
2  Remove the valve covers (see Section 4).
3  Rotate the crankshaft until the number one piston is at Top Dead Center on the compression stroke (see Section 3).
4  Remove the rocker arms for the number 1 piston.
5  Thread an adapter into the spark plug hole and connect an air

6.5 Thread the air hose adapter into the spark plug hole - adapters are commonly available from auto parts stores

6.8 Once the spring is depressed, the keepers can be removed with a small magnet or needle-nose pliers (a magnet is preferred to prevent dropping the keepers)

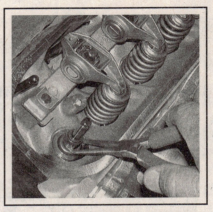
6.10 Use a pair of needle-nose pliers to remove the valve seals

hose from a compressed air source to it (see illustration). Most auto parts stores can supply the air hose adapter.

→Note: *Many cylinder compression gauges utilize a screw-in fitting that may work with your air hose quick-disconnect fitting. If a cylinder compression gauge fitting is used, it will be necessary to remove the Schrader valve from the end of the fitting before using it in this procedure.*

6  Apply compressed air to the cylinder. The valves should be held in place by the air pressure.

### ※※ WARNING:

**If the cylinder isn't exactly at TDC, air pressure may force the piston down, causing the engine to quickly rotate. DO NOT leave a wrench on the crankshaft balancer bolt or you may be injured by the tool.**

7  Stuff shop rags into the cylinder head holes around the valves to prevent parts and tools from falling into the engine.

8  Using a socket and a hammer, gently tap on the top of each valve spring retainer several times (this will break the seal between the valve keeper and the spring retainer and allow the keeper to separate from the valve spring retainer as the valve spring is compressed), then use a valve-spring compressor to compress the spring. Remove the keepers with small needle-nose pliers or a magnet (see illustration).

→Note: *Several different types of tools are available for compressing the valve springs with the head in place. One type grips the lower spring coils and presses on the retainer as the knob is turned, while the lever-type shown here utilizes the rocker arm bolt for leverage. Both types work very well, although the lever type is usually less expensive.*

9  Remove the valve spring and retainer.

→Note: *If air pressure fails to retain the valve in the closed position during this operation, the valve face or seat may be damaged. If so, the cylinder head will have to be removed for repair.*

10  Remove the old valve stem seals, noting differences between the intake and exhaust seals (see illustration).

11  Wrap a rubber band or tape around the top of the valve stem so the valve won't fall into the combustion chamber, then release the air pressure.

12  Inspect the valve stem for damage. Rotate the valve in the guide and check the end for eccentric movement, which would indicate that the valve is bent.

13  Move the valve up-and-down in the guide and make sure it does not bind. If the valve stem binds, either the valve is bent or the guide is damaged. In either case, the head will have to be removed for repair.

14  Reapply air pressure to the cylinder to retain the valve in the closed position, then remove the tape or rubber band from the valve stem.

15  If you're working on an exhaust valve, install the new exhaust valve seal on the valve stem and press it down over the valve guide to the specified depth. Don't force the seal against the top of the guide (see illustrations).

→Note: *On aluminum heads, be sure to take this measurement from the steel spring seat to the top edge of the intake and exhaust valve seals, not from the aluminum seat on the head!*

16  If you're working on an intake valve, install a new intake valve stem seal over the valve stem and press it down over the valve guide to the specified depth. Don't force the intake valve seal against the top of the guide.

### ※※ CAUTION:

**Do not install an exhaust valve seal on an intake valve, as high oil consumption will result.**

17  Install the spring and retainer in position over the valve.

18  Compress the valve spring assembly only enough to install the keepers in the valve stem.

19  Position the keepers in the valve stem groove. Apply a small dab of grease to the inside of each keeper to hold it in place if necessary (see illustration). Remove the pressure from the spring tool and make sure the keepers are seated.

20  Disconnect the air hose and remove the adapter from the spark plug hole.

21  Repeat the above procedure on the remaining cylinders, following the firing order sequence (see this Chapter's Specifications). Bring each

# V8 ENGINES 2B-7

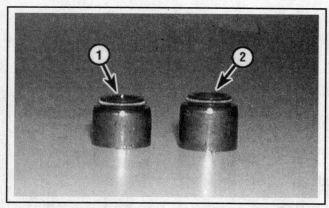

**6.15a  Be sure to install the seals on the correct valve stems**

1   Intake valve seal  
2   Exhaust valve seal

**6.15b  Install the intake and exhaust valve seals to the specified depth - measure from the spring seat to the top edge of the valve seal**

piston to Top Dead Center on the compression stroke before applying air pressure (see Section 3).

22  Reinstall the rocker arm assemblies and the valve covers (see Sections 4 and 5).

23  Start the engine, then check for oil leaks and unusual sounds coming from the valve cover area. Allow the engine to idle for at least five minutes before revving the engine.

**6.19  Apply a small dab of grease to each keeper as shown here before installation - it'll hold them in place on the valve stem as the spring is released**

## 7   Intake manifold - removal and installation

### ❄❄ WARNING:

**Wait until the engine is completely cool before starting this procedure.**

### REMOVAL

▸ **Refer to illustrations 7.9a and 7.9b**

1  Disconnect the cable from the negative terminal of the battery (see Chapter 5, Section 1). Relieve the fuel system pressure (see Chapter 4).

2  Refer to Chapter 4 and remove the air filter housing outlet duct.

3  Refer to Chapter 5 and remove the alternator.

4  Label, then disconnect all of the engine wiring harnesses that interfere with intake manifold removal. Also detach all the harness retainers and clips. Move the detached harnesses aside and secure them out of the way.

➡ **Note:** Clear labeling will make the assembly procedure go smoothly and quickly.

5  Disconnect the power brake vacuum hose from the booster.

6  Disconnect the hose from the canister purge solenoid (see Chapter 6). Disconnect the fuel line from the fuel rail (see Chapter 4). Disconnect the PCV hose from the intake manifold.

7  Disconnect any remaining electrical connectors or vacuum hoses connected to the intake manifold or throttle body.

8  Loosen the intake manifold mounting bolts in 1/4-turn increments in the reverse order of the tightening sequence until they can be removed by hand (see illustration 7.14). The manifold will probably be stuck to the cylinder heads and force may be required to break the gasket seal. A pry bar can be positioned between the front of the manifold and the valley tray to break the bond made by the gasket.

### ❄❄ CAUTION:

**Do not pry between the manifold and the heads or damage to the gasket sealing surfaces may result and vacuum leaks could develop. Also, don't use too much force - the manifold is made of a plastic composite and could crack.**

## 2B-8  V8 ENGINES

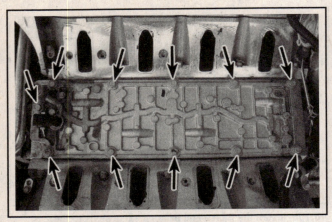

**7.9a** Valve Lifter Oil Manifold (VLOM) assembly mounting bolts

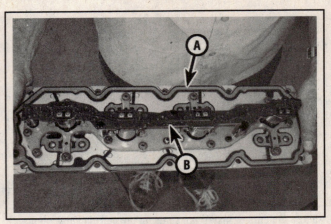

**7.9b** The Valve Lifter Oil Manifold (VLOM) gasket (A) can be re-used if it's in good condition. Don't pick up the assembly by the electrical lead frame (B)

9  Remove the intake manifold. As the manifold is lifted from the engine, be sure to check for and disconnect anything still attached to the manifold.

➥**Note:** *Removing the manifold will be easier if you have an assistant hold the wiring harnesses out of the way.*

The Valve Lifter Oil Manifold (VLOM) assembly can be removed at this time if your engine is equipped with this feature (see illustrations).

### INSTALLATION

♦ *Refer to illustrations 7.12 and 7.14*

➥**Note:** *The mating surfaces of the cylinder heads, block and manifold must be perfectly clean when the manifold is installed.*

10  Carefully remove all traces of old gasket material. Note that the intake manifold is made of a composite material and the cylinder heads on some engines are made of aluminum, therefore aggressive scraping is not suggested and will damage the sealing surfaces. After the gasket surfaces are cleaned and free of any gasket material, wipe the mating surfaces with a cloth saturated with safety solvent. If there is old sealant or oil on the mating surfaces when the manifold is installed, oil or vacuum leaks may develop. Use a vacuum cleaner to remove any gasket material that falls into the intake ports in the heads.

11  Use a tap of the correct size to chase the threads in the bolt holes, then use compressed air (if available) to remove the debris from the holes.

### ❋❋ WARNING:

**Wear safety glasses or a face shield to protect your eyes when using compressed air.**

12  Position the new gaskets on the intake manifold (see illustration). Note that the gaskets are equipped with installation tabs that must snap into place on the intake manifold. The words "Manifold Side" may appear on the gasket. If so, this will ensure proper installation. Make sure the gaskets snap into place and all intake port openings align.

13  Carefully set the manifold in place.

14  Install the bolts and tighten them following the recommended sequence (see illustration) to the torque listed in this Chapter's Specifications. Do not overtighten the bolts or gasket leaks may develop.

15  The remainder of installation is the reverse of removal. Start the engine and check carefully for vacuum leaks at the intake manifold joints.

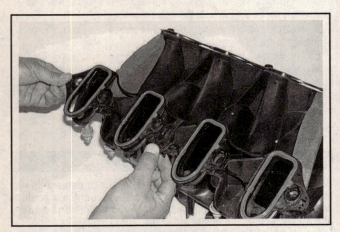

**7.12** Align the tabs on the intake gaskets with the tabs on the manifold and snap the gasket into place

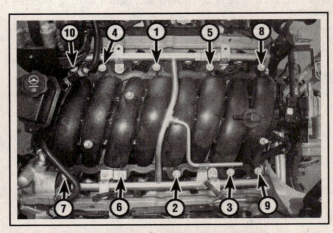

**7.14** Intake manifold bolt tightening sequence - all V8 engines

# V8 ENGINES  2B-9

## 8  Exhaust manifolds - removal and installation

### REMOVAL

▶ Refer to illustrations 8.3, 8.7, 8.9a and 8.9b

> **WARNING:**
> Use caution when working around the exhaust manifolds - the sheetmetal heat shields can be sharp on the edges. Also, the engine should be cold when this procedure is followed.

1  Disconnect the cable from the negative terminal of the battery (see Chapter 5, Section 1).
2  Raise the vehicle and support it securely on jackstands.
3  Working under the vehicle, apply penetrating oil to the exhaust pipe-to-manifold studs and nuts (they're usually rusty) (see illustration).
4  Refer to Chapter 11 and remove the inner fender splash shield.
5  Disconnect the exhaust pipe/catalytic converter pipe from the exhaust manifold.
6  Remove the spark plug wires.

> **CAUTION:**
> Twist the boots to loosen them, then pull on the boots only - not the wires.

### Left side

7  Use a dab of paint to mark the alignment of the upper section of the steering intermediate shaft to the lower part of the steering column (see illustration). Remove the fasteners, then separate the shafts so they can be moved aside.

> **CAUTION:**
> Don't move the wheels or the steering wheel after disconnecting the shafts. The airbag system clockspring could be damaged.

### Right side

8  Remove the oil dipstick tube.

### Both sides

9  If a new manifold is to be installed, remove the heat shield (see illustration). Remove the manifold mounting bolts, then lift off the exhaust manifold and its gasket (see illustration).

8.3  Typical exhaust pipe-to-manifold nuts

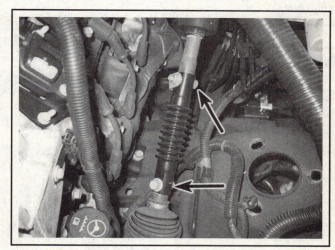

8.7  The steering shaft should be disconnected for access to the left exhaust manifold; don't turn the steering wheel or the front wheels after separating it

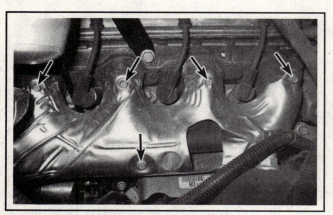

8.9a  Exhaust manifold - heat shield bolts

8.9b  Exhaust manifold mounting bolts

# 2B-10 V8 ENGINES

## INSTALLATION

10  Check the manifold for cracks and make sure the bolt threads are clean and undamaged. The manifold and cylinder head mating surfaces must be clean before the manifolds are reinstalled - use a gasket scraper to remove all carbon deposits and gasket material.

→ **Note: The cylinder heads on some engines are made of aluminum, therefore aggressive scraping is not suggested and will damage the sealing surfaces.**

11  Install the heat shields, then install the bolts and gaskets onto the manifold. Retaining tabs surrounding the gasket bolt holes should hold the assembly together as the manifold is installed.

12  Starting at the fourth thread, apply a 1/4-inch wide band of medium-strength threadlocking compound to the threads of the bolts.

→ **Note: The manufacturer recommends not applying threadlocking compound on the first three threads.**

13  Place the manifold on the cylinder head and install the mounting bolts finger tight.

14  When tightening the mounting bolts, work from the center to the ends. Tighten the bolts in two steps to the torque listed in this Chapter's Specifications. If required, bend the exposed end of the exhaust manifold gasket back against the cylinder head.

15  The remainder of installation is the reverse of removal.

16  Start the engine and check for exhaust leaks.

## 9  Cylinder heads - removal and installation

**※※ WARNING:**

**Wait until the engine is completely cool before beginning this procedure.**

→ **Note: It will be necessary to purchase a new set of 11 mm cylinder head bolts for this procedure.**

### REMOVAL

◆ Refer to illustrations 9.7, 9.10 and 9.12

1  Disconnect the cable from the negative terminal of the battery (see Chapter 5, Section 1) and drain the cooling system (see Chapter 1). Relieve the fuel system pressure (see Chapter 4).

2  Remove the intake manifold (see Section 7) and the coolant pipe.

3  Detach the exhaust manifold(s) from the cylinder head(s) (see Section 8). It is not necessary to disconnect the manifold(s) from the exhaust pipe(s).

4  Remove the valve cover(s) (see Section 4).

5  Remove the rocker arms and pushrods (see Section 5).

**※※ CAUTION:**

**Again, as mentioned in Section 5, keep all the parts in order so they can be reinstalled in the same locations.**

6  Remove the coolant air bleed pipe.

### Left cylinder head

7  Remove the alternator and its bracket (see Chapter 5) (see illustration).

8  Disconnect the ground strap at the back of the cylinder head.

### Right cylinder head

9  Remove the dipstick tube.

10  Disconnect the ground cables from the front of the cylinder head (see illustration).

### Both cylinder heads

11  Loosen the head bolts in 1/4-turn increments in the reverse

**9.7  The alternator bracket can be unbolted and moved forward as a complete assembly**

**9.10  Label then disconnect the ground wires that are bolted to the right cylinder head**

order of the tightening sequence (see illustration 9.21) until they can be removed by hand.

➡ **Note:** There will be different length and size head bolts for different locations. Make a note of the different sizes and lengths and where they go when removing the bolts to ensure correct installation of the new bolts.

12 Lift the head(s) off the engine. If resistance is felt, do not pry between the head and block as damage to the mating surfaces will result. To dislodge the head, place a pry bar or long screwdriver into the intake port and carefully pry the head off the engine (see illustration).

13 Store the heads on blocks of wood to prevent damage to the gasket sealing surfaces.

## INSTALLATION

◆ Refer to illustrations 9.18 and 9.21

14 The mating surfaces of the cylinder heads and block must be perfectly clean when the heads are installed. Gasket removal solvents are available at auto parts stores and may prove helpful.

15 Use a gasket scraper to remove all traces of carbon and old gasket material, then wipe the mating surfaces with a cloth saturated with brake system cleaner.

➡ **Note:** The cylinder heads on some engines are made of aluminum, therefore aggressive scraping is not suggested and will damage the sealing surfaces.

If there is oil on the mating surfaces when the heads are installed, the gaskets may not seal correctly and leaks may develop. When working on the block, use a vacuum cleaner to remove any debris that falls into the cylinders.

16 Check the block and head mating surfaces for nicks, deep scratches and other damage. If damage is slight, it can be removed with emery cloth. If it is excessive, machining may be the only alternative.

17 Use a tap of the correct size to chase the threads in the head bolt holes in the block. If a tap is not available, spray a liberal amount of brake cleaner into each hole. Use compressed air (if available) to remove the debris from the holes.

### ❋❋ WARNING:

Wear safety glasses or a face shield to protect your eyes when using compressed air.

All cylinder head bolts should be replaced with new bolts.

18 Position the new gaskets over the dowels in the block (see illustration).

19 Carefully position the heads on the block without disturbing the gaskets.

20 Before installing the 8mm head bolts, coat the threads with a medium-strength threadlocking compound. Then install the new 8mm head bolts (bolts 11 through 15).

21 Install new 11 mm head bolts (bolts 1 through 10) and tighten them finger tight. Following the recommended sequence (see illustration), tighten the bolts in four steps to the torque listed in this Chapter's Specifications.

### ❋❋ WARNING:

DO NOT reuse head bolts - always replace them with new ones.

22 Install the coolant pipe, using new gaskets, onto the cylinder heads. Tighten the bolts to the torque listed in this Chapter's Specifications.

23 The remainder of installation is the reverse of removal.

24 Add coolant and change the oil and filter (see Chapter 1). Start the engine and check for proper operation and coolant or oil leaks.

9.12 Using a prybar inserted into an intake port to break the head loose - do not use excessive force or damage to the head may result

9.18 Position the head gasket over the dowels at each end of the cylinder head with the mark facing the front of the vehicle

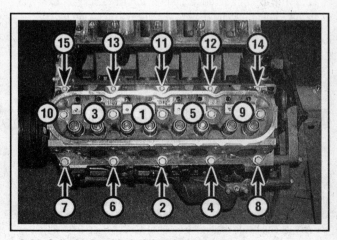

9.21 Cylinder head bolt tightening sequence

## 10 Crankshaft balancer - removal and installation

◆ Refer to illustrations 10.6 and 10.9

➡ **Note:** *This procedure requires a special balancer installation tool and a new crankshaft balancer bolt. Read through the entire procedure and obtain the tool and materials before proceeding.*

1 Disconnect the cable from the negative terminal of the battery (see Chapter 5, Section 1).

2 Raise the front of the vehicle and support it securely on jackstands. Then apply the parking brake.

3 Remove the drivebelt (see Chapter 1). Remove the cooling fan and shroud assembly (see Chapter 3).

4 Working under the vehicle, remove the stone shield from below the engine (if equipped).

5 Remove the starter motor (see Chapter 5). Have an assistant wedge a large screwdriver or prybar into the driveplate ring gear teeth, then loosen the crankshaft pulley center bolt.

6 Pull the balancer off the crankshaft with a puller (see illustration).

※※ **CAUTION:**

**The jaws of the puller must only contact the hub of the balancer - not the outer ring.**

➡ **Note:** *The proper adapter or a long Allen-head bolt should be inserted into the crankshaft nose for the puller's tapered tip to push against to prevent damage to the crankshaft threads.*

7 Position the crankshaft pulley/balancer on the crankshaft and slide it on as far as it will go.

8 Using the crankshaft balancer installation tool, press the crankshaft pulley/balancer onto the crankshaft.

9 Install the old crankshaft balancer bolt and tighten the crankshaft bolt to the Step 1 torque listed in this Chapter's Specifications. Remove the old bolt and measure the distance from the snout of the crankshaft to the balancer hub (see illustration). When properly installed, the balancer hub should extend 3/32 to 11/64-inch past the crankshaft snout. If the measurement is incorrect, reinstall the balancer installation tool and press the balancer on the crankshaft until the measurement is correct.

10 Install a new crankshaft balancer bolt and tighten it to the torque and angle of rotation listed in this Chapter's Specifications.

11 The remainder of installation is the reverse of removal.

**10.6** The use of a three jaw puller will be necessary to remove the crankshaft balancer - always place the puller jaws around the hub, not the outer ring

**10.9** Before the new crankshaft bolt is installed and tightened, the balancer must be measured for proper installation - when properly installed, the balancer hub should extend 3/32 to 11/64-inch past the crankshaft snout

## 11 Crankshaft front oil seal - removal and installation

◆ Refer to illustrations 11.2 and 11.4

1 Remove the crankshaft balancer (see Section 10).

2 Note how the seal is installed - the new one must be installed to the same depth and facing the same way. Carefully pry the oil seal out of the cover with a seal puller or a large screwdriver (see illustration). Be very careful not to distort the cover or scratch the crankshaft! Wrap electrician's tape around the tip of the screwdriver to avoid damage to the crankshaft.

3 If the seal is being replaced with the timing chain cover removed, support the cover on top of two blocks of wood and drive the seal out from the rear with a hammer and punch.

※※ **CAUTION:**

**Be careful not to scratch, gouge or distort the area that the seal fits into or a leak will develop.**

# V8 ENGINES   2B-13

**11.2 Carefully pry the old seal out of the timing chain cover - don't damage the crankshaft in the process**

**11.4 Drive the new seal into place with a large socket and hammer**

4  Apply clean engine oil or multi-purpose grease to the outer edge of the new seal, then install it in the cover with the lip (spring side) facing IN. Drive the seal into place (see illustration) with a large socket and a hammer (if a large socket isn't available, a piece of pipe will also work). Make sure the seal enters the bore squarely and stop when the front face is at the proper depth.

5  Check the surface on the balancer hub that the oil seal rides on. If the surface has been grooved from long-time contact with the seal, replace the crankshaft balancer.
6  Lubricate the balancer hub with clean engine oil and reinstall the crankshaft balancer as described in Section 10.
7  The remainder of installation is the reverse of removal.

## 12  Timing chain - removal, inspection and installation

**✳✳ WARNING:**

**Wait until the engine is completely cool before beginning this procedure.**

**✳✳ CAUTION:**

The timing system is complex, and severe engine damage will occur if you make any mistakes. Do not attempt this procedure unless you are highly experienced with this type of repair. If you are at all unsure of your abilities, be sure to consult an expert. Double-check all your work and be sure everything is correct before you attempt to start the engine.

### REMOVAL AND INSPECTION

▶ **Refer to illustrations 12.3, 12.6, 12.8, 12.9, 12.11, 12.12, 12.14 and 12.17**

➥ **Note 1:** Refer to Step 10 for information on the camshaft actuator components used on engines with VINs with K, Y, 2 or 8 as the eighth digit.

➥ **Note 2:** A special tool is recommended for aligning the timing chain cover during installation (see Step 29).

1  Disconnect the cable from the negative terminal of the battery (see Chapter 5, Section 1).
2  Refer to Chapter 1 and drain the cooling system and engine oil.
3  Refer to Chapter 3 and remove the water pump (see illustration).
4  Remove the crankshaft balancer (see Section 10).
5  Disconnect the wiring from the camshaft position sensor and the camshaft position actuator magnet (if equipped).

**12.3 Detach the hoses, then remove the bolts and lift off the water pump manifold (engine removed for clarity)**

## 2B-14 V8 ENGINES

12.6 Timing chain cover mounting bolts (arrows)

12.8 Timing chain alignment marks - when properly aligned, the crankshaft sprocket should be in the 12 o'clock position and the camshaft sprocket should be in the 6 o'clock position

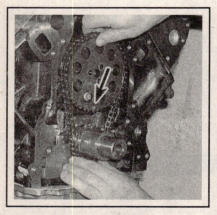

12.9 Remove the upper sprocket along with the sprocket (engine without camshaft position actuator system shown)

12.11 Camshaft position actuator system components - the "bolt" is the solenoid valve (it must be replaced with a new one whenever it's removed); don't try to disassemble the actuator mechanism

12.12 Place your fingers as shown when pulling the actuator from the end of the camshaft - don't pull on the reluctor ring as it could pop apart and be destroyed

6  Remove the timing chain cover mounting bolts and separate the timing chain cover from the block (see illustration). The cover may be stuck; if so, use a putty knife to break the gasket seal. The cover can easily be damaged, so DO NOT attempt to pry it off.

7  Remove the oil pump (see Section 15).

### Engines without camshaft position actuator

8  Loosen the camshaft sprocket bolts one turn, then screw the crankshaft balancer bolt into the end of the crankshaft and rotate the crankshaft in the normal direction of rotation (clockwise) until the timing marks align (see illustration).

9  Remove the bolt from the end of the camshaft, then detach the camshaft sprocket and chain as an assembly (see illustration).

### ✱✱ CAUTION:

Do not turn the crankshaft or the camshaft while the chain is off. Damage to the pistons and/or the valves could result.

### Engines with camshaft position actuator

10  Screw the crankshaft balancer bolt into the end of the crankshaft and rotate the crankshaft in the normal direction of rotation (clockwise) until the timing marks align (see illustration 12.8).

11  Remove the solenoid valve from the center of the upper sprocket assembly and discard it (see illustration). It must be replaced with a new one.

12  Loosen the actuator from the front of the camshaft by putting your fingers around the rear of the chain and sprocket while pulling it off (see illustration).

### ✱✱ WARNING:

Don't grasp the reluctor wheel or any other parts of the assembly. The actuator could pop apart, injure you and become damaged.

13 Remove the actuator assembly along with the chain, then separate the chain.

> **CAUTION:**
> Do not turn the crankshaft or the camshaft while the chain is off. Damage to the pistons and/or the valves could result.

14 Tie the actuator assembly together for safety (see illustration).

## All engines

15 Remove the timing chain tensioner and inspect it for wear and damage.
16 Also inspect the camshaft and crankshaft sprockets for wear and damage.
17 If replacement of the timing chain is necessary, remove the sprocket from the crankshaft with a two-or three-jaw puller, but be careful not to damage the threads in the end of the crankshaft (see illustration).

## INSTALLATION

▶ Refer to illustration 12.28

➥Note: Timing chains must be replaced as a set with the camshaft and crankshaft sprockets. Never put a new chain on old sprockets.

18 Use a gasket scraper to remove all traces of old gasket material and sealant from the cover and engine block.
19 Align the crankshaft sprocket with the Woodruff key and press the sprocket onto the crankshaft (if removed) with the vibration damper bolt, a large socket and some washers or tap it gently into place until it is completely seated.

> **CAUTION:**
> If resistance is encountered, do not hammer the sprocket onto the crankshaft. It may eventually move onto the shaft, but it may be cracked in the process and fail later, causing extensive engine damage.

20 Loop the new chain over the camshaft sprocket, then turn the sprocket until the timing mark is at the bottom (if you're working on an engine with the camshaft position actuator system, you'll be installing the actuator/sprocket assembly rather than the sprocket). Mesh the chain with the crankshaft sprocket and position the camshaft sprocket on the end of the camshaft. If necessary, turn the camshaft so the dowel in the camshaft fits into the hole in the sprocket with the timing mark in the 6 o'clock position (see illustration 12.8). When the chain is installed, the timing marks MUST align as shown.
21 For engines using a camshaft position actuator, lay a straightedge across the front of the block and verify that the timing chain does not protrude in front of the face of the engine block.
22 Tighten either the camshaft sprocket bolt or the NEW camshaft position actuator valve to the torque listed in this Chapter's Specifications.
23 Compress the timing chain tensioner and temporarily insert a pin in the hole to secure it in the retracted position.

**12.14 Secure the actuator with a wire tie or something similar to ensure that it doesn't come apart**

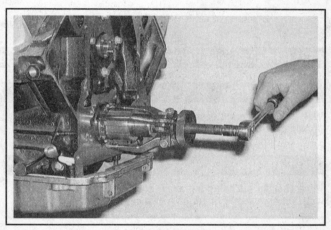

**12.17 The sprocket on the crankshaft can be removed with a two or three-jaw puller**

24 Install the tensioner and tighten the bolts to the torque listed in this Chapter's Specifications. Remove the pin to allow the tensioner to extend.

> **CAUTION:**
> Carefully and slowly rotate the crankshaft by hand through at least two full revolutions (use a socket and breaker bar on the crankshaft pulley center bolt). If you feel any resistance, STOP! There is something wrong - most likely valves are contacting the pistons. You must find the problem before proceeding.

25 Install the oil pump onto the engine (see Section 15).
26 Clean all sealing areas of the timing chain cover and remove all traces of old sealant. Clean the area with brake system cleaner to remove oily residue. Remove the crankshaft oil seal and purchase a new one.
27 Apply a bead of RTV sealant to the corners where the oil pan meets the engine block.

## 2B-16  V8 ENGINES

28  Install the timing chain cover on the engine loosely using a new gasket (see illustration). The bolts should be snug but not tight.

29  Align the timing chain cover as follows:
  a)  Obtain a cover alignment tool GM part no. J41476. This tool is also available from other tool manufacturers.
  b)  Place the tool over the crankshaft snout with the legs registered into the slots on the front cover.
  c)  Secure the tool using the crankshaft balancer bolt but don't over-tighten it.
  d)  With the timing chain cover properly aligned, tighten the cover bolts and oil pan bolts to the torque listed in this Chapter's Specifications.
  e)  Remove the tool.

30  Install a new crankshaft oil seal (see Section 11).
31  The remainder of installation is the reverse of removal.
32  Refill the cooling system and engine oil, and replace the engine oil filter (see Chapter 1). Run the engine and check for oil and coolant leaks.

**12.28  Install the front cover with a new gasket onto the engine block LOOSELY - the cover must be aligned properly before final installation**

## 13  Camshaft and lifters - removal, inspection and installation

**✶✶ WARNING:**

**Wait until the engine is completely cool before beginning this procedure.**

**✶✶ CAUTION:**

**If the camshaft is being replaced, always install new lifters as well. Do not use old lifters on a new camshaft.**

### REMOVAL

▶ **Refer to illustrations 13.13a, 13.13b and 13.15**

1  Disconnect the cable from the negative battery terminal (see Chapter 5, Section 1).
2  Relieve the fuel system pressure (see Chapter 4).
3  Refer to Chapter 3 and remove the radiator.
4  Remove the auxiliary automatic transmission fluid cooler, if equipped.
5  Disconnect the power steering cooler and move it aside.
6  Refer to Chapter 3 and remove the air conditioning condenser.
7  Remove the valve covers (see Section 4).
8  Remove the intake manifold (see Section 7).
9  Remove the rocker arms and pushrods (see Section 5).
10  Remove the timing chain and sprockets (see Section 12).
11  Remove the cylinder heads (see Section 9).
12  Before removing the lifters, arrange to store them in a clearly labeled box to ensure that they're reinstalled in their original locations.
13  Remove the lifter retainers and lifters and store them where they won't get dirty (see illustrations). DO NOT attempt to withdraw the camshaft with the lifters in place.
14  If the lifters are built up with gum and varnish they may not come out with the retainer. If so, there are several ways to extract the lifters from the bores. A special tool designed to grip and remove lifters is manufactured by many tool companies and is widely available, but it may not be required in every case. On newer engines without a lot of varnish buildup, the lifters can often be removed with a small magnet or even with your fingers. A machinist's scribe with a bent end can be used to pull the lifters out by positioning the point under the retainer ring in the top of each lifter.

**✶✶ CAUTION:**

**Don't use pliers to remove the lifters unless you intend to replace them with new ones. The pliers will damage the precision machined and hardened lifters, rendering them useless.**

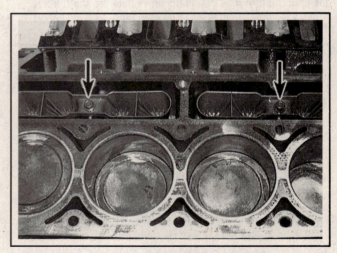

**13.13a  The roller lifters are held in place by retainers - remove the retainer bolts and remove the retainers and the lifters as an assembly - note that each retainer houses four individual lifters and they must be installed back in their original locations if they're going to be reused**

# V8 ENGINES  2B-17

13.13b  Once the lifters and retainers are removed from the block they can be marked (for location and installation purposes) and inspected

13.15  Remove the bolts and take off the camshaft retainer plate, noting which side faces the block

13.18a  If the camshaft is removed from the engine, lobe lift can be obtained by measuring camshaft lobe height . . .

15  Remove the bolts and the camshaft retainer plate, noting which direction faces the block (see illustration).

16  Thread a bolt into the camshaft sprocket bolt hole to use as a handle when removing the camshaft from the block.

17  Carefully and slowly pull the camshaft out. Support the cam near the block so the lobes don't nick or gouge the bearings as it's withdrawn.

## INSPECTION

### Camshaft lobe lift check

▶ Refer to illustrations 13.18a and 13.18b

18  Measure the camshaft lobe height and the base circle (see illustrations). The difference between the two measurements is the lobe lift (lobe height - base circle = lobe lift). Record this figure for future reference and repeat the check on the remaining camshaft lobes.

19  After the lobe lift check is complete, compare the results to the values listed in this Chapter's Specifications. If the lobe lift is 0.002 inch less than specified, cam lobe wear has occurred and a new camshaft should be installed.

### Camshaft bearing journals, lobes and bearings

▶ Refer to illustration 13.21

20  After the camshaft has been removed from the engine, cleaned with solvent and dried, inspect the bearing journals for uneven wear, pitting and evidence of seizure. If the journals are damaged, the bearing inserts in the block are probably damaged as well. Both the camshaft and bearings will have to be replaced.

➡ Note: Camshaft bearing replacement requires special tools and expertise that place it beyond the scope of the average home mechanic. The tools for bearing removal and installation are available at stores that carry automotive tools, possibly even found at a tool rental business. It is advisable though, if the bearings are bad and the procedure is beyond your ability, take the engine block to an automotive machine shop to ensure that the job is done correctly.

21  Measure the bearing journals with a micrometer to determine if they are excessively worn or out-of-round (see illustration).

22  Check the camshaft lobes for heat discoloration, score marks, chipped areas, pitting and uneven wear. If the lobes are in good condition and if the lobe lift measurements recorded earlier are as specified, the camshaft can be reused.

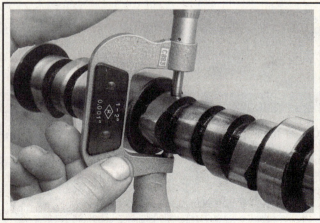

13.18b  . . . and by measuring the camshaft base circle - the difference between the two measurements equals lobe lift

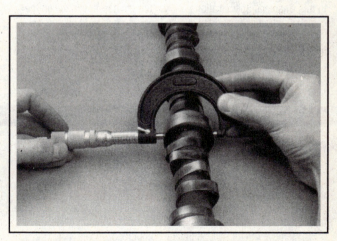

13.21  Check the diameter of each camshaft bearing journal to pinpoint excessive wear and out-of-round conditions

# 2B-18 V8 ENGINES

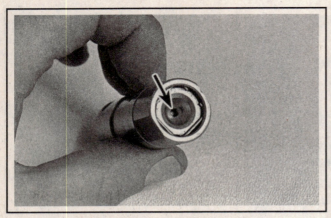

**13.23 Check the pushrod seat in the top of each lifter for wear**

**13.24 The roller on hydraulic roller lifters must turn freely - check for wear and excessive play as well**

### Lifters

▸ *Refer to illustrations 13.23 and 13.24*

23  Clean the lifters with solvent and dry them thoroughly without mixing them up. Check each lifter wall and pushrod seat and for score marks and uneven wear (see illustration). If the lifter walls are damaged or worn (which is not very likely), inspect the lifter bores in the engine block as well. If the pushrod seats are worn, check the pushrod ends.

24  Check the rollers carefully for wear and damage and make sure they turn freely without excessive play (see illustration).

25  Used roller lifters can not be reinstalled with a new camshaft, but the original camshaft can be used if new lifters are installed. Always use new lifters when installing a new camshaft.

## INSTALLATION

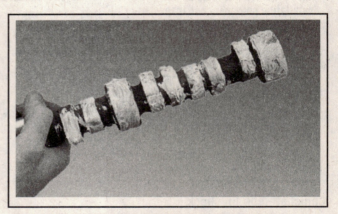

**13.26 Be sure to apply camshaft assembly lube to the cam lobes and bearing journals before installing the camshaft**

▸ *Refer to illustration 13.26*

26  Lubricate the camshaft bearing journals and cam lobes with camshaft and lifter assembly lube (see illustration).

27  Slide the camshaft into the engine. Support the cam near the block and be careful not to scrape or nick the bearings.

28  Turn the camshaft until the dowel pin is in the 3 o'clock position, install the camshaft thrust plate and tighten the bolts to the torque listed in this Chapter's Specifications. Make sure the gasket surface on the camshaft thrust plate and the engine block are free from oil and dirt.

29  Install the timing chain and sprockets (see Section 12). Also install the camshaft position sensor using a new O-ring (see Chapter 6).

30  Lubricate the lifters with clean engine oil and install them in the lifter retainers. Be sure to align the flats on the lifters with the flats in the lifter retainers. Install the retainer and lifters into the engine block as an assembly. If the original lifters are being reinstalled, be sure to return them to their original locations. If a new camshaft is being installed, install new lifters as well. Tighten the lifter retainer bolts to the torque listed in this Chapter's Specifications.

31  The remainder of installation is the reverse of removal.

32  Before starting and running the engine, refill the cooling system, change the oil and install a new oil filter (see Chapter 1).

## 14  Oil pan - removal and installation

### REMOVAL

1  Disconnect the cable from the negative terminal of the battery (see Chapter 5, Section 1).

2  Raise the vehicle and support it securely on jackstands, then refer to Chapter 1 and drain the engine oil and remove the oil filter.

3  Remove the oil pan skid plate, if equipped.

4  Unbolt the steering rack and allow it to hang out of the way (see Chapter 8).

5  On 4WD vehicles, remove the front differential carrier (see Chapter 8).

6  Remove the transmission bellhousing covers from the rear of the engine block.

7  Remove the crossmember.

8  Remove the lower transmission-to-engine bolt(s). These bolts vary depending on the model of transmission used but all screw into the oil pan.

# V8 ENGINES 2B-19

9  Disconnect the wiring from the engine oil level sensor.
10  Disconnect the wiring harness retainers from the oil pan.
11  Disconnect the oil cooler lines from the oil pan on models with automatic transmissions.
12  Remove all the oil pan bolts, then lower the pan from the engine. The pan will probably stick to the engine, so strike the pan with a rubber mallet until it breaks the gasket seal.

### ❄ CAUTION:
**Before using force on the oil pan, be sure all the bolts have been removed. Carefully slide the oil pan down and out, to the rear.**

## INSTALLATION

▸ **Refer to illustration 14.13**

13  Drill out the rivets securing the oil pan gasket to the oil pan and remove the old gasket (see illustration). Wash out the oil pan with solvent.
14  Thoroughly clean the mounting surfaces of the oil pan and engine block of old gasket material and sealant. Wipe the gasket surfaces clean with a rag soaked in brake system cleaner.
15  Apply a 3/16-inch wide, one-inch long bead of RTV sealant to the corners of the block where the front cover and the rear cover meet the engine block. The gasket tabs protrude at these points and must have sealant around them. Attach the new gasket to the pan, install the pan and tighten the bolts finger-tight. Be sure the oil gallery passages in the pan and the gasket are aligned properly.

➡ **Note: Oil pan gasket rivets do not need to be installed on assembly.**

16  Install all of the oil pan mounting bolts snug.
17  Install the lower transmission mounting bolts snug.
18  Tighten the oil pan bolts and the transmission bolts to the torques listed in this Chapter's Specifications.
19  The remainder of installation is the reverse of removal.
20  Add the proper type and quantity of oil (see Chapter 1), start the engine and check for leaks before placing the vehicle back in service.

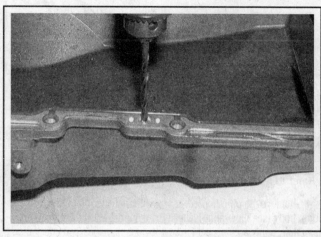

**14.13  The manufacturer uses rivets to hold the gasket to the oil pan during assembly - carefully drill them out (it isn't necessary to rivet the new gasket to the oil pan)**

## 15  Oil pump - removal, inspection and installation

### REMOVAL

▸ **Refer to illustrations 15.2 and 15.3**

1  Refer to the Sections 12 and 14 and remove the timing chain cover and oil pan.
2  Remove the oil pump pick-up tube mounting nuts and bolts and lower the pick-up tube and screen assembly from the vehicle (see illustration).
3  Remove the oil pump retaining bolts and slide the pump off the end of the crankshaft (see illustration).

**15.2  Remove the bolt securing the oil pick-up tube to the oil pump and remove the pick-up tube from the engine**

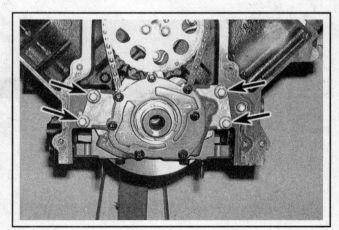

**15.3  Oil pump mounting bolts**

# 2B-20 V8 ENGINES

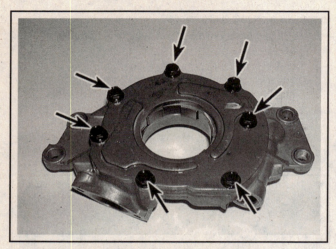

**15.4 Oil pump cover-to-oil pump housing mounting bolts (arrows)**

**15.8 Always install a new O-ring on the oil pump pick up tube**

## INSPECTION

♦ **Refer to illustration 15.4**

4  Remove the oil pump cover and withdraw the rotors from the pump body (see illustration). Clean the components with solvent, dry them thoroughly and inspect for any obvious damage. Also check the bolt holes for damaged threads and the splined surfaces on the crankshaft sprocket for any apparent damage. If any of the components are scored, scratched or worn, replace the entire oil pump assembly. There are no serviceable parts currently available.

## INSTALLATION

♦ **Refer to illustration 15.8**

5  Prime the pump by pouring clean engine oil into the pick-up tube hole, while turning the pump by hand.
6  Position the oil pump over the end of the crankshaft and align the teeth on the crank-shaft sprocket with the teeth on the oil pump drive gear. Making sure the pump is fully seated against the block.
7  Install the oil pump mounting bolts and tighten them to the torque listed in this Chapter's Specifications.
8  Install a new O-ring on the oil pump pick-up tube, then fasten it to the oil pump and the engine block main studs (see illustration).

**✱✱✱ CAUTION:**
Be absolutely certain that the pick-up tube-to-oil pump bolts are properly tightened so that no air can be sucked into the oiling system at this connection.

9  Install and align the timing chain cover, then install the oil pan. Refer to Sections 12 and 14 for the installation procedures.
10  The remainder of installation is the reverse of removal.
11  Add oil and coolant as necessary. Run the engine and check for oil and coolant leaks. Also check the oil pressure as described in Chapter 2C.

## 16  Driveplate - removal and installation

The driveplate replacement for V8 engines is identical to the driveplate replacement procedure for the V6 engines. Refer to Chapter 2 Part A for the procedure and use the torque figures in this Chapter's Specifications.

➡ **Note:** If the spacer between the driveplate and the crankshaft must be removed and it's stuck, insert bolts (M11x1.5 mm) into the two threaded holes in the spacer. Tightening the bolts will force the spacer off the crankshaft.

# V8 ENGINES  2B-21

## 17  Rear main oil seal - replacement

▶ Refer to illustrations 17.3 and 17.4

**Note:** *If the rear main seal housing is removed from the engine block, special tools must be used to properly center the housing over the end of the crankshaft. This alignment procedure is critical because a seal that is slightly off-center will leak. Refer to Chapter 2C, Section 10 for information on rear main seal housing alignment.*

1  Remove the transmission (see Chapter 7A).
2  Remove the driveplate (see Section 16).
3  Pry the oil seal from the rear cover with a screwdriver (see illustration). Be careful not to nick or scratch the crankshaft or the seal bore. Be sure to note how far it's recessed into the housing bore before removal so the new seal can be installed to the same depth. Thoroughly clean the seal bore in the block with a shop towel. Remove all traces of oil and dirt.
4  Don't lubricate the seal or touch the sealing lip. Make sure the seal is installed in the correct orientation (it has a reverse-lip design). The seal part number must be visible when it's installed. Preferably, a seal installation tool (available at most auto parts store) should be used to press the new seal back into place. If the proper seal installation tool is unavailable, use a large drift and carefully drive the new seal squarely into the seal bore and flush with the rear cover (see illustration).
5  Install the driveplate (see Section 16).
6  Install the transmission (see Chapter 7A).

**17.3  Carefully pry the old seal out with a screwdriver at the notches provided in the rear cover**

**17.4  The rear oil seal can be pressed into place with a seal installation tool, a section of pipe or a blunt object shown here - in any case be sure the seal is installed squarely into the seal bore and flush with the rear cover**

## 18  Engine mounts - check and replacement

1  Engine mounts seldom require attention, but broken or deteriorated mounts should be replaced immediately or the added strain placed on the driveline components may cause damage.

### CHECK

2  During the check, the engine must be raised slightly to remove the weight from the mounts.
3  Raise the vehicle and support it securely on jackstands.
4  Check the mounts to see if the rubber is cracked, hardened or separated from the metal plates. Sometimes the rubber will split right down the center.
5  Check for relative movement between the mount plates and the engine or frame (use a large prybar to attempt to move the engine away from the mounts). If movement is noted, check the tightness of the mount fasteners first before condemning the mounts. Usually when engine mounts are broken, they are very obvious as the engine will easily move away from the mount when pried or under load.

### REPLACEMENT

▶ Refer to illustrations 18.9 and 18.11

6  Disconnect the cable from the negative terminal of the battery, then raise the vehicle and support it securely on jackstands.
7  On 4WD models, remove the interfering driveaxle (see Chapter 8).
8  If you're working on the left mount, it is necessary to remove the left inner fender splash shield and the exhaust manifold heat shield on most models.

**18.9  One of the three engine mount-to-frame bolts, seen from above**

**18.11  Three of the four engine mount-to-block bolts**

9  Remove the engine mount-to-frame bracket bolts (see illustration). There are three bolts on each side securing the mounts to the frame bracket.

10  Attach an engine hoist to the top of the engine for lifting; do not use a jack under the oil pan to support the entire weight of the engine or the oil pump pick-up could be damaged. If a hoist is not available, a jack can be placed under the large casting lugs on each side of the engine block at the rear (near the transmission) to support the weight of the engine while the engine mounts are being replaced. To access this lug on the right side of the engine, the starter motor will have to be removed (see Chapter 5).

11  Unbolt the mount from the engine block (see illustration), then raise the engine far enough to remove the mount from between the engine and the frame.

12  Installation is the reverse of removal. Use non-hardening thread-locking compound on the mount bolts and be sure to tighten them to the torque listed in this Chapter's Specifications.

# V8 ENGINES 2B-23

## Specifications

### General

| | |
|---|---|
| Displacement | |
| 4.8L engines | 293 cubic inches |
| 5.3L engines | 325 cubic inches |
| 6.0L engines | 364 cubic inches |
| 6.2L engines | 376 cubic inches |
| Bore and stroke | |
| 4.8L engines | 3.779 x 3.268 inches |
| 5.3L engines | 3.779 x 3.622 inches |
| 6.0L engines | 4.001 x 3.622 inches |
| 6.2L engines | 4.065 x 3.622 inches |
| Cylinder numbers (front-to-rear) | |
| Left (driver's) side | 1-3-5-7 |
| Right side | 2-4-6-8 |
| Firing order | 1-8-7-2-6-5-4-3 |

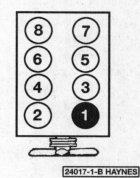

**Cylinder numbering - V8 engines**

### Camshaft

| | |
|---|---|
| Journal diameters | 2.164 to 2.166 inches |
| Camshaft endplay | 0.001 to 0.012 inch |
| Lobe lift | |
| 4.8L engines | 0.283 inch |
| 5.3L engines | |
|   Active fuel management cylinders | 0.289 inch |
|   Non-active fuel management cylinders | 0.283 inch |
| 6.0L VIN "Y" engines | |
|   Active fuel management cylinders | |
|     Intake | 0.283 inch |
|     Exhaust | 0.287 inch |
|   Non-active fuel management cylinders | |
|     Intake | 0.279 inch |
|     Exhaust | 0.282 inch |
| 6.0L VIN "K" engines | |
|   Intake | 0.274 inch |
|   Exhaust | 0.281 inch |
| 6.2L engines | 0.294 inch |

### Torque specifications*  Ft-lbs (unless otherwise indicated)

➡ **Note:** One foot-pound (ft-lb) of torque is equivalent to 12 inch-pounds (in-lbs) of torque. Torque values below approximately 15 foot-pounds are expressed in inch-pounds, because most foot-pound torque wrenches are not accurate at these smaller values.

| | |
|---|---|
| Camshaft sprocket bolt | |
|   Step 1 | 55 |
|   Step 2 (2007 through 2009 only) | Tighten an additional 50 degrees |
| Camshaft position actuator magnet bolts | 106 in-lbs |
| Camshaft position actuator solenoid valve | |
|   Step 1 | 48 |
|   Step 2 | Tighten an additional 90 degrees |

## V8 ENGINES

**Torque specifications (continued)***     **Ft-lbs (unless otherwise indicated)**

➡ **Note:** One foot-pound (ft-lb) of torque is equivalent to 12 inch-pounds (in-lbs) of torque. Torque values below approximately 15 foot-pounds are expressed in inch-pounds, because most foot-pound torque wrenches are not accurate at these smaller values.

| | |
|---|---|
| Camshaft retainer bolts | |
|    Hex head | 18 |
|    Torx head | 132 in-lbs |
| Crankshaft balancer bolt | |
|    2007 and 2008 models | |
|       Step 1 (use old bolt) | 240 |
|       Step 2 (use new bolt) | 37 |
|       Step 3 | Tighten an additional 140 degrees |
|    2009 and later models (use new bolt) | |
|       Step 1 | 110 |
|       Step 2 | Loosen 360 degrees |
|       Step 3 | 59 |
|       Step 4 | Tighten an additional 125 degrees |
| Crankshaft rear oil seal retainer bolts | 22 |
| Crossmember mounting bolts | |
|    1500 Series | 74 |
|    2500 Series | 89 |
| Cylinder head bolts (use new bolts) (in sequence - see illustration 9.21) | |
|    Step 1 | |
|       All 11 mm bolts (1 through 10) | 22 |
|    Step 2 | |
|       All 11 mm bolts (1 through 10) | Tighten an additional 90 degrees |
|    Step 3 | |
|       All 11 mm bolts (1 through 10) | Tighten an additional 70 degrees |
|    Step 4 | |
|       All 8 mm bolts (11 through 15) | 22 |
| Driveplate bolts | |
|    Step 1 | 15 |
|    Step 2 | 37 |
|    Step 3 | 74 |
| Engine mount fasteners | |
|    To engine | 37 |
|    To frame | 48 |
| Exhaust manifold bolts | |
|    Step 1 | 132 in-lbs |
|    Step 2 | 15 |
| Exhaust manifold heat shield bolt | 80 in-lbs |
| Exhaust pipe flange nuts | 20 to 25 |
| Intake manifold bolts | |
|    Step 1 | 44 in-lbs |
|    Step 2 | 89 in-lbs |
| Oil pan baffle bolts | 89 in-lbs |
| Oil pan drain plug | 18 |
| Oil pan rear access plugs | 80 in-lbs |

# V8 ENGINES 2B-25

## Torque specifications (continued)* — Ft-lbs (unless otherwise indicated)

➥**Note:** One foot-pound (ft-lb) of torque is equivalent to 12 inch-pounds (in-lbs) of torque. Torque values below approximately 15 foot-pounds are expressed in inch-pounds, because most foot-pound torque wrenches are not accurate at these smaller values.

| | |
|---|---|
| Oil pan bolts | |
|     To engine and front cover | 18 |
|     To rear cover | 106 in-lbs |
|     To bellhousing, converter cover and transmission | 37 |
| Oil pump cover bolts | 106 in-lbs |
| Oil pump mounting bolts | 18 |
| Rocker arm bolts | 22 |
| Timing chain cover bolts | 18 |
| Transmission-to-oil pan bolts | 37 |
| Valley cover bolts | 18 |
| Valve cover bolts | 106 in-lbs |
| Valve lifter guide bolts | 106 in-lbs |

*Note: Refer to Part C for additional specifications.

**Notes**

# 2C GENERAL ENGINE OVERHAUL PROCEDURES

**Section**

1. General information - engine overhaul
2. Oil pressure check
3. Cylinder compression check
4. Vacuum gauge diagnostic checks
5. Engine rebuilding alternatives
6. Engine removal - methods and precautions
7. Engine - removal and installation
8. Engine overhaul - disassembly sequence
9. Pistons and connecting rods - removal and installation
10. Crankshaft - removal and installation
11. Engine overhaul - reassembly sequence
12. Initial start-up and break-in after overhaul

**Reference to other Chapters**

CHECK ENGINE light on - See Chapter 6

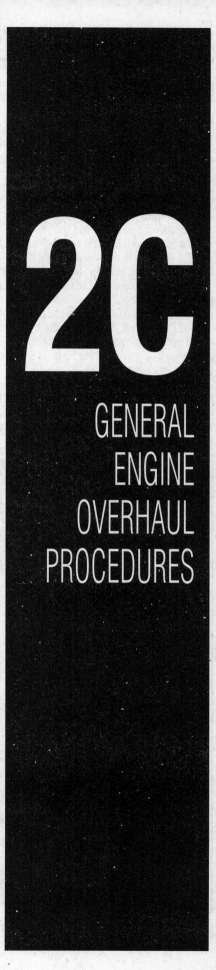

## 2C-2 GENERAL ENGINE OVERHAUL PROCEDURES

### 1 General information - engine overhaul

▶ **Refer to illustrations 1.1, 1.2, 1.3, 1.4, 1.5 and 1.6**

Included in this portion of Chapter 2 are general information and diagnostic testing procedures for determining the overall mechanical condition of your engine.

The information ranges from advice concerning preparation for an overhaul and the purchase of replacement parts and/or components to detailed, step-by-step procedures covering removal and installation.

The following Sections have been written to help you determine whether your engine needs to be overhauled and how to remove and install it once you've determined it needs to be rebuilt. For information concerning in-vehicle engine repair, see Chapter 2A or 2B.

The Specifications included in this Part are general in nature and include only those necessary for testing the oil pressure, checking the engine compression, and bottom-end torque specifications. Refer to Chapter 2A or 2B for additional engine Specifications.

It's not always easy to determine when, or if, an engine should be completely overhauled, because a number of factors must be considered.

High mileage is not necessarily an indication that an overhaul is needed, while low mileage doesn't preclude the need for an overhaul. Frequency of servicing is probably the most important consideration. An engine that's had regular and frequent oil and filter changes, as well as other required maintenance, will most likely give many thousands of miles of reliable service. Conversely, a neglected engine may require an overhaul very early in its service life.

Excessive oil consumption is an indication that piston rings, valve seals and/or valve guides are in need of attention. Make sure that oil leaks aren't responsible before deciding that the rings and/or guides are bad. Perform a cylinder compression check to determine the extent of the work required (see Section 3). Also check the vacuum readings under various conditions (see Section 4).

Check the oil pressure with a gauge installed in place of the oil pressure sending unit and compare it to this Chapter's Specifications (see Section 2). If it's extremely low, the bearings and/or oil pump are probably worn out.

Loss of power, rough running, knocking or metallic engine noises, excessive valve train noise and high fuel consumption rates may also point to the need for an overhaul, especially if they're all present at the same time. If a complete tune-up doesn't remedy the situation, major mechanical work is the only solution.

An engine overhaul involves restoring the internal parts to the specifications of a new engine. During an overhaul, the piston rings are replaced and the cylinder walls are reconditioned (rebored and/or honed) (see illustrations 1.1 and 1.2). If a rebore is done by an automotive machine shop, new oversize pistons will also be installed. The main

**1.1 An engine block being bored - an engine rebuilder will use special machinery to recondition the cylinder bores**

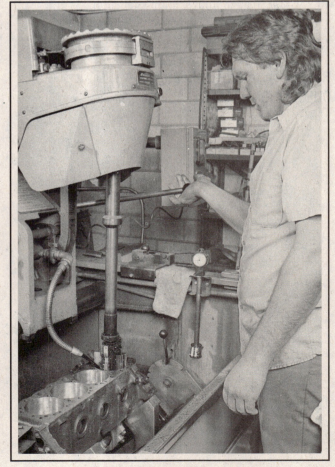

**1.2 If the cylinders are bored, the machine shop will normally hone the engine on a machine like this**

# General Engine Overhaul Procedures 2C-3

1.3 A crankshaft having a main bearing journal ground

1.4 A machinist checks for a bent connecting rod, using specialized equipment

1.5 A bore gauge being used to check the main bearing bore

1.6 Uneven piston wear like this indicates a bent connecting rod

bearings and connecting rod bearings are generally replaced with new ones and, if necessary, the crankshaft may be reground to restore the journals (see illustration 1.3). Generally, the valves are serviced as well, since they're usually in less-than-perfect condition at this point. While the engine is being overhauled; other components, such as the starter and alternator, can be rebuilt as well. The end result should be a like-new engine that will give many trouble-free miles.

➡ Note: Critical cooling system components such as the hoses, drivebelts, thermostat and water pump should be replaced with new parts when an engine is overhauled. The radiator should be checked carefully to ensure that it isn't clogged or leaking (see Chapter 3). If you purchase a rebuilt engine or short block, some rebuilders will not warranty their engines unless the radiator has been professionally flushed. Also, we don't recommend overhauling the oil pump - always install a new one when an engine is rebuilt.

Overhauling the internal components on today's engines is a difficult and time-consuming task that requires a significant amount of specialty tools and is best left to a professional engine rebuilder (see illustrations 1.4, 1.5 and 1.6). A competent engine rebuilder will handle the inspection of your old parts and offer advice concerning the reconditioning or replacement of the original engine. Never purchase parts or have machine work done on other components until the block has been thoroughly inspected by a professional machine shop. As a general rule, time is the primary cost of an overhaul, especially since the vehicle may be tied up for a minimum of two weeks or more. Be aware that some engine builders only have the capability to rebuild the engine you bring them while other rebuilders have a large inventory of rebuilt exchange engines in stock. Also be aware that many machine shops could take as much as two weeks time to completely rebuild your engine depending on shop workload. Sometimes it makes more sense to simply exchange your engine for another engine that's already rebuilt to save time.

# 2C-4 GENERAL ENGINE OVERHAUL PROCEDURES

## 2  Oil pressure check

♦ **Refer to illustration 2.2**

1  Low engine oil pressure can be a sign of an engine in need of rebuilding. A low oil pressure indicator (often called an "idiot light") is not a test of the oiling system. Such indicators only come on when the oil pressure is dangerously low. Even a factory oil pressure gauge in the instrument panel is only a relative indication, although much better for driver information than a warning light. A better test is with a mechanical (not electrical) oil pressure gauge.

2  Locate the engine oil pressure sending unit:

   a) *On V8 engines, the oil pressure sender is located in the top of the valley cover or Valve Lifter Oil Manifold (VLOM) near the rear of the intake manifold. However, since access to the oil pressure sender is limited, the manufacturer recommends the use of an adapter that screws in place of the oil filter.*

   b) *On V6 engines, the oil pressure sending unit is located at the top rear of the engine (see illustration).*

3  On V6 engines, unscrew and remove the oil pressure sending unit and screw in the hose for your oil pressure gauge. On V8 engines, install the adapter in place of the oil filter, then connect the gauge hose to the adapter.

4  Connect an accurate tachometer to the engine, according to the tachometer manufacturer's instructions.

5  Check the oil pressure with the engine running (normal operating temperature) at the specified engine speed, and compare it to this Chapter's Specifications. If it's extremely low, the bearings and/or oil pump are probably worn out.

**2.2  Location of the oil pressure sending unit on V6 engines - on V8 engines you'll need an adapter that attaches to the oil filter mount**

## 3  Cylinder compression check

♦ **Refer to illustration 3.6**

1  A compression check will tell you what mechanical condition the upper end of your engine (pistons, rings, valves, head gaskets) is in. Specifically, it can tell you if the compression is down due to leakage caused by worn piston rings, defective valves and seats or a blown head gasket.

➡ **Note: The engine must be at normal operating temperature and the battery must be fully charged for this check.**

2  Begin by cleaning the area around the spark plugs before you remove them (compressed air should be used, if available). The idea is to prevent dirt from getting into the cylinders as the compression check is being done.

3  Remove all of the spark plugs from the engine (see Chapter 1).

4  Remove the air intake duct (see Chapter 4) and block the throttle plate wide open.

5  **V6 engine:** Disable the ignition system by disconnecting the primary (low voltage) wiring from the ignition coil. Disable the fuel system by disconnecting the electrical connector from the fuel meter body (see Chapter 4, Section 14).

**V8 engines:** Disable the ignition system by disconnecting the primary (low voltage) electrical connector from each ignition coil (see Chapter 5). Disable the fuel injectors by removing the INJ A and INJ B fuses from the underhood fuse/relay box.

# GENERAL ENGINE OVERHAUL PROCEDURES   2C-5

6  Install a compression gauge in the number one cylinder spark plug hole (see illustration).

7  Crank the engine over at least seven compression strokes and watch the gauge. The compression should build up quickly in a healthy engine. Low compression on the first stroke, followed by gradually increasing pressure on successive strokes, indicates worn piston rings. A low compression reading on the first stroke, which doesn't build up during successive strokes, indicates leaking valves or a blown head gasket (a cracked head could also be the cause). Deposits on the undersides of the valve heads can also cause low compression. Record the highest gauge reading obtained.

8  Repeat the procedure for the remaining cylinders and compare the results to this Chapter's Specifications.

9  Add some engine oil (about three squirts from a plunger-type oil can) to each cylinder, through the spark plug hole, and repeat the test.

10  If the compression increases after the oil is added, the piston rings are definitely worn. If the compression doesn't increase significantly, the leakage is occurring at the valves or head gasket. Leakage past the valves may be caused by burned valve seats and/or faces or warped, cracked or bent valves.

11  If two adjacent cylinders have equally low compression, there's a strong possibility that the head gasket between them is blown. The appearance of coolant in the combustion chambers or the crankcase would verify this condition.

12  If one cylinder is slightly lower than the others, and the engine has a slightly rough idle, a worn lobe on the camshaft could be the cause.

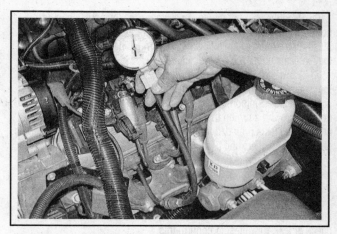

**3.6  Use a compression gauge with a threaded fitting for the spark plug hole, not the type that requires hand pressure to maintain the seal**

13  If the compression is unusually high, the combustion chambers are probably coated with carbon deposits. If that's the case, the cylinder head(s) should be removed and decarbonized.

14  If compression is way down or varies greatly between cylinders, it would be a good idea to have a leak-down test performed by an automotive repair shop. This test will pinpoint exactly where the leakage is occurring and how severe it is.

## 4  Vacuum gauge diagnostic checks

♦ **Refer to illustration 4.6**

1  A vacuum gauge provides inexpensive but valuable information about what is going on in the engine. You can check for worn rings or cylinder walls, leaking head or intake manifold gaskets, incorrect carburetor adjustments, restricted exhaust, stuck or burned valves, weak valve springs, improper ignition or valve timing and ignition problems.

2  Unfortunately, vacuum gauge readings are easy to misinterpret, so they should be used in conjunction with other tests to confirm the diagnosis.

3  Both the absolute readings and the rate of needle movement are important for accurate interpretation. Most gauges measure vacuum in inches of mercury (in-Hg). The following references to vacuum assume the diagnosis is being performed at sea level. As elevation increases (or atmospheric pressure decreases), the reading will decrease. For every 1,000 foot increase in elevation above approximately 2,000 feet, the gauge readings will decrease about one inch of mercury.

4  Connect the vacuum gauge directly to the intake manifold vacuum, not to ported (throttle body) vacuum. Be sure no hoses are left disconnected during the test or false readings will result.

5  Before you begin the test, allow the engine to warm up completely. Block the wheels and set the parking brake. With the transaxle in Park, start the engine and allow it to run at normal idle speed.

**✷✷ WARNING:**

**Keep your hands and the vacuum gauge clear of the fans.**

# 2C-6 GENERAL ENGINE OVERHAUL PROCEDURES

6  Read the vacuum gauge; an average, healthy engine should normally produce about 17 to 22 in-Hg with a fairly steady needle (see illustration). Refer to the following vacuum gauge readings and what they indicate about the engine's condition:

7  A low steady reading usually indicates a leaking gasket between the intake manifold and cylinder head(s) or throttle body, a leaky vacuum hose, late ignition timing or incorrect camshaft timing. Check ignition timing with a timing light and eliminate all other possible causes, utilizing the tests provided in this Chapter before you remove the timing chain cover to check the timing marks.

8  If the reading is three to eight inches below normal and it fluctuates at that low reading, suspect an intake manifold gasket leak at an intake port or a faulty fuel injector.

9  If the needle has regular drops of about two-to-four inches at a steady rate, the valves are probably leaking. Perform a compression check or leak-down test to confirm this.

10  An irregular drop or down-flick of the needle can be caused by a sticking valve or an ignition misfire. Perform a compression check or leak-down test and read the spark plugs.

11  A rapid vibration of about four in-Hg vibration at idle combined with exhaust smoke indicates worn valve guides. Perform a leak-down test to confirm this. If the rapid vibration occurs with an increase in engine speed, check for a leaking intake manifold gasket or head gasket, weak valve springs, burned valves or ignition misfire.

12  A slight fluctuation, say one inch up and down, may mean ignition problems. Check all the usual tune-up items and, if necessary, run the engine on an ignition analyzer.

13  If there is a large fluctuation, perform a compression or leak-down test to look for a weak or dead cylinder or a blown head gasket.

14  If the needle moves slowly through a wide range, check for a clogged PCV system, incorrect idle fuel mixture, throttle body or intake manifold gasket leaks.

15  Check for a slow return after revving the engine by quickly snapping the throttle open until the engine reaches about 2,500 rpm and let it shut. Normally the reading should drop to near zero, rise above normal idle reading (about 5 in-Hg over) and return to the previous idle reading. If the vacuum returns slowly and doesn't peak when the throttle is snapped shut, the rings may be worn. If there is a long delay, look for a restricted exhaust system (often the muffler or catalytic converter). An easy way to check this is to temporarily disconnect the exhaust ahead of the suspected part and redo the test.

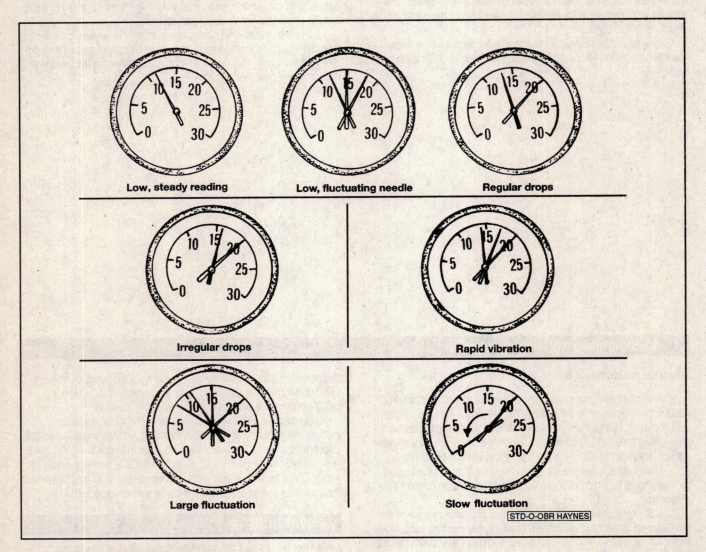

4.6  Typical vacuum gauge readings

# GENERAL ENGINE OVERHAUL PROCEDURES    2C-7

## 5  Engine rebuilding alternatives

The do-it-yourselfer is faced with a number of options when purchasing a rebuilt engine. The major considerations are cost, warranty, parts availability and the time required for the rebuilder to complete the project. The decision to replace the engine block, piston/connecting rod assemblies and crankshaft depends on the final inspection results of your engine. Only then can you make a cost effective decision whether to have your engine overhauled or simply purchase an exchange engine for your vehicle.

Some of the rebuilding alternatives include:

**Individual parts** - If the inspection procedures reveal that the engine block and most engine components are in reusable condition, purchasing individual parts and having a rebuilder rebuild your engine may be the most economical alternative. The block, crankshaft and piston/connecting rod assemblies should all be inspected carefully by a machine shop first.

**Short block** - A short block consists of an engine block with a crankshaft and piston/connecting rod assemblies already installed. All new bearings are incorporated and all clearances will be correct. The existing camshafts, valve train components, cylinder head and external parts can be bolted to the short block with little or no machine shop work necessary.

**Long block** - A long block consists of a short block plus an oil pump, oil pan, cylinder head, valve cover, camshaft and valve train components, timing sprockets and chain or gears and timing cover. All components are installed with new bearings, seals and gaskets incorporated throughout. The installation of manifolds and external parts is all that's necessary.

**Low mileage used engines** - Some companies now offer low mileage used engines which is a very cost effective way to get your vehicle up and running again. These engines often come from vehicles which have been in totaled in accidents or come from other countries which have a higher vehicle turn-over rate. A low mileage used engine also usually has a similar warranty like the newly remanufactured engines.

Give careful thought to which alternative is best for you and discuss the situation with local automotive machine shops, auto parts dealers and experienced rebuilders before ordering or purchasing replacement parts.

## 6  Engine removal - methods and precautions

♦ **Refer to illustrations 6.1, 6.2, 6.3 and 6.4**

If you've decided that an engine must be removed for overhaul or major repair work, several preliminary steps should be taken. Read all removal and installation procedures carefully prior to committing to this job.

Locating a suitable place to work is extremely important. Adequate work space, along with storage space for the vehicle, will be needed. If a shop or garage isn't available, at the very least a flat, level, clean work surface made of concrete or asphalt is required.

Cleaning the engine compartment and engine before beginning the removal procedure will help keep tools clean and organized (see illustrations 6.1 and 6.2).

An engine hoist will also be necessary. Make sure the hoist is rated in excess of the combined weight of the engine and its accessories. Safety is of primary importance, considering the potential hazards involved in removing the engine from the vehicle.

If you're a novice at engine removal, get at least one helper. One person cannot easily do all the things you need to do to remove a big heavy engine assembly from the engine compartment. Also helpful is to seek advice and assistance from someone who's experienced in engine removal.

6.1  After tightly wrapping water-vulnerable components, use a spray cleaner on everything, with particular concentration on the greasiest areas, usually around the valve cover and lower edges of the block. If one section dries out, apply more cleaner

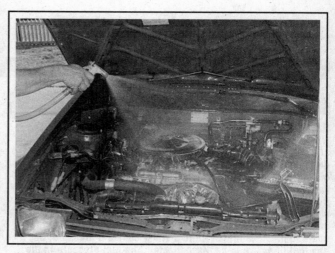

6.2  Depending on how dirty the engine is, let the cleaner soak in according to the directions and hose off the grime and cleaner. Get the rinse water down into every area you can get at; then dry important components with a hair dryer or paper towels

## 2C-8  GENERAL ENGINE OVERHAUL PROCEDURES

Plan the operation ahead of time. Arrange for or obtain all of the tools and equipment you'll need prior to beginning the job (see illustrations 6.3 and 6.4). Some of the equipment necessary to perform engine removal and installation safely and with relative ease are (in addition to a vehicle hoist and an engine hoist) a heavy duty floor jack (preferably fitted with a transaxle jack head adapter), complete sets of wrenches and sockets as described in the front of this manual, wooden blocks, plenty of rags and cleaning solvent for mopping up spilled oil, coolant and gasoline.

Plan for the vehicle to be out of use for quite a while. A machine shop can do the work that is beyond the scope of the home mechanic. Machine shops often have a busy schedule, so before removing the engine, consult the shop for an estimate of how long it will take to rebuild or repair the components that may need work.

**6.3  Get an engine stand sturdy enough to firmly support the engine while you're working on it. Stay away from three-wheeled models; they have a tendency to tip over more easily, so get a four-wheeled unit**

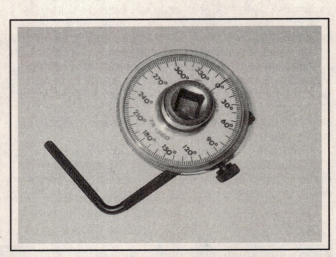

**6.4  Since many of the fasteners on these engines are tightened using the angle torque method, a torque angle gauge is essential for proper assembly**

## 7  Engine - removal and installation

### ✳✳ WARNING 1:

Gasoline is extremely flammable, so take extra precautions when you work on any part of the fuel system. Don't smoke or allow open flames or bare light bulbs near the work area, and don't work in a garage where a gas-type appliance (such as a water heater or clothes dryer) is present. Since gasoline is carcinogenic, wear fuel-resistant gloves when there's a possibility of being exposed to fuel, and, if you spill any fuel on your skin, rinse it off immediately with soap and water. Mop up any spills immediately and do not store fuel-soaked rags where they could ignite. The fuel system is under constant pressure, so, if any fuel lines are to be disconnected, the fuel pressure in the system must be relieved first (see Chapter 4 for more information). When you perform any kind of work on the fuel system, wear safety glasses and have a Class B type fire extinguisher on hand.

### ✳✳ WARNING 2:

The engine must be completely cool before beginning this procedure.

### REMOVAL

▸ Refer to illustrations 7.25, 7.32, 7.34 and 7.35

➥ Note: Keep in mind that during this procedure you'll have to adjust the height of the vehicle to perform certain operations.

1  Have the air conditioning system discharged by an approved station before beginning work if you're working on a V6 model.

2  Disconnect the cable from the negative terminal of the battery (see Chapter 5, Section 1).

# GENERAL ENGINE OVERHAUL PROCEDURES  2C-9

**7.25 The upper radiator support bar must be removed for clearance on V8 models**

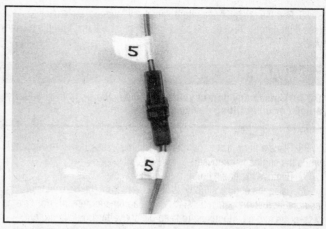

**7.32 Label both ends of each wire and hose before disconnecting it**

3  Relieve the fuel system pressure (see Chapter 4).
4  Refer to Chapter 11 and remove the hood.
5  Remove the engine cover.
6  Remove the entire air cleaner assembly (see Chapter 4).
7  Remove the drivebelt (see Chapter 1).
8  Raise the vehicle and support it securely on jackstands. Remove the engine lower shield and the skid plate on 4WD models.
9  Drain the coolant and the engine oil (see Chapter 1). Remove the radiator hoses and the heater hoses.
10  Disconnect the catalytic converter(s) from the exhaust manifolds. Wire them out of the way or remove them completely.
11  Remove the starter (see Chapter 5).
12  Remove the flywheel cover. Remove the torque converter bolts. You'll have to rotate the crankshaft with a socket and a breaker bar to access all of the bolts.
13  Remove the transmission-to-engine bolts (see Chapter 7A), then lower the vehicle.

## V6 models

14  Refer to Chapter 4 and remove the fuel pipes and hoses.
15  Refer to Chapter 3 and remove the lower fan shroud.
16  Remove the air conditioning hoses.
17  Remove the power brake booster vacuum hose.
18  Loosen the power steering pump rear bracket nut, then remove the rear bracket front nut.
19  Remove the remaining bracket fasteners and slide the bracket (with the air conditioning compressor and power steering pump attached) off of the bracket. Set it out of the way.

## V8 models

20  Remove the inner diagonal brace rod from the left front of the engine compartment.
21  Remove the hood latch (see Chapter 11).
22  Remove the upper fan shroud bolts, upper radiator mount bolts and upper condenser bolts.
23  Remove the radiator and air conditioning condenser (see Chapter 3).
24  Remove the headlights (see Chapter 12).
25  Remove the upper radiator support tie bar that connects the upper front sections of the fenders (see illustration).
26  Disconnect the transmission cooler lines from the oil pan.
27  Remove the rear mounting bolt from the power steering pump.

**7.34 On V8 engines, it will be necessary to attach the chains to the cylinder heads; most V6 engines have lifting brackets installed**

28  Unbolt the alternator bracket and move the assembly aside.
29  Remove the automatic transmission dipstick tube.
30  Remove the EVAP pipe bracket from the stud on the transmission.
31  Remove the transfer case vent hose if you're working on a 4WD model.

## All models

32  Label, then disconnect each wiring connection from the engine (see illustration). Also detach all wiring clips, brackets and retainers. Make notes of the positions of wiring harnesses as necessary.
33  Check all sides of the engine to make sure that all components are disconnected and that hoses as well as wires are labeled.
34  Attach lengths of chain to the engine (see illustration). There are lifting brackets on some of the engines but not all. If you can't find a lifting bracket, make sure to attach the chains to secure points such as the cylinder heads.

**✷✷ CAUTION:**

**DO NOT lift the engine by the intake manifold.**

# 2C-10 GENERAL ENGINE OVERHAUL PROCEDURES

35 Roll the engine hoist into position and connect the chains to it (see illustration). Take up the slack in the chain, but don't lift the engine yet.

> ※ **WARNING:**
>
> **DO NOT place any part of your body under the engine when only a hoist or other lifting device supports it.**

36 Place a floor jack under the front of the transmission to support it while the engine is removed.

37 Remove the engine mount-to-frame bracket bolts.

38 Raise the engine slightly. Work it away from the transmission, being sure that the converter stays in the automatic transmission (clamp vise-grips to the bellhousing to keep the converter from sliding out).

39 Carefully raise the engine from the engine compartment. Check as you go to make sure nothing is hanging up. Remove the driveplate and mount the engine on an engine stand.

## INSTALLATION

40 Installation is the reverse of the removal procedure, noting the following points:

a) *Check the engine and transmission mounts. If they're worn, replace them.*
b) *Tighten the torque converter bolts to the torque listed in this Chapter's Specifications.*

**7.35 Pull the engine forward as far as possible to clear the transmission and the cowl, then lift the engine high enough to clear the body**

c) *Refill the cooling system with the proper mixture of coolant and refill the engine with the recommended oil (see Chapter 1).*
d) *Check the transmission fluid level and add fluid as necessary.*
e) *Have the air conditioning system recharged by the shop that discharged it.*

## 8 Engine overhaul - disassembly sequence

1 It's much easier to disassemble the engine if it's mounted on a portable engine stand. A stand can often be rented quite cheaply from an equipment rental yard. Before the engine is mounted on a stand, the driveplate should be removed from the engine.

2 If a stand isn't available, it's possible to remove the external engine components with it blocked up on the floor. Be extra careful not to tip or drop the engine when working without a stand.

3 If you're going to obtain a rebuilt engine, all external components must come off first, to be transferred to the replacement engine. These components include:

*Driveplate*
*Ignition system components*
*Emissions-related components*
*Engine mounts and mount brackets*
*Fuel injection components*
*Intake/exhaust manifolds*
*Oil filter*
*Thermostat and housing assembly*
*Water pump*

➡**Note: When removing the external components from the engine, pay close attention to details that may be helpful or important during installation. Note the installed position of gaskets, seals, spacers, pins, brackets, washers, bolts and other small items.**

4 If you're going to obtain a short block (assembled engine block, crankshaft, pistons and connecting rods), then you should remove the timing belt, cylinder head, oil pan, oil pump pick-up tube, oil pump and water pump from your engine so that you can turn in your old short block to the rebuilder as a core. See *Engine rebuilding alternatives* for additional information regarding the different possibilities to be considered.

# GENERAL ENGINE OVERHAUL PROCEDURES

## 9  Pistons and connecting rods - removal and installation

### REMOVAL

♦ **Refer to illustrations 9.1, 9.3 and 9.4**

➡ **Note: Prior to removing the piston/connecting rod assemblies, remove the cylinder head and oil pan (see Chapter 2A or 2B).**

1  Use your fingernail to feel if a ridge has formed at the upper limit of ring travel (about 1/4-inch down from the top of each cylinder). If carbon deposits or cylinder wear have produced ridges, they must be completely removed with a special tool (see illustration). Follow the manufacturer's instructions provided with the tool. Failure to remove the ridges before attempting to remove the piston/connecting rod assemblies may result in piston breakage.

2  After the cylinder ridges have been removed, turn the engine so the crankshaft is facing up.

3  Before the connecting rods are removed, check the connecting rod endplay with feeler gauges. Slide them between the first connecting rod and the crankshaft throw until the play is removed (see illustration). Repeat this procedure for each connecting rod. The endplay is equal to the thickness of the feeler gauge(s). Check with an automotive machine shop for the endplay service limit (a typical endplay limit should measure between 0.005 to 0.015 inch [0.127 to 0.396 mm]). If the play exceeds the service limit, new connecting rods will be required. If new rods (or a new crankshaft) are installed, the endplay may fall under the minimum allowable. If it does, the rods will have to be machined to restore it. If necessary, consult an automotive machine shop for advice.

4  Check the connecting rods and caps for identification marks. If they aren't plainly marked, use paint or marker to clearly identify each rod and cap (1, 2, 3, etc., depending on the cylinder they're associated with) (see illustration).

5  Remove the connecting rod cap fasteners.

6  Remove the number one connecting rod cap and bearing insert. Don't drop the bearing insert out of the cap.

7  Remove the bearing insert and push the connecting rod/piston assembly out through the top of the engine. Use a wooden or plastic hammer handle to push on the upper bearing surface in the connecting rod. If resistance is felt, double-check to make sure that all of the ridge was removed from the cylinder.

8  Repeat the procedure for the remaining cylinders.

9  After removal, reassemble the connecting rod caps and bearing inserts in their respective connecting rods and install the fasteners finger tight. Leaving the old bearing inserts in place until reassembly will help prevent the connecting rod bearing surfaces from being accidentally nicked or gouged.

10  The pistons and connecting rods are now ready for inspection and overhaul at an automotive machine shop.

### PISTON RING INSTALLATION

♦ **Refer to illustrations 9.13, 9.14, 9.15, 9.19a, 9.19b and 9.22**

11  Before installing the new piston rings, the ring end gaps must be checked. It's assumed that the piston ring side clearance has been checked and verified correct.

**9.1  Before you try to remove the pistons, use a ridge reamer to remove the raised material (ridge) from the top of the cylinders**

**9.3  Checking the connecting rod endplay (side clearance)**

**9.4  If the connecting rods and caps are not marked, use paint to mark the caps to the rods by cylinder number (for example, this would be the No. 4 connecting rod)**

➡ **Note:** V6 engines are equipped with two different brand pistons. The correct piston rings must match the piston. To identify the piston brand, turn the piston over and look between the pin bores for an FM mark (Federal Mogul) or a no mark (Mahle) brand type. Use only the correct type of piston rings that match the piston.

12 Lay out the piston/connecting rod assemblies and the new ring sets so the ring sets will be matched with the same piston and cylinder during the end gap measurement and engine assembly.

13 Insert the top (number one) ring into the first cylinder and square it up with the cylinder walls by pushing it in with the top of the piston (see illustration). The ring should be near the bottom of the cylinder, at the lower limit of ring travel.

14 To measure the end gap, slip feeler gauges between the ends of the ring until a gauge equal to the gap width is found (see illustration). The feeler gauge should slide between the ring ends with a slight amount of drag. A typical ring gap should fall between 0.010 and 0.020 inch [0.25 to 0.50 mm] for compression rings and up to 0.030 inch [0.76 mm] for the oil ring steel rails. If the gap is larger or smaller than specified, double-check to make sure you have the correct rings before proceeding.

15 If the gap is too small, it must be enlarged or the ring ends may come in contact with each other during engine operation, which can cause serious damage to the engine. If necessary, increase the end gaps by filing the ring ends very carefully with a fine file. Mount the file in a vise equipped with soft jaws, slip the ring over the file with the ends contacting the file face and slowly move the ring to remove material from the ends. When performing this operation, file only by pushing the ring from the outside end of the file towards the vise (see illustration).

16 Excess end gap isn't critical unless it's greater than 0.040 inch (1 mm). Again, double-check to make sure you have the correct ring type.

17 Repeat the procedure for each ring that will be installed in the first cylinder and for each ring in the remaining cylinders. Remember to keep rings, pistons and cylinders matched up.

18 Once the ring end gaps have been checked/corrected, the rings can be installed on the pistons.

19 The oil control ring (lowest one on the piston) is usually installed first. It's composed of three separate components. Slip the spacer/expander into the groove (see illustration). If an anti-rotation tang is used, make sure it's inserted into the drilled hole in the ring groove. Next, install the upper side rail in the same manner (see illustration). Don't use a piston ring installation tool on the oil ring side rails, as

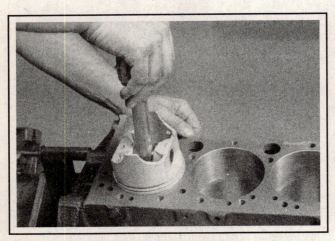

**9.13 Install the piston ring into the cylinder then push it down into position using a piston so the ring will be square in the cylinder**

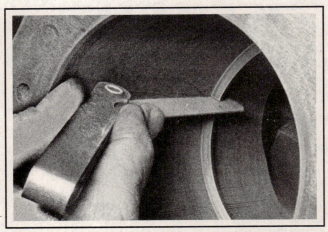

**9.14 With the ring square in the cylinder, measure the ring end gap with a feeler gauge**

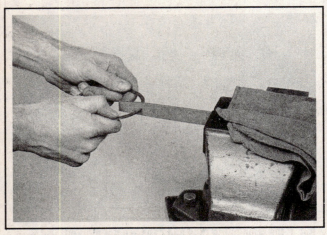

**9.15 If the ring end gap is too small, clamp a file in a vise as shown and file the piston ring ends - be sure to remove all raised material**

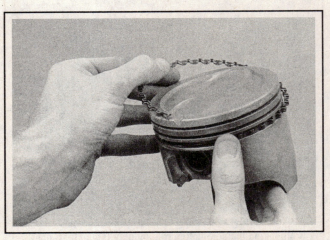

**9.19a Installing the spacer/expander in the oil ring groove**

# GENERAL ENGINE OVERHAUL PROCEDURES  2C-13

they may be damaged. Instead, place one end of the side rail into the groove between the spacer/expander and the ring land, hold it firmly in place and slide a finger around the piston while pushing the rail into the groove. Finally, install the lower side rail.

20  After the three oil ring components have been installed, check to make sure that both the upper and lower side rails can be rotated smoothly inside the ring grooves.

21  The number two (middle) ring is installed next. It's usually stamped with a mark which must face up, toward the top of the piston. Do not mix up the top and middle rings, as they have different cross-sections.

➡ **Note 1: Always follow the instructions printed on the ring package or box - different manufacturers may require different approaches.**

➡ **Note 2: On V6 engines, original equipment compression rings are marked as follows: The top compression ring has a green stripe on the top side of the ring, 180-degrees from the gap. the second compression ring has a green stripe 90-degrees from the gap.**

22  Use a piston ring installation tool and make sure the identification mark is facing the top of the piston, then slip the ring into the middle groove on the piston (see illustration). Don't expand the ring any more than necessary to slide it over the piston.

23  Install the number one (top) ring in the same manner. Make sure the mark is facing up. Be careful not to confuse the number one and number two rings.

24  Repeat the procedure for the remaining pistons and rings.

## INSTALLATION

25  Before installing the piston/connecting rod assemblies, the cylinder walls must be perfectly clean, the top edge of each cylinder bore must be chamfered, and the crankshaft must be in place.

26  Remove the cap from the end of the number one connecting rod (refer to the marks made during removal). Remove the original bearing inserts and wipe the bearing surfaces of the connecting rod and cap with a clean, lint-free cloth. They must be kept spotlessly clean.

### Connecting rod bearing oil clearance check
▶ Refer to illustrations 9.30, 9.35, 9.37 and 9.41

27  Clean the back side of the new upper bearing insert, then lay it in place in the connecting rod.

28  Make sure the tab on the bearing fits into the recess in the rod. Don't hammer the bearing insert into place and be very careful not to nick or gouge the bearing face. Don't lubricate the bearing at this time.

29  Clean the back side of the other bearing insert and install it in the rod cap. Again, make sure the tab on the bearing fits into the recess in the cap, and don't apply any lubricant. It's critically important that the mating surfaces of the bearing and connecting rod are perfectly clean and oil free when they're assembled.

30  Position the piston ring gaps at the specified intervals around the piston as shown (see illustration).

31  Lubricate the piston and rings with clean engine oil and attach a piston ring compressor to the piston. Leave the skirt protruding about 1/4-inch to guide the piston into the cylinder. The rings must be compressed until they're flush with the piston.

**9.19b  DO NOT use a piston ring installation tool when installing the oil control side rails**

**9.22  Use a piston ring installation tool to install the number 2 and the number 1 (top) rings - be sure the directional mark on the piston ring(s) is facing toward the top of the piston**

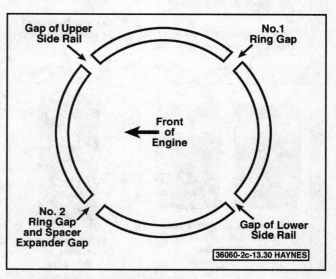

**9.30  Piston ring end gap positions**

# ENGINE BEARING ANALYSIS

## Debris

**Babbitt bearing embedded with debris from machinings**
**Microscopic detail of debris**

**Microscopic detail of gouges**
**Overplated copper alloy bearing gouged by cast iron debris**

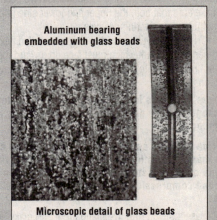

**Aluminum bearing embedded with glass beads**
**Microscopic detail of glass beads**

**Damaged lining caused by dirt left on the bearing back**

## Misassembly

**Result of a lower half assembled as an upper - blocking the oil flow**

**Excessive oil clearance is indicated by a short contact arc**

**Polished and oil-stained backs are a result of a poor fit in the housing bore**

**Result of a wrong, reversed, or shifted cap**

## Overloading

**Damage from excessive idling which resulted in an oil film unable to support the load imposed**

**Damaged upper connecting rod bearings caused by engine lugging; the lower main bearings (not shown) were similarly affected**

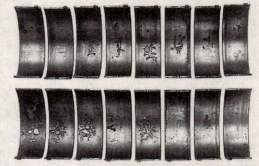

**The damage shown in these upper and lower connecting rod bearings was caused by engine operation at a higher-than-rated speed under load**

# Misalignment

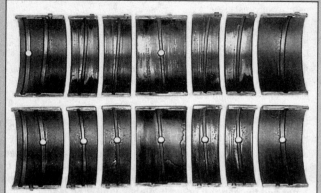
A warped crankshaft caused this pattern of severe wear in the center, diminishing toward the ends

A poorly finished crankshaft caused the equally spaced scoring shown

A tapered housing bore caused the damage along one edge of this pair

A bent connecting rod led to the damage in the "V" pattern

# Lubrication

Result of dry start: The bearings on the left, farthest from the oil pump, show more damage

Result of a low oil supply or oil starvation

Severe wear as a result of inadequate oil clearance

Damage from excessive thrust or insufficient axial clearance

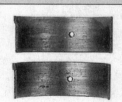
Bearing affected by oil dilution caused by excessive blow-by or a rich mixture

# Corrosion

Microscopic detail of corrosion

Corrosion is an acid attack on the bearing lining generally caused by inadequate maintenance, extremely hot or cold operation, or inferior oils or fuels

Microscopic detail of cavitation

Example of cavitation - a surface erosion caused by pressure changes in the oil film

© 1986 Federal-Mogul Corporation
Copy and photographs courtesy of Federal Mogul Corporation

# 2C-16 GENERAL ENGINE OVERHAUL PROCEDURES

32 Rotate the crankshaft until the number one connecting rod journal is at BDC (Bottom Dead Center) and apply a liberal coat of engine oil to the cylinder walls.

33 With the arrow on top of the piston facing the front (timing chain) of the engine, gently insert the piston/connecting rod assembly into the number one cylinder bore and rest the bottom edge of the ring compressor on the engine block.

34 Tap the top edge of the ring compressor to make sure it's contacting the block around its entire circumference.

35 Gently tap on the top of the piston with the end of a wooden or plastic hammer handle (see illustration) while guiding the end of the connecting rod into place on the crankshaft journal. The piston rings may try to pop out of the ring compressor just before entering the cylinder bore, so keep some downward pressure on the ring compressor. Work slowly, and if any resistance is felt as the piston enters the cylinder, stop immediately. Find out what's hanging up and fix it before proceeding. Do not, for any reason, force the piston into the cylinder - you might break a ring and/or the piston.

36 Once the piston/connecting rod assembly is installed, the connecting rod bearing oil clearance must be checked before the rod cap is permanently installed.

37 Cut a piece of the appropriate size Plastigage slightly shorter than the width of the connecting rod bearing and lay it in place on the number one connecting rod journal, parallel with the journal axis (see illustration).

38 Clean the connecting rod cap bearing face and install the rod cap. Make sure the mating mark on the cap is on the same side as the mark on the connecting rod (see illustration 9.4).

39 Install the rod nuts or bolts, and tighten them to the torque listed in this Chapter's Specifications.

➡ **Note: Use a thin-wall socket to avoid erroneous torque readings that can result if the socket is wedged between the rod cap and the bolt. If the socket tends to wedge itself between the fastener and the cap, lift up on it slightly until it no longer contacts the cap. DO NOT rotate the crankshaft at any time during this operation.**

40 Remove the fasteners and detach the rod cap, being very careful not to disturb the Plastigage.

41 Compare the width of the crushed Plastigage to the scale printed on the Plastigage envelope to obtain the oil clearance (see illustration). The connecting rod oil clearance is usually about 0.001 to 0.002 inch. Consult an automotive machine shop for the clearance specified for the rod bearings on your engine.

42 If the clearance is not as specified, the bearing inserts may be the wrong size (which means different ones will be required). Before deciding that different inserts are needed, make sure that no dirt or oil was between the bearing inserts and the connecting rod or cap when the clearance was measured. Also, recheck the journal diameter. If the Plastigage was wider at one end than the other, the journal may be tapered. If the clearance still exceeds the limit specified, the bearing will have to be replaced with an undersize bearing.

❄ **CAUTION:**

**When installing a new crankshaft, always use a standard size bearing.**

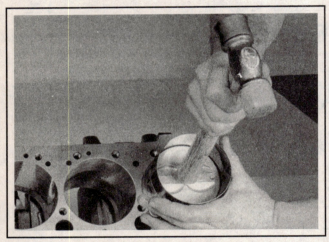

9.35 Use a plastic or wooden hammer handle to push the piston into the cylinder

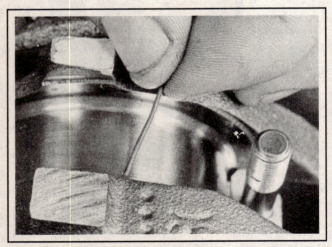

9.37 Place Plastigage on each connecting rod bearing journal parallel to the crankshaft centerline

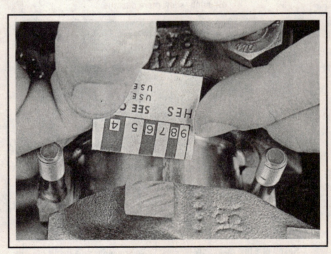

9.41 Use the scale on the Plastigage package to determine the bearing oil clearance - be sure to measure the widest part of the Plastigage and use the correct scale; it comes with both standard and metric scales

# General Engine Overhaul Procedures 2C-17

## Final installation

43 Carefully scrape all traces of the Plastigage material off the rod journal and/or bearing face. Be very careful not to scratch the bearing - use your fingernail or the edge of a plastic card.

44 Make sure the bearing faces are perfectly clean, then apply a uniform layer of clean moly-base grease or engine assembly lube to both of them. You'll have to push the piston into the cylinder to expose the face of the bearing insert in the connecting rod.

45 Slide the connecting rod back into place on the journal, install the rod cap, install the bolts and tighten them to the torque listed in this Chapter's Specifications.

46 Repeat the entire procedure for the remaining pistons/connecting rods.

47 The important points to remember are:
a) Keep the back sides of the bearing inserts and the insides of the connecting rods and caps perfectly clean when assembling them.
b) Make sure you have the correct piston/rod assembly for each cylinder.
c) The arrow on the piston must face the front (timing chain end) of the engine.
d) Lubricate the cylinder walls liberally with clean oil.
e) Lubricate the bearing faces when installing the rod caps after the oil clearance has been checked.

48 After all the piston/connecting rod assemblies have been correctly installed, rotate the crankshaft a number of times by hand to check for any obvious binding.

49 As a final step, check the connecting rod endplay as described in Step 3. If it was correct before disassembly and the original crankshaft and rods were reinstalled, it should still be correct. If new rods or a new crankshaft were installed, the endplay may be inadequate. If so, the rods will have to be removed and taken to an automotive machine shop for resizing.

## 10 Crankshaft - removal and installation

### REMOVAL

♦ Refer to illustrations 10.1 and 10.3

➡ Note: The crankshaft can be removed only after the engine has been removed from the vehicle. It's assumed that the driveplate, crankshaft pulley, timing chain, oil pan, oil pump body, oil filter and piston/connecting rod assemblies have already been removed. The rear main oil seal retainer must be unbolted and separated from the block before proceeding with crankshaft removal.

1 Before the crankshaft is removed, measure the endplay. Mount a dial indicator with the indicator in line with the crankshaft and just touching the end of the crankshaft as shown (see illustration).

2 Pry the crankshaft all the way to the rear and zero the dial indicator. Next, pry the crankshaft to the front as far as possible and check the reading on the dial indicator. The distance traveled is the endplay. A typical crankshaft endplay will fall between 0.003 to 0.010 inch (0.076 to 0.254 mm). If it is greater than that, check the crankshaft thrust surfaces for wear after it's removed. If no wear is evident, new main bearings should correct the endplay.

3 If a dial indicator isn't available, feeler gauges can be used. Gently pry the crankshaft all the way to the front of the engine. Slip feeler gauges between the crankshaft and the front face of the thrust bearing or washer to determine the clearance (see illustration).

4 Loosen the main bearing cap bolts 1/4-turn at a time each, until they can be removed by hand.

5 Remove the main bearing caps. Pull the main bearing cap straight up and off the cylinder block. Removing the caps from a V8 engine may require a special slide-hammer removal tool that threads into the caps.

**10.1 Checking crankshaft endplay with a dial indicator**

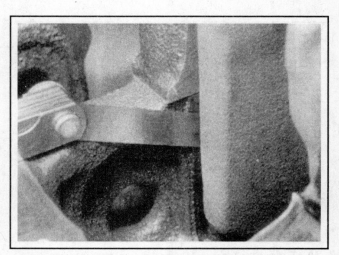

**10.3 Checking the crankshaft endplay with feeler gauges at the thrust bearing journal**

# 2C-18  GENERAL ENGINE OVERHAUL PROCEDURES

Gently tap the main bearing cap with a soft-face hammer, if necessary.

6  Carefully lift the crankshaft out of the engine. It may be a good idea to have an assistant available, since the crankshaft is quite heavy and awkward to handle. With the bearing inserts in place inside the engine block and main bearing caps, reinstall the main bearing caps onto the engine block and tighten the bolts finger tight. Make sure the caps are in the exact order they were removed with the arrow pointing toward the front (timing chain and front cover) of the engine.

## INSTALLATION

7  Crankshaft installation is the first step in engine reassembly. It's assumed at this point that the engine block and crankshaft have been cleaned, inspected and repaired or reconditioned.

8  Position the engine block with the bottom facing up.

9  Remove the bolts and lift off the main bearing caps.

10  If they're still in place, remove the original bearing inserts from the block and from the main bearing caps. Wipe the bearing surfaces of the block and main bearing cap assembly with a clean, lint-free cloth. They must be kept spotlessly clean. This is critical for determining the correct bearing oil clearance.

## MAIN BEARING OIL CLEARANCE CHECK

▶ Refer to illustrations 10.17 and 10.21

11  Without mixing them up, clean the back sides of the new upper main bearing inserts (with grooves and oil holes) and lay one in each main bearing saddle in the engine block. Each upper bearing (engine block) has an oil groove and oil hole in it.

### ✳✳ CAUTION:

**The oil holes in the block must line up with the oil holes in the engine block inserts. The thrust washer or thrust bearing insert must be installed in the correct location.**

➡ Note: The thrust bearing is located on the third journal in the main bearing cap journals (counting from the front).

Clean the back sides of the lower main bearing inserts and lay them in the corresponding location in the main bearing caps. Make sure the tab on the bearing insert fits into the recess in the block or main bearing caps.

### ✳✳ CAUTION:

**Do not hammer the bearing insert into place and don't nick or gouge the bearing faces. DO NOT apply any lubrication at this time.**

12  Clean the faces of the bearing inserts in the block and the crankshaft main bearing journals with a clean, lint-free cloth.

13  Check or clean the oil holes in the crankshaft, as any dirt here can go only one way - straight through the new bearings.

14  Once you're certain the crankshaft is clean, carefully lay it in position in the cylinder block.

15  Before the crankshaft can be permanently installed, the main bearing oil clearance must be checked.

16  Cut several strips of the appropriate size of Plastigage. They must be slightly shorter than the width of the main bearing journal.

17  Place one piece on each crankshaft main bearing journal, parallel with the journal axis as shown (see illustration).

18  Clean the faces of the bearing inserts in the main bearing caps or lower crankcase. Install the caps without disturbing the Plastigage.

19  Apply clean engine oil to all bolt threads prior to installation, install all bolts finger-tight, then tighten them to the torque listed in this Chapter's Specifications. DO NOT rotate the crankshaft at any time during this operation.

➡ Note: On V8 engines, follow the correct torque sequence (see illustration 10.30), but note that it isn't necessary to install the side bolts at this time (just bolts 1 through 20).

20  Remove the bolts and carefully lift the main bearing caps straight up and off the block. Do not disturb the Plastigage or rotate the crankshaft.

21  Compare the width of the crushed Plastigage on each journal to the scale printed on the Plastigage envelope to determine the main bearing oil clearance (see illustration). Check with an automotive machine shop for the crankshaft bearing oil clearance for your engine.

**10.17  Place the Plastigage onto the crankshaft bearing journal as shown**

**10.21  Use the scale on the Plastigage package to determine the bearing oil clearance - be sure to measure the widest part of the Plastigage and use the correct scale; it comes with both standard and metric scales**

# GENERAL ENGINE OVERHAUL PROCEDURES  2C-19

22 If the clearance is not as specified, the bearing inserts may be the wrong size (which means different ones will be required). Before deciding if different inserts are needed, make sure that no dirt or oil was between the bearing inserts and the caps or block when the clearance was measured. If the Plastigage was wider at one end than the other, the crankshaft journal may be tapered. If the clearance still exceeds the limit specified, the bearing insert(s) will have to be replaced with an undersize bearing insert(s).

> **CAUTION:**
> **When installing a new crankshaft, always install a standard bearing insert set.**

23 Carefully scrape all traces of the Plastigage material off the main bearing journals and/or the bearing insert faces. Be sure to remove all residue from the oil holes. Use your fingernail or the edge of a plastic card - don't nick or scratch the bearing faces.

## FINAL INSTALLATION

▸ **Refer to illustration 10.30**

24 Carefully lift the crankshaft out of the cylinder block.
25 Clean the bearing insert faces in the cylinder block, then apply a thin, uniform layer of moly-base grease or engine assembly lube to each of the bearing surfaces. Be sure to coat the thrust faces as well as the journal face of the thrust bearing.
26 Make sure the crankshaft journals are clean, then lay the crankshaft back in place in the cylinder block.
27 Clean the bearing insert faces and apply the same lubricant to them. Clean the engine block and the bearing cap mating surfaces thoroughly. The surfaces must be free of oil residue.
28 Prior to installation, apply clean engine oil to all bolt threads, wiping off any excess, then install all bolts finger-tight.
29 Tighten the main bearing cap bolts of V6 engines to the torque listed in this Chapter's Specifications.
30 On V8 engines, tighten bolts one through ten to the torque listed in Step 1 of this Chapter's Specifications. Be sure to follow the correct sequence (see illustration). Tap the crankshaft to the rear and then to the front to align the thrust bearing on V8 engines. The last tap must be in the forward direction.
31 Proceed with the rest of the fastener tightening steps listed in this Chapter's Specifications for the V8 engine, ending with the tightening of the NEW side bolts.
➡ **Note: Using old side bolts may result in oil leaks as the bolts have a sealing compound on them.**
32 Recheck the crankshaft endplay with a feeler gauge or a dial indicator. The endplay should be correct if the crankshaft thrust faces aren't worn or damaged and if new bearings have been installed.
33 Rotate the crankshaft a number of times by hand to check for any obvious binding. It should rotate with a running torque of 50 in-lbs or less. If the running torque is too high, correct the problem at this time.

## REAR MAIN OIL SEAL RETAINER INSTALLATION

### V6 models

34 Install the rear seal retainer with a new gasket. Tighten the fasteners to the torque listed in this Chapter's Specifications

35 Refer to Chapter 2A for information regarding installing the seal into the retainer.

### V8 models

#### Without a seal in the retainer

36 Install the rear seal retainer with a new gasket but only tighten the bolts snug.
37 Install the proper alignment tool (GM part J 41480 or an equivalent) to the rear of the engine block oil pan rail. Tighten the bolts to 18 ft-lbs.
38 Install the proper alignment tool (GM part J 41476 or an equivalent) to the rear of the crankshaft. The two mounting bolts should be parallel to the oil pan rail. The legs of the tool will register into the seal bore.
39 Snug the J 41476 mounting bolts. Tighten the J 41480 bolts that go into the retainer to 106 in-lbs.
40 Tighten the rear oil seal retainer mounting bolts to the torque listed in this Chapter's Specifications.
41 Remove the tools and lay a straightedge across the rear of the oil pan rail. Use feeler gauges to measure the clearance between the oil seal retainer and the straightedge. The clearance must be between 0 and 0.02-inch. If the clearance is not correct, repeat the installation procedure.
42 Refer to Chapter 2B for information regarding seal installation.

#### With a seal in the retainer

43 Check the rear oil seal retainer for two alignment tabs in the bore at the 3 o'clock and 9 o'clock positions. If the retainer doesn't have these tabs, it must be installed without the seal - refer to Step 36. If it does have the tabs, continue to the next Step.
44 Install the proper alignment tool (GM part J 41479-2A or an equivalent) to the rear of the crankshaft. Tighten the bolts snug.
45 Install the oil seal retainer (with a seal in it) with a new gasket and finger-tighten the bolts.
46 Remove the alignment tool.
47 Install the proper alignment tool (GM part J 41480 or an equivalent) to the rear of the engine block. Tighten the bolts to 18 ft-lbs.
48 Tighten the tool bolts that go into the seal retainer to 106 in-lbs.
49 Tighten the oil seal retainer bolts to the torque listed in this Chapter's Specifications.

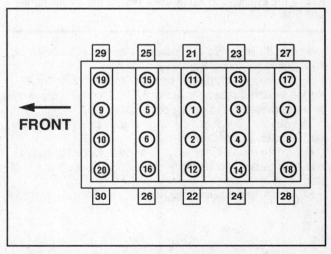

**10.30 Main bearing cap tightening sequence - V8 engine**

# 2C-20 GENERAL ENGINE OVERHAUL PROCEDURES

50  Remove the tool and lay a straightedge across the rear of the oil pan rail. Use feeler gauges to measure the clearance between the oil seal retainer and the straightedge. The clearance must be between 0 and 0.02-inch. If the clearance is not correct, repeat the installation procedure.

51  Refer to Chapter 2B for information regarding seal installation.

## 11  Engine overhaul - reassembly sequence

1  Before beginning engine reassembly, make sure you have all the necessary new parts, gaskets and seals as well as the following items on hand:

   Common hand tools
   A 1/2-inch drive torque wrench
   New engine oil
   Gasket sealant
   Thread locking compound

2  If you obtained a short block it will be necessary to install the cylinder head, the oil pump and pick-up tube, the oil pan, the water pump, the timing belt and timing cover, and the valve cover (see Chapter 2A or 2B). In order to save time and avoid problems, the external components must be installed in the following general order:

   Thermostat and housing cover
   Water pump
   Intake and exhaust manifolds
   Fuel injection components
   Emission control components
   Spark plugs
   Ignition coils or coil pack and spark plug wires
   Oil filter
   Engine mounts and mount brackets
   Driveplate

## 12  Initial start-up and break-in after overhaul

**WARNING:**

**Have a fire extinguisher handy when starting the engine for the first time.**

1  Once the engine has been installed in the vehicle, double-check the engine oil and coolant levels.

2  With the spark plugs out of the engine and the ignition system and fuel injectors disabled (see Section 3, Step 5) crank the engine until oil pressure registers on the gauge or the light goes out.

3  Install the spark plugs, hook up the plug wires and restore the ignition system and fuel injector functions.

4  Start the engine. It may take a few moments for the fuel system to build up pressure, but the engine should start without a great deal of effort.

5  After the engine starts, it should be allowed to warm up to normal operating temperature. While the engine is warming up, make a thorough check for fuel, oil and coolant leaks.

6  Shut the engine off and recheck the engine oil and coolant levels.

7  Drive the vehicle to an area with minimum traffic, accelerate from 30 to 50 mph, then allow the vehicle to slow to 30 mph with the throttle closed. Repeat the procedure 10 or 12 times. This will load the piston rings and cause them to seat properly against the cylinder walls. Check again for oil and coolant leaks.

8  Drive the vehicle gently for the first 500 miles (no sustained high speeds) and keep a constant check on the oil level. It is not unusual for an engine to use oil during the break-in period.

9  At approximately 500 to 600 miles, change the oil and filter.

10  For the next few hundred miles, drive the vehicle normally. Do not pamper it or abuse it.

11  After 2,000 miles, change the oil and filter again and consider the engine broken in.

# GENERAL ENGINE OVERHAUL PROCEDURES 2C-21

## Specifications

### General

Displacement
- 4.3L V6 models — 262 cubic inches
- 4.8L V8 models — 293 cubic inches
- 5.3L V8 models — 325 cubic inches
- 6.0L V8 models — 364 cubic inches
- 6.2L V8 models — 376 cubic inches

Bore and stroke
- 4.3L V6 models — 4.012 x 3.480 inches
- 4.8L V8 models — 3.779 x 3.268 inches
- 5.3L V8 models — 3.779 x 3.622 inches
- 6.0L V8 models — 4.001 x 3.622 inches
- 6.2L V8 models — 4.065 x 3.622 inches

Cylinder compression pressure
- Minimum — 100 psi
- Maximum variation — 25 percent from highest reading

Oil pressure (minimum, at operating temperature)
- 1000 rpm — 6 psi
- 2000 rpm — 18 psi
- 4000 rpm — 24 psi

## Torque specifications — Ft-lbs (unless otherwise indicated)

Connecting rod bearing cap fasteners
- V6 models
  - 2007 (nuts)
    - Step 1 — 20
    - Step 2 — Tighten an additional 70 degrees
  - 2008 and later (bolts)
    - Step 1 — 15
    - Step 2 — Tighten an additional 100 degrees
- V8 models
  - Step 1 — 15
  - Step 2 — Tighten an additional 85 degrees

Main bearing cap bolts
- V6 models
  - Step 1 — 15
  - Step 2 — Tighten an additional 73 degrees
- V8 models
  - Bolts 1 through 10
    - Step 1 — 15
    - Step 2 — Tighten an additional 80 degrees
  - Studs 11 through 20
    - Step 1 — 15
    - Step 2 — Tighten an additional 51 degrees
  - Bolts 21 through 30* — 18

Transmission-to-engine bolts — 37

Driveplate-to-torque converter bolts
- 4L80-E transmission — 44
- All other transmissions — 47

*Must be replaced with new bolts*

# GLOSSARY

## B

**Backlash** - The amount of play between two parts. Usually refers to how much one gear can be moved back and forth without moving the gear with which it's meshed.

**Bearing Caps** - The caps held in place by nuts or bolts which, in turn, hold the bearing surface. This space is for lubricating oil to enter.

**Bearing clearance** - The amount of space left between shaft and bearing surface. This space is for lubricating oil to enter.

**Bearing crush** - The additional height which is purposely manufactured into each bearing half to ensure complete contact of the bearing back with the housing bore when the engine is assembled.

**Bearing knock** - The noise created by movement of a part in a loose or worn bearing.

**Blueprinting** - Dismantling an engine and reassembling it to EXACT specifications.

**Bore** - An engine cylinder, or any cylindrical hole; also used to describe the process of enlarging or accurately refinishing a hole with a cutting tool, as to bore an engine cylinder. The bore size is the diameter of the hole.

**Boring** - Renewing the cylinders by cutting them out to a specified size. A boring bar is used to make the cut.

**Bottom end** - A term which refers collectively to the engine block, crankshaft, main bearings and the big ends of the connecting rods.

**Break-in** - The period of operation between installation of new or rebuilt parts and time in which parts are worn to the correct fit. Driving at reduced and varying speed for a specified mileage to permit parts to wear to the correct fit.

**Bushing** - A one-piece sleeve placed in a bore to serve as a bearing surface for shaft, piston pin, etc. Usually replaceable.

## C

**Camshaft** - The shaft in the engine, on which a series of lobes are located for operating the valve mechanisms. The camshaft is driven by gears or sprockets and a timing chain. Usually referred to simply as the cam.

**Carbon** - Hard, or soft, black deposits found in combustion chamber, on plugs, under rings, on and under valve heads.

**Cast iron** - An alloy of iron and more than two percent carbon, used for engine blocks and heads because it's relatively inexpensive and easy to mold into complex shapes.

**Chamfer** - To bevel across (or a bevel on) the sharp edge of an object.

**Chase** - To repair damaged threads with a tap or die.

**Combustion chamber** - The space between the piston and the cylinder head, with the piston at top dead center, in which air-fuel mixture is burned.

**Compression ratio** - The relationship between cylinder volume (clearance volume) when the piston is at top dead center and cylinder volume when the piston is at bottom dead center.

**Connecting rod** - The rod that connects the crank on the crankshaft with the piston. Sometimes called a con rod.

**Connecting rod cap** - The part of the connecting rod assembly that attaches the rod to the crankpin.

**Core plug** - Soft metal plug used to plug the casting holes for the coolant passages in the block.

**Crankcase** - The lower part of the engine in which the crankshaft rotates; includes the lower section of the cylinder block and the oil pan.

**Crank kit** - A reground or reconditioned crankshaft and new main and connecting rod bearings.

**Crankpin** - The part of a crankshaft to which a connecting rod is attached.

**Crankshaft** - The main rotating member, or shaft, running the length of the crankcase, with offset throws to which the connecting rods are attached; changes the reciprocating motion of the pistons into rotating motion.

**Cylinder sleeve** - A replaceable sleeve, or liner, pressed into the cylinder block to form the cylinder bore.

## D

**Deburring** - Removing the burrs (rough edges or areas) from a bearing.

**Deglazer** - A tool, rotated by an electric motor, used to remove glaze from cylinder walls so a new set of rings will seat.

## E

**Endplay** - The amount of lengthwise movement between two parts. As applied to a crankshaft, the distance that the crankshaft can move forward and back in the cylinder block.

## F

**Face** - A machinist's term that refers to removing metal from the end of a shaft or the face of a larger part, such as a flywheel.

**Fatigue** - A breakdown of material through a large number of loading and unloading cycles. The first signs are cracks followed shortly by breaks.

**Feeler gauge** - A thin strip of hardened steel, ground to an exact thickness, used to check clearances between parts.

**Free height** - The unloaded length or height of a spring.

**Freeplay** - The looseness in a linkage, or an assembly of parts, between the initial application of force and actual movement. Usually perceived as slop or slight delay.

**Freeze plug** - See Core plug.

## G

**Gallery** - A large passage in the block that forms a reservoir for engine oil pressure.

**Glaze** - The very smooth, glassy finish that develops on cylinder walls while an engine is in service.

## H

**Heli-Coil** - A rethreading device used when threads are worn or damaged. The device is installed in a retapped hole to reduce the thread size to the original size.

## I

**Installed height** - The spring's measured length or height, as installed on the cylinder head. Installed height is measured from the spring seat to the underside of the spring retainer.

## J

**Journal** - The surface of a rotating shaft which turns in a bearing.

## K

**Keeper** - The split lock that holds the valve spring retainer in position on the valve stem.

**Key** - A small piece of metal inserted into matching grooves machined into two parts fitted together - such as a gear pressed onto a shaft - which prevents slippage between the two parts.

**Knock** - The heavy metallic engine sound, produced in the combustion chamber as a result of abnormal combustion - usually detonation. Knock is usually caused by a loose or worn bearing. Also referred to as detonation, pinging and spark knock. Connecting rod or main bearing knocks are created by too much oil clearance or insufficient lubrication.

## L

**Lands** - The portions of metal between the piston ring grooves.

**Lapping the valves** - Grinding a valve face and its seat together with lapping compound.

**Lash** - The amount of free motion in a gear train, between gears, or in a mechanical assembly, that occurs before movement can begin. Usually refers to the lash in a valve train.

**Lifter** - The part that rides against the cam to transfer motion to the rest of the valve train.

## M

**Machining** - The process of using a machine to remove metal from a metal part.

**Main bearings** - The plain, or babbitt, bearings that support the crankshaft.

**Main bearing caps** - The cast iron caps, bolted to the bottom of the block, that support the main bearings.

## O

**O.D.** - Outside diameter.

**Oil gallery** - A pipe or drilled passageway in the engine used to carry engine oil from one area to another.

**Oil ring** - The lower ring, or rings, of a piston; designed to prevent excessive amounts of oil from working up the cylinder walls and into the combustion chamber. Also called an oil-control ring.

**Oil seal** - A seal which keeps oil from leaking out of a compartment. Usually refers to a dynamic seal around a rotating shaft or other moving part.

**O-ring** - A type of sealing ring made of a special rubberlike material; in use, the O-ring is compressed into a groove to provide the sealing action.

**Overhaul** - To completely disassemble a unit, clean and inspect all parts, reassemble it with the original or new parts and make all adjustments necessary for proper operation.

## P

**Pilot bearing** - A small bearing installed in the center of the flywheel (or the rear end of the crankshaft) to support the front end of the input shaft of the transmission.

**Pip mark** - A little dot or indentation which indicates the top side of a compression ring.

**Piston** - The cylindrical part, attached to the connecting rod, that moves up and down in the cylinder as the crankshaft rotates. When the fuel charge is fired, the piston transfers the force of the explosion to the connecting rod, then to the crankshaft.

**Piston pin (or wrist pin)** - The cylindrical and usually hollow steel pin that passes through the piston. The piston pin fastens the piston to the upper end of the connecting rod.

**Piston ring** - The split ring fitted to the groove in a piston. The ring contacts the sides of the ring groove and also rubs against the cylinder wall, thus sealing space between piston and wall. There are two types of rings: Compression rings seal the compression pressure in the combustion chamber; oil rings scrape excessive oil off the cylinder wall.

**Piston ring groove** - The slots or grooves cut in piston heads to hold piston rings in position.

**Piston skirt** - The portion of the piston below the rings and the piston pin hole.

**Plastigage** - A thin strip of plastic thread, available in different sizes, used for measuring clearances. For example, a strip of plastigage is laid across a bearing journal and mashed as parts are assembled. Then parts are disassembled and the width of the strip is measured to determine clearance between journal and bearing. Commonly used to measure crankshaft main-bearing and connecting rod bearing clearances.

**Press-fit** - A tight fit between two parts that requires pressure to force the parts together. Also referred to as drive, or force, fit.

**Prussian blue** - A blue pigment; in solution, useful in determining the area of contact between two surfaces. Prussian blue is commonly used to determine the width and location of the contact area between the valve face and the valve seat.

## R

**Race (bearing)** - The inner or outer ring that provides a contact surface for balls or rollers in bearing.

**Ream** - To size, enlarge or smooth a hole by using a round cutting tool with fluted edges.

**Ring job** - The process of reconditioning the cylinders and installing new rings.

**Runout** - Wobble. The amount a shaft rotates out-of-true.

## S

**Saddle** - The upper main bearing seat.

**Scored** - Scratched or grooved, as a cylinder wall may be scored by abrasive particles moved up and down by the piston rings.

**Scuffing** - A type of wear in which there's a transfer of material between parts moving against each other; shows up as pits or grooves in the mating surfaces.

**Seat** - The surface upon which another part rests or seats. For example, the valve seat is the matched surface upon which the valve face rests. Also used to refer to wearing into a good fit; for example, piston rings seat after a few miles of driving.

**Short block** - An engine block complete with crankshaft and piston and, usually, camshaft assemblies.

**Static balance** - The balance of an object while it's stationary.

**Step** - The wear on the lower portion of a ring land caused by excessive side and back-clearance. The height of the step indicates the ring's extra side clearance and the length of the step projecting from the back wall of the groove represents the ring's back clearance.

**Stroke** - The distance the piston moves when traveling from top dead center to bottom dead center, or from bottom dead center to top dead center.

**Stud** - A metal rod with threads on both ends.

## T

**Tang** - A lip on the end of a plain bearing used to align the bearing during assembly.

**Tap** - To cut threads in a hole. Also refers to the fluted tool used to cut threads.

**Taper** - A gradual reduction in the width of a shaft or hole; in an engine cylinder, taper usually takes the form of uneven wear, more pronounced at the top than at the bottom.

**Throws** - The offset portions of the crankshaft to which the connecting rods are affixed.

**Thrust bearing** - The main bearing that has thrust faces to prevent excessive endplay, or forward and backward movement of the crankshaft.

**Thrust washer** - A bronze or hardened steel washer placed between two moving parts. The washer prevents longitudinal movement and provides a bearing surface for thrust surfaces of parts.

**Tolerance** - The amount of variation permitted from an exact size of measurement. Actual amount from smallest acceptable dimension to largest acceptable dimension.

## U

**Umbrella** - An oil deflector placed near the valve tip to throw oil from the valve stem area.

**Undercut** - A machined groove below the normal surface.

**Undersize bearings** - Smaller diameter bearings used with re-ground crankshaft journals.

## V

**Valve grinding** - Refacing a valve in a valve-refacing machine.

**Valve train** - The valve-operating mechanism of an engine; includes all components from the camshaft to the valve.

**Vibration damper** - A cylindrical weight attached to the front of the crankshaft to minimize torsional vibration (the twist-untwist actions of the crankshaft caused by the cylinder firing impulses). Also called a harmonic balancer.

## W

**Water jacket** - The spaces around the cylinders, between the inner and outer shells of the cylinder block or head, through which coolant circulates.

**Web** - A supporting structure across a cavity.

**Woodruff key** - A key with a radiused backside (viewed from the side).

## Notes

**Section**

1 General information
2 Antifreeze - general information
3 Thermostat - check and replacement
4 Engine cooling fans - check and replacement
5 Radiator and coolant expansion tank - removal and installation
6 Water pump - check and replacement
7 Coolant temperature gauge sending unit - check and replacement
8 Blower motor and control module - removal and installation
9 Heater and air conditioning control assembly - removal and installation
10 Heater core - removal and installation
11 Air conditioning and heating system - check and maintenance
12 Air conditioning accumulator - removal and installation
13 Air conditioning compressor - removal and installation
14 Air conditioning condenser - removal and installation
15 Air conditioning expansion (orifice) tube - removal and installation
16 Air conditioning low pressure switch and pressure sensor - removal and installation

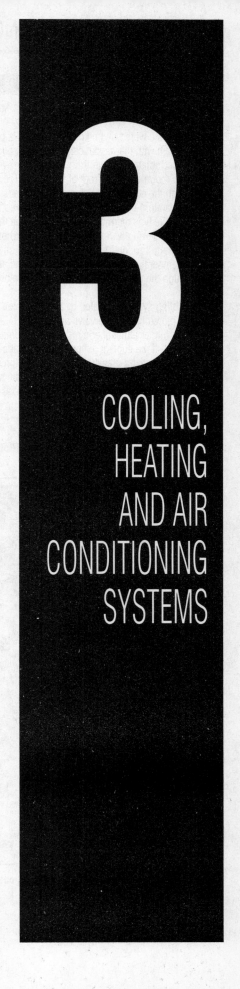

# 3

# COOLING, HEATING AND AIR CONDITIONING SYSTEMS

# 3-2 COOLING, HEATING AND AIR CONDITIONING SYSTEMS

## 1  General information

▶ Refer to illustration 1.1

All vehicles covered by this manual employ a pressurized engine cooling system with thermostatically controlled coolant circulation (see illustration). Coolant is drawn from the radiator by an impeller-type water pump mounted at the front of the block. The coolant is then circulated through the engine block, the cylinder heads and, on V6 engines, the intake manifold before it's redirected back into the radiator.

A wax pellet type thermostat is located in the thermostat housing on the engine. During warm up, the closed thermostat prevents coolant from circulating through the radiator. When the engine reaches normal operating temperature, the thermostat opens and allows hot coolant to travel through the radiator, where it is cooled before returning to the engine.

The cooling system is sealed by the expansion tank cap, which contains a blow-off pressure valve and a vacuum atmospheric valve. By maintaining higher atmospheric pressure, it increases the boiling point of the coolant. If the coolant temperature goes above this increased boiling point, the extra pressure in the system forces the cap valve off its seat and allows excess pressure to escape the system.

The expansion tank serves as both the point at which fresh coolant is added to the cooling system to maintain the proper fluid level and as the point where coolant is circulated to allow air to bubbles out of the system.

V8 engines are equipped with an Overheat Protection Mode operating system to protect the engine from damage caused by severe overheating. When the computer senses an overheat condition, an instrument panel warning light comes on that says "REDUCED ENGINE POWER." In this mode, the computer switches the firing of the individual coils On and Off at each cylinder to allow cooling cycles between the firing cycles. The engine will have a dramatic loss of power, but will allow vehicle operation in an emergency. If this light is On, find a safe place to get off the road as soon as possible, and allow the engine to cool thoroughly. Check the coolant level and inspect for a split hose or other obvious signs of coolant leakage. The engine oil will be ruined after this mode has operated, since unburned fuel will get into the oil. After fixing the overheating problem, change the oil and filter right away and reset the Oil Life Monitor (see Chapter 1).

The heating system works by directing air through the heater core mounted in the dash and to the interior of the vehicle by a system of ducts. Rear heating systems are equipped with a separate heater core located at the rear of the vehicle. Temperature is controlled by mixing heated air with fresh air, using a system of doors in the ducts, and a blower motor.

Air conditioning is an optional accessory, consisting of an evaporator core located under the dash, a condenser in front of the radiator, an accumulator in the engine compartment and a belt-driven compressor mounted at the front of the engine. Rear air conditioning systems are equipped with a separate evaporator core located at the rear of the vehicle.

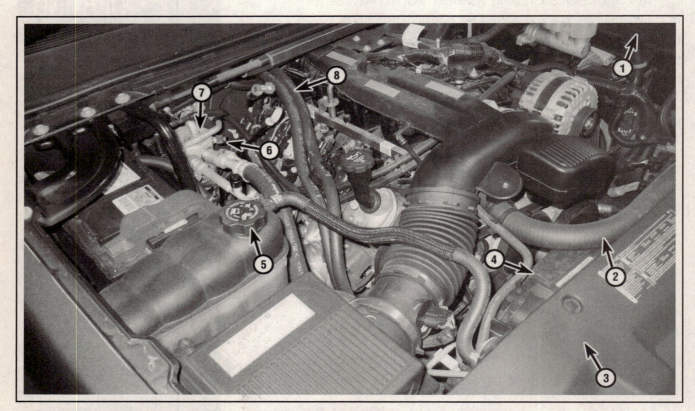

**1.1 Cooling and air conditioning system components - V8 Silverado model (engine cover removed)**

1. Fuse/relay box
2. Upper radiator hose
3. Radiator support/condenser cover
4. Engine cooling fans
5. Coolant expansion tank cap
6. Low side refrigerant charging port
7. Accumulator
8. Heater hoses

# COOLING, HEATING AND AIR CONDITIONING SYSTEMS

## 2 Antifreeze - general information

▶ Refer to illustration 2.4

**WARNING:**

Do not allow antifreeze to come in contact with your skin or painted surfaces of the vehicle. Rinse off spills immediately with plenty of water. Antifreeze is highly toxic if ingested. Never leave antifreeze lying around in an open container or in puddles on the floor; children and pets are attracted by it's sweet smell and may drink it. Check with local authorities about disposing of used antifreeze. Many communities have collection centers which will see that antifreeze is disposed of safely. Never dump used antifreeze on the ground or pour it into drains.

**CAUTION:**

The manufacturer recommends using only DEX-COOL coolant for these systems. DEX-COOL is a long-lasting coolant designed for 100,000 miles or 5 years. Never mix green-colored ethylene glycol antifreeze and orange-colored DEX-COOL silicate-free coolant because doing so will destroy the efficiency of the DEX-COOL.

The cooling system should be filled with a water/ethylene glycol based antifreeze solution which will prevent freezing down to at least -20-degrees F (even lower in cold climates). It also provides protection against corrosion and increases the coolant boiling point.

The cooling system should be drained, flushed and refilled at least every other year (see Chapter 1). The use of antifreeze solutions for periods of longer than two years is likely to cause damage and encourage the formation of rust and scale in the system. However, these models are filled with a new, long-life DEX-COOL coolant, which the manufacturer claims is good for five years.

Before adding antifreeze, check all hose connections, because antifreeze tends to leak through very minute openings. Engines don't normally consume coolant, so if the level goes down, find the cause and correct it.

The exact mixture of antifreeze to water which you should use depends on the relative weather conditions. The mixture should contain at least 50-percent antifreeze, but should never contain more than 70-percent antifreeze. Consult the mixture ratio chart on the antifreeze container before adding coolant. Hydrometers are available at most auto parts stores to test the coolant (see illustration). Always use antifreeze which meets the vehicle manufacturer's specifications.

2.4 An inexpensive hydrometer can be used to test the condition of your coolant

## 3 Thermostat - check and replacement

### CHECK

1  Before assuming the thermostat is to blame for a cooling system problem, check the coolant level, drivebelt tension (see Chapter 1) and temperature gauge (or light) operation.

2  If the engine seems to be taking a long time to warm up (based on heater output or temperature gauge operation), the thermostat is probably stuck open. Replace the thermostat with a new one.

3  If the engine runs hot, use your hand to check the temperature of the upper radiator hose (V6 engine) or the lower radiator hose (V8 engine). If the hose isn't hot, but the engine is, the thermostat is probably stuck closed, preventing the coolant inside the engine from escaping to the radiator. Replace the thermostat.

**CAUTION:**

Don't drive the vehicle without a thermostat. The computer may stay in open loop and emissions and fuel economy will suffer.

4  If the hose is hot, it means the coolant is flowing and the thermostat is open. Consult the *Troubleshooting* Section at the front of this manual for cooling system diagnosis.

# 3-4 COOLING, HEATING AND AIR CONDITIONING SYSTEMS

3.18 Thermostat housing cover bolts - V8 engines

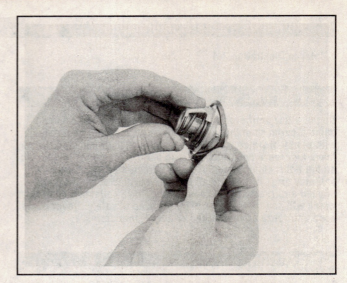

3.21 Install a new rubber seal around the thermostat

## REPLACEMENT

### ⚠ WARNING:

**Wait until the engine is completely cool before beginning this procedure.**

5 Disconnect the cable from the negative terminal of the battery (see Chapter 5, Section 1).

6 Drain the cooling system (see Chapter 1). If the coolant is relatively new or in good condition, save it and reuse it. If it is to be replaced, see Section 2 for cautions about proper disposal of used anti-freeze.

### V6 engines

7 Remove the engine cover (see Chapter 2A) and the air intake duct and resonator assembly (see Chapter 4).

8 Remove the hose clamp, then detach the radiator hose from the thermostat housing cover. If the hose sticks, grasp it near the end with a pair of adjustable pliers and twist it to break the seal, then pull it off. If the hose is old or deteriorated, cut it off and install a new one.

9 If the outer surface of the thermostat housing cover that mates with the hose is deteriorated (corroded, pitted, etc.), it may be damaged further by hose removal. If it is, the thermostat housing cover will have to be replaced.

10 Remove the bolts and detach the thermostat cover. If the cover is stuck, tap it with a soft-face hammer to jar it loose. Be prepared for some coolant to spill as the gasket seal is broken.

11 Note how it's installed (which end is facing up), then remove the thermostat.

12 Clean the mating surfaces of the intake manifold and thermostat housing cover.

13 Install a new rubber O-ring around the thermostat (see illustration 3.21).

14 Install the thermostat into the intake manifold and make sure the correct end faces out - the spring is directed toward the engine.

15 Reattach the thermostat housing cover to the intake manifold and tighten the bolts to the torque listed in this Chapter's Specifications.

### V8 engines

♦ Refer to illustrations 3.18 and 3.21

16 Remove the air intake duct and resonator assembly (see Chapter 4).

17 Follow the lower radiator hose to the engine, then disconnect it. If the outer surface of the thermostat housing cover that mates with the hose is deteriorated (corroded, pitted, etc.) it may be damaged further by hose removal. If it is, the thermostat housing cover will have to be replaced.

18 Remove the thermostat housing cover from the engine (see illustration).

19 Note how it's installed (which end is facing out), then remove the thermostat.

20 Clean the sealing surfaces on the water pump housing and the thermostat housing cover.

21 Install a new rubber O-ring around the thermostat (see illustration). Install the thermostat into the water pump housing, spring-end first.

22 Place the thermostat housing onto the water pump housing and install the bolts. Tighten the bolts to the torque listed in this Chapter's Specifications.

### All engines

23 The remainder of installation is the reverse of the removal procedure. Now would be a good time to check and replace the hoses and clamps (see Chapter 1).

24 Refer to Chapter 1 and refill the cooling system, then run the engine and check carefully for leaks.

25 Repeat Steps 1 through 4 to be sure the repair corrected the problem.

# COOLING, HEATING AND AIR CONDITIONING SYSTEMS 3-5

## 4  Engine cooling fans - check and replacement

### ※ WARNING:

To avoid possible injury or damage, DO NOT operate the engine with a damaged fan. Do not attempt to repair fan blades - replace a damaged fan with a new one.

### CHECK

▶ Refer to illustrations 4.2 and 4.3

1  These models are equipped with two cooling fans mounted side-by-side behind the radiator. The PCM (Powertrain Control Module - engine computer) and three relays are used to operate the fans at Low or High speeds depending on requirements. The fans are protected by fuses inside the engine compartment's fuse/relay box.

2  If the engine is overheating and neither of the cooling fans operate, locate the fuses in the engine compartment fuse/relay box (see illustration). Remove the fuse(s) and check for continuity (see Chapter 12). If a fuse is blown, replace it and see if it blows again. If it does, trace and repair the source of the short circuit.

3  If the fuses are okay, check each fan by unplugging the fan motor electrical connector and applying battery power directly to the motor terminals with fused jumper wires (see illustration). When done correctly, the fan should come on. If a fan motor doesn't work, replace the motor.

4  If the fan motors are okay but are still not coming on when the engine gets hot, the fan relays might be defective. Locate the relays in the engine compartment's fuse/relay box (see illustration 4.2). You can pull each relay out and test it individually (see Chapter 12).

➡ Note: Relays are used to control a circuit by turning it on and off in response to a signal from the PCM. It's also likely that a failure in one of these circuits will result in a warning light in the instrument display.

5  If no obvious problems are found, have the cooling fan system diagnosed by a dealer service department or other qualified repair shop.

### REPLACEMENT

▶ Refer to illustrations 4.14, 4.15, 4.16 and 4.17

### ※ WARNING:

Wait until the engine is completely cool before beginning this procedure.

6  Disconnect the cable from the negative terminal of the battery (see Chapter 5, Section 1).

7  Drain the cooling system (see Chapter 1). If the coolant is relatively new or in good condition, save it and reuse it. If it is to be replaced, see Section 2 for cautions about proper disposal of used antifreeze.

8  Remove the intake air duct and resonator assembly (see Chapter 4).

9  Disconnect the hose clip from the fan shroud and remove the upper radiator hose from the radiator.

10  Detach the coolant air bleed hose from its clip on the fan shroud (see illustration 5.5).

11  Disconnect the fan motor electrical connectors (see illustration 4.3). Remove any hoses or wiring harnesses that may be attached to the fan shroud.

12  Disconnect the transmission cooler line bracket bolts from the fan shroud.

13  Disconnect the engine oil cooler line mounting clips, if equipped, and position the engine oil cooler lines off to the side.

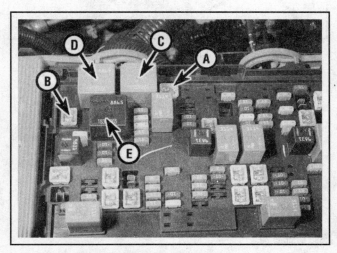

4.2  The engine compartment fuse/relay box with fan relay and fuse details (V8 engine model shown):

- A  Fan fuse 1
- B  Fan fuse 2
- C  Fan relay LO
- D  Fan relay HI
- E  Fan relay CNTRL

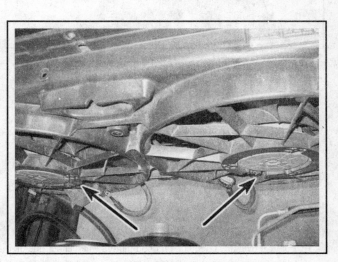

4.3  To test either fan motor, disconnect the electrical connector and use jumper wires to connect the fan directly to the battery and ground - if the fan still doesn't work, replace the motor

## 3-6 COOLING, HEATING AND AIR CONDITIONING SYSTEMS

4.14 Location of the engine cooling fan/shroud mounting bolts

4.15 Separate the cooling fan/shroud from the radiator and lift it out of the engine compartment

4.16 Slide the retaining clip off the motor shaft and lift the fan off the motor

4.17 Remove the cooling fan motor mounting bolts

14 Remove the cooling fan/shroud assembly mounting bolts (see illustration).

15 Remove the cooling fan/shroud assembly by lifting it up and out of the engine compartment (see illustration).

16 Remove the cooling fan retainer clips (see illustration) and separate the fan blade(s) from the fan motor(s).

17 Remove the mounting bolts and separate the cooling fan motor(s) from the shroud (see illustration).

18 Put tape over any openings on good motors when either one of them is removed to protect them from debris.

19 After thoroughly cleaning all debris from the shroud, remove the protective tape from the motor(s).

20 Install the fan motor mounting bolts and tighten them securely.

21 Turn the fan blade and confirm that it is correctly installed.

22 The remainder of installation is the reverse of the removal procedure. Tighten all fasteners securely.

23 Refill the cooling system and check the transmission fluid level (see Chapter 1).

# COOLING, HEATING AND AIR CONDITIONING SYSTEMS 3-7

## 5 Radiator and coolant expansion tank - removal and installation

**WARNING:**
Wait until the engine is completely cool before beginning this procedure.

### RADIATOR

♦ Refer to illustrations 5.3a, 5.3b, 5.4, 5.5 and 5.6

➥Note: *The vehicles covered by this manual use spring type radiator hose clamps. If you decide to reuse them, make sure that the hose is installed on a connection that is clean and dry.* Don't try to reuse these clamps on aftermarket hoses. Replace them with conventional worm drive type clamps.

1 Disconnect the cable from the negative terminal of the battery (see Chapter 5, Section 1).
2 Drain the cooling system (see Chapter 1). If the coolant is relatively new or in good condition, save it and reuse it. If it is to be replaced, see Section 2 for cautions about proper handling of used antifreeze.
3 Remove the radiator support cover (see illustrations).
4 Disconnect the upper radiator hose from the radiator (see illustration).
5 Disconnect the coolant air bleed pipe and the lower radiator hose from the radiator (see illustration).

5.3a Remove the plastic fasteners to lift off the radiator support cover (pick-up models)

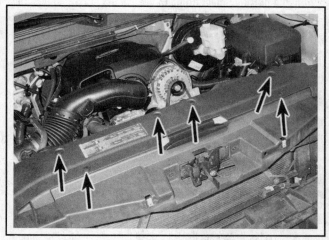

5.3b On SUV models, remove these fasteners to remove the radiator support cover

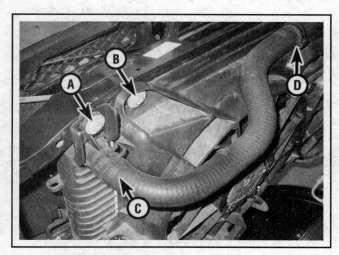

5.4 Radiator details - left side

- A  Radiator mounting bolt
- B  Fan shroud mounting bolt
- C  Upper radiator hose
- D  Hose support clamp

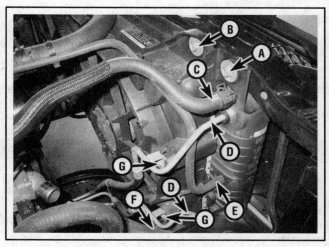

5.5 Radiator details - right side

- A  Radiator mounting bolt
- B  Fan shroud mounting bolt
- C  Expansion tank hose
- D  Transmission fluid cooler lines
- E  Coolant air bleed hose
- F  Lower radiator hose
- G  Transmission cooler line bracket

# 3-8 COOLING, HEATING AND AIR CONDITIONING SYSTEMS

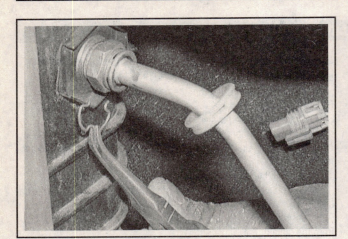

**5.6 Unsnap the plastic collar, then pry off the retaining clip from the quick-connect fitting to disconnect the transmission and engine oil cooler lines**

**5.16 Expansion tank mounting details (V8 shown, V6 similar)**

A  Expansion tank coolant hoses
B  Expansion tank mounting bolts and brackets

6  Disconnect the transmission cooler lines from the right side of the radiator (see illustration). To disconnect the lines from the radiator, simply unsnap the plastic collar from the quick-connect fitting, then pry off the quick-connect fitting retaining clip and remove the lines. Plug the ends of the lines to prevent fluid from leaking out after you disconnect them. Have a drip pan ready to catch any spills. Always be sure to inspect the O-rings on the cooler lines before reinstallation.

➥**Note: Do not remove the clips by pulling straight out. Hold one side in with your fingers while using a pick (with a bent tip) to pull the other side out, then rotate the clip off. Install clips the same way, not straight on.**

7  If equipped with an external engine oil cooler, disconnect the engine oil cooler lines from the left side of the radiator and the shroud. The quick-connect fittings are identical to the transmission cooler line fittings (see illustration 5.6).

8  Remove the engine cooling fans (see Section 4).

9  Remove the radiator mounting bolts (see illustrations 5.4 and 5.5) and lift the radiator from the engine compartment.

10  Prior to installation of the radiator, replace any damaged radiator hoses and hose clamps.

11  Radiator installation is the reverse of removal. Tighten the bolts to the torque listed in this Chapter's Specifications. When installing the radiator, make sure that the radiator seats properly in the lower saddles and that the upper brackets are secure. Install the cooler line retaining clips onto the quick connect fitting before installing the lines, then snap the cooler lines into place on the quick connect fittings. Be sure to reinstall the plastic collars on the quick connect fittings as they lock the retaining clip in place.

12  After installation, refill the cooling system (see Chapter 1), then check the engine oil and automatic transmission fluid levels.

## COOLANT EXPANSION TANK

▶ Refer to illustration 5.16

13  Disconnect the cable from the negative terminal of the battery (see Chapter 5, Section 1).

14  Drain the cooling system as described in Chapter 1 until the expansion tank is empty. Refer to the coolant **Warning** in Section 2.

15  Remove the air filter housing (see Chapter 4).

16  Detach the coolant hoses from the expansion tank. Unscrew the mounting fasteners, then remove the expansion tank (see illustration).

17  Prior to installation, make sure the tank is clean and free of debris which could be drawn into the radiator (wash it with soapy water and a brush if necessary, then rinse thoroughly).

18  Installation is the reverse of removal. Refill the cooling system (see Chapter 1) and check for leaks.

## 6  Water pump - check and replacement

**❄❄ WARNING:**

**Wait until the engine is completely cool before beginning this procedure.**

### CHECK

1  A failure in the water pump can cause serious engine damage due to overheating.

2  If a failure occurs in the shaft seal on the water pump, a coolant leak can usually be seen coming from that area.

3  Most water pumps are equipped with small weep or vent holes. Although it might seem difficult, it is possible to check these holes for leaks using a flashlight and a mirror. In most cases, the leak can be detected from coolant streams or drips near the bottom of the water pump housing.

4  If the water pump shaft bearings fail, there may be a howling sound near the water pump while it's running. With the engine off, shaft wear can be felt if the water pump pulley is rocked up-and-down. Don't mistake drivebelt slippage, which causes a squealing sound, for water

# COOLING, HEATING AND AIR CONDITIONING SYSTEMS

6.11 Location of the upper radiator hose on the water pump (V8 engine)

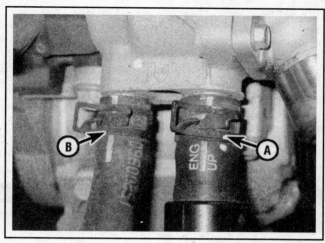

6.12 Disconnect the expansion tank hose (A) and the heater hose (B) on the V8 engine

6.15a Water pump mounting bolts - V6 engine

6.15b Water pump mounting bolts - V8 engines

pump bearing failure.

5  A quick water pump performance check is to turn the heater on. If the pump is failing, it might not be able to efficiently circulate hot water all the way to the heater core as it should.

## REPLACEMENT

6  Disconnect the cable from the negative terminal of the battery (see Chapter 5, Section 1).

7  Drain the coolant (see Chapter 1). Remove the air intake duct and resonator assembly (see Chapter 4).

8  Remove the drivebelt (see Chapter 1).

### V6 engines

9  Remove the water pump pulley.

➡ **Note:** *To detach the pulley on V6 engines, you will need a strap wrench or two-pin spanner to hold the pulley while the bolts are removed.*

10  Remove the radiator hoses and the expansion tank hose from the water pump.

### V8 engines

▶ **Refer to illustrations 6.11 and 6.12**

11  Remove the water pump inlet hose (see illustration).

12  Remove the water pump outlet hose at the thermostat (see Section 3) and the heater and expansion tank hoses (see illustration).

13  If necessary for access, detach the coolant air bleed hose from the engine.

14  If a new water pump is to be installed, remove the thermostat housing (see Section 3).

### All models

▶ **Refer to illustrations 6.15a and 6.15b**

15  Unbolt the water pump (see illustrations). It may be necessary to tap the pump with a soft-face hammer to break the gasket seal.

# 3-10 COOLING, HEATING AND AIR CONDITIONING SYSTEMS

16 Remove all traces of the old gasket seal from the mounting surface on the engine. Do the same on the water pump if the same pump is going to be re-installed.

17 Clean the mounting bolt threads and threaded holes on the mounting surface to remove corrosion and sealant, if necessary.

18 Compare the replacement pump with the old one to make sure that they're identical.

19 Place the gaskets and water pump into position. Use caution to ensure that the gasket doesn't slip out of position. Install the mounting bolts until they are all finger tight.

➡ **Note: On V6 engines, use RTV sealant on the water pump bolt threads.**

20 Tighten the bolts to the torque listed in this Chapter's Specifications.

21 The remainder of installation is the reverse of the removal procedure.

22 Refill the engine with coolant (see Chapter 1). Run the engine and check for leaks.

## 7  Coolant temperature gauge sending unit - check and replacement

### CHECK

1 The coolant temperature indicator system is composed of a temperature gauge mounted in the dash and a coolant temperature sensor mounted on the engine. This coolant temperature sensor doubles as an information sensor for the fuel and emissions systems (see Chapter 6).

2 If an overheating indication occurs, check the coolant level in the system (see Chapter 1).

3 Check the operation of the coolant temperature sensor (see Chapter 6). If the sensor is defective, replace it with a new one.

4 If the coolant temperature sensor is good, have the temperature gauge checked by a dealer service department. This test will require a scan tool to access the information as it is processed by the on-board computer.

### REPLACEMENT

5 Refer to Chapter 6 for the Engine Coolant Temperature (ECT) sensor replacement procedure.

## 8  Blower motor and control module - removal and installation

### ⚠ WARNING:

**These models have airbags. Always disable the airbag system before working in the vicinity of any airbag system component to avoid the possibility of accidental deployment of the airbag, which could cause personal injury (see Chapter 12).**

➡ **Note: These models are equipped with a control module that controls activation and fan blower speeds. The control module is mounted on the HVAC housing near the blower fan. Some models are equipped with rear auxiliary heating and air conditioning systems that have the control module mounted on the HVAC housing in the rear of the vehicle. Fuses for this system can be checked, but due to the use of an integrated electronic control module and Body Control Module that can only be tested with specialized equipment, it will be necessary to take these vehicles to dealer service department or other qualified repair shop to have the blower motor circuit checked in the event of a problem.**

### BLOWER MOTOR

#### Front blower motor (all models)

▸ **Refer to illustration 8.2 and 8.3**

1 Disconnect the cable from the negative terminal of the battery (see Chapter 5, Section 1). Working under the instrument panel on the passenger's side, remove the trim panel (see Chapter 11, illustration 26.44).

2 Disconnect the electrical connector from the blower motor (see illustration). Remove the mounting screw.

3 Pull down the retaining tab and rotate the blower motor to remove it (see illustration).

4 Installation is the reverse of removal.

#### Rear auxiliary blower motor

▸ **Refer to illustrations 8.5a, 8.5b, 8.5c, 8.5d, 8.5e, 8.5f, 8.5g, 8.5h and 8.6**

5 Remove the interior trim panels from the right rear of the vehicle (from behind the right rear door opening to the liftgate) for access to the rear HVAC unit (see illustrations).

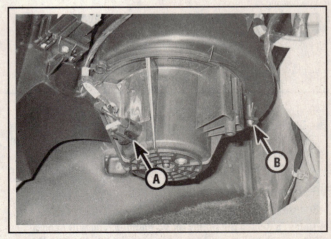

**8.2 Blower motor electrical connector (A) and mounting screw (B)**

# COOLING, HEATING AND AIR CONDITIONING SYSTEMS   3-11

8.3  Pull down the retaining tab and rotate the blower motor

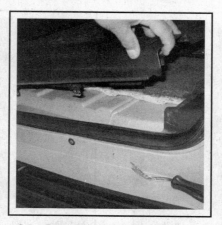

8.5a  Remove the liftgate sill plate

8.5b  Peel off the weatherstripping from the liftgate opening

8.5c  Carefully pry off the headliner trim panel

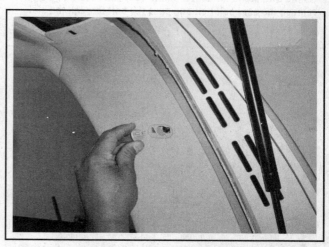

8.5d  Remove the cover and the screw underneath . . .

8.5e  . . . then separate the liftgate opening trim panel from the body

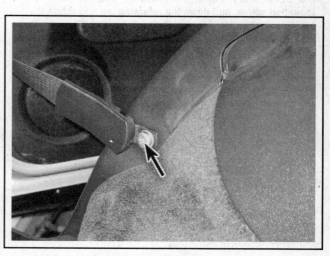

8.5f  Remove the seat belt mounting bolt

## 3-12 COOLING, HEATING AND AIR CONDITIONING SYSTEMS

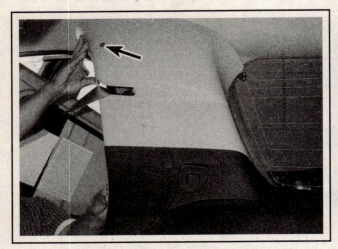

8.5g  Remove the right rear door sill plate, remove the cover and screw from the top of the "C" pillar trim panel, then pry off the trim panel for access to the seat belt mounting bolt; remove the seat belt bolt and pass the belt through the slot

8.5h  Remove the right rear quarter trim panel for access to the rear HVAC unit

8.6  Disconnect the electrical connector, remove the mounting screw, pull down the retaining tab and rotate the blower motor counterclockwise

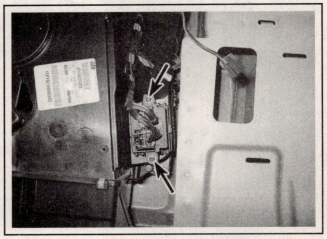

8.17  Cooling fan control module mounting screws on the rear A/C heater assembly

6   Disconnect the electrical connector from the blower motor (see illustration).
7   Remove the retaining screw.
8   Pull down the retaining tab and rotate the blower motor counterclockwise to remove it.
9   Installation is the reverse of removal.

### CONTROL MODULE

#### Front blower motor module (all models)

10   Disconnect the cable from the negative terminal of the battery (see Chapter 5, Section 1).
11   Working under the instrument panel on the passenger's side, remove the trim panel (see Chapter 11, illustration 26.44).
12   Disconnect the control module harness connector.
13   Remove the control module mounting screws (see illustration 8.4) and separate the module from the blower housing.
14   Installation is the reverse of removal.

#### Rear auxiliary blower motor module

▶ Refer to illustration 8.17

15   Remove the right rear quarter trim panel (see illustrations 8.5a through 8.5h)
16   Disconnect the control module harness connector.
17   Remove the control module mounting screws (see illustration) and separate the module from the blower housing.
18   Installation is the reverse of removal.

# COOLING, HEATING AND AIR CONDITIONING SYSTEMS 3-13

## 9  Heater and air conditioning control assembly - removal and installation

### ※※ WARNING:

These models have airbags. Always disable the airbag system before working in the vicinity of any airbag system component to avoid the possibility of accidental deployment of the airbag, which could cause personal injury (see Chapter 12).

1  Disconnect the cable from the negative terminal of the battery (see Chapter 5, Section 1).

### FRONT HEATER CONTROLS

▶ Refer to illustration 9.3

2  Remove the instrument panel center trim panel on the Standard Dash or Premium Dash to access to the heater/air conditioning control mounting bolts (see Chapter 11).

3  Remove the mounting bolts (see illustration) from the heater and air conditioning control assembly.

4  Pull the unit from the dash. It can be pulled out just far enough to allow disconnection of the electrical connectors. Use a small screwdriver to release the clips.

5  Installation is the reverse of removal.

### REAR AUXILIARY HEATER CONTROLS

▶ Refer to illustration 9.6

6  Pry out the trim panel from the back of the console, then disconnect the electrical connectors and remove the control mounting screws (see illustration).

7  Installation is the reverse of removal.

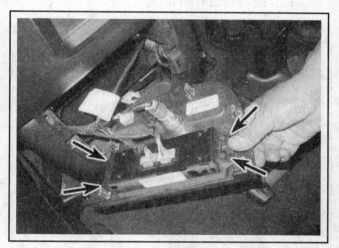

9.3  Heater and A/C control assembly mounting bolts (standard dash shown, premium dash similar)

9.6  Remove the mounting screws and remove the rear heater and A/C control assembly from the console

## 10  Heater core - removal and installation

### ※※ WARNING 1:

These models have airbags. Always disable the airbag system before working in the vicinity of any airbag system component to avoid the possibility of accidental deployment of the airbag, which could cause personal injury (see Chapter 12).

### ※※ WARNING 2:

The air conditioning system is under high pressure. DO NOT loosen any fittings or remove any components until after the system has been discharged. Air conditioning refrigerant must be properly discharged into an EPA-approved container at a dealership service department or an automotive air conditioning facility. Always wear eye protection when disconnecting air conditioning system fittings.

### ※※ WARNING 3:

Wait until the engine is completely cool before beginning this procedure.

1  Have the air conditioning system discharged by a dealership service department or an automotive air conditioning facility (see **Warning** 2 above).

# 3-14 COOLING, HEATING AND AIR CONDITIONING SYSTEMS

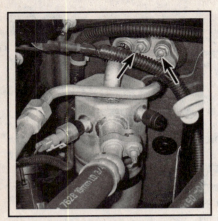

**10.3** Have the air conditioning system discharged by a dealership service department or an automotive air conditioning facility, then detach the air conditioning lines from the evaporator core fittings at the firewall

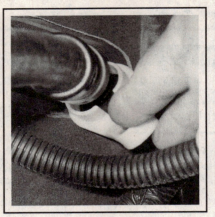

**10.4a** Disconnect the heater core hoses at the engine compartment firewall using a special offset disconnect tool

**10.4b** After the special disconnect tool is slid into the spring lock mechanism, pull the heater core hose back and off the heater core pipe

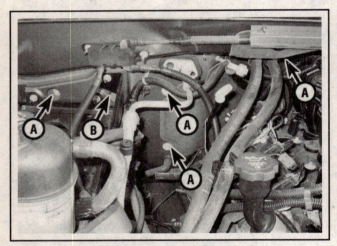

**10.10a** Heater/air conditioning unit mounting details from inside the right side (passenger side) engine compartment

| A | Mounting bolts | B | Mounting stud and nut |

**10.10b** Heater/air conditioning unit mounting details from inside the left side (driver's side) engine compartment

| A | Mounting bolt | B | Mounting stud and nut |

## FRONT HEATER CORE

▶ Refer to illustrations 10.3, 10.4a, 10.4b, 10.10a, 10.10b, 10.10c and 10.12

2  Disconnect the cable from the negative terminal of the battery (see Chapter 5, Section 1).

3  Drain the cooling system (see Chapter 1). Detach the air conditioning lines from the evaporator core fittings at the firewall (see illustration). Be sure to plug each refrigerant line to avoid contamination of the air conditioning system.

4  Disconnect the heater hoses from the heater core pipes on the engine side of the firewall (see illustrations) and plug the open fittings.

5  Remove the engine cover (see Chapter 2A or 2B).
6  Remove the battery (see Chapter 5).
7  Remove the accumulator (see Section 12).
8  Place the instrument panel in the service position (see Chapter 11, Section 28).
9  Disconnect the drain tube at the firewall (see illustration 11.1).
10  Working in the engine compartment, remove the fasteners securing the heating/air conditioning unit to the firewall (see illustrations).
11  Working back inside the passenger compartment, remove the electrical connectors and any ground straps, then remove the heating/air conditioning unit from the vehicle.
12  Remove the heater core cover (see illustration) and slide the heater core out of the housing.

# COOLING, HEATING AND AIR CONDITIONING SYSTEMS 3-15

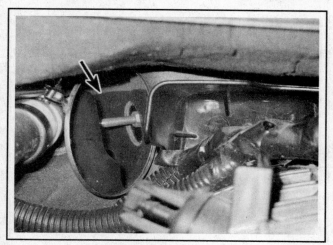

10.10c The stud and nut (viewed in mirror) can be accessed from behind the intake manifold sight shield using an open-end wrench. Note here that the manufacturer recommends using a hole saw in the sight shield to access the stud and nut for the heater and A/C unit removal

10.12 The heater core cover is at the top rear of the heater/air conditioning unit

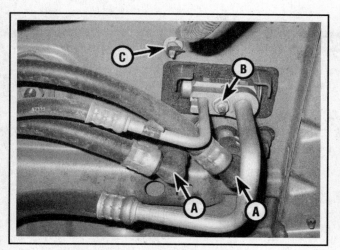

10.17 The heater hoses (A) and the fastener for the refrigerant line fittings (B) on SUV models are accessible from under the vehicle. C is the lower mounting nut for the heater/air conditioning unit

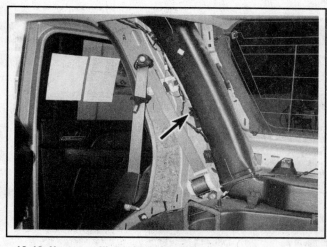

10.19 Upper auxiliary air duct fastener

13  Installation is the reverse of removal.

→Note: *When reinstalling the heater core, make sure any original insulating/sealing materials are in place around the heater core pipes, heater core and heating/air conditioning unit.*

14  Refill the cooling system (see Chapter 1). Have the air conditioning system charged by the shop that discharged it.

15  Start the engine and check for proper operation and leaks.

## REAR (AUXILIARY) HEATER CORE

▸ Refer to illustrations 10.17, 10.19 and 10.21

16  Remove the right rear quarter trim panel (see illustrations 8.5a through 8.5h)

17  Detach the air conditioning lines from the rear evaporator core fittings on the underside of the vehicle (see illustration). Be sure to plug each refrigerant line to avoid contamination of the air conditioning system. Use special release tools to remove the quick-connect fittings.

18  Also working from the underside of the vehicle, clamp off the heater hoses leading to the rear heater core, then disconnect the hoses from the inlet and outlet fittings and plug the open fittings. To disconnect the hose fittings, squeeze the plastic retainer tabs, then pull off the hoses.

19  Remove the upper auxiliary air duct (see illustration).

20  Remove the rear heating/air conditioning unit lower mounting nut from under the vehicle (see illustration 10.17).

# 3-16 COOLING, HEATING AND AIR CONDITIONING SYSTEMS

21 Remove the electrical connectors and wiring harness retaining straps, then detach the rear heating/air conditioning unit mounting upper screws and remove it from the vehicle (see illustration). Follow the remaining steps to remove the heater core from the rear heating/air conditioning unit.

22 Remove the heater core cover located on the bottom of the heater/air conditioning unit.

23 Remove the heater core from the housing.

24 Installation is the reverse of removal.

➥ **Note: When reinstalling the heater core, make sure any original insulating/sealing materials are in place around the heater core pipes, heater core and housing.**

25 Check the coolant level, adding coolant as necessary (see Chapter 1).

26 Start the engine and check for proper operation. Recheck the coolant level. Have the air conditioning system charged by the shop that discharged it.

**10.21 Location of the rear heater and A/C unit upper mounting bolts (the lower mounting nut is accessible from under the vehicle)**

## 11 Air conditioning and heating system - check and maintenance

▸ Refer to illustration 11.1

**✻✻ WARNING:**

**The air conditioning system is under high pressure. Do not loosen any hose fittings or remove any components until after the system has been discharged by an air conditioning technician. Always wear eye protection when disconnecting air conditioning system fittings.**

1 The following maintenance checks should be performed on a regular basis to ensure the air conditioner continues to operate at peak efficiency.

  a) Check the compressor drivebelt. If it's worn or deteriorated, replace it (see Chapter 1).
  b) Check the system hoses. Look for cracks, bubbles, hard spots and deterioration. Inspect the hoses and all fittings for oil bubbles and seepage. If there's any evidence of wear, damage or leaks, replace the hose(s).
  c) Inspect the condenser fins for leaves, bugs and other debris. Use a fin comb or compressed air to clean the condenser.
  d) Make sure the system has the correct refrigerant charge.
  e) Check the evaporator housing drain tube (see illustration) for blockage.

2 It's a good idea to operate the system for about 10 minutes at least once a month, particularly during the winter. Long term non-use can cause hardening, and subsequent failure, of the seals.

3 Because of the complexity of the air conditioning system and the special equipment necessary to service it, in-depth troubleshooting and repairs are not included in this manual. However, simple checks and component replacement procedures are provided in this Chapter.

4 The most common cause of poor cooling is simply a low system refrigerant charge. If a noticeable drop in cool air output occurs, the following quick check will help you determine if the refrigerant level is low.

### CHECKING THE REFRIGERANT CHARGE

▸ Refer to illustration 11.7 and 11.8

5 Warm the engine up to normal operating temperature.

6 Place the air conditioning temperature selector at the coldest setting and the blower at the highest setting. Open the vehicle doors (to make sure the air conditioning system doesn't cycle off as soon as it cools the passenger compartment).

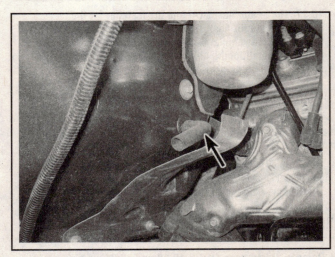

**11.1 Check that the evaporator housing drain tube at the firewall is clear of any blockage - the view here is from the right fenderwell**

# COOLING, HEATING AND AIR CONDITIONING SYSTEMS

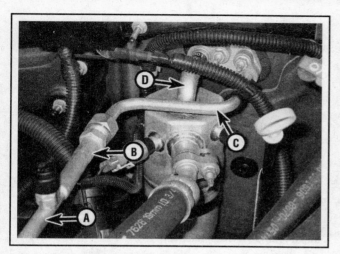

11.7 You can get a fairly good idea of the refrigerant charge by feeling the lines running to and from the evaporator:

- A  Liquid line - should be warm
- B  Orifice tube location
- C  Evaporator inlet line - should be cold
- D  Evaporator outlet line - should be cold (but won't be as cold as "C")

7  Feel the refrigerant lines before and after the orifice tube (expansion valve), and at the evaporator outlet. The liquid line (small diameter pipe) before the expansion valve should be warm, but directly after the orifice tube (the inlet to the evaporator) should be cold. The outlet (large diameter) pipe from the evaporator should also be cold (but it won't be quite as cold as the inlet) (see illustration).

8  Place a thermometer in the dashboard vent nearest the evaporator (see illustration) and operate the system until the indicated temperature is around 40 to 45-degrees F. If the ambient (outside) air temperature is very high, say 110-degrees F, the duct air temperature may be as high as 60-degrees F, but generally the air conditioning is 35 to 40-degrees F cooler than the ambient air.

➡ **Note: Humidity of the ambient air also affects the cooling capacity of the system. Higher ambient humidity lowers the effectiveness of the air conditioning system.**

## ADDING REFRIGERANT

♦ **Refer to illustrations 11.9 and 11.12**

9  Buy an air conditioning charging kit at an auto parts store. A charging kit includes a can of refrigerant, a tap valve and a short section of hose that can be attached between the tap valve and the system low side service valve (see illustration).

➡ **Note: Leak detection kits are also available and can help you pinpoint leaks in your air conditioning system.**

### ※※ CAUTION:

**There are two types of refrigerant used in automotive air conditioning systems; R-12 - which has been widely used on earlier models - and the more environmentally-friendly R-134a used in all models covered by this manual. These two refrigerants (and their appropriate refrigerant oils) are not compatible with each other and must never be mixed or components will be damaged. Use only R-134a refrigerant in the models covered by this manual.**

11.8 Insert a thermometer in the center duct while operating the air conditioning system - the output air should be 35 to 40 degrees F less than the ambient temperature, depending on humidity (but not lower than 40-degrees F)

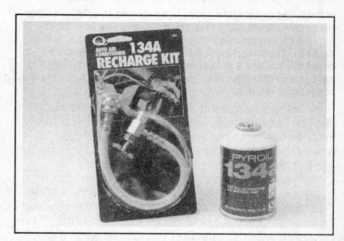

11.9 A basic charging kit for 134a systems is available at most auto parts stores - it must say 134a (not R-12) and so must the can of refrigerant

10  Hook up the charging kit by following the kit manufacturer's instructions.

### ※※ WARNING:

**DO NOT attempt to connect the charging kit hose to the system high side! The fittings on the charging kit are designed to fit only on the low side of the system.**

11  Back off the valve handle on the charging kit and screw the kit onto the refrigerant can, making sure first that the O-ring or rubber seal inside the threaded portion of the kit is in place.

### ※※ WARNING:

**Wear protective eyewear when dealing with pressurized refrigerant cans.**

## 3-18 COOLING, HEATING AND AIR CONDITIONING SYSTEMS

**11.12 Attach the refrigerant kit to the low-side charging port - the cap should be marked with an "L"**

**11.23 Insert the disinfectant can's nozzle through the RECIRC door above the blower motor at this location, and aim it toward the evaporator core**

12 Remove the dust cap from the low-side charging connection and attach the quick-connect fitting on the kit hose (see illustration).

13 Warm up the engine and turn on the air conditioner. Keep the charging kit hose away from the fan and other moving parts.

➡ **Note:** *The compressor needs to be running in order to charge the system. However, if your system is very low on refrigerant, the compressor may turn off or not come on at all. If this happens, disconnect the A/C low pressure switch electrical connector and use a jumper wire between the two terminals in the harness connector (see illustration 12.2). This will keep the compressor running.*

14 Turn the valve handle on the kit until the stem pierces the can, then back the handle out to release the refrigerant. You should be able to hear the rush of gas. Add refrigerant to the low side of the system until the compressor discharge line feels warm and the compressor inlet pipe feels cool. Allow stabilization time between each addition.

15 If you have an accurate thermometer, place it in the center air conditioning vent (see illustration 11.8) and note the temperature of the air coming out of the vent. A fully-charged system which is working correctly should cool down to about 40-degrees F. Generally, an air conditioning system will put out air that is 35 to 40-degrees F cooler than the ambient air. For example, if the ambient (outside) air temperature is very high (over 100 degrees F), the temperature of air coming out of the registers should be 60 to 70 degrees F.

16 When the can is empty, turn the valve handle to the closed position and release the connection from the low-side port. Replace the dust cap.

### ❈❈ CAUTION:
**Never add more than one can of refrigerant to the system. If more refrigerant than that is required, the system should be evacuated and leak tested.**

17 Remove the charging kit from the can and store the kit for future use with the piercing valve in the UP position, to prevent inadvertently piercing the can on the next use.

## HEATING SYSTEMS

18 If the carpet under the heater core is damp, or if antifreeze vapor or steam is coming through the vents, the heater core is leaking. Remove it (see Section 10) and install a new unit (most radiator shops will not repair a leaking heater core).

19 If the air coming out of the heater vents isn't hot, the problem could stem from any of the following causes:

a) *The thermostat is stuck open, preventing the engine coolant from warming up enough to carry heat to the heater core. Replace the thermostat (see Section 3).*
b) *There is a blockage in the system, preventing the flow of coolant through the heater core. Feel both heater hoses at the firewall. They should be hot. If one of them is cold, there is an obstruction in one of the hoses or in the heater core. Detach the hoses and back flush the heater core with a water hose. If the heater core is clear but circulation is impeded, remove the two hoses and flush them out with a water hose.*
c) *If flushing fails to remove the blockage from the heater core, the core must be replaced (see Section 10).*

## ELIMINATING AIR CONDITIONING ODORS

**Refer to illustration 11.23**

20 Unpleasant odors that often develop in air conditioning systems are caused by the growth of a fungus, usually on the surface of the evaporator core. The warm, humid environment there is a perfect breeding ground for mildew to develop.

21 The evaporator core on most vehicles is difficult to access, and dealerships have a lengthy, expensive process for eliminating the fungus by opening up the evaporator case and using a powerful disinfectant and rinse on the core until the fungus is gone. You can service your own system at home, but it takes something much stronger than basic household germ-killers or deodorizers.

22 Aerosol disinfectants for automotive air conditioning systems are available in most auto parts stores, but remember when shopping for them that the most effective treatments are also the most expensive. Make sure that the disinfectant can comes with a long spray hose. The basic procedure for using these sprays is to start by running the system in the RECIRC mode for ten minutes with the blower on its highest speed. Use the highest heat mode to dry out the system and keep the compressor from engaging by disconnecting the wiring connector at the compressor (see Section 13).

23 Change the blower motor setting to low and the temperature to the middle setting. Open the glove box door, push up on the door stop and pivot the door down. Guide the nozzle into the RECIRC door and towards the evaporator housing (see illustration), then spray according

# COOLING, HEATING AND AIR CONDITIONING SYSTEMS    3-19

to the manufacturer's recommendations. Follow the manufacturer's recommendations for the length of spray and waiting time between applications.

> **CAUTION:**
> Be careful not to get the hose caught in the blower motor fan that is just below the opening.

24 Once the evaporator has been cleaned, the best way to prevent the mildew from coming back again is to make sure your evaporator housing drain tube is clear (see illustration 11.1).

## 12  Air conditioning accumulator - removal and installation

### REMAL

▶ Refer to illustrations 12.2 and 12.4

> **WARNING:**
> The air conditioning system is under high pressure. DO NOT loosen any fittings or remove any components until after the system has been discharged. Air conditioning refrigerant must be properly discharged into an EPA-approved container at a dealership service department or an automotive air conditioning repair facility. Always wear eye protection when disconnecting air conditioning system fittings.

1  Have the air conditioning system discharged by an automotive air conditioning technician (see **Warning** above).
2  Disconnect the refrigerant inlet and outlet lines (see illustration). Cap or plug the open lines immediately to prevent the entry of dirt or moisture.
3  Disconnect the electrical connector from the A/C low pressure switch.
4  Remove the mounting bracket nut (see illustration) and maneuver the accumulator out of the engine compartment.

### INSTALLATION

5  If you are replacing the accumulator with a new one, add two ounces of fresh refrigerant oil to the new unit (oil must be R-134a compatible).
6  Place the new accumulator into position in the bracket. Install the accumulator assembly onto the bulkhead.
7  Install the inlet and outlet lines, using clean refrigerant oil on the new O-rings. Tighten the mounting nuts securely.
8  Connect the electrical connector to the A/C low pressure switch.
9  Have the system evacuated, recharged and leak tested by the shop that discharged it.

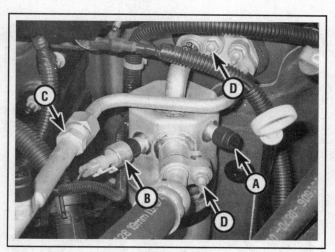

**12.2  Accumulator and related component details (Silverado shown, others similar)**

   A   Low side charging port
   B   A/C low pressure switch
   C   Expansion (orifice) tube fitting
   D   Outlet line fitting nut

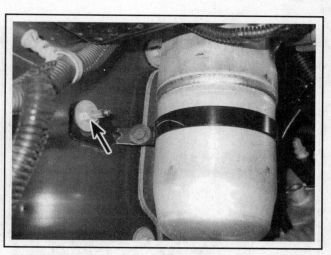

**12.4  Accumulator bracket mounting nut (viewed from the fenderwell with the inner fender splash shield removed)**

# 3-20 COOLING, HEATING AND AIR CONDITIONING SYSTEMS

## 13 Air conditioning compressor - removal and installation

### REMOVAL

▸ **Refer to illustrations 13.6a and 13.6b**

**※ WARNING:**

The air conditioning system is under high pressure. DO NOT loosen any fittings or remove any components until after the system has been discharged. Air conditioning refrigerant must be properly discharged into an EPA-approved container at a dealership service department or an automotive air conditioning repair facility. Always wear eye protection when disconnecting air conditioning system fittings.

➡ **Note 1:** The compressor on V6 engines is mounted on top of the engine and is driven by the serpentine drivebelt. On V8 engines, the compressor is mounted on the right side (passenger side) below the thermostat housing and is driven by a separate drivebelt that can be accessed after the engine drivebelt is removed.

➡ **Note 2:** The accumulator (see Section 12) should be replaced whenever the compressor is replaced.

➡ **Note 3:** Whenever the compressor is replaced because of internal damage, the expansion (orifice) tube should also be replaced (see Section 15).

1  Have the air conditioning system discharged by an automotive air conditioning technician (see **Warning** above).
2  Disconnect the cable from the negative terminal of the battery (see Chapter 5, Section 1).
3  On V8 engines, remove the air intake duct and resonator assembly (see Chapter 4).
4  Remove the drivebelt, and on models with a V8 engine, the air conditioning compressor drivebelt (see Chapter 1).
5  Disconnect the electrical connector(s) from the air conditioning compressor.
6  Clean the compressor thoroughly around the refrigerant line fittings. Disconnect the suction and discharge lines from the compressor (see illustrations). Plug the open fittings to prevent the entry of dirt and moisture, and discard the O-rings between the refrigerant lines and compressor.
7  Remove the compressor mounting bolts. Detach the compressor from the mounting bracket and remove the compressor from the engine compartment.

### INSTALLATION

8  If a new compressor is being installed, pour two ounces of R-134a-compatible refrigerant oil into it prior to installation. Also follow any directions included with the new compressor.

➡ **Note:** Some replacement compressors come with refrigerant oil in them. Follow the directions with the compressor regarding the draining of excess oil prior to installation.

**※ CAUTION:**

The oil used must be labeled as compatible with R-134a refrigerant systems.

9  Installation is the reverse of removal. When installing the line fitting bolt to the compressor, use new seals lubricated with clean refrigerant oil, and tighten the bolt securely.
10  Reconnect the cable to the negative terminal of the battery.
11  Have the system evacuated, recharged and leak tested by the shop that discharged it.

**13.6a Air conditioning compressor mounting details - V8 engine**

A  Compressor clutch electrical connector
B  Refrigerant line fitting bolts
C  Compressor upper mounting bolts

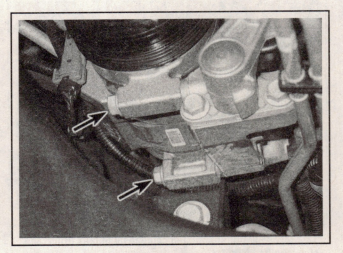

**13.6b Location of the compressor lower mounting bolts**

# COOLING, HEATING AND AIR CONDITIONING SYSTEMS    3-21

## 14  Air conditioning condenser - removal and installation

▸ Refer to illustrations 14.3, 14.4, 14.5, 14.6, 14.7a, 14.7b, 14.8a and 14.8b

### ✲✲ WARNING:

The air conditioning system is under high pressure. DO NOT loosen any fittings or remove any components until after the system has been discharged. Air conditioning refrigerant must be properly discharged into an EPA-approved container at a dealership service department or an automotive air conditioning repair facility. Always wear eye protection when disconnecting air conditioning system fittings.

➡ **Note:** The accumulator should be replaced if the condenser was damaged, causing the system to be open for some time (see Section 12).

1  Have the air conditioning system discharged by an automotive air conditioning technician (see **Warning** above).
2  If you're working on a pickup model, remove the radiator support cover (see illustrations 5.3a).
3  Remove the radiator grille (see Chapter 11). If you're working on an SUV model, also remove the hood latch/condenser cover (see illustration 5.3b and the accompanying illustration).
4  Remove the radiator air baffles (see illustration).
5  Remove the power steering fluid cooler (see illustration).
6  Remove the hood latch (see Chapter 11) and the hood latch support brace (see illustration).

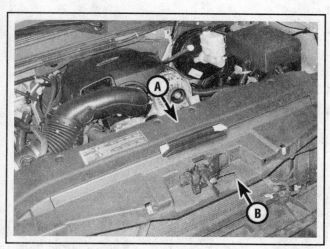

14.3  Remove the fasteners and take off the radiator support cover (A) and the hood latch/condenser cover (B) (SUV models)

14.4  Remove the radiator air baffle from each side

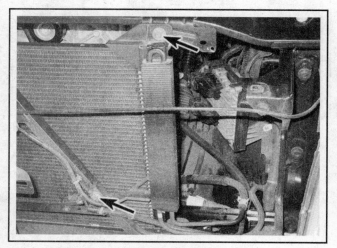

14.5  Power steering fluid cooler mounting bolts

14.6  Unscrew these bolts and remove the latch support brace

7  Remove the radiator support bracket (see illustrations).

8  Disconnect the refrigerant lines from the condenser and the remaining condenser mounting bolt (see illustration). Plug the open ends of the condenser and the disconnected refrigerant lines to prevent entry of dirt or moisture. Remove the condenser from the vehicle (see illustration).

9  Inspect the rubber insulator pads (on the lower crossmember) on which the condenser sits. Replace them if they're dried or cracked.

10  If the original condenser will be reinstalled, store it with the line fittings on top to prevent oil from draining out. If a new condenser is being installed, pour one ounce of R-134a-compatible refrigerant oil into it prior to installation.

11  Reinstall the components in the reverse order of removal. Be sure the rubber pads are in place under the condenser.

12  Have the system evacuated, recharged and leak tested by the shop that discharged it.

14.7a  Remove the radiator support bracket mounting bolts

14.7b  The upper mounting bolts are accessed from the backside of the support bracket

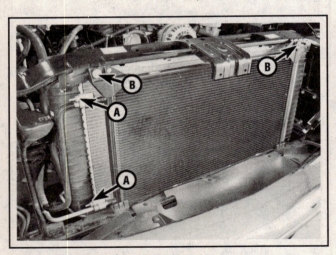

14.8a  Condenser mounting details

A  Condenser refrigerant line fittings
B  Condenser mounting bolts

14.8b  Carefully move the condenser clear of the upper bracket, then lift it from the vehicle

# COOLING, HEATING AND AIR CONDITIONING SYSTEMS  3-23

## 15  Air conditioning expansion (orifice) tube - removal and installation

▶ Refer to illustrations 15.2 and 15.5

### ❄ WARNING:

The air conditioning system is under high pressure. DO NOT loosen any fittings or remove any components until after the system has been discharged. Air conditioning refrigerant must be properly discharged into an EPA-approved container at a dealership service department or an automotive air conditioning repair facility. Always wear eye protection when disconnecting air conditioning system fittings.

1  Have the air conditioning system discharged by an automotive air conditioning technician (see **Warning** above).
2  Locate the expansion (orifice) tube fitting in the liquid (high-pressure) line near the accumulator (see illustration).
3  Hold the stationary fitting (on the line coming from the condenser) with one wrench, then loosen the other fitting with another wrench.
4  The expansion tube is a tube with a fixed-diameter orifice and a mesh filter at each end. When you separate the pipe at the fitting you will see one end of the orifice tube inside the pipe leading to the evaporator. Use needle-nose pliers to remove the orifice tube. If it's stuck or breaks off, a special tool (available at some auto parts stores or specialty tool suppliers) may be required to extract it.
5  The orifice tube acts to meter the refrigerant, changing it from a high-pressure liquid to a low-pressure gas. It is possible to reuse the orifice tube if (see illustration):
  a) The screens aren't plugged with grit or foreign material
  b) Neither screen is torn
  c) The plastic housing over the screens is intact
  d) The brass orifice inside the plastic housing is unrestricted
6  Installation is the reverse of removal. Be sure to insert the expansion tube with the shorter end pointing toward the evaporator.

### ❄ CAUTION:

Always use a new O-ring when installing the expansion (orifice) tube.

7  Reconnect the refrigerant line and tighten the fitting securely, then have the system evacuated, recharged and leak-tested by the shop that discharged it.

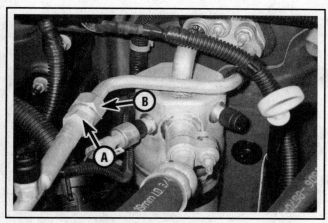

15.2  The expansion (orifice) tube is located in the small line leading to the evaporator core. Hold the stationary fitting (A) with a wrench while loosening the tube nut (B)

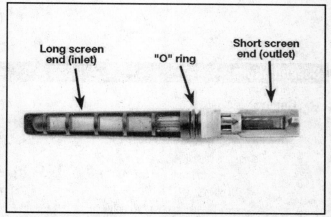

15.5  The expansion tube is equipped with a tapered mesh screen that must be clean and must not have any holes or damage

## 16  Air conditioning low pressure switch and pressure sensor - removal and installation

### ❄ WARNING:

The air conditioning system is under high pressure. Do not loosen any hose fittings or remove any components until the system has been discharged. Air conditioning refrigerant should be properly discharged into an EPA-approved recovery/recycling unit by a dealer service department or an automotive air conditioning repair facility. Always wear eye protection when disconnecting air conditioning system fittings.

### LOW PRESSURE SWITCH

1  Unplug the electrical connector from the low pressure switch (see illustration 12.2).
2  Unscrew the low pressure switch using an open-end wrench.
3  Screw the new switch onto the threads until hand tight, then tighten it securely.
4  Reconnect the electrical connector.

# 3-24  COOLING, HEATING AND AIR CONDITIONING SYSTEMS

## PRESSURE SENSOR

▸ **Refer to illustration 16.5**

5  Working near the air conditioning compressor, unplug the electrical connector from the pressure sensor (see illustration).

6  Remove the sensor from the air conditioning line fitting. Be sure to hold the fitting using an open-end wrench to prevent deforming the pressure line.

7  Install the new pressure sensor, then tighten it securely.

8  Reconnect the electrical connector.

**16.5  Location of the low pressure sensor**

## Specifications

### General

| | |
|---|---|
| Coolant capacity | See Chapter 1 |
| Refrigerant type | R-134a |
| Refrigerant capacity | |
|   Suburban (front and rear A/C) | 3.0 pounds |
|   All others | |
|     Standard (front A/C only) | 1.6 pounds |
|     With rear air conditioning | 2.5 pounds |

## Torque specifications      Ft-lbs (unless otherwise indicated)

➡ **Note:** One foot-pound (ft-lb) of torque is equivalent to 12 inch-pounds (in-lbs) of torque. Torque values below approximately 15 foot-pounds are expressed in inch-pounds, because most foot-pound torque wrenches are not accurate at these smaller values.

| | |
|---|---|
| Air conditioning low pressure switch | 53 in-lbs |
| Air conditioning pressure sensor | 53 in-lbs |
| Radiator mounting bolts | 18 |
| Thermostat housing nuts/bolts | 18 |
| Water pump bolts | |
|   V6 engine | 33 |
|   V8 engines | |
|     Step 1 | 132 in-lbs |
|     Step 2 | 22 |
| Water pump pulley bolts (V6 engine) | 18 |

# 4

# FUEL AND EXHAUST SYSTEMS

**Section**

1. General information and precautions
2. Fuel pressure relief procedure
3. Fuel pump/fuel pressure - check
4. Fuel lines and fittings - general information
5. Fuel tank - removal and installation
6. Fuel tank cleaning and repair - general information
7. Fuel pump/fuel level sending unit - removal and installation
8. Fuel pump/fuel level sending unit - component replacement
9. Fuel pressure sensor and fuel pump flow control module - replacement
10. Air filter housing - removal and installation
11. Sequential Fuel Injection (SFI) system - general information
12. Sequential Fuel Injection (SFI) system - general check
13. Throttle body - inspection, removal and installation
14. Fuel meter body and injectors (V6 models) - removal and installation
15. Fuel rail and injectors (V8 models) - removal and installation
16. Exhaust system servicing - general information

**Reference to other Chapters**

Air filter check and replacement - See Chapter 1
Catalytic converter - See Chapter 6
Exhaust system check - See Chapter 1
Fuel filter replacement - See Chapter 1
Underhood hose check and replacement - See Chapter 1

# 4-2 FUEL AND EXHAUST SYSTEMS

## 1 General information and precautions

This Chapter covers the removal and installation procedures for the important parts of the air intake, fuel and exhaust systems. Because emission-control systems are integral parts of the engine management system, there are many cross-references to Chapter 6. Information on the engine management system, information sensors and output actuators is in Chapter 6.

The air intake system consists of the air filter housing, the air intake duct, the throttle body and the intake manifold. Incoming air passes through the air filter element, the Mass Air Flow (MAF) sensor, the air intake duct, the throttle body, the intake manifold plenum and the intake manifold runners before being mixed with fuel sprayed into the intake ports by the fuel injectors.

The Sequential Fuel Injection (SFI) system consists of the fuel tank, an electric fuel pump/fuel level sending unit module mounted inside the tank, the fuel rail, the fuel injectors, the fuel pressure regulator (V6 models), the fuel pressure sensor and Fuel Pressure Control Module (FPCM) (V8 models) and the metal and flexible fuel lines that connect the various components of the SFI system.

The exhaust system consists of the exhaust manifold(s), the catalytic converter(s), the resonator, the muffler and the exhaust pipes connecting these components. The system is suspended from the vehicle pan by rubber hangers. You'll find the removal and installation procedures for the exhaust manifold(s) in Chapter 2, and for the rest of the exhaust system in this Chapter. There is more information about - and the replacement procedures for - the catalytic converter(s) in Chapter 6.

## 2 Fuel pressure relief procedure

### ✽✽ WARNING:

**Gasoline is extremely flammable, so take extra precautions when you work on any part of the fuel system. Don't smoke or allow open flames or bare light bulbs near the work area, and don't work in a garage where a gas-type appliance (such as a water heater or clothes dryer) is present. Since gasoline is carcinogenic, wear fuel-resistant gloves when there's a possibility of being exposed to fuel, and, if you spill any fuel on your skin, rinse it off immediately with soap and water. Mop up any spills immediately and do not store fuel-soaked rags where they could ignite. The fuel system is under constant pressure, so, if any fuel lines are to be disconnected, the fuel pressure in the system must be relieved first. When you perform any kind of work on the fuel system, wear safety glasses and have a Class B type fire extinguisher on hand.**

### ✽✽ CAUTION:

**After the fuel pressure has been relieved, it's a good idea to lay a shop towel over any fuel connection to be disassembled, to absorb the residual fuel that may leak out when servicing the fuel system.**

1  Remove the fuel filler cap to relieve any pressure built-up in the fuel tank.
2  Disconnect the cable from the negative terminal of the battery (see Chapter 5).
3  Unscrew the cap on the fuel pressure test port (see illustration 3.4a).
4  Surround and cover the test port with shop rags, then depress the Schrader valve inside the test port with a small screwdriver until the pressure in the fuel system is relieved. Properly dispose of the rags.

## 3 Fuel pump/fuel pressure - check

### ✽✽ WARNING:

**Gasoline is extremely flammable, so take extra precautions when you work on any part of the fuel system. See the Warning in Section 2.**

### FUEL PUMP OPERATION CHECK

▸ Refer to illustration 3.2

1  Sit inside the vehicle with the windows closed, turn the ignition key to ON (not START) and listen carefully for the sound made by the fuel pump as it's briefly turned on by the PCM to pressurize the fuel system prior to starting the engine. You will only hear a whirring sound for a second or two, but that sound tells you that the pump is working. If you can't hear the pump, remove the fuel filler cap, then have an assistant turn the ignition switch to ON while you listen for the sound of the pump operating for a couple of seconds.
2  If the pump does not come on when the ignition key is turned to ON, check the fuel pump fuse and relay (see illustration 3.2). If the fuse and relay are okay, check the wiring back to the fuel pump. If the fuse, relay and wiring are okay, the fuel pump is probably defective. If the pump runs continuously with the ignition key in its ON position, the Powertrain Control Module (PCM) is probably defective. Have the PCM checked by a dealer service department or other qualified repair shop.

### FUEL PRESSURE CHECK

▸ Refer to illustrations 3.3, 3.4a and 3.4b

3  To measure the fuel pressure, you'll need a fuel pressure gauge compatible with high-pressure fuel injection systems, and a hose and fitting suitable for connecting the gauge to the Schrader valve-type test port on the fuel feed line (see illustration).

## FUEL AND EXHAUST SYSTEMS    4-3

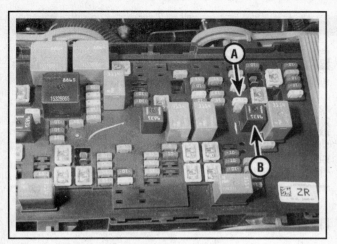

3.2  The fuel pump fuse (A) is located on the engine compartment fuse and relay panel, which is located on the left side of the engine compartment. The fuel pump relay (B) is also located on this fuse and relay panel

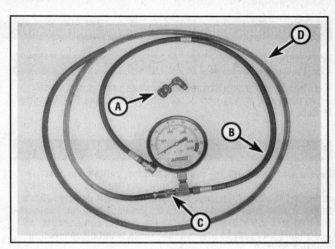

3.3  A typical fuel pressure gauge set up

- A   Screw-on adapter for the Schrader valve on the fuel rail
- B   Hose with fitting to connect to the adapter
- C   Bleeder valve (optional)
- D   Bleeder hose (optional)

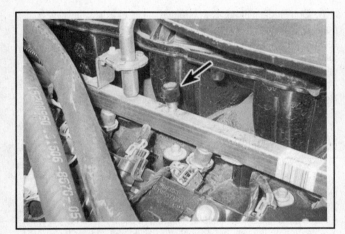

3.4a  The fuel pressure test port on V8 models is located on the right fuel rail

3.4b  The fuel pressure gauge connected to the test port on the fuel rail (V8 model shown)

4   Locate the fuel pressure test port, unscrew the cap and connect a fuel pressure gauge (see illustrations).

5   Start the engine and allow it to idle. Note the gauge reading as soon as the pressure stabilizes, and compare it with the pressure listed in this Chapter's Specifications.

- a) If the pressure is lower than specified, suspect a restricted fuel filter (the fuel filter is an integral part of the fuel pump/fuel level sensor module; to replace it you must replace the fuel pump). Also check the fuel lines and hoses for kinks, blockages and leaks.
- b) V6 models: If there are no restrictions and you changed the fuel filter but the pressure is still lower than specified, the regulator (which is an integral component of the fuel pump module) might be stuck in the open position, which will prevent the fuel system from reaching its normal operating pressure range. The regulator is not serviceable separately; if it's defective you must replace the fuel pump module.
- c) It's possible that one or more of the fuel injectors might be leaking or stuck open (but if this is happening the engine will be running very poorly and will most likely have set a trouble code). On V6 models, remove the fuel meter body and inspect the injectors (see Section 14); on V8 models, remove the fuel rail and inspect the injectors (see Section 15).
- d) V6 models: If the fuel pressure is higher than specified, the fuel pressure regulator might be stuck in the closed position. The regulator is an integral component of the fuel pump module, and is not serviceable separately; if it's defective, you must replace the fuel pump module.

6   Turn off the engine. Verify that the fuel pressure loses no more than 5 psi for one minute after the engine is turned off.

7   Disconnect the fuel pressure gauge and screw on the test port cap. Clean up any spilled gasoline.

8   Start the engine and verify that there are no fuel leaks.

# 4-4 FUEL AND EXHAUST SYSTEMS

## 4 Fuel lines and fittings - general information

### ✺✺ WARNING 1:

**Gasoline is extremely flammable, so take extra precautions when you work on any part of the fuel system. See the Warning in Section 2.**

### ✺✺ WARNING 2:

**Before disconnecting any fuel line fittings, relieve the fuel system pressure (see Section 2) and equalize tank pressure by removing the fuel filler cap. This procedure will merely relieve the increased pressure necessary for the engine to run - remember that fuel will still be present in the system components, so you should be ready to mop up fuel spills when disconnecting fuel line fittings.**

1  Always disconnect the cable from the negative battery terminal (see Chapter 5, Section 1) and relieve the fuel pressure (see Section 2) before servicing fuel lines or fittings.

2  Whenever you're working under the vehicle, be sure to inspect all fuel and evaporative emission lines for leaks, kinks, dents and other damage. Always replace a damaged fuel or EVAP line immediately. Leaking fuel and EVAP lines will result in loss of fuel and excessive air pollution (the leaking raw fuel emits unburned hydrocarbon vapors into the atmosphere).

3  If you find signs of dirt in the lines during disassembly, disconnect all lines and blow them out with compressed air. If the fuel inlet strainer or fuel filter are clogged or dirty you will not be able to inspect either one because they're inside the fuel pump module, a sealed unit that cannot be disassembled and must therefore be replaced (see Section 7). So consider this option only if, after cleaning all the fuel lines, the fuel system is still clogged.

4  The fuel supply line connects the fuel pump in the fuel tank to the fuel rail on the engine. The Evaporative Emission (EVAP) system vapor lines connect the fuel tank to the EVAP canister and the canister to the intake manifold. The fuel and EVAP lines are secured to the underbody with plastic clips.

## STEEL TUBING

5  Because fuel lines used on fuel-injected vehicles are under fairly high pressure, it is critical that they be replaced with lines of equivalent specification. Never use copper or aluminum tubing to replace steel tubing. These materials cannot withstand normal vehicle vibration.

6  Some steel fuel lines have threaded fittings. When loosening these fittings to service or replace components:

  a) *Hold the stationary fitting with one wrench while loosening or tightening the tubing nut with another.*
  b) *If you're going to replace one of these fittings, use original equipment parts or parts that meet original equipment standards.*

## PLASTIC TUBING

7  Most of the fuel (and EVAP) lines on the vehicles covered in this manual are plastic. If you ever have to replace a plastic line, use only plastic tubing meeting original equipment standards.

### ✺✺ CAUTION:

**When removing or installing plastic fuel line tubing, be careful not to bend or twist it too much, which can damage it. And damaged fuel lines MUST be replaced! Also, be aware that the plastic fuel tubing is NOT heat resistant, so keep it away from excessive heat. Nor is it acid-proof, so don't wipe it off with a shop rag that has been used to wipe off battery electrolyte. If you accidentally spill or wipe electrolyte on plastic fuel tubing, replace the tubing.**

## FLEXIBLE HOSES

### ✺✺ WARNING:

**Use only original equipment replacement hoses or their equivalent. Unapproved hoses might fail when subjected to the high operating pressures of the fuel system.**

8  Don't route fuel hoses within four inches of exhaust system components or within ten inches of a catalytic converter. Make sure that no rubber hoses are installed directly against the vehicle, particularly in places where there is any vibration. If allowed to touch some vibrating part of the vehicle, a hose can easily become chafed and it might start leaking. A good rule of thumb is to maintain a minimum of 1/4-inch clearance around a hose (or metal line) to prevent contact with the vehicle underbody.

## FUEL LINE AND EVAP LINE FITTINGS

9  The vehicles covered in this manual use two kinds of fuel line quick-connect fittings (metal or plastic) for most connections at the fuel pump, the fuel tank, under the vehicle and in the engine compartment. A third type of plastic quick-connect fitting is used only at the EVAP canister and on the vent hose connection at the fuel tank for the EVAP canister vent solenoid.

10  The procedure for releasing each type of fuel line fitting is different. But a few rules of thumb apply to all fittings:

  a) *Inspect the fitting for dirt. If the fitting is dirty, clean it off before disassembling it. The seals in the fitting will stick to the fuel line as they age. Twist the fitting on the line, then push and pull the fitting until it moves freely.*
  b) *Always disconnect all fuel line fittings from a fuel system component before removing the component.*
  c) *When disconnecting a quick-connect fitting, inspect the condition of the retainer before reconnecting the fitting. The best strategy with respect to retainers is to simply replace the retainer every time that you disconnect the fitting.*
  d) *When you disconnect a fitting with an O-ring inside, inspect the O-ring before reconnecting the fitting. If it has been leaking or looks deteriorated, replace the line and fitting (in most cases, the fitting itself is a non-removable part of the fuel line, so you might have to replace an entire fuel line if a fitting is damaged or defective).*

# FUEL AND EXHAUST SYSTEMS   4-5

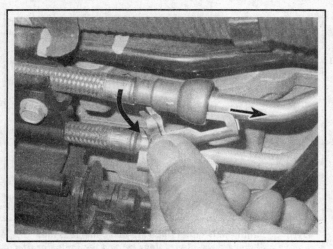

4.12  Pull the end of the retainer off the fuel line, then disengage the other end from the female side of the fitting

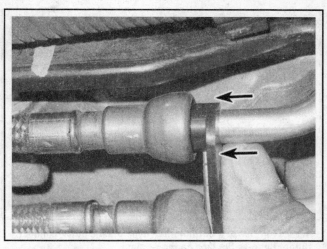

4.13a  Insert the fuel line separator tool into the female side of the fitting, push it into the fitting until it releases the locking tabs inside the fitting . . .

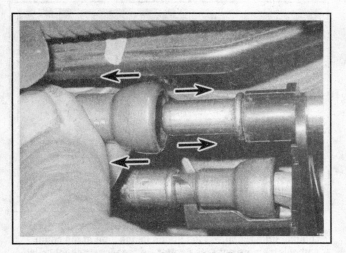

4.13b  . . . then pull the two halves of the fitting apart

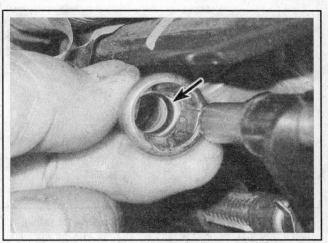

4.14  Inspect the old O-ring inside the female side of the fitting; if it's cracked, torn or deteriorated, replace the line

## METAL COLLAR QUICK-CONNECT FITTINGS

### Disconnection

▸ Refer to illustrations 4.12, 4.13a and 4.13b

➡ Note 1: You'll find these fittings at the connections between the fuel supply and return lines in the engine compartment.

➡ Note 2: You'll need a special tool set (available at most auto parts stores) to disconnect these fittings.

➡ Note 3: The photos accompanying the disconnection procedure depicted here shows the metal collar quick-connect fittings at a four-cylinder fuel rail, but the metal quick-connect fittings on a V6 or V8 fuel rail are identical.

11  Relieve the fuel system pressure (see Section 2).

12  Pull off the clip end of the retainer, then remove it from the fitting (see illustration).

13  Using a fuel line separator tool of the proper size (available at most auto parts stores), insert the tool into the female side of the fitting, then push it into the fitting to release the locking tabs and pull the fitting apart (see illustrations).

### Reconnection

▸ Refer to illustration 4.14

14  Inspect the O-ring inside the fitting (see illustration).
15  Apply a few drops of clean engine oil to the male pipe end.
16  Push both sides of the fitting together until the retaining tabs snap into place. Pull on both sides of the fitting to verify that it's securely connected.
17  Install the retainer, making sure it clips into place.
18  Start the engine and check for fuel leaks.

## 4-6 FUEL AND EXHAUST SYSTEMS

4.19a To release a Bartholomew-type plastic quick-connect fitting, depress the tabs on the retainer . . .

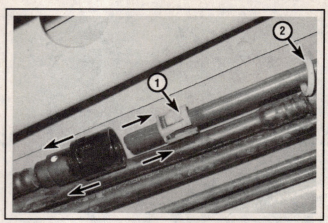

4.19b . . . until the two fuel lines are disconnected, then remove and discard the old retainer (1) and the indicator ring (2) (the indicator ring is used only during factory assembly; there is no need to reinstall it)

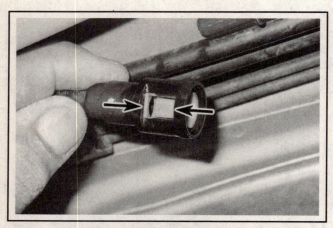

4.23 Install a new retainer in the female side of the fitting; make sure that the release tabs are aligned with the windows in the connector

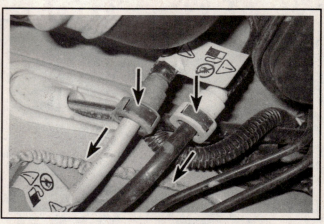

4.29 To disconnect a Push Down TI type plastic collar quick-connect fitting, depress the button on the side of the collar and pull the pipe out of the collar. To reconnect, simply push the pipe into the collar until it clicks into place

### PLASTIC COLLAR QUICK-CONNECT FITTINGS

#### Bartholomew-type

**Disconnection**

▸ Refer to illustrations 4.19a and 4.19b

19  To release a Bartholomew-type quick-connect fitting, depress the tabs of the retainer (see illustration) and pull the two fuel lines apart (see illustration).

20  Remove and discard the old retainer from the male side of the fitting.

21  Remove and discard the indicator ring from the male side of the fitting.

**Reconnection**

▸ Refer to illustration 4.23

22  Inspect the old O-ring inside the female side of the fitting.
23  Insert a new retainer in the female side of the fitting. Make sure that the release tabs are aligned with the windows of the connector (see illustration).
24  Apply a few drops of engine oil to the tip of the male fuel line.
25  Push both sides of the fitting together until the retainer release tabs snap into place.
26  Pull on both sides of the fitting to verify that it's securely connected.
27  Start the engine and check for fuel leaks.

#### Push Down TI type

▸ Refer to illustration 4.29

28  Push Down TI type fittings are typically used as chassis fuel and EVAP line connections. They're frequently used to connect the fuel and EVAP lines that are coming from components in, on or near the fuel tank to the fuel/EVAP lines going forward to the engine compartment.
29  To disconnect a Push Down TI type fitting, simply depress the button on the female, or collar, side of the fitting (see illustration) and pull the fuel or EVAP pipe out of the collar.
30  Be sure to wipe off the end of the pipe with a clean cloth before

# FUEL AND EXHAUST SYSTEMS   4-7

4.32  To disconnect a pull-up type plastic collar quick-connect fitting, squeeze the tangs on the bottom of the retainer, then carefully pry the retainer from the fitting

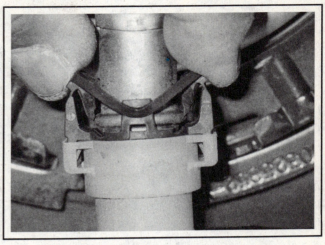

4.34  To disconnect a squeeze-type fitting, squeeze the plastic retainer with your fingers to release it, then pull the fitting off the line

reconnecting it to a Push Down TI fitting, then simply push it into the collar until it clicks into place.

### Pull-up type

▶ Refer to illustration 4.32

31  Pull-up type fittings are also used on chassis fuel line connections, and resemble Push Down TI type fittings. If you can't push down the button on the fitting, it's a pull-up type.

32  To disconnect a pull-up type fitting, squeeze the tangs on the underside of the fitting, then carefully pry up the retainer from the fitting (see illustration).

33  Be sure to wipe off the end of the pipe with a clean cloth before

reconnecting it to the fitting, then push the female side of the fitting onto the fuel line until it bottoms, then push the retainer in until it clicks into place.

### Squeeze type

▶ Refer to illustration 4.34

34  To disconnect a squeeze-type fitting, squeeze the sides of the retainer with your fingers until it releases, then pull the fitting off the line (see illustration).

35  Be sure to wipe off the end of the pipe with a clean cloth before reconnecting it to the fitting, then push the female side of the fitting onto the fuel line until the retainer clicks into place.

## 5  Fuel tank - removal and installation

Refer to illustrations 5.4, 5.5, 5.11 and 5.12

**WARNING 1:**
oline is extremely flammable, so take extra precautions  you work on any part of the fuel system. See the Warning ction 2.

**WARNING 2:**
disconnecting or opening any part of the fuel system, the fuel system pressure (see Section 2), and equalize ssure inside the fuel tank by removing the fuel filler cap.

sconnect the cable from the negative battery terminal (see , Section 1).
ve the fuel system pressure (see Section 2).
e the vehicle and place it securely on jackstands.
ove the fuel tank rock guard, if equipped (see illustration).

5.4  The outer edge of the rock guard, if equipped, is secured by these three bolts. After removing the bolts, allow the rock guard to swing down, then disengage the four slots on the inner edge of the rock guard from their mounting tabs

## 4-8 FUEL AND EXHAUST SYSTEMS

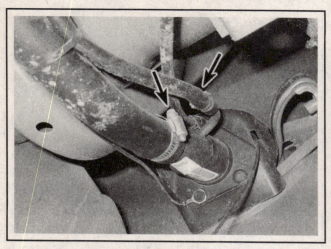

5.5 Loosen the hose clamps on the fuel tank filler neck hose and the EVAP system vent hose, then work the two hoses off the pipes

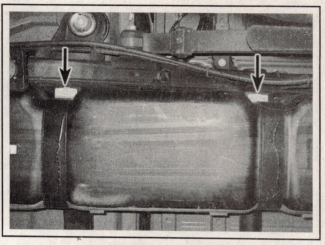

5.11 To detach the fuel tank from the vehicle, remove these fuel tank strap bolts

5  Locate the fuel tank filler neck hose and the EVAP system vent hose (see illustration), loosen both hose clamp screws and disconnect the filler neck hose from the tank pipe.

6  Locate the EVAP line that goes to the EVAP canister and disconnect it from the EVAP canister.

7  Locate the fuel supply line quick-connect fitting at the front of the fuel tank and disconnect it.

8  It's easier to remove the fuel tank when it's nearly empty. But there is no fuel tank drain plug, so if there's still a lot of fuel in the tank, siphon or hand-pump the remaining fuel from the tank through the tank filler neck pipe.

### ✺✺ WARNING:

Don't start the siphoning action by mouth! Use a siphoning kit (available at most auto parts stores).

9  Detach any clips that secure the fuel pump/fuel level sending unit harness to the crossmember or frame rail and detach it/them.

### ✺✺ CAUTION:

Failure to detach these clips will result in damage to the fuel pump/fuel level sending unit harness when the tank is lowered.

10  Support the fuel tank.
11  The fuel tank is supported by two transverse straps. The outer end of each fuel tank strap is secured to the vehicle by a bolt (see illustration). The inner end of each strap is secured to a mounting bracket by a hinge. Remove the two fuel tank strap bolts, then disengage each strap

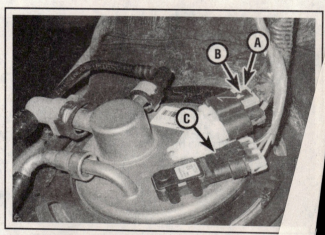

5.12 To disconnect the electrical connector from the fuel pump/fuel level sending unit module, slide the retainer out (A), then depress the release tab (B) and pull off the connector. To disconnect the electrical connector from the fuel tank pressure sensor, pull up the retainer (C) and the connector

hinge from its mounting bracket and remove the straps.
12  Carefully lower the tank just far enough to access the fuel level sensor module, then disconnect the electrical from the pump module and from the EVAP fuel pressure sensor (see illustration).
13  Lower the tank the rest of the way.
14  Installation is the reverse of removal.

# FUEL AND EXHAUST SYSTEMS 4-9

## 6  Fuel tank cleaning and repair - general information

**⁕⁕ WARNING:**
**Gasoline is extremely flammable, so take extra precautions when you work on any part of the fuel system. See the Warning in Section 2.**

1  The fuel tank is plastic and cannot be repaired. No reliable repair procedures are available to correct leaks or damage. Fuel tank replacement is the only approved service.

2  To remove sediment from the bottom of the tank, have the fuel tank steam-cleaned. Remove the fuel pump/level sending unit (see Section 7) and all EVAP system components (see Chapter 6) prior to cleaning. Allow plenty of time for the tank to air-dry before returning it to service.

## 7  Fuel pump/fuel level sending unit - removal and installation

♦ Refer to illustrations 7.5, 7.6, 7.7 and 7.8

**⁕⁕ WARNING:**
**Gasoline is extremely flammable, so take extra precautions when you work on any part of the fuel system. See the Warning in Section 2.**

1  Disconnect the cable from the negative battery terminal (see Chapter 5, Section 1).

2  Relieve the system fuel pressure (see Section 2), and equalize tank pressure by removing the fuel filler cap.

3  Remove the fuel tank (see Section 5).

4  Disconnect the EVAP vent line, the fuel tank vent line fitting and the fuel supply line quick-connect fittings from the fuel pump/fuel level sending unit mounting flange.

5  Using a pair of large water pump pliers, unscrew the fuel pump/fuel level sending unit module lock ring by turning it counterclockwise (see illustration). If the lock ring is tight, use a hammer and a brass punch to loosen it (don't use a steel punch, which could produce sparks when struck by the hammer).

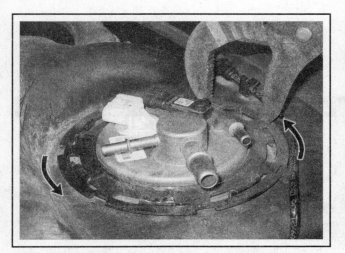

7.5  Use a large pair of water pump pliers to unscrew the fuel pump lock ring; if the lock ring is too tight to loosen this way, carefully tap it loose with a hammer and a brass punch

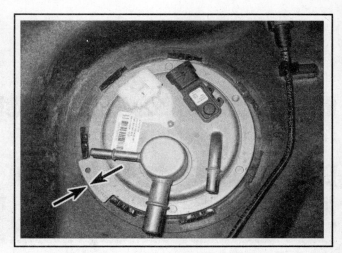

7.6  To ensure that the pipes for the EVAP vent line, fuel tank vent line and fuel supply line are oriented correctly when you install the fuel pump/fuel level sending unit, make alignment marks on the tank and the pump flange at this wider part of the flange

# 4-10 FUEL AND EXHAUST SYSTEMS

7.7 Carefully remove the fuel pump/fuel level sensor module from the fuel tank; once the module has cleared the mounting hole in the tank, angle it as shown to work the float arm through the mounting hole without damaging anything

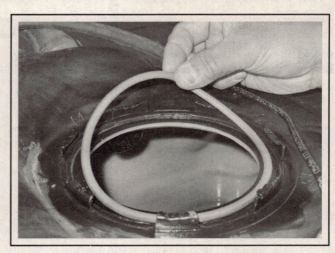

7.8 Remove and inspect the O-ring seal for the fuel pump mounting flange; if it's cracked, torn or deteriorated, replace it

6  Note the orientation of the pump mounting flange and mark the orientation of the fuel pump in relation to the fuel tank (see illustration) to ensure that the fuel pump is correctly realigned when you install it again. If you're going to install a new pump, note the location of your alignment mark on the old pump and make a mark at the same spot on the new unit.

7  Remove the fuel pump/fuel level sensor module high enough to disconnect the quick-connect fitting for the ventilation harness, then remove the pump assembly (see illustration), taking care not to damage the fuel level sensor float arm and float.

8  Before installing the pump, inspect the O-ring (see illustration) for cracks, tears and deterioration. If it isn't in perfect condition, replace it.

9  Insert the fuel pump/fuel level sensor module into the fuel tank, align the mark that you made on the fuel pump/fuel level sending unit flange with the mark you made on the tank and carefully seat the pump flange on the tank. Make sure that you don't damage the float arm or the float during installation. If the float arm is bent, the fuel level that is indicated on the fuel level gauge on the instrument cluster will be incorrect.

10  Installation is otherwise the reverse of removal.

## 8  Fuel pump/fuel level sending unit - component replacement

♦ Refer to illustrations 8.3 and 8.5

**※ WARNING:**

**Gasoline is extremely flammable, so take extra precautions when you work on any part of the fuel system. See the Warning in Section 2.**

➡ Note: You can purchase the complete fuel pump/fuel level sending unit module, or you can purchase either the fuel pump or the fuel level sending unit separately. If only one component fails, use this procedure to separate the two components, then reassemble the good component and the new replacement component.

1  Remove the fuel pump/fuel level sending unit (see Section 7).

2  Place the fuel pump/fuel level sending unit on a clean workbench surface.

3  Disconnect the fuel level sending unit electrical connector from the fuel pump mounting flange (see illustration).

4  Disengage the fuel level sensor wiring harness from the molded-in harness guide.

5  Remove the fuel level sensor unit from the fuel pump module (see illustration).

6  No further disassembly of the fuel pump module is possible. If any component of the fuel pump module is defective, replace the fuel pump module.

7  To install the fuel level sending unit on the fuel pump module, slide the sensor unit into place until you hear a click, then pull on the sensor unit to verify that it's locked into place.

8  Installation is otherwise the reverse of removal.

# FUEL AND EXHAUST SYSTEMS    4-11

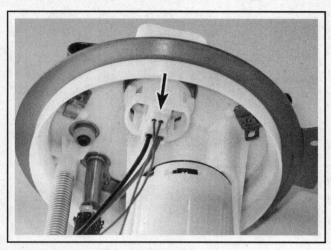

8.3 Disconnect the fuel level sending unit electrical connector from its terminal on the underside of the fuel pump mounting flange

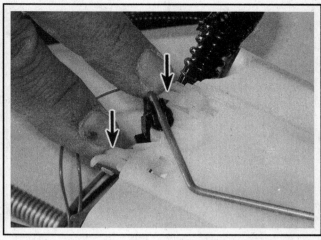

8.5 Depress these two levers on the locking lugs on the sending unit, then slide the sending unit up and remove it from the pump module

## 9   Fuel pressure sensor and fuel pump flow control module - replacement

### FUEL PRESSURE SENSOR

➤ Refer to illustration 9.3

**✲✲ WARNING:**

**Gasoline is extremely flammable, so take extra precautions when you work on any part of the fuel system. See the Warning in Section 2.**

➡ Note: *The fuel pressure sensor is located under the vehicle, on the fuel supply line.*

1   Relieve the system fuel pressure (see Section 2).
2   Raise the vehicle and place it securely on jackstands.
3   Disconnect the electrical connector from the fuel pressure sensor (see illustration).
4   Unscrew the fuel pressure sensor from the fuel supply line.

**✲✲ WARNING:**

**Be prepared for fuel spillage.**

5   Remove and discard the old fuel pressure sensor O-ring. Coat a new O-ring with a little clean engine oil, then install it on the fuel pressure sensor.
6   Installation is otherwise the reverse of removal. Be sure to tighten the fuel pressure sensor securely.

7   Before lowering the vehicle, test for leaks as follows:
*Turn the ignition switch key to the ON position (don't start the engine) for two seconds.*
*Turn the ignition switch key to the OFF position for ten seconds.*
*Turn the ignition switch key to the ON position (don't start the engine).*
*Inspect the area around the fuel pressure sensor for leaks.*

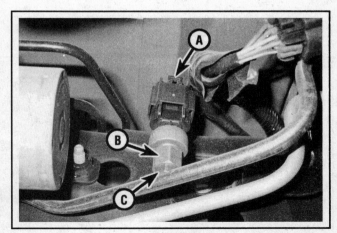

9.3 To disconnect the electrical connector from the fuel pressure sensor, depress the release tab (A). To remove the sensor (B) from the fuel supply line, hold the flats on the fuel line (C) with a wrench (to prevent twisting the line), then unscrew the sensor

# 4-12  FUEL AND EXHAUST SYSTEMS

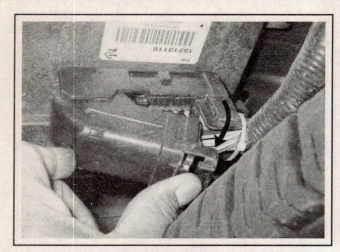

**9.9  To disconnect the electrical connector from the fuel pump flow control module, depress the release tab, flip the lever down and pull down the connector**

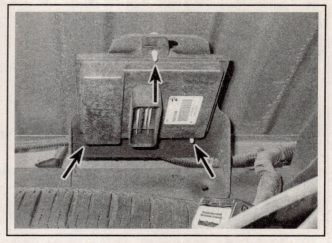

**9.10  To detach the fuel pump flow control module from its mounting bracket, remove the three mounting bolts from the top**

## FUEL PUMP FLOW CONTROL MODULE

▸ Refer to illustrations 9.9 and 9.10

➡ Note: *The fuel pump flow control module is located under the vehicle, above the front part of the spare tire. It's not necessary to remove the spare tire to replace the module.*

8  Raise the vehicle and place it securely on jackstands.
9  Disconnect the electrical connector from the fuel pump flow control module (see illustration).
10  Remove the three control module mounting bolts (see illustration) and remove the module from its mounting bracket.
11  Installation is the reverse of removal.

## 10  Air filter housing - removal and installation

### AIR INTAKE DUCT

▸ Refer to illustrations 10.2, 10.3, 10.4, 10.5 and 10.6

1  On V8 models, remove the intake manifold cover.

2  Disconnect the PCV fresh air inlet hose from the air intake duct (see illustration).
3  On V8 models, detach the radiator hose clip from the air intake duct (see illustration).
4  Loosen the hose clamp screw at the Mass Air Flow (MAF) sensor (see illustration).

**10.2  To disconnect the PCV fresh air inlet hose from the air intake duct, pull it out (V8 model shown)**

**10.3  On V8 models, detach the radiator hose clip from the air intake duct**

**10.4  MAF sensor hose clamps (V8 model shown)**

# FUEL AND EXHAUST SYSTEMS  4-13

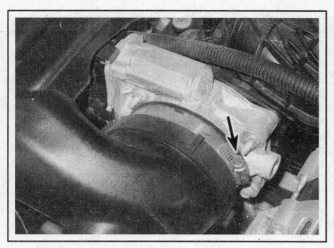

10.5  To detach the air intake duct from the throttle body, loosen this hose clamp (V8 model shown)

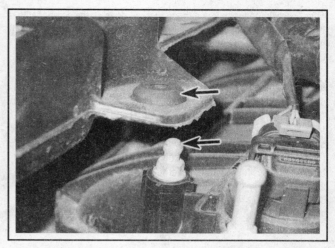

10.6  To remove the air intake duct from a V8 model, grasp it firmly and lift it straight up to disengage the locator pin from its grommet

5  Loosen the hose clamp screw at the throttle body (see illustration).
6  To remove the air intake duct, grasp it firmly and lift it straight up to disengage the locator pin from its insulator grommet (see illustration).
7  Installation is the reverse of removal.

## AIR FILTER HOUSING

8  Remove the air intake duct (see Steps 1 through 5).

9  Disconnect the electrical connector from the Mass Air Flow (MAF) sensor.
10  Grasp the air filter housing firmly and pull it straight up to disengage the two locator pins from their insulator grommets.
11  While the air filter housing is out, inspect the rubber mounting grommets. If a grommet is cracked, torn or otherwise damaged, replace it.
12  Installation is the reverse of removal.

## 11  Sequential Fuel Injection (SFI) system - general information

These models are equipped with a Sequential Fuel Injection (SFI) system. The SFI system consists of three basic sub-systems: the air induction system, the fuel system and the electronic control system.

➡ **Note:** *Refer to Chapter 6 for more information on the components of the electronic control system.*

### AIR INDUCTION SYSTEM

The air induction system consists of the air filter housing, the Mass Air Flow (MAF) sensor, the Intake Air Temperature (IAT) sensor (an integral part of the MAF sensor), the air intake duct, the throttle body, the Throttle Position (TP) sensor, the air intake plenum and the intake manifold.

➡ **Note:** *For more information about the MAF/IAT sensor, refer to Chapter 6. The TP sensor is an integral component of the throttle body and cannot be serviced separately.*

All models are equipped with an electronic throttle body. The electronic throttle control system consists of the Accelerator Pedal Position (APP) sensor, the Powertrain Control Module (PCM) and the motor inside the throttle body that controls the angle of the throttle plate. The APP sensor, which is an integral component of the accelerator pedal assembly, is a potentiometer that monitors the position (or angle) of the accelerator pedal. As you depress the accelerator pedal, the APP sensor outputs a variable voltage signal to the PCM, which sends a command to the throttle body's motor, which opens the throttle plate in proportion to the position (angle) of the accelerator pedal. As the throttle plate opens or closes, the amount of air that can pass through the system increases or decreases accordingly. As the throttle plate opens or closes, the Throttle Position (TP) sensor, which is located on the end of the throttle plate shaft, opens or closes with it. And as more or less air enters the engine, the MAF/IAT sensor signal to the PCM also changes. In response to these two signals from the TP and MAF/IAT sensors, the PCM opens each injector for a longer or shorter duration to increase the amount of fuel delivered to the inlet ports. The interval during which an injector is open is known as pulse width.

### FUEL SYSTEM

An electric fuel pump located inside the fuel tank supplies fuel under pressure to the fuel rail, which distributes fuel evenly to all injectors. A

## FUEL AND EXHAUST SYSTEMS

filter between the fuel pump and the fuel rail protects the components of the system. From the fuel rail, fuel is injected into the intake ports, just above the intake valves, by a fuel injector.

The amount of fuel supplied by the injectors is precisely controlled by injector drivers inside the PCM. The injector drivers, which are turned on and off by the PCM, control the ground side of each injector circuit: When the ground path is closed, the injectors are on; when the ground path is open, the injectors are off. The PCM uses signals from the Crankshaft Position (CKP) sensor and the Camshaft Position (CMP) sensor to determine when to trigger each injector in cylinder firing order (hence the term "sequential injection"). This precise control of injector timing produces more power, better fuel economy and lower exhaust emissions.

### ELECTRONIC CONTROL SYSTEM

The PCM controls the SFI system and the engine management system. It receives signals from an array of information sensors that monitor such variables as intake air mass and temperature, coolant temperature, engine speed, crankshaft position, acceleration/deceleration, and exhaust gas oxygen content. These signals help the PCM determine the injection duration necessary for the optimal air/fuel ratio. These sensors and various PCM-controlled output actuators (relays, solenoids, etc.) are located throughout the engine compartment. For further information regarding the engine management system, see Chapter 6.

## 12 Sequential Fuel Injection (SFI) system - general check

▶ Refer to illustration 12.7

### ※※ WARNING:

**Gasoline is extremely flammable, so take extra precautions when you work on any part of the fuel system. See the Warning in Section 2.**

1  Inspect the SFI system electrical connectors. Verify that all ground wire connections are tight. Loose connectors and poor grounds can cause many problems that resemble more serious malfunctions.

2  Verify that the battery is fully charged (see Chapters 1 and 5 for help with the battery). The PCM, information sensors and output actuators depend on a steady and adequate voltage to function correctly.

3  Inspect the air filter element (see Chapter 1). A dirty or partially blocked filter will severely impede performance and economy.

4  Inspect any fuses for the circuit you're checking. If you find a blown fuse, replace it and note whether it blows again. If it does, look for a short in the circuit.

5  Inspect the air intake duct, the throttle body and the intake manifold for leaks, which will cause an excessively lean mixture. Also inspect all vacuum hoses connected to the intake manifold and to the throttle body.

6  Remove the air intake duct (see Section 10) and inspect the throttle body for dirt, carbon, varnish or other residue inside the bore of the throttle body, particularly around the throttle plate. If it's dirty, clean it with carburetor cleaner spray and a shop towel.

### V8 MODELS ONLY

7  With the engine running, place an automotive stethoscope against each injector (see illustration), one at a time, and listen for a clicking sound that indicates operation. If you don't have a stethoscope, you can place the tip of a long screwdriver against the injector and listen through the handle. If you can hear all of the injectors operating, but there is a misfire condition present, the electrical circuits are functioning, but the injectors might be dirty or fouled from carbon deposits. Commercial cleaning products might help. If not, then you might have to replace the injectors (see Section 15).

8  If you can't hear an injector operating, disconnect its electrical connector and note whether it makes any difference in the way the engine runs (engine rpm should drop if the injector is working). Unplugging a non-operational injector shouldn't make any difference in engine rpm.

9  If the rpm doesn't drop when you disconnect an injector, the injector is probably defective (provided that the cylinder has compression and spark is present). Disconnect the electrical connector and measure the resistance of the injector coil with an ohmmeter, then compare your measurement with the injector resistance listed in this Chapter's Specifications.

10  If the injector coil resistance is okay, but the injector is not operating, the circuit between the PCM and the injector might be open somewhere. It might also indicate that the driver inside the PCM is defective. The signal to the injector can be tested with a 'noid light (short for solenoid), available at most auto parts stores. If the light doesn't flash when connected to the injector harness connector and the engine is cranked, have the circuit between the PCM and the injectors tested by a dealer service department before replacing the PCM or the injector.

**12.7 Use an automotive stethoscope to listen to each injector. If an injector sounds different from the other injectors, or isn't making any sound at all, disconnect it and see if there is any difference in the way the engine runs**

# FUEL AND EXHAUST SYSTEMS 4-15

## 13  Throttle body - inspection, removal and installation

### INSPECTION

1  Remove the air intake duct from the throttle body, open the throttle plate and inspect the throttle body bore for carbon and residue build-up. If it's dirty, clean it with solvent or carburetor cleaner. Make sure that the solvent or carb cleaner is safe for oxygen sensor systems and catalytic converters.

**CAUTION:**

Do not clean the Throttle Position (TP) sensor or the throttle motor with solvent. Also, do NOT use a metal brush to clean the bore of the throttle body, which is protected by a special coating. Scrubbing the bore with a stiff brush could ruin the coating. Instead, wipe out the bore with a clean shop rag and a little carburetor cleaner.

### REMOVAL AND INSTALLATION

▶ Refer to illustrations 13.3, 13.4 and 13.5

**WARNING:**

Wait until the engine is completely cool before beginning this procedure.

➡Note: The photos accompanying this section depict the throttle body on a V8 model, but the throttle body used on V6 models is virtually identical, except that it's mounted on top of the intake manifold instead of the front of the manifold.

2  Remove the air intake duct (see Section 10).
3  Disconnect the electrical connector from the throttle body (see illustration).
4  Remove the throttle body mounting fasteners (see illustration) and remove the throttle body.
5  Remove the old throttle body gasket (see illustration) and discard it.
6  Installation is the reverse of removal. Be sure to use a new gasket and tighten the throttle body mounting fasteners to the torque listed in this Chapter's Specifications.
7  Start the engine, then verify that the throttle body operates correctly and that there are no air leaks.

13.3  To disconnect the throttle body electrical connector, pull out the lock, then depress the release tab and pull off the connector (V8 model shown, V6 models similar)

13.4  To detach the throttle body from the intake manifold, remove the four fasteners

13.5  Remove and inspect the throttle body's O-ring type gasket and discard it. Always use a new O-ring gasket when installing the throttle body, to prevent air leaks

## 14  Fuel meter body and injectors (V6 models) - removal and installation

### UPPER INTAKE MANIFOLD

▶ **Refer to illustrations 14.5 and 14.10**

1  Disconnect the cable from the negative battery terminal (see Chapter 2, Section 1).
2  Relieve the fuel system pressure (see Section 2).
3  Disconnect the quick-connect fitting between the under-chassis fuel supply line and the engine compartment fuel supply line.
**Note:** *Refer to Section 4 if you're unfamiliar with quick-connect fittings. Cap the two open fuel lines to prevent contaminants from entering the fuel system.*
4  If the vehicle is equipped with air conditioning, disconnect the electrical connectors from the air conditioning pressure switch and the compressor clutch (see Chapter 3).
5  Trace each lead in the engine harness to its corresponding electrical connector and disconnect all connectors, including the electrical connector for the port injector module on top of the fuel meter body (see illustration) and the connectors for the EVAP canister purge solenoid and the Manifold Absolute Pressure (MAP) sensor (see Chapter 6). Remove the bolts that secure the engine wiring harness rear bracket and remove the harness bracket, then detach all harness clips.
6  Disconnect the EVAP purge line quick-connect fitting from the EVAP canister purge solenoid (see Chapter 6) and push the purge line aside.
7  Remove the nut that secures the engine wiring harness ground wire to the stud on the rear of the right cylinder head, then remove the stud and detach the engine wiring harness bracket. Push the engine wiring harness and bracket aside.
8  Remove the PCV crankcase ventilation hose.
9  Disconnect the power brake booster vacuum hose from the vacuum fitting.
10  Remove the upper intake manifold mounting bolts and the two throttle body mounting studs (see illustration) and remove the upper intake manifold.
11  Remove and discard the old upper intake manifold gasket.
12  Installation is the reverse of removal. Be sure to use a new gasket and tighten the upper intake manifold fasteners to the torque listed in this Chapter's Specifications.

### FUEL PULSATION DAMPER

▶ **Refer to illustration 14.14**

13  Remove the upper intake manifold (see Steps 1 through 11).
14  Remove the fuel pulsation damper retaining clip (see illustration).
15  Pull the fuel pulsation damper out of the fuel meter body. Cover the cavity for the pulsation damper to prevent contamination from entering the fuel meter body.
16  Remove and discard the old fuel pulsation damper O-ring.
17  Installation is the reverse of removal. Be sure to use a new damper O-ring. Lubricate the new O-ring with clean engine oil.

**14.5  Remove the retaining clip and disconnect the electrical connector from the port injector module on top of the fuel meter body**

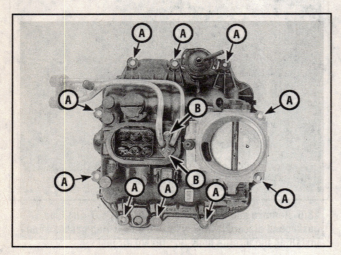

**14.10  Upper intake manifold bolt locations**

A  Mounting bolts     B  Fuel line retainer bolts

**14.14  To remove the fuel pulsation damper retaining clip, spread the ends of the retainer apart, then pull the damper out of the fuel meter body (fuel meter body removed for clarity)**

# FUEL AND EXHAUST SYSTEMS    4-17

## FUEL METER BODY AND INJECTORS

♦ Refer to illustrations 14.21, 14.22a, 14.22b, 14.23 and 14.24

18  Remove the upper intake manifold (see Steps 1 through 11).
19  Remove the fuel supply line bracket bolt, remove the fuel supply line retainer bolts (see illustration 14.10) and remove the fuel supply line from the fuel meter body.
20  Remove and discard the fuel seal retainer, the upper fuel seal, the spacer ring and the lower fuel seal from the hole in the top of the fuel meter body into which the fuel supply line is installed.
21  Detach the injector poppet nozzles from the lower intake manifold (see illustration).
22  Pry the fuel meter body mounting bracket retaining tabs away from the mounting lugs, then lift the fuel meter body, injector lines and poppet nozzles off the upper intake manifold as a single assembly (see illustrations).
23  Remove the injector retainer nuts and the retainer (see illustration).
24  Remove the injectors from the fuel meter body (see illustration).
25  Installation is the reverse of removal.

14.21  Squeeze the lock tabs together and pull each poppet nozzle out of its bore in the lower intake manifold

14.22a  Using two screwdrivers, pry the locking tabs away from the fuel meter body . . .

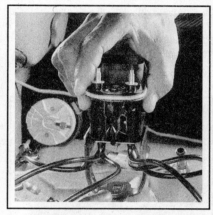

14.22b  . . . and remove the fuel meter body, injector lines and poppet nozzles from the lower intake manifold

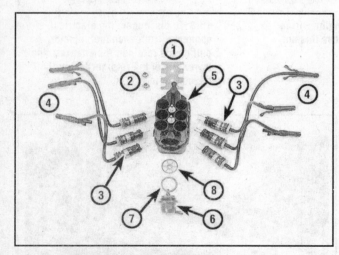

14.23  Fuel meter body disassembled:

1  Fuel injector retaining plate
2  Retaining plate nuts
3  Fuel injectors
4  Poppet nozzles
5  Fuel meter body
6  Fuel pulsation damper
7  O-ring
8  Filter screen (if equipped)

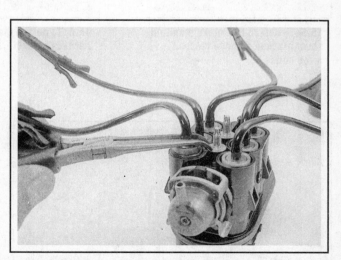

14.24  To remove each injector from the fuel meter body, pull on the tube fitting while pushing on the injector from the other side

## 15 Fuel rail and injectors (V8 models) - removal and installation

◆ Refer to illustrations 15.5, 15.6, 15.8, 15.10, 15.11, 15.13, 15.14, 15.15, 15.16 and 15.18

1  Remove the intake manifold cover.
2  Relieve the fuel system pressure (see Section 2).
3  Disconnect the cable from the negative battery terminal (see Chapter 5, Section 1).
4  Remove the air intake duct (see Section 10).
5  Remove the intake manifold cover bracket bolts (see illustration) and remove the manifold cover bracket.
6  Remove the nut that secures the engine wiring harness bracket to the intake manifold (see illustration) and detach the harness bracket.
7  Trace each electrical lead in the engine wiring harness to its corresponding electrical connector and disconnect the electrical connectors from the following components:

Alternator (see Chapter 5) - also detach the engine wiring harness clip from the alternator
EVAP purge solenoid (see Chapter 6)
Ignition coils (see Chapter 5)
Manifold Absolute Pressure (MAP) sensor (see Chapter 6)
Throttle body (see Section 13)

8  Disconnect the electrical connectors from the fuel injectors (see illustration).
9  Detach all engine wiring harness clips (including the clip on the alternator) and set the engine wiring harness aside.
10  Disconnect the PCV crankcase ventilation hose (PCV hose) from the intake manifold (see illustration) and set the PCV hose aside.
11  Disconnect the fuel line from the fuel rail (see illustration).
12  Disconnect the EVAP canister purge solenoid from the intake manifold and remove the EVAP canister from the fuel rail (see Chapter 6).

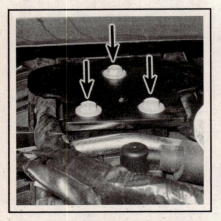

**15.5  To detach the intake manifold cover bracket, remove these three bolts**

**15.6  To detach the engine wiring harness bracket, remove this nut**

**15.8  To disconnect the electrical connector from each fuel injector, pull up the retainer, then depress the lower end of the retainer and pull off the connector**

**15.10  To detach the PCV hose from the intake manifold, grasp it firmly and pull it straight up**

**15.11  Disconnect the fuel supply line quick-connect fitting at the fuel rail (if you're unfamiliar with metal-collar type quick-connect fittings, see Section 4).**

## FUEL AND EXHAUST SYSTEMS    4-19

13  Remove the fuel rail mounting bolts (see illustration).

14  Carefully disengage the injectors from the intake manifold and lift the fuel rail and all six injectors from the engine as a single assembly (see illustration).

15  Remove each injector retainer (see illustration) and remove the injectors from the fuel rail.

16  Remove and discard the old injector O-rings (see illustration). Always install new O-rings on the injectors before reassembling the injectors and the fuel rail.

17  To ensure that the new injector O-rings are not damaged when the injectors are installed into the fuel rail and into the intake manifold, lubricate them with clean engine oil.

18  When installing the injector retainers, make sure that they're aligned properly and fit tightly (see illustration). If a retainer is damaged, replace it.

19  Installation is otherwise the reverse of removal. Be sure to tighten the fuel rail mounting bolts securely.

20  Turn the ignition key to the On position to pressurize the fuel system and verify that there are no fuel leaks.

**15.13  To detach the fuel rail, remove the four mounting bolts (two bolts on right side fuel rail shown)**

**15.14  Disengage the fuel injectors from the intake manifold, then lift the fuel rail and the injectors as a single assembly**

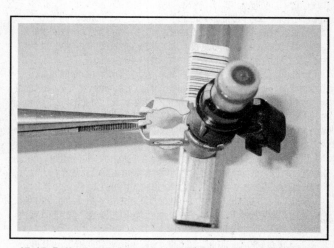

**15.15  Pull off each injector retaining clip with needle nose pliers, then pull the injector out of the fuel rail**

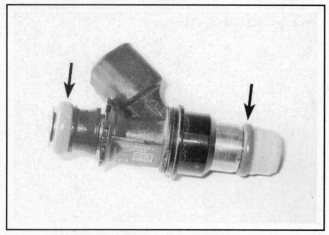

**15.16  Remove the O-rings from each fuel injector, then install new O-rings. Coat the O-rings with a little clean engine oil to protect them from damage**

**15.18  When installing each injector retainer, make sure that it's aligned properly and fits tightly**

# 4-20 FUEL AND EXHAUST SYSTEMS

## 16 Exhaust system servicing - general information

### INSPECTION

**⁕⁕ WARNING:**

Inspect and repair exhaust system components only after allowing the exhaust components to cool completely. This applies particularly to the catalytic converter, which operates at very high temperatures. Also, when working under the vehicle, make sure it is securely supported on jackstands.

1  The exhaust system consists of the exhaust manifold(s), the catalytic converter(s), the exhaust pipes, the muffler, and all brackets, hangers and clamps that support the exhaust system. Inspect the exhaust system regularly to ensure that it remains safe and quiet. Look for any damaged or bent parts, open seams, holes, loose connections, excessive corrosion or other defects which could allow exhaust fumes to enter the vehicle. Also check the catalytic converter(s) when you inspect the exhaust system. Inspect the catalytic converter heat shield(s) for cracks, dents and loose or missing fasteners. If a heat shield is damaged, the converter might also be damaged. Damaged or deteriorated exhaust system components should not be repaired; they should be replaced with new parts.

2  Before trying to disassemble any exhaust components, spray the fasteners with a penetrating oil to help ease removal. If the exhaust system components are extremely corroded or rusted together, welding equipment will probably be required to remove them. The convenient way to accomplish this is to have a muffler repair shop remove the corroded sections with a cutting torch. If, however, you want to save money by doing it yourself (and you don't have a welding outfit with a cutting torch), simply cut off the old components with a hacksaw. If you have compressed air, special pneumatic cutting chisels can also be used. If you decide to tackle the job at home, be sure to wear safety goggles to protect your eyes from metal chips and work gloves to protect your hands.

3  Here are some simple guidelines to follow when repairing the exhaust system:

a) Work from the back to the front when removing exhaust system components.
b) Apply penetrating oil to the exhaust system component fasteners to make them easier to remove.
c) Use new gaskets, hangers and clamps when installing exhaust systems components.
d) Apply anti-seize compound to the threads of all exhaust system fasteners at reassembly.
e) Be sure to allow sufficient clearance between newly installed parts and all points on the underbody to avoid overheating the floor pan and possibly damaging the interior carpet and insulation. Pay particularly close attention to the catalytic converter and heat shield.

### COMPONENT REPLACEMENT

#### Rubber hangers

▶ Refer to illustration 16.4

4  The exhaust system is attached to the body with rubber hangers (see illustration). Anytime you must raise the vehicle to perform any under-vehicle service, make sure that you inspect the exhaust hangers. Look for cracks, tears and deterioration. If a rubber hanger is worn or damaged, replace it.

#### Catalytic converters

5  The procedure for replacing the catalytic converters is in Chapter 6.

**16.4 Exhaust system rubber hanger**

# FUEL AND EXHAUST SYSTEMS 4-21

## Specifications

| | |
|---|---|
| Fuel pressure | |
|     Ignition key turned to ON, engine not running | 50 to 60 psi |
|     Engine idling | |
|         V6 engine | Not available |
|         V8 engine | 43 to 45 psi |
| Fuel pressure leakdown | |
|     One minute after turning key to OFF | No more than 5 psi drop |
| Fuel injector coil resistance (V8 models only) | 11 to 14 ohms (approximate) |

## Torque specifications

➡ **Note: One foot-pound (ft-lb) of torque is equivalent to 12 inch-pounds (in-lbs) of torque. Torque values below approximately 15 ft-lbs are expressed in inch-pounds, since most foot-pound torque wrenches are not accurate at these smaller values.**

| | |
|---|---|
| Throttle body mounting fasteners | 89 in-lbs |

# 4-22 FUEL AND EXHAUST SYSTEMS

## Notes

**Section**

1. General information, precautions and battery disconnection
2. Battery - emergency jump starting
3. Battery - check and replacement
4. Battery cables - check and replacement
5. Ignition system - general information
6. Ignition system - check
7. Ignition coil pack - removal and installation
8. Charging system - general information and precautions
9. Charging system - check
10. Alternator - removal and installation
11. Starting system - general information and precautions
12. Starter motor and circuit - check
13. Starter motor - removal and installation

# 5
## ENGINE ELECTRICAL SYSTEMS

# 5-2 ENGINE ELECTRICAL SYSTEMS

## 1  General information, precautions and battery disconnection

The engine electrical systems include all ignition, charging and starting components. Because of their engine-related functions, these components are discussed separately from chassis electrical devices such as the lights, the instruments, etc. (see Chapter 12).

### PRECAUTIONS

Always observe the following precautions when working on the electrical system:

a) Be extremely careful when servicing engine electrical components. They are easily damaged if checked, connected or handled improperly.
b) Never leave the ignition switched on for long periods of time when the engine is not running.
c) Never disconnect the battery cables while the engine is running.
d) Maintain correct polarity when connecting battery cables from another vehicle during jump starting - see the "Booster battery (jump) starting" Section at the front of this manual.
e) Always disconnect the negative cable from the battery before working on the electrical system, but read the following battery disconnection procedure first.
f) The ignition coils generate extremely high voltage. Never touch an ignition coil or a spark plug wire with the engine running.

It's also a good idea to review the safety-related information regarding the engine electrical systems located in the "Safety first!" Section at the front of this manual before beginning any operation included in this Chapter.

### BATTERY DISCONNECTION

#### ✶✶ WARNING:

On models with OnStar, make absolutely sure the ignition key is in the Off position and Retained Accessory Power (RAP) has been depleted before disconnecting the cable from the negative battery terminal. Also, never remove the OnStar fuse with the ignition key in any position other than Off. If these precautions are not taken, the OnStar system's back-up battery will be activated, and remain activated, until it goes dead. If this happens, the OnStar system will not function as it should in the event that the main vehicle battery power is cut off (as might happen during a collision).

➡ Note: To disconnect the battery for service procedures requiring power to be cut from the vehicle, first open the driver's door to disable Retained Accessory Power (RAP), then loosen the cable end bolt and disconnect the cable from the negative battery terminal. Isolate the cable end to prevent it from coming into accidental contact with the battery terminal.

The battery is located in the engine compartment on all vehicles covered by this manual. To disconnect the battery for service procedures that require battery disconnection, simply disconnect the cable from the negative battery terminal. Make sure that you isolate the cable to prevent it from coming into contact with the battery negative terminal.

Some vehicle systems (radio, alarm system, power door locks, etc.) require battery power all the time, either to enable their operation or to maintain control unit memory (Powertrain Control Module, Transmission Control Module, etc.), which would be lost if the battery were to be disconnected. So before you disconnect the battery, note the following points:

a) Before connecting or disconnecting the cable from the negative battery terminal, make sure that you turn the ignition key and the lighting switch to their OFF positions. Failure to do so could damage semiconductor components.
b) On a vehicle with power door locks, it is a wise precaution to remove the key from the ignition and to keep it with you, so that it does not get locked inside if the power door locks should engage accidentally when the battery is reconnected!
c) After the battery has been disconnected, then reconnected (or a new battery has been installed), the Transmission Control Module (TCM) will need some time to relearn its adaptive strategy. As a result, shifting might feel firmer than usual. This is a normal condition and will not adversely affect the operation or service life of the transmission. Eventually, the TCM will complete its adaptive learning process and the shift feel of the transmission will return to normal.
d) The engine management system's PCM has some learning capabilities that allow it to adapt or make corrections in response to minor variations in the fuel system in order to optimize driveability and idle characteristics. However, the PCM might lose some or all of this information when the battery is disconnected. The PCM must go through a relearning process before it can regain its former driveability and performance characteristics. Until it relearns this lost data, you might notice a difference in driveability, idle and/or shift feel.

### MEMORY SAVERS

Devices known as "memory savers" (typically, small 9-volt batteries) can be used to avoid some of the above problems. A memory saver is usually plugged into the cigarette lighter, then you can disconnect the vehicle battery from the electrical system. The memory saver will deliver sufficient current to maintain security alarm codes and maybe PCM memory. It will also run unswitched (always on) circuits such as the clock and radio memory, while isolating the car battery in the event that a short circuit occurs while the vehicle is being serviced.

#### ✶✶ WARNING:

If you're going to work around any airbag system components, disconnect the battery and do not use a memory saver. If you do, the airbag could accidentally deploy and cause personal injury.

#### ✶✶ CAUTION:

Because memory savers deliver current to operate unswitched circuits when the battery is disconnected, make sure that the circuit that you're going to service is actually open before working on it!

## 2  Battery - emergency jump starting

Refer to the *Booster battery (jump) starting* procedure at the front of this manual.

# ENGINE ELECTRICAL SYSTEMS 5-3

## 3  Battery - check and replacement

**※※ WARNING:**
Hydrogen gas is produced by the battery, so keep open flames and lighted cigarettes away from it at all times. Always wear eye protection when working around a battery. Rinse off spilled electrolyte immediately with large amounts of water.

## CHECK

▶ Refer to illustrations 3.1a, 3.1b and 3.1c

1  A battery cannot be accurately tested until it is at or near a fully charged state. Disconnect the negative battery cable from the battery and perform the following tests:

   a) **Battery state of charge test** - Visually inspect the indicator eye (if equipped) on the top of the battery. If the indicator eye is dark in color, charge the battery as described in Chapter 1. If the battery is equipped with removable caps, check the battery electrolyte. The electrolyte level should be above the upper edge of the plates. If the level is low, add distilled water. DO NOT OVERFILL. The excess electrolyte may spill over during periods of heavy charging. Test the specific gravity of the electrolyte using a hydrometer (see illustration). Remove the caps and extract a sample of the electrolyte and observe the float inside the barrel of the hydrometer. Follow the instructions from the tool manufacturer and determine the specific gravity of the electrolyte for each cell. A fully charged battery will indicate approximately 1.270 (green zone) at 68-degrees F (20-degrees C). If the specific gravity of the electrolyte is low (red zone), charge the battery as described in Chapter 1.

   b) **Open circuit voltage test** - Using a digital voltmeter, perform an open circuit voltage test (see illustration). Connect the negative probe of the voltmeter to the negative battery post and the positive probe to the positive battery post. The battery voltage should be greater than 12.5 volts. If the battery is less than the specified voltage, charge the battery before proceeding to the next test. Do not proceed with the battery load test until the battery is fully charged.

   c) **Battery load test** - An accurate check of the battery condition can only be performed with a load tester (available at most auto parts stores). This test evaluates the ability of the battery to operate the starter and other accessories during periods of heavy amperage draw (load). Connect a battery load-testing tool to the battery terminals (see illustration). Load test the battery according to the tool manufacturer's instructions. This tool increases the load demand (amperage draw) on the battery. Maintain the load on the battery for 15 seconds and observe that the battery voltage does not drop below 9.6 volts. If the battery condition is weak or defective, the tool will indicate this condition immediately. Note: Cold temperatures will cause the minimum voltage reading to drop slightly. Follow the chart given in the tool manufacturer's instructions to compensate for cold climates. Minimum load voltage for freezing temperatures (32-degrees F/0-degrees C) should be approximately 9.1 volts.

   d) **Battery drain test** - This test will indicate whether there's a constant drain on the vehicle's electrical system that can cause the battery to discharge. Make sure all accessories are turned off. If the vehicle has an underhood light, verify that it's working properly, then disconnect it. Connect one lead of a digital ammeter to the disconnected negative battery cable clamp and the other lead to the negative battery post. A drain of approximately 100 milliamps or less is considered normal (due to the engine control computers, clocks, digital radios and other components that normally cause a key-off battery drain). An excessive drain (approximately 500 milliamps or more) will cause the battery to discharge. The problem circuit or component can be located by removing the fuses, one at a time, until the excessive drain stops.

3.1a  Checking the specific gravity of the battery electrolyte with a hydrometer; this hydrometer is equipped with a thermometer to make temperature corrections

3.1b  To test the open-circuit voltage of the battery, connect the black probe of a voltmeter to the negative terminal and the red probe to the positive terminal of the battery; if the battery is fully charged, the voltmeter should indicate at least 12.6 volts (depending on the outside air temperature)

3.1c  Some battery load testers are equipped with an ammeter, which enables you to impose a precise load on the battery (less expensive testers, like this one, have only a load switch and a voltmeter)

## 5-4 ENGINE ELECTRICAL SYSTEMS

3.2  To detach the front fender rear upper brace, remove these bolts

3.3  Disconnect the negative battery cable (1) first, then the positive cable (2)

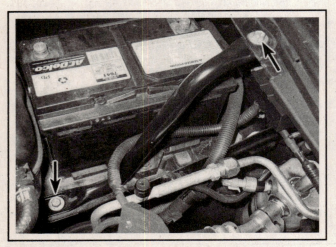

3.5  To detach the battery tray support bracket, remove these bolts

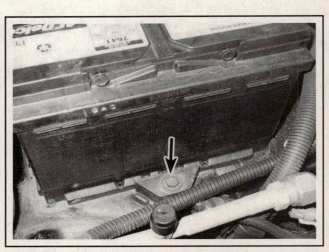

3.6  To detach the battery from the battery tray, remove the battery hold-down clamp bolt and clamp

### REPLACEMENT

#### Battery

♦ Refer to illustrations 3.2, 3.3, 3.5 and 3.6

2  Remove the front fender rear upper brace (see illustration).
3  Disconnect the cable from the negative battery terminal (see illustration).

**※※ WARNING:**

**ALWAYS disconnect the negative cable FIRST and connect it last, or you might accidentally short the battery with the tool you're using to loosen the cable clamps.**

4  Open the red plastic cover protecting the positive battery terminal and disconnect the cable from the positive terminal.
5  Remove the battery tray support bracket (see illustration).
6  Remove the hold-down clamp bolt and clamp (see illustration).
7  Lift out the battery.
8  While the battery is removed, clean and inspect the battery tray. If the tray has any corrosion on it, be sure to neutralize it with baking soda and water. If the tray is damaged, replace it (see Steps 10 through 12).
9  Installation is the reverse of removal.

**※※ WARNING:**

**When connecting the battery cables, always connect the positive cable first and the negative cable last.**

## ENGINE ELECTRICAL SYSTEMS 5-5

### Battery tray

▶ **Refer to illustration 3.11**

10  Remove the battery (see Steps 2 through 7).
11  Remove the battery tray (see illustration).
12  Installation is the reverse of removal.

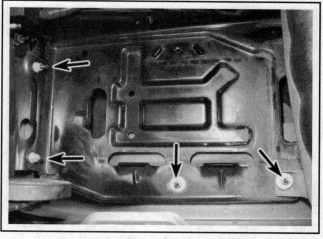

3.11  To detach the battery tray, remove these nuts

### 4  Battery cables - check and replacement

#### CHECK

▶ **Refer to illustration 4.2**

1  Periodically inspect the entire length of each battery cable for damage, cracked or burned insulation and corrosion. Poor battery cable connections can cause starting problems and decreased engine performance.
2  Check the cable-to-terminal connections at the ends of the cables for cracks, loose wire strands and corrosion (see illustration). The presence of white, fluffy deposits under the insulation at the cable terminal connection is a sign that the cable is corroded and should be replaced.

Check the terminals for distortion, missing mounting bolts and corrosion (see Chapter 1 for further information regarding battery cable maintenance).

#### REPLACEMENT

3  Before disconnecting anything, review the following procedure and photos, and note the routing of the cables to ensure correct installation.
4  Disconnect the cables from the battery terminals (negative first, positive last), then disconnect the negative or positive battery cable as follows:

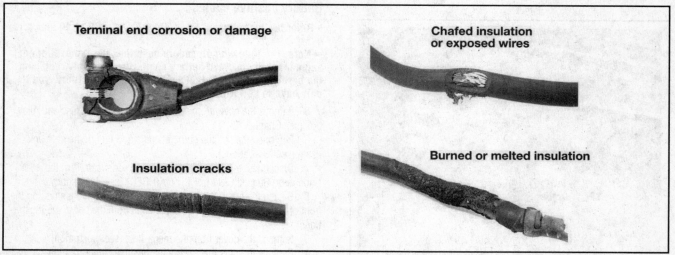

4.2  Typical battery cable problems

## 5-6 ENGINE ELECTRICAL SYSTEMS

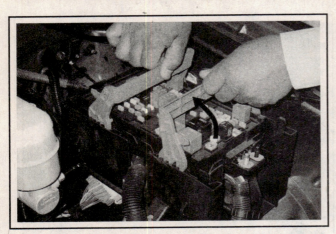

4.6a To remove the fuse panel from the engine compartment fuse and relay box, flip up the two fuse box retainers . . .

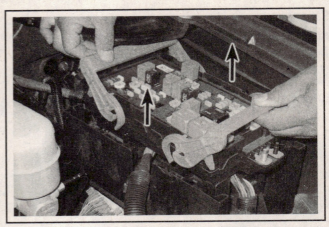

4.6b . . . then carefully work the fuse panel up and out of the fuse box

4.7 To disengage the positive cable from the Universal Bussed Electrical Center (UBEC) on the fuse panel, carefully spread these four retainers away from the terminal cover with a pair of small screwdrivers and pull up the cover

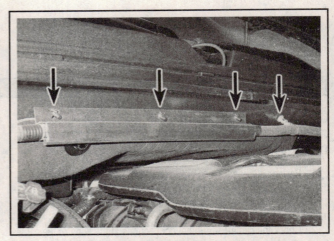

4.8 Carefully pry off the positive cable protector and detach the cable clips from the firewall

### Battery positive cables

♦ Refer to illustrations 4.6a, 4.6b, 4.7, 4.8, 4.9, 4.10 and 4.11

→ Note: The following procedure describes the removal of each segment of the positive battery cable. If you're only replacing the portion of the cable that's connected to the battery, you'll only have to perform Steps 9 and 10.

5  Remove the cover from the engine compartment fuse and relay box (see Chapter 12).

6  Flip open the two fuse box retainers and lift out the fuse box panel (see illustrations).

7  Disconnect the positive battery cable connector from the Underhood Bussed Electrical Center (UBEC) (see illustration).

8  Detach the positive battery cable protector from the studs on the front of the cowl and detach the cable clips from the cowl (see illustrations).

9  Remove the cover from the mega-fuse (see illustration).

10  Remove the nuts that secure the positive cable(s) to the alternator, battery positive terminal and fuse and relay box, as required (see illustration) and disconnect the cable(s) being replaced from the mega-fuse.

11  Trace the cable to the alternator and from the stud on the mega-

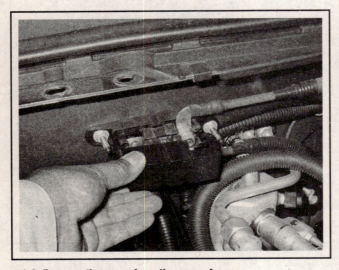

4.9 Remove the cover from the mega-fuse

# ENGINE ELECTRICAL SYSTEMS  5-7

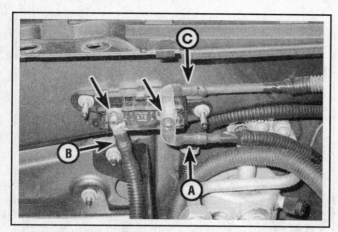

**4.10 To detach the three positive cables from the mega-fuse, remove the nuts from both stud terminals and disconnect the cables:**

- A  Positive cable to alternator B+ (battery) terminal
- B  Positive cable to battery positive terminal
- C  Positive cable to engine compartment fuse and relay box

**4.11 To disconnect the other end of the alternator positive cable, pull back the weather boot and remove this nut**

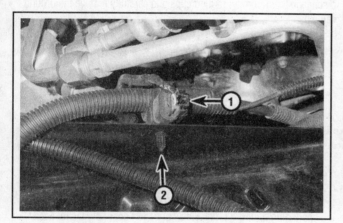

**4.13 To detach the current sensor (if equipped) from the battery tray, depress the release tab (1) and disconnect the electrical connector, then carefully push out the retaining clip (2) (battery removed for clarity)**

**4.14 To detach the ground cable (A) from the stud on the front of the right cylinder head, remove this nut. When reattaching the ground cable to the stud, don't forget to also reattach the engine harness terminal (B) to the stud**

fuse to the alternator, detach any cable clips or disengage the cable from the clips and disconnect the terminal from the B+ stud on the alternator (see illustration).

12  Installation is the reverse of removal.

## Battery ground cable

▶ **Refer to illustrations 4.13, 4.14, 4.16, 4.17a and 4.17b**

13  Trace the ground cable down to the battery current sensor (see illustration), which, if equipped, will be secured to the left side of the battery tray by a clip. Disconnect the electrical connector from the current sensor, then detach the sensor clip from the battery tray.

14  Disconnect the ground cable from the stud on the front end of the right cylinder head (see illustration).

15  Raise the front of the vehicle and place it securely on jackstands.

➡ **Note: This may or may not be necessary, depending on the ground clearance of the vehicle.**

16  Remove the under-engine rock guard, if equipped (see illustration).

**4.16 To detach the under-engine rock guard (if equipped), remove these four bolts**

# 5-8  ENGINE ELECTRICAL SYSTEMS

**4.17a  Locate the left lower ground cable terminal on the front side of the outer left end of the front frame crossmember, remove this bolt and detach the ground cable from the crossmember**

**4.17b  Locate the right lower ground cable terminal on the front side of the outer right end of the front frame crossmember, remove this bolt and detach the ground cable from the crossmember**

17  Detach the lower ground cable terminals from the left and right ends of the front frame crossmember (see illustrations).

18  Be sure to note how the two lower ground cables are routed, then disengage them from all routing clips and carefully remove the ground cable harness.

19  Installation is the reverse of removal.

## 5  Ignition system - general information

The ignition system consists of the ignition control module (V6 models), the ignition coil pack(s), the spark plugs, the Camshaft Position (CMP) sensor (V6 models only), the Crankshaft Position (CKP) sensor, the knock sensor(s) and the Powertrain Control Module (PCM).

### V6 MODELS

V6 models use three ignition coils; each coil fires two spark plugs via spark plug wires. The three coils and the ignition control module are combined into one integral coil pack unit mounted on top of the engine behind the throttle body. The ignition control module contains an ignition coil driver circuit for each ignition coil. If a coil pack or the ignition module fails, you must replace the entire coil pack unit.

Each coil fires two spark plugs. One coil fires the No. 1 and No. 4 cylinders; the second coil fires the No. 2 and No. 5 cylinders; the third coil fires the No. 3 and No. 6 cylinders. When the piston in cylinder No. 1 is on the compression stroke, the piston in cylinder No. 4 is on the exhaust stroke. When the coil fires cylinder No. 1, most of the spark voltage goes to that cylinder because the pressure - and therefore the resistance - is high in that cylinder (the higher the resistance, the higher the voltage needed to jump the gap from the spark plug's center electrode to ground). Conversely, the piston in cylinder No. 4, which is on the exhaust stroke, produces no pressure, and therefore no resistance, so little voltage is needed to jump the gap from the spark plug's center electrode to ground. The ignition coils for cylinders 2 and 5 and 3 and 6 work the same way. This design is known as a waste spark ignition.

### V8 MODELS

V8 models have eight ignition coils; each coil fires its corresponding spark plug, in firing order, via a very short spark plug wire. Each coil is located on the valve cover, directly above the plug that it fires. Each gang of four coils is mounted on a bracket that is bolted to the valve cover. If you need to remove the valve cover, you can remove all four coils and the mounting bracket as a single assembly. If a coil fails, you can remove and replace it without having to replace the entire coil assembly for that cylinder head. There is no separate ignition control module; each coil has an integral module.

### ALL MODELS

Each ignition coil has a terminal for voltage supply (12 volts) and a terminal for ground. The Powertrain Control Module (PCM) supplies a low reference circuit and an ignition control (IC) circuit, which controls the ground path for the primary winding in the coil.

The PCM uses inputs from a variety of information sensors to control ignition timing: the Camshaft Position (CMP) sensor (V6 models), the Crankshaft Position (CKP) sensor, the Engine Coolant Temperature (ECT) sensor, the Intake Air Temperature (IAT) sensor, the knock sensors, the Mass Air Flow (MAF) sensor, the Transmission Range (TR) sensor and the Vehicle Speed Sensor (VSS). For more information on these other sensors, refer to Chapter 6.

## 6  Ignition system - check

♦ Refer to illustration 6.2

**WARNING:**
Because of the very high voltage generated by the ignition system, use extreme care when you're servicing ignition components such as the ignition coil pack and spark plugs.

**Note 1:** The ignition system components are difficult to diagnose. In the event of ignition system failure, if the checks do not clearly indicate the source of the ignition system problem, have the vehicle tested by a dealer service department or other qualified auto repair facility.

**Note 2:** For the following test, you'll need a calibrated spark tester (available at auto parts stores). Don't use the old-fashioned type of spark tester that looks like a spark plug with an alligator clip on it. This type of tester disables the spark plug for the plug wire that you're testing, which might set a Diagnostic Trouble Code for a misfire. Instead, use the type of spark tester shown in the photo that accompanies this Section. This newer type of tester enables you to check for spark without disabling the spark plug for the plug wire that you're testing.

1  If a malfunction occurs and the vehicle won't start, do not immediately assume that the ignition system is causing the problem. First, check the following items:
  a) Make sure the battery cable clamps, where they connect to the battery, are clean and tight.
  b) Test the condition of the battery (see Section 3). If it does not pass all the tests, replace it with a new battery.
  c) Check the wiring and connections for the ignition control module and for the ignition coil pack. On V6 models, inspect the spark plug wires; make sure that they're in good shape and tightly connected to the coil pack and to the spark plugs.
  d) Check the related fuses inside the fuse box (see Chapter 12). If they're burned, determine the cause and repair the circuit.

2  If the engine turns over but won't start, disconnect a spark plug wire from a spark plug and attach a spark tester between the ignition

6.2  Checking an ignition coil with a spark tester. If the coil is delivering power to the plug, the tester will flash (this type of spark tester is available at most auto parts stores)

coil high-tension terminal and the spark plug (see illustration). Spark testers are available at most auto parts stores. Crank the engine and note whether or not the tester flashes.

3  If the tester flashes during cranking, the coil is delivering sufficient voltage to the spark plug to fire it. Repeat this test for each cylinder to verify that the other coils are OK.

4  If the tester doesn't flash, remove a coil from another cylinder and swap it for the one being tested. If the tester now flashes, you know that the original coil is bad. If the tester still doesn't flash, the PCM or wiring harness is probably defective. Have the PCM checked out by a dealer service department or other qualified repair shop (testing the PCM is beyond the scope of the do-it-yourselfer because it requires expensive special tools).

5  If the tester flashes during cranking but a misfire code (related to the cylinder being tested) has been stored, the spark plug could be fouled or defective, or the fuel injector for that cylinder could be defective.

## 7  Ignition coil pack - removal and installation

### V6 MODELS

1  Relieve the fuel system pressure (see Chapter 4), then disconnect the cable from the negative battery terminal (see Section 3).

2  Remove the air filter housing (see Chapter 4).

3  Disconnect the quick-connect fitting between the fuel supply line and the fuel rail (see Chapter 4).

4  Disconnect the electrical connectors from the Manifold Absolute Pressure (MAP) sensor electrical connector and from the EVAP canister purge solenoid valve (see Chapter 6).

5  Disconnect the throttle body electrical connector (see Chapter 4).

6  Detach the engine wiring harness clips from their respective brackets and set the harness aside, then remove the engine harness rear bracket bolts and the bracket.

7  Mark the positions of the spark plug wires, then disconnect the plug wires from the ignition coil (see Chapter 1) and set the plug wires aside.

8  Disconnect the electrical connector from the ignition coil pack.

9  Remove the four ignition coil pack mounting bolts and remove the ignition coil pack assembly.

10  Before installing the boots on the ignition coil pack, coat the interior of each boot with silicone dielectric compound. Installation is otherwise the reverse of removal.

## 5-10  ENGINE ELECTRICAL SYSTEMS

### V8 MODELS

**Refer to illustration 7.13**

11 Disconnect the cable from the negative battery terminal (see Section 3).

12 Remove the intake manifold cover (see Chapter 2B).

13 Disconnect the electrical connector from the ignition coil (see illustration).

14 Disconnect the spark plug wire boot from the spark plug (see Chapter 1).

15 Remove the two ignition coil mounting bolts and remove the coil.

16 Installation is the reverse of removal.

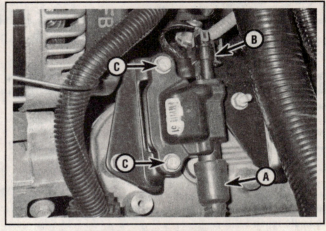

**7.13  Coil mounting details (V8 engine)**

A   Spark plug wire     C   Mounting bolts
B   Electrical connector

### 8  Charging system - general information and precautions

The charging system supplies electrical power for the ignition system, the lights, the radio, the electronic control systems and all other electrical components on the car. The charging system consists of the battery, the alternator (with an integral voltage regulator), the Powertrain Control Module (PCM), the charge indicator lamp on the instrument panel cluster, the mega-fuse (located on the left side of the firewall) and the wiring between all the components.

The alternator generates alternating current (AC), which is rectified to direct current (DC) to charge the battery and supply power to other electrical systems. The alternator is driven by a drivebelt at the front of the engine. The voltage regulator limits the alternator charging voltage by regulating the current supplied to the alternator field circuit. The regulator is a solid-state electronic assembly mounted inside the alternator. The regulator is not separately replaceable on these vehicles. If it's defective, you must replace the alternator.

Inspect the alternator drivebelt, battery and all charging system wires and connections at the intervals listed in Chapter 1.

Be very careful when making any circuit connections and note the following:

a) Never start the engine with a battery charger connected.
b) Never disconnect a battery cable with the engine running.
c) Always disconnect both battery cables before using a battery charger: negative cable first, positive cable last.

### 9  Charging system - check

**Refer to illustration 9.3**

1  If the charging system malfunctions, don't immediately assume that the alternator is causing the problem. First check the following items:

a) Ensure that the battery cable connections at the battery are clean and tight.
b) If the battery is not a maintenance-free type, check the electrolyte level and specific gravity. If the electrolyte level is low, add clean, mineral-free tap water. If the specific gravity is low, charge the battery.
c) Check the alternator wiring and connections.
d) Check the drivebelt condition and tension (see Chapter 1).
e) Check the alternator mounting bolts for looseness.
f) Run the engine and check the alternator for abnormal noise.

2  Use a voltmeter to check the battery voltage with the engine off. It should be at least 12.6 volts (see illustration 3.1b).

3  Start the engine and check the battery voltage again. It should now be approximately 13.5 to 15.5 volts (see illustration)

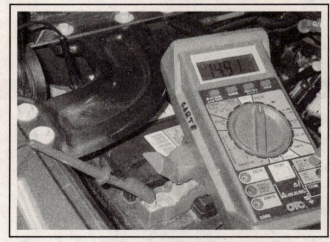

**9.3  To check charging voltage, hook up a voltmeter to the battery terminals and note the voltage with the engine running - it should be 13.5 to 15.5 volts**

# ENGINE ELECTRICAL SYSTEMS 5-11

4  If the charging voltage reading is zero, inspect the condition of the mega-fuse located on the left side of the firewall (see illustration 4.9).

5  If the voltage reading is more or less than the specified charging voltage, the voltage regulator is defective. Replace the alternator (the voltage regulator cannot be replaced separately).

## 10 Alternator - removal and installation

1  Disconnect the cable from the negative battery terminal (see Section 3).
2  Remove the air intake duct (see Chapter 4, Section 10).
3  Remove the accessory drivebelt (see Chapter 1).

### V6 MODELS

4  Detach the clip for the engine wiring harness from the air conditioning compressor and condenser hose bracket and remove the bolt from the bracket that secures the air conditioning compressor and condenser hose.
5  Unbolt the clamp that secures the heater outlet hose to the alternator mounting bracket.
6  Disconnect the electrical connector from the alternator.
7  Peel back the weather boot from the B+ terminal, remove the nut and disconnect the battery cable from the alternator.
8  Remove the alternator mounting bolts.
9  Push aside the heater outlet hose and the air conditioning compressor and condenser hoses and remove the alternator.
10  Installation is the reverse of removal.

### V8 MODELS

▶ **Refer to illustration 10.13**

11  Peel back the weather boot from the B+ terminal, remove the nut and disconnect the battery cable from the alternator (see illustration 4.11).

**10.13  To detach the alternator from a V8 model, remove these two bolts**

12  Disconnect the electrical connector from the alternator.
13  Remove the two alternator mounting bolts (see illustration).
14  Remove the alternator.
15  Installation is the reverse of removal. Be sure to tighten the alternator mounting bolts securely.

## 11 Starting system - general information and precautions

The starting system consists of the battery, the ignition switch, the Transmission Range (TR) switch, the starter motor solenoid, the starter motor and the wires that connect these components. The solenoid is located on top of and is an integral part of the starter motor.

The starter motor can be operated only when the shift lever is in PARK or NEUTRAL. When the ignition key is turned to the START position, it sends a signal to the Body Control Module (BCM), which sends a signal to the Powertrain Control Module (PCM). The PCM uses the Transmission Range (TR) switch to verify that the transmission is in PARK or NEUTRAL, then sends battery voltage to the control circuit of the starter relay, which closes the starter circuit and supplies battery voltage the S terminal on the starter solenoid.

# 5-12 ENGINE ELECTRICAL SYSTEMS

Always observe the following precautions when working on the starting system:

a) Excessive cranking of the starter motor can overheat it and cause serious damage. Never operate the starter motor for more than 15 seconds at a time without pausing for at least two minutes to allow it to cool.
b) The starter is connected directly to the battery and could arc or cause a fire if mishandled, overloaded or short-circuited.
c) Always detach the cable from the negative battery terminal before working on the starting system.

## 12 Starter motor and circuit - check

♦ **Refer to illustrations 12.3 and 12.4**

1 If a malfunction occurs in the starting circuit, do not immediately assume that the starter is causing the problem. First, check the following items:

a) Make sure that the battery cable clamps are clean and tight where they connect to the battery.
b) Check the condition of the battery cables (see Section 4). Replace any defective battery cables with new parts.
c) Test the condition of the battery (see Section 3). If it does not pass all the tests, replace it with a new battery.
d) Check the starter solenoid wiring and connections. Refer to the wiring diagrams at the end of Chapter 12.
e) Check the starter mounting bolts for tightness.
f) Make sure that the shift lever is in PARK or NEUTRAL.

2 If the starter motor does not operate when the ignition switch is turned to the START position, check for battery voltage to the solenoid. Connect a test light or voltmeter to the starter solenoid switched terminal (the small wire) while an assistant turns the ignition switch to the START position. If voltage is not available, check the starting system circuit (see the wiring diagrams at the end of Chapter 12). If voltage is available but the starter motor does not operate, remove the starter (see Section 13) and bench test it (see Step 4).

3 If the starter turns over slowly, check the starter cranking voltage and the current draw from the battery. This test must be performed with the starter on the engine. Crank the engine over (for 10 seconds or less) and observe the battery voltage. It should not drop below 8.5 volts. Also, observe the current draw using an inductive type ammeter (see illustration). It should not exceed 400 amps or drop below 250 amps.

※※ **CAUTION:**

**The battery cables might overheat because of the large amount of current being drawn from the battery. Discontinue the testing until the starting system has cooled down.**

If the starter motor cranking amp values are not within the correct range, replace it with a new unit. There are several conditions that may affect the starter cranking potential. The battery must be in good condition and the battery cold-cranking rating must not be under-rated for the particular application. The battery terminals and cables must be clean and not corroded. Also, in cases of extreme cold temperatures, make sure the battery and/or engine block is warmed before performing the tests.

4 If the starter is receiving voltage but does not activate, remove and check the starter/solenoid assembly on the bench (see illustration). Most likely the solenoid is defective. In some rare cases, the engine may be seized so be sure to try and rotate the crankshaft pulley (see Chapter 2) before proceeding. With the starter/solenoid assembly mounted in a vise on the bench, connect one jumper cable from the negative battery terminal to the body of the starter. Connect the other jumper cable from the positive battery terminal to the B+ terminal on the starter. Connect a starter switch and apply battery voltage to the solenoid S terminal (for 10 seconds or less) and see if the solenoid plunger, shift lever and overrunning clutch extends and rotates the pinion drive. If the pinion drive extends but does not rotate, the solenoid is operating but the starter motor is defective. If there is no movement but the solenoid clicks, the solenoid and/or the starter motor is defective. If the solenoid plunger extends and rotates the pinion drive, the starter/solenoid assembly is working properly.

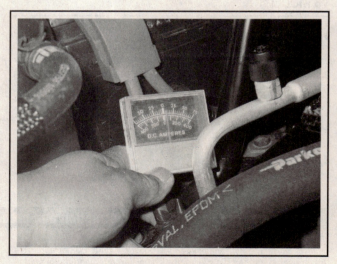

**12.3 Use an inductive-type ammeter to measure the current draw. It can be placed directly on the positive or negative (shown) battery cable, whichever is easier to access (there's no reading on this ammeter because the engine is not being cranked)**

# ENGINE ELECTRICAL SYSTEMS 5-13

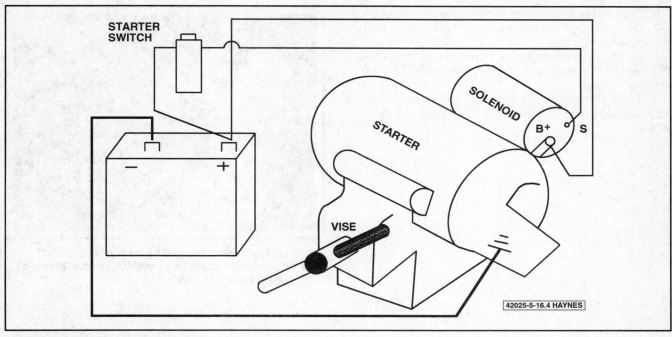

12.4 Starter motor bench testing details

## 13 Starter motor - removal and installation

▶ Refer to illustrations 13.4, 13.5 and 13.6

➡ **Note:** *The photos accompanying this Section depict the starter motor assembly on a V8 model, but the starter motor used on V6 models is in the same location and is similar to what is shown here.*

1  Disconnect the cable from the negative battery terminal (see Section 3).

2  On V8 models, loosen the wheel lug nuts for the right front wheel. On all models, raise the front of the vehicle and place it securely on jackstands. On V8 models, remove the right front wheel.

3  On V6 models, remove the oil pan skid plate, if equipped. On V8 models, remove the right front wheel well splash shield (see Chapter 11).

4  Remove the starter motor heat shield, if equipped (see illustration).

5  Disconnect the battery cable (the larger cable) and the starter control cable (the smaller cable) from the starter motor solenoid terminals (see illustration).

➡ **Note:** *If you're unable to remove the nuts that secure the starter cable and starter control cable nuts with the starter bolted in place (particularly on V8 models), skip this step for now and proceed to Steps 6 and 7, then remove the nuts and disconnect the two electrical cables when you have better access. But be sure to support the starter; don't allow the starter to hang by the cables!*

13.4  To remove the heat shield (if equipped) from the starter motor, simply pull it off

13.5  Remove the smaller nut (A) and disconnect the cable from the starter solenoid terminal, then remove the larger nut (B) and disconnect the battery cable from the starter

## 5-14 ENGINE ELECTRICAL SYSTEMS

6  On V8 models, remove the transmission cover (see illustration).

7  Remove the starter motor mounting bolts (see illustration 13.6) and remove the starter motor.

8  You can remove the starter motor from underneath the vehicle on V6 models. On V8 engines, remove the starter motor through the right wheel well opening.

9  Installation is the reverse of removal. Be sure to tighten the starter motor mounting bolts to the torque listed in this Chapter's Specifications.

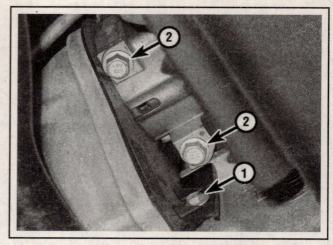

13.6  On V8 engines, remove the transmission cover bolt (1) and remove the cover. On all engines, remove these two bolts (2) to remove the starter

## Specifications

### General

Firing order
- V6 engines ............ 1-6-5-4-3-2
- V8 engines ............ 1-8-7-2-6-5-4-3

Cylinder numbering
- V6 engines
  - Left bank ............ 1-3-5
  - Right bank ........... 2-4-6
- V8 engines
  - Left bank ............ 1-3-5-7
  - Right bank ........... 2-4-6-8

Ignition timing ............ Not adjustable

## Torque specifications

Starter motor mounting bolts (all engines) ............ 37 ft-lbs

**Section**

1 General information
2 On-Board Diagnostic (OBD) system and Diagnostic Trouble Codes (DTCs)
3 Accelerator Pedal Position (APP) sensor - replacement
4 Camshaft Position (CMP) sensor - replacement
5 Crankshaft Position (CKP) sensor - replacement
6 Engine Coolant Temperature (ECT) sensor - replacement
7 Intake Air Temperature (IAT) sensor - replacement
8 Knock sensor - replacement
9 Manifold Absolute Pressure (MAP) sensor - replacement
10 Mass Air Flow (MAF) sensor - replacement
11 Oxygen sensors - general information and replacement and component replacement
12 Transmission Range (TR) switch - replacement and adjustment
13 Vehicle Speed Sensor (VSS) - replacement
14 Powertrain Control Module (PCM) - removal and installation
15 Catalytic converters - description, check and replacement
16 Evaporative emissions control (EVAP) system - description and component replacement
17 Positive Crankcase Ventilation (PCV) system - description
18 Camshaft Position (CMP) Actuator System - description
19 Cylinder Deactivation System - description and component replacement

# 6

# EMISSIONS AND ENGINE CONTROL SYSTEMS

# 6-2 EMISSIONS AND ENGINE CONTROL SYSTEMS

## 1 General information

◆ Refer to illustration 1.7

The emission control systems and components are an integral part of the engine management system, which is called the Sequential Fuel Injection (SFI) system (see Chapter 4 for more information on the SFI system). The SFI system also includes all the government-mandated diagnostic features of the second generation of on-board diagnostics, which is known as On-Board Diagnostics II (OBD-II).

At the center of the SFI and OBD-II systems is the on-board computer, which is known as the Powertrain Control Module (PCM). Using a variety of information sensors, the PCM monitors all of the important engine operating parameters (temperature, speed, load, etc.). It also uses an array of output actuators (such as the ignition coils, fuel injectors and various solenoids and relays) to respond to and alter these parameters as necessary to maintain optimal performance, economy and emissions. The principal emission control systems used on the vehicles covered in this manual include the:

Camshaft Actuator System (V6 models)
Catalytic converter
Displacement on Demand (DOD) system (some V8 models)
Evaporative Emission Control (EVAP) system
Exhaust Gas Recirculation (EGR) system
Positive Crankcase Ventilation (PCV) system
Throttle Actuator Control (TAC) system

The Sections in this Chapter include general descriptions and component replacement procedures for most of the information sensors and output actuators, as well as the important components that are part of the systems listed above. Refer to Chapter 4 for more information on the air intake, fuel and exhaust systems, and to Chapter 5 for information on the ignition system. Refer to Chapter 1 for any scheduled maintenance for emission-related systems and components.

The procedures in this Chapter are intended to be practical, affordable and within the capabilities of the home mechanic. The diagnosis of most engine and emission control functions and driveability problems requires specialized tools, equipment and training. When servicing emission devices or systems becomes too difficult or requires special test equipment, consult a dealer service department or other qualified repair shop.

Although engine and emission control systems are very sophisticated on late-model vehicles, you can do most of the regular maintenance and some servicing at home with common tune-up and hand tools and relatively inexpensive meters. Because of the Federally mandated warranty that covers the emission control system, check with a dealer about warranty coverage before working on any emission-related systems. After the warranty has expired, you may wish to perform some of the component replacement procedures in this Chapter to save money. Remember that the most frequent cause of emission and driveability problems is a loose electrical connector or a broken wire or vacuum hose, so always check the electrical connections, the electrical wiring and the vacuum hoses first.

Pay close attention to any special precautions given in this Chapter. Remember that illustrations of various systems might not exactly match the system installed on the vehicle on which you're working because of changes made by the manufacturer during production or from year to year.

A Vehicle Emission Control Information (VECI) label (see illustration) is located in the engine compartment. This label contains emission-control and engine tune-up specifications and adjustment information for emission-control components. When servicing the engine or emission systems, always check the VECI label in your vehicle. If any information in this manual contradicts what you read on the VECI label on your vehicle, always defer to the information on the VECI label.

1.7 The Vehicle Emissions Control Information (VECI) label, which is located on the air filter housing, specifies what emission control systems are installed on the vehicle

## 2 On-Board Diagnostic (OBD) system and Diagnostic Trouble Codes (DTCs)

### SCAN TOOL INFORMATION

◆ Refer to illustrations 2.1 and 2.2

1  Hand-held scanners are handy for analyzing the engine management systems used on late-model vehicles. Because extracting the Diagnostic Trouble Codes (DTCs) from an engine management system is now the first step in troubleshooting many computer-controlled systems and components, even the most basic generic code readers are capable of accessing a computer's DTCs (see illustration). More powerful scan tools can also perform many of the diagnostics once associated with expensive factory scan tools. If you're planning to obtain a generic scan tool for your vehicle, make sure that it's compatible with OBD-II systems. If you don't plan to purchase a code reader or scan tool and don't have access to one, you can have the codes extracted by a dealer service department or by an independent repair shop.

# EMISSIONS AND ENGINE CONTROL SYSTEMS  6-3

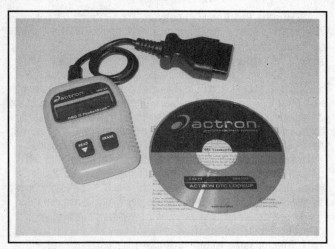

**2.1  Simple code readers are an economical way to extract trouble codes when the CHECK ENGINE light comes on**

**2.2  Scanners like these from Actron and AutoXray are powerful diagnostic aids - they can tell you just about anything that you want to know about your engine management system**

2  With the advent of the Federally mandated emission control system known as On-Board Diagnostics-II (OBD-II), specially designed scanners were developed. Several tool manufacturers have released OBD-II scan tools for the home mechanic (see illustration).

➥**Note: An aftermarket generic scanner should work with any model covered by this manual. Before purchasing a generic scan tool, verify that it will work properly with the OBD-II system you want to scan. If necessary, of course, you can always have the codes extracted by a dealer service department or an independent repair shop with a professional scan tool. Some auto parts stores even provide this service for free.**

## OBD-II SYSTEM GENERAL DESCRIPTION

3  All vehicles covered by this manual are equipped with the OBD-II system. This system consists of the on-board computer, known as the Powertrain Control Module (PCM), and information sensors that monitor various functions of the engine and send a constant stream of data to the PCM during engine operation. Unlike earlier on-board diagnostics systems, the OBD-II system doesn't just monitor everything, store Diagnostic Trouble Codes (DTCs) and illuminate a Malfunction Indicator Light (MIL) when there's a problem. It also predicts the probable failure of systems and components when their data starts to become suspicious!

4  The PCM is the brain of the electronically controlled OBD-II system. It receives data from a number of information sensors and switches. Based on the data that it receives from the sensors, the PCM constantly alters engine operating conditions to optimize driveability, performance, emissions and fuel economy. It does so by turning on and off and by controlling various output actuators such as relays, solenoids, valves and other devices. The PCM is mounted in the left front corner of the engine compartment.

5  Virtually every fuel, ignition and emission control component in the OBD-II system is covered by a Federally mandated extended emissions warranty that is longer than the warranty covering the rest of the vehicle. Read your owner's manual for the terms of the warranty protecting the emission-control systems on your vehicle. It isn't a good idea to do-it-yourself at home while the vehicle emission systems are still under warranty because owner-induced damage to the PCM, the sensors and/or the control devices might void this warranty. So as long as the emission systems are still warranted, take the vehicle to a dealer service department if there's a problem.

## INFORMATION SENSORS

6  **Accelerator Pedal Position (APP) sensor** - The APP sensor, which is an integral component of the accelerator pedal assembly, is part of the electronic throttle control system. None of the vehicles covered in this manual use a conventional accelerator cable. The APP sensor constantly monitors the angle of the accelerator pedal and sends this data to the PCM, which controls an output actuator known as the throttle motor (located in the throttle body) to open or close the throttle plate inside the throttle body to the correct position. The APP sensor actually consists of two identical sensors known as potentiometers (variable resistors), each of which receives a 5-volt reference and a low reference voltage from the PCM and returns a signal voltage to the PCM that's proportional to the angle of the accelerator pedal. One of the potentiometers is redundant, and serves as a back-up in the event that the primary potentiometer fails. The PCM compares the signal outputs from both potentiometers to assess the accuracy of the primary potentiometer's signal. The APP sensor is an integral component of the accelerator pedal assembly. If it's defective, replace the accelerator pedal assembly.

7  **Camshaft Position (CMP) sensor** - The CMP sensor is a Hall effect switching device that produces a signal that the PCM uses to monitor the position of the camshaft. This data enables the CMP sensor to identify the number 1 cylinder so that it can time the firing order of the spark plugs and the firing sequence of the fuel injectors.

8  **Crankshaft Position (CKP) sensor** - The CKP sensor is a permanent magnet generator (also known as a variable reluctance sensor) that produces a variable AC voltage signal that the PCM uses to determine the position of the crankshaft. The PCM uses data from the CKP sensor (and from the CMP sensor on V6 models) to synchronize ignition timing with fuel injector timing, to control spark knock and to detect misfires.

**9 Engine Coolant Temperature (ECT) sensor** - The ECT sensor is a thermistor (temperature-sensitive variable resistor) that sends a voltage signal to the PCM, which uses this data to determine the temperature of the engine coolant. The ECT sensor tells the PCM when the engine is sufficiently warmed up to go into closed loop, helps the PCM control the air/fuel mixture ratio and ignition timing, and also helps the PCM determine when to turn the Exhaust Gas Recirculation (EGR) system on and off.

**10 Fuel tank pressure sensor** - The fuel tank pressure sensor measures the fuel tank pressure when the PCM tests the EVAP system. It's also used to control fuel tank pressure by signaling the EVAP system to purge the tank when the pressure becomes excessive.

**11 Intake Air Temperature (IAT) sensor** - The IAT sensor monitors the temperature of the air entering the engine and sends a signal to the PCM. The IAT sensor is an integral component of the Mass Air Flow (MAF) sensor. For more information, see Step 14.

**12 Knock sensor** - The knock sensor is a piezoelectric crystal that oscillates in proportion to engine vibration. The term piezoelectric refers to the property of certain crystals that produce a voltage when subjected to a mechanical stress. The oscillation of the piezoelectric crystal produces a voltage output that is monitored by the PCM, which retards the ignition timing when the oscillation exceeds a certain threshold. When the engine is operating normally, the knock sensor oscillates consistently and its voltage signal is steady. When detonation occurs, engine vibration increases, and the oscillation of the knock sensor exceeds a design threshold. Detonation is an uncontrolled explosion, after the spark occurs at the spark plug, which spontaneously combusts the remaining air/fuel mixture, resulting in a pinging or knocking sound. If allowed to continue, the engine can be damaged.

**13 Manifold Absolute Pressure (MAP) sensor** - The MAP sensor monitors the pressure or vacuum downstream from the throttle plate, inside the intake manifold. The MAP sensor measures intake manifold pressure and vacuum on the absolute scale - from zero instead of from sea-level atmospheric pressure (14.7 psi). The MAP sensor converts the absolute pressure into a variable voltage signal that changes with the pressure. The PCM uses this data to determine engine load so that it can alter the ignition advance and fuel enrichment.

**14 Mass Air Flow (MAF) sensor** - The MAF sensor uses a hot-wire sensing element to enable the PCM to measure the amount of intake air drawn into the engine. The wire is constantly maintained at a specified temperature above the ambient temperature of the incoming air by electrical current. As intake air passes through the MAF sensor and over the hot wire, it cools the wire, and the control system immediately corrects the temperature back to its constant value. The current required to maintain the constant value is used by the PCM to determine the amount of air flowing through the MAF sensor. The MAF sensor also includes an integral Intake Air Temperature (IAT) sensor. The two components cannot be serviced separately; if either sensor is defective, replace the MAF sensor. On all models, the MAF sensor is located between the air filter housing and the air intake duct. The MAF sensor is also referred to an MAF/IAT sensor because it also incorporates the Intake Air Temperature (IAT) sensor.

**15 Oxygen sensors** - An oxygen sensor is a galvanic battery that generates a small variable voltage signal in proportion to the difference between the oxygen content in the exhaust stream and the oxygen content in the ambient air. The PCM uses the voltage signal from the upstream oxygen sensor to maintain a stoichiometric air/fuel ratio of 14.7:1 by constantly adjusting the on-time of the fuel injectors. There are four oxygen sensors on all models: two upstream and two downstream.

**16 Throttle Position (TP) sensor** - The TP sensor is a potentiometer that receives a constant voltage input from the PCM and sends back a proportional voltage signal that varies in relation to the opening angle of the throttle plate inside the throttle body. This voltage signal tells the PCM when the throttle is closed, half-open, wide open or anywhere in between. The PCM uses this data, along with information from other sensors, to calculate injector pulse width (the interval of time during which an injector solenoid is energized by the PCM). There are actually two TP sensors on the electronic throttle body. One TP sensor has a signal that's above four volts when the throttle plate is closed, and which decreases as the throttle plate opens. The other sensor has a signal that's below one volt when the throttle plate is closed, and which increases as the throttle plate opens. The PCM compares the two signals to make sure that the primary sensor is functioning correctly. The TP sensors are an integral part of the electronic throttle body and cannot be serviced separately.

**17 Transmission speed sensors** - The Input Shaft Speed (ISS) sensor is located inside the transmission, so you cannot replace it at home. The Vehicle Speed Sensor (VSS), a magnetic pick-up coil, is located on the outside of automatic transmissions, so you can replace it. The VSS provides the PCM with information about the rotational speed of the output shaft in the transmission. The PCM uses this information to control the torque converter and to calculate speed scheduling and the correct operating pressure for the transmission.

**18 Valve Lifter Oil Manifold (VLOM) oil pressure sensor** - The VLOM is a component of the Cylinder Deactivation (or Active Fuel Management) System. The VLOM oil pressure sensor is a sub-component of the VLOM. The VLOM oil pressure sensor monitors the oil pressure inside the system and sends this information to the PCM, which controls the system. The Cylinder Deactivation System is used on all 5.3L V8 engines and on L76 6.0L V8 engines; it is not used on 4.3L V6 engines, 4.8L V8 engines and L92 6.2L V8 engines. For more information about the Cylinder Deactivation System, see Section 19.

## POWERTRAIN CONTROL MODULE (PCM)

19 The PCM, which is located at the left front corner of the engine compartment, inside the air filter housing, is the brain of the engine management system. It monitors the data input of all the information sensors, processes this information by comparing it to the operational map (program) for the engine management system, then sends command decisions to the output actuators, which execute those commands. These decisions are made so quickly (in milliseconds) that the entire process is virtually undetectable. As a result, driveability is smooth and seamless.

## OUTPUT ACTUATORS

**20 Camshaft actuator solenoid** - The PCM-controlled camshaft actuator solenoid controls the flow of oil into the camshaft actuator system. When the PCM sends a pulse-width modulated voltage signal to the camshaft actuator solenoid, oil flows through one of two passages into the cam phaser. When oil is directed through one passage, the cam phaser advances the camshaft; when it's directed through the other passage, the cam phaser retards the camshaft. The camshaft actuator solenoid and the cam phaser are components of the camshaft actuator system, which lowers emissions, increases fuel economy and improves engine idle stability. The camshaft actuator system is used on 6.0L and 6.2L V8 engines.

# EMISSIONS AND ENGINE CONTROL SYSTEMS

21 **EVAP canister purge solenoid** - The EVAP canister purge solenoid (or purge valve) controls the flow of fuel vapors (unburned hydrocarbons) from the EVAP canister to the intake manifold. The EVAP canister purge solenoid is normally closed. But when ordered to do so by the PCM, it allows intake manifold vacuum to draw the fuel vapors that are stored in the EVAP canister into the intake manifold, where they're mixed with intake air, then burned along with the normal air/fuel mixture.

22 **EVAP canister vent solenoid valve** - The EVAP canister vent solenoid valve is normally open, to allow fresh outside air to flow through the vent and into the EVAP canister. But when energized by the PCM, the vent solenoid closes and seals off the EVAP system for inspection and maintenance tests and for OBD-II leak and pressure tests.

23 **Fuel injectors** - The fuel injectors spray a fine mist of fuel into the intake ports, where it is mixed with incoming air. The injectors are PCM-controlled inductive coils that open and close small valves in firing order to administer precise amounts of pressurized fuel through their nozzles. For more information about the injectors, see Chapter 4.

24 **Ignition coils** - The PCM-controlled ignition coils are under the control of the Powertrain Control Module (PCM). V6 models use a coil pack, which consists of three ignition coils; each coil fires two cylinders. V8 models employ eight coils, one for each cylinder. For more information about the ignition coils, see Chapter 5.

25 **Throttle actuator motor** - The throttle actuator motor, which is an electric motor that opens and closes the throttle plate in response to commands from the PCM, is located inside the electronic throttle body. It cannot be serviced separately from the throttle body.

26 **Valve Lifter Oil Manifold (VLOM) solenoids** - The VLOM solenoids are components of the Cylinder Deactivation System, or Active Fuel Management Control System, which improves fuel economy on V8 engines by turning off four cylinders under certain driving conditions. This system is used on all 5.3L V8 engines and on L76 (aluminum block) 6.0L V8 engines. The PCM-controlled VLOM solenoids control the flow of engine oil to special hydraulic lifters used to activate and deactivate cylinders 4 and 6 on the rear cylinder bank and cylinders 1 and 7 on the front bank.

## OBTAINING AND CLEARING DIAGNOSTIC TROUBLE CODES (DTCS)

27 All models covered by this manual are equipped with on-board diagnostics. When the PCM recognizes a malfunction in a monitored emission control system, component or circuit, it turns on the Malfunction Indicator Light (MIL) on the dash. The PCM will continue to display the MIL until the problem is fixed and the Diagnostic Trouble Code (DTC) is cleared from the PCM's memory. You'll need a scan tool to access any DTCs stored in the PCM.

28 Before outputting any DTCs stored in the PCM, thoroughly inspect ALL electrical connectors and hoses. Make sure that all electrical connections are tight, clean and free of corrosion. Make sure that all hoses are correctly connected, fit tightly and are in good condition (no cracks or tears). Also, make sure that the engine is tuned up. A poorly running engine is probably one of the biggest causes of emission-related malfunctions. Often, simply giving the engine a good tune-up will correct the problem.

2.29 The 16-pin Data Link Connector (DLC), also referred to as the diagnostic connector, is located under the left part of the dash

### Accessing the DTCs

▶ Refer to illustration 2.29

29 On these models, all of which are equipped with On-Board Diagnostic II (OBD-II) systems, the Diagnostic Trouble Codes (DTCs) can only be accessed with a code reader or a scan tool (see illustrations 2.1 and 2.2). Simply plug the connector of the tool into the Data Link Connector (DLC) or diagnostic connector (see illustration), which is located under the lower edge of the dash, just to the right of the steering column. Then follow the instructions included with the scan tool to extract the DTCs.

30 Once you have outputted all of the stored DTCs, look them up on the accompanying DTC chart.

31 After troubleshooting the source of each DTC, make any necessary repairs or replace the defective component(s).

### Clearing the DTCs

32 Clear the DTCs with the scan tool in accordance with the instructions provided by the scan tool's manufacturer.

## DIAGNOSTIC TROUBLE CODES

33 The accompanying tables are a list of the Diagnostic Trouble Codes (DTCs) that can be accessed by a do-it-yourselfer working at home (there are many, many more DTCs available to dealerships with proprietary scan tools and software, but those codes cannot be accessed by a generic scan tool). If, after you have checked and repaired the connectors, wire harness and vacuum hoses (if applicable) for an emission-related system, component or circuit, the problem persists, have the vehicle checked by a dealer service department.

## OBD-II DIAGNOSTIC TROUBLE CODES (DTCS)

➡ **Note: Not all trouble codes apply to all models.**

| Code | Probable cause |
|---|---|
| P0010 | Camshaft Position (CMP) actuator solenoid control circuit |
| P0011 | Camshaft Position (CMP) system performance |
| P0016 | Crankshaft Position (CKP) sensor/Camshaft Position (CMP) sensor correlation |
| P0030 | Oxygen sensor heater control circuit (bank 1, sensor 1) |
| P0036 | Oxygen sensor heater control circuit (bank 1, sensor 2) |
| P0050 | Oxygen sensor heater control circuit (bank 2, sensor 1) |
| P0053 | Oxygen sensor heater resistance (bank 1, sensor 1) |
| P0054 | Oxygen sensor heater resistance (bank 1, sensor 2) |
| P0056 | Oxygen sensor heater control circuit (bank 2, sensor 2) |
| P0059 | Oxygen sensor heater resistance (bank 2, sensor 1) |
| P0060 | Oxygen sensor heater resistance (bank 2, sensor 2) |
| P0068 | Throttle body air flow performance |
| P0087 | Fuel rail pressure too low |
| P0088 | Fuel rail pressure too high |
| P0090 | Fuel pressure regulator control circuit |
| P0097 | Intake Air Temperature (IAT) sensor 2, low circuit voltage |
| P0098 | Intake Air Temperature (IAT) sensor 2, high circuit voltage |
| P0101 | Mass Air Flow (MAF) sensor performance |
| P0102 | Mass Air Flow (MAF) sensor circuit, low frequency |
| P0103 | Mass Air Flow (MAF) sensor circuit, high frequency |
| P0106 | Manifold Absolute Pressure (MAP) sensor performance |
| P0107 | Manifold Absolute Pressure (MAP) sensor circuit, low voltage |
| P0108 | Manifold Absolute Pressure (MAP) sensor circuit, high voltage |
| P0112 | Intake Air Temperature (IAT) sensor circuit, low voltage |
| P0113 | Intake Air Temperature (IAT) sensor circuit, high voltage |
| P0116 | Engine Coolant Temperature (ECT) sensor performance |
| P0117 | Engine Coolant Temperature (ECT) sensor circuit, low voltage |
| P0118 | Engine Coolant Temperature (ECT) sensor circuit, high voltage |
| P0120 | Throttle Position (TP) sensor 1 circuit |

## EMISSIONS AND ENGINE CONTROL SYSTEMS

| Code | Probable cause |
|---|---|
| P0121 | Throttle Position (TP) sensor 1 performance |
| P0122 | Throttle Position (TP) sensor 1 circuit, low voltage |
| P0123 | Throttle Position (TP) sensor 1 circuit, high voltage |
| P0128 | Engine coolant temperature below thermostat-regulating temperature |
| P0131 | Oxygen sensor circuit, low voltage (bank 1, sensor 1) |
| P0132 | Oxygen sensor circuit, high voltage (bank 1, sensor 1) |
| P0133 | Oxygen sensor circuit, slow response (bank 1, sensor 1) |
| P0134 | Oxygen sensor circuit, insufficient activity (bank 1, sensor 1) |
| P0135 | Oxygen sensor heater performance (bank 1, sensor 1) |
| P0136 | Oxygen sensor circuit (bank 1, sensor 2) |
| P0137 | Oxygen sensor circuit, low voltage (bank 1, sensor 2) |
| P0138 | Oxygen sensor circuit, high voltage (bank 1, sensor 2) |
| P013A | Oxygen sensor, rich-to-lean response too slow (bank 1, sensor 2) |
| P013B | Oxygen sensor, lean-to-rich response too slow (bank 1, sensor 2) |
| P013C | Oxygen sensor, lean-to-rich response too slow (bank 2, sensor 2) |
| P013D | Oxygen sensor, lean-to-rich response too slow (bank 2, sensor 2) |
| P013E | Oxygen sensor, delayed rich-to-lean response (bank 1, sensor 2) |
| P013F | Oxygen sensor, delayed lean-to-rich response (bank 1, sensor 2) |
| P0140 | Oxygen sensor circuit, insufficient activity (bank 1, sensor 2) |
| P0141 | Oxygen sensor heater performance (bank 1, sensor 2) |
| P014A | Oxygen sensor, delayed rich-to-lean response (bank 2, sensor 2) |
| P014B | Oxygen sensor, delayed lean-to-rich response (bank 2, sensor 2) |
| P0151 | Oxygen sensor circuit, low voltage (bank 2, sensor 1) |
| P0152 | Oxygen sensor circuit, high voltage (bank 2, sensor 1) |
| P0153 | Oxygen sensor, slow response (bank 2, sensor 1) |
| P0154 | Oxygen sensor circuit, insufficient activity (bank 2, sensor 1) |
| P0155 | Oxygen sensor heater performance (bank 2, sensor 1) |
| P0156 | Oxygen sensor performance (bank 2, sensor 2) |
| P0157 | Oxygen sensor circuit, low voltage (bank 2, sensor 2) |
| P0158 | Oxygen sensor circuit, high voltage (bank 2, sensor 2) |
| P0160 | Oxygen sensor circuit, insufficient activity (bank 2, sensor 2) |

## OBD-II DIAGNOSTIC TROUBLE CODES (DTCS) (CONTINUED)

➡ Note: Not all trouble codes apply to all models.

| Code | Probable cause |
|---|---|
| P0161 | Oxygen sensor heater performance (bank 2, sensor 2) |
| P0168 | Engine fuel temperature too high |
| P0171 | Fuel trim system lean (bank 1) |
| P0172 | Fuel trim system rich (bank 1) |
| P0174 | Fuel trim system lean (bank 2) |
| P0175 | Fuel trim system rich (bank 2) |
| P0181 | Fuel temperature sensor performance |
| P0182 | Fuel temperature sensor circuit, low voltage |
| P0183 | Fuel temperature sensor circuit, high voltage |
| P0191 | Fuel Rail Pressure (FRP) sensor performance |
| P0192 | Fuel Rail Pressure (FRP) sensor circuit, low voltage |
| P0193 | Fuel Rail Pressure (FRP) sensor circuit, high voltage |
| P0200 | Injector control circuit |
| P0201 | Injector no. 1 circuit malfunction |
| P0202 | Injector no. 2 circuit malfunction |
| P0203 | Injector no. 3 circuit malfunction |
| P0204 | Injector no. 4 circuit malfunction |
| P0205 | Injector no. 5 circuit malfunction |
| P0206 | Injector no. 6 circuit malfunction |
| P0207 | Injector no. 7 circuit malfunction |
| P0208 | Injector no. 8 circuit malfunction |
| P0218 | Transmission fluid temperature too high |
| P0220 | Throttle Position (TP) sensor 2 circuit |
| P0222 | Throttle Position (TP) sensor 2 circuit, low voltage |
| P0223 | Throttle Position (TP) sensor 2 circuit, high voltage |
| P0230 | Fuel pump relay control circuit |
| P0231 | Fuel pump control circuit, low voltage |
| P0232 | Fuel pump control circuit, high voltage |
| P023F | Fuel pump control circuit |

## EMISSIONS AND ENGINE CONTROL SYSTEMS 6-9

| Code | Probable cause |
|---|---|
| P025A | Fuel pump control module enable circuit |
| P029D | Injector no. 1 leak |
| P02A1 | Injector no. 2 leak |
| P02A5 | Injector no. 3. Leak |
| P02A9 | Injector no. 4 leak |
| P02AD | Injector no. 5 leak |
| P02B1 | Injector no. 6 leak |
| P02B5 | Injector no. 7 leak |
| P02B9 | Injector no. 8 leak |
| P0300 | Engine misfire detected |
| P0301 | Engine misfire detected (cylinder 1) |
| P0302 | Engine misfire detected (cylinder 2) |
| P0303 | Engine misfire detected (cylinder 3) |
| P0304 | Engine misfire detected (cylinder 4) |
| P0305 | Engine misfire detected (cylinder 5) |

| Code | Probable cause |
|---|---|
| P0306 | Engine misfire detected (cylinder 6) |
| P0307 | Engine misfire detected (cylinder 7) |
| P0308 | Engine misfire detected (cylinder 8) |
| P0315 | Crankshaft Position (CKP) system, system variation not learned |
| P0324 | Knock Sensor (KS) module performance |
| P0325 | Knock Sensor (KS) circuit (bank 1) |
| P0326 | Knock Sensor (KS) performance |
| P0327 | Knock Sensor (KS) circuit, low frequency or low voltage (bank 1) |
| P0328 | Knock Sensor (KS) circuit, high frequency or high voltage (bank 1) |
| P0330 | Knock Sensor (KS) circuit (bank 2) |
| P0332 | Knock Sensor (KS) circuit, low frequency or low voltage (bank 2) |
| P0333 | Knock Sensor (KS) circuit, high frequency or high voltage (bank 2) |
| P0335 | Crankshaft Position (CKP) sensor circuit |
| P0336 | Crankshaft Position (CKP) sensor performance |
| P0340 | Camshaft Position (CMP) sensor circuit |

## 6-10 EMISSIONS AND ENGINE CONTROL SYSTEMS

### OBD-II DIAGNOSTIC TROUBLE CODES (DTCS) (CONTINUED)

➡ **Note: Not all trouble codes apply to all models.**

| Code | Probable cause |
| --- | --- |
| P0341 | Camshaft Position (CMP) sensor performance |
| P0342 | Camshaft Position (CMP) sensor circuit, low voltage |
| P0343 | Camshaft Position (CMP) sensor circuit, high voltage |
| P0351 | Ignition coil 1 control circuit |
| P0352 | Ignition coil 2 control circuit |
| P0353 | Ignition coil 3 control circuit |
| P0354 | Ignition coil 4 control circuit |
| P0355 | Ignition coil 5 control circuit |
| P0356 | Ignition coil 6 control circuit |
| P0357 | Ignition coil 7 control circuit |
| P0358 | Ignition coil 8 control circuit |
| P0401 | Exhaust Gas Recirculation (EGR) flow insufficient |
| P0402 | Exhaust Gas Recirculation (EGR) flow excessive |
| P0403 | Exhaust Gas Recirculation (EGR) solenoid control circuit |
| P0405 | Exhaust Gas Recirculation (EGR) position sensor circuit, low voltage |
| P0406 | Exhaust Gas Recirculation (EGR) position sensor circuit, high voltage |
| P040C | Exhaust Gas Recirculation (EGR) temperature sensor 1 circuit, low voltage |
| P040D | Exhaust Gas Recirculation (EGR) temperature sensor 1 circuit, high voltage |
| P040F | Exhaust Gas Recirculation (EGR) temperature sensor 1 circuit-to-EGR temperature sensor 2 circuit correlation |
| P041C | Exhaust Gas Recirculation (EGR) temperature sensor 2 circuit, low voltage |
| P041D | Exhaust Gas Recirculation (EGR) temperature sensor 2 circuit, high voltage |
| P0420 | Catalyst system, low efficiency (bank 1) |
| P0430 | Catalyst system, high efficiency (bank 2) |
| P0442 | Evaporative Emission (EVAP) system, small leak detected |
| P0443 | Evaporative Emission (EVAP) system, purge solenoid control circuit |
| P0446 | Evaporative Emission (EVAP) vent system performance |
| P0449 | Evaporative Emission (EVAP) vent solenoid control circuit |
| P0451 | Fuel Tank Pressure (FTP) sensor performance |
| P0452 | Fuel Tank Pressure (FTP) sensor circuit, low voltage |

## EMISSIONS AND ENGINE CONTROL SYSTEMS

| Code | Probable cause |
|---|---|
| P0453 | Fuel Tank Pressure (FTP) sensor circuit, high voltage |
| P0454 | Fuel Tank Pressure (FTP) sensor circuit intermittent |
| P0455 | Evaporative Emission (EVAP) system, large leak detected |
| P0461 | Fuel level sensor circuit performance |
| P0462 | Fuel level sensor circuit, low voltage |
| P0463 | Fuel level sensor circuit, high voltage |
| P0464 | Fuel level sensor circuit intermittent |
| P046C | Exhaust Gas Recirculation (EGR) position sensor performance |
| P0480 | Cooling fan relay 1 control circuit |
| P0481 | Cooling fan relay 2 control circuit |
| P0496 | EVAP system flow during non-purge |
| P0500 | Vehicle Speed Sensor (VSS) circuit |
| P0502 | Vehicle Speed Sensor (VSS) circuit, low voltage |
| P0503 | Vehicle Speed Sensor (VSS) circuit intermittent |
| P0506 | Idle speed low |
| P0507 | Idle speed high |
| P0513 | Theft deterrent key incorrect |
| P0521 | Engine Oil Pressure (EOP) sensor, performance |
| P0522 | Engine Oil Pressure (EOP) sensor circuit, low voltage |
| P0523 | Engine Oil Pressure (EOP) sensor circuit, high voltage |
| P0530 | Air conditioning refrigerant |
| P0532 | Air conditioning refrigerant pressure sensor circuit, low voltage |
| P0533 | Air conditioning refrigerant pressure sensor circuit, high voltage |
| P0540 | Intake air heater feedback circuit |
| P0545 | Exhaust Gas Temperature (EGT) sensor 1 circuit, low voltage |
| P0546 | Exhaust Gas Temperature (EGT) sensor 1 circuit, high voltage |
| P0561 | System voltage performance unstable |
| P0562 | System voltage low |
| P0563 | System voltage high |
| P0567 | Cruise control resume switch circuit |
| P0568 | Cruise control set switch circuit |

## OBD-II DIAGNOSTIC TROUBLE CODES (DTCS) (CONTINUED)

➡ Note: Not all trouble codes apply to all models.

| Code | Probable cause |
|---|---|
| P0571 | Cruise control brake switch circuit |
| P0572 | Brake switch circuit 1, low voltage |
| P0573 | Brake switch circuit 1, high voltage |
| P0575 | Cruise control switch signal circuit |
| P0601 | Powertrain or Transmission Control Module Read Only Memory (ROM) |
| P0602 | Powertrain or Transmission Control Module not programmed |
| P0603 | Powertrain or Transmission Control Module long-term memory reset |
| P0604 | Powertrain or Transmission Control Module Random Access Memory |
| P0606 | Powertrain Control Module (PCM) internal performance |
| P0607 | Powertrain Control Module (PCM) performance |
| P0608 | Vehicle speed output circuit |
| P0609 | Rear wheel speed sensor |
| P060B | Powertrain Control Module (PCM) analog-to-digital performance |
| P060D | Powertrain Control Module (PCM), Accelerator Pedal Position (APP) system performance |
| P0613 | Transmission Control Module (TCM) processor |
| P0615 | Starter relay control circuit |
| P0616 | Starter relay control circuit, low voltage |
| P0617 | Starter relay control circuit, high voltage |
| P061C | Powertrain Control Module (PCM), engine speed performance |
| P0621 | Alternator L-terminal circuit |
| P0622 | Alternator F-terminal circuit |
| P062C | Powertrain Control Module (PCM), vehicle speed performance |
| P062F | Powertrain Control Module (PCM) long-term memory performance |
| P062F | Transmission Control Module (TCM) EEPROM error |
| P0633 | Theft deterrent key not programmed |
| P0634 | Transmission Control Module (TCM), internal temperature too high |
| P0641 | 5-volt reference 1 circuit |
| P0642 | 5-volt reference 1 circuit, low voltage |
| P0643 | 5-volt reference 1 circuit, high voltage |

# EMISSIONS AND ENGINE CONTROL SYSTEMS

| Code | Probable cause |
|---|---|
| P0645 | Air conditioning clutch relay control circuit |
| P0646 | Air conditioning clutch relay control circuit, low voltage |
| P0647 | Air conditioning clutch relay control circuit, high voltage |
| P064A | Fuel pump control module driver temperature too high |
| P0650 | Malfunction Indicator Light (MIL) control circuit |
| P0651 | 5-volt reference 2 circuit |
| P0652 | 5-volt reference 2 circuit, low voltage |
| P0653 | 5-volt reference 2 circuit, high voltage |
| P0654 | Engine speed output circuit |
| P0658 | Actuator supply voltage 1 low |
| P0659 | Actuator supply voltage 1 high |
| P0667 | Transmission Control Module (TCM) temperature sensor performance |
| P0668 | Transmission Control Module (TCM) temperature sensor circuit, low voltage |
| P0669 | Transmission Control Module (TCM) temperature sensor circuit, high voltage |
| P0685 | Engine controls ignition relay control circuit |
| P0689 | Engine controls ignition relay feedback circuit, low voltage |
| P0690 | Engine controls ignition relay feedback circuit, high voltage |
| P0698 | 5-volt reference 3 circuit, low voltage |
| P0699 | 5-volt reference 3 circuit, high voltage |
| P069E | Fuel Pump Control Module (FPCM) requested MIL illumination |
| P06A6 | 5-volt reference circuit performance |
| P0700 | Transmission Control Module (TCM) requested MIL illumination |
| P0701 | Transmission Control Module (TCM) performance |
| P0703 | Brake switch circuit 2 |
| P0705 | Transmission Range (TR) switch circuit |
| P0706 | Transmission Range (TR) sensor circuit, PRNDL input |
| P0708 | Transmission Range (TR) sensor circuit, high voltage |
| P0711 | Transmission Fluid Temperature (TFT) sensor performance or sensor circuit performance |
| P0712 | Transmission Fluid Temperature (TFT) sensor circuit, low voltage or low input |
| P0713 | Transmission Fluid Temperature (TFT) sensor circuit, high voltage or high input |
| P0716 | Input Speed Sensor (ISS) performance or ISS circuit performance |

## OBD-II DIAGNOSTIC TROUBLE CODES (DTCS) (CONTINUED)

➡ Note: **Not all trouble codes apply to all models.**

| Code | Probable cause |
|---|---|
| P0717 | Input Speed Sensor (ISS) circuit, low voltage |
| P0719 | Brake switch circuit, low voltage |
| P071A | Transmission tow mode switch circuit |
| P0721 | Output Speed Sensor (OSS) performance |
| P0722 | Output Speed Sensor (OSS) circuit, low voltage or no signal |
| P0723 | Output Speed Sensor (OSS) circuit, intermittent |
| P0724 | Brake switch circuit, high voltage |
| P0725 | Engine speed circuit |
| P0726 | Engine speed circuit performance |
| P0727 | No engine speed signal |
| P0729 | Incorrect 6th gear ratio |
| P0730 | Incorrect gear ratio |
| P0731 | Incorrect 1st gear ratio |
| P0732 | Incorrect 2nd gear ratio |
| P0733 | Incorrect 3rd gear ratio |
| P0734 | Incorrect 4th gear ratio |
| P0735 | Incorrect 5th gear ratio |
| P0736 | Incorrect reverse ratio |
| P0740 | Torque Converter Clutch (TCC) enable solenoid control circuit |
| P0741 | Torque Converter Clutch (TCC) system, stuck off |
| P0742 | Torque Converter Clutch (TCC) system, stuck on |
| P0748 | Pressure Control (PC) solenoid control circuit |
| P0776 | Clutch Pressure Control (PC) solenoid 2, stuck off |
| P0777 | Clutch Pressure Control (PC) solenoid 2, stuck on |
| P0796 | Clutch Pressure Control (PC) solenoid 3 stuck off |
| P0797 | Clutch Pressure Control (PC) solenoid 3 stuck on |
| P0806 | Clutch Pedal Position (CPP) sensor circuit performance |
| P0807 | Clutch Pedal Position (CPP) sensor circuit, low voltage |
| P0808 | Clutch Pedal Position (CPP) sensor circuit, high voltage |

# EMISSIONS AND ENGINE CONTROL SYSTEMS   6-15

| Code | Probable cause |
|---|---|
| P080A | Clutch Pedal Position (CPP) not learned |
| P0815 | Upshift switch circuit |
| P0816 | Downshift switch circuit |
| P0826 | TAP up and down shift switch circuit |
| P0827 | TAP up and down shift switch circuit low |
| P0828 | TAP up and down shift switch circuit high |
| P0833 | Clutch pedal switch 2 circuit |
| P0851 | Park/Neutral Position (PNP) switch circuit, low voltage |
| P0852 | Park/Neutral Position (PNP) switch circuit, high voltage |

## 3  Accelerator Pedal Position (APP) sensor - replacement

♦ Refer to illustration 3.2

➡ Note: The APP sensor is located at the top of the accelerator pedal arm. The APP sensor and the accelerator pedal are a one-piece assembly and are replaced as a unit.

1  Remove the knee bolster (see Chapter 11).
2  Locate the APP sensor at the top of the accelerator pedal assembly (see illustration).
3  Disconnect the electrical connector from the APP sensor.
4  Unscrew the mounting bolts and remove the APP sensor and accelerator pedal as a single assembly.
5  Installation is the reverse of removal. Tighten the APP sensor assembly mounting bolts securely.

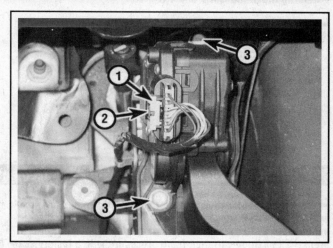

3.2  Pull the connector lock (1) out, then depress the release tab (2) and disconnect the electrical connector. To detach the APP sensor/pedal assembly, remove the two mounting bolts (3)

## 4  Camshaft Position (CMP) sensor - replacement

### V6 MODELS

➡ Note: The CMP sensor is located on the engine front cover, at about 11 o'clock in relation to the crankshaft pulley.

1  Disconnect the CMP sensor electrical connector.
2  Remove the water pump (see Chapter 3).
3  Remove the CMP sensor retaining bolt and remove the CMP sensor.
4  Remove and discard the old CMP sensor O-ring.
5  Installation is the reverse of removal. Be sure to use a new O-ring, lubricate it with a film of clean engine oil, and tighten the CMP sensor bolt securely.

# 6-16 EMISSIONS AND ENGINE CONTROL SYSTEMS

**4.7 On V8 engines, the Camshaft Position (CMP) sensor is located on the front of the timing chain cover. Pry the release tab with a small screwdriver and disconnect the electrical connector (1), remove the harness mounting bolt (2) . . .**

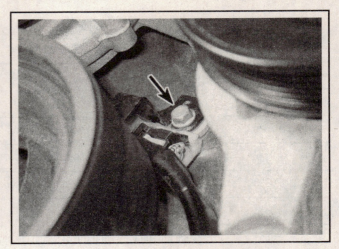

**4.8 . . . then follow the harness up to the sensor and remove the sensor mounting bolt**

## V8 MODELS

▶ Refer to illustrations 4.7 and 4.8

➡ **Note: The CMP sensor is located on the timing chain cover, to the left of the crankshaft pulley.**

6  Raise the front of the vehicle and place it securely on jackstands.
7  Disconnect the CMP sensor harness electrical connector and remove the sensor harness bolt (see illustration).
8  Remove the mounting bolt (see illustration) and remove the CMP sensor.
9  Remove the CMP sensor O-ring and inspect it for cracks, tears and deterioration. If the O-ring is still in good shape you can reuse it. If not, replace it. Install the O-ring and lubricate it with clean engine oil.
10  Installation is the reverse of removal. Be sure to tighten the mounting bolts securely.

## 5  Crankshaft Position (CKP) sensor - replacement

➡ **Note: If the Malfunction Indicator Light (MIL) comes on after replacing the CKP sensor and starting the engine, drive the vehicle to a dealer and have the service department perform a crankshaft position variation learn procedure with a factory scan tool.**

## V6 MODELS

▶ Refer to illustrations 5.2 and 5.4

1  Raise the vehicle and place it securely on jackstands. Remove the oil pan skid plate, if equipped.
2  Disconnect the electrical connector from the CKP sensor (see illustration).
3  Remove the CKP sensor mounting bolt.
4  Remove and discard the old CKP sensor O-ring (see illustration). Install a new O-ring and lubricate it with clean engine oil.
5  Installation is the reverse of removal. Be sure to use a new O-ring and tighten the CKP sensor mounting bolt securely.

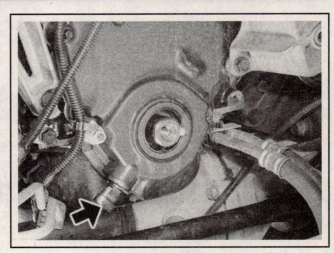

**5.2 On V6 engines, the CKP sensor is located at about 7 o'clock in relation to the crankshaft pulley (removed for clarity in this photo). To remove the CKP sensor, disconnect the electrical connector and remove the sensor mounting bolt**

# EMISSIONS AND ENGINE CONTROL SYSTEMS

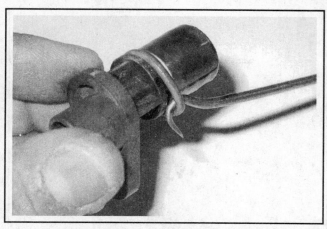

5.4 On V6 engines, be sure to remove the old O-ring from the CKP sensor; always install the sensor with a new O-ring (even if you're installing the old sensor)

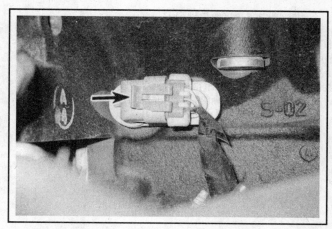

5.8 To remove the CKP sensor from a V8 engine, pry the retaining tab away from the electrical connector and disconnect the connector, then remove the sensor mounting bolt

## V8 MODELS

♦ Refer to illustration 5.8

➡ Note: The CKP sensor is located on the right side of the engine block, near the flywheel.

6   Loosen the right front wheel lug nuts. Raise the vehicle and place it securely on jackstands. Remove the right front wheel. Remove the right front wheel well splash shield (see Chapter 11).
7   Remove the starter motor (see Chapter 5).
8   Disconnect the electrical connector from the CKP sensor (see illustration).
9   Remove the CKP sensor mounting bolt and remove the CKP sensor.
10  Installation is the reverse of removal. Be sure to tighten the CKP sensor mounting bolt securely. Tighten the wheel lug nuts to the torque listed in the Chapter 1 Specifications.

## 6   Engine Coolant Temperature (ECT) sensor - replacement

♦ Refer to illustrations 6.2a, 6.2b and 6.4

**WARNING:**

Wait until the engine is completely cool before beginning this procedure.

➡ Note: On V6 models, the ECT sensor is located on the left side of the left cylinder head, between the spark plugs for cylinders 3 and 5, just below the valve cover. On V8 models, the ECT sensor is located on the left side of the left cylinder head, just ahead of the exhaust manifold.

1   Drain the cooling system to a level that's below the level of the ECT sensor (see Chapter 1).
2   Disconnect the electrical connector from the ECT sensor (see illustrations).

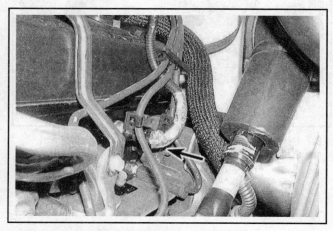

6.2a  Location of the ECT sensor on the V6 engine

6.2b  Location of the ECT sensor on V8 engines

# 6-18 EMISSIONS AND ENGINE CONTROL SYSTEMS

3  Unscrew and remove the ECT sensor.

4  If you're installing the old sensor, wrap the threads of the new ECT sensor with Teflon tape (see illustration). The threads of a new sensor should already be coated with sealant.

5  Installation is the reverse of removal. Be sure to tighten the ECT sensor to the torque listed in this Chapter's Specifications.

6  Refill the cooling system (see Chapter 1).

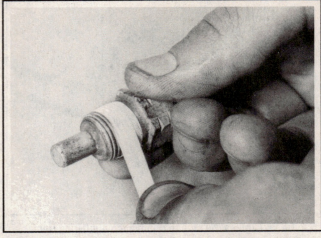

**6.4  Before installing the ECT sensor, wrap the threads of the sensor with Teflon tape to prevent leaks**

## 7  Intake Air Temperature (IAT) sensor - replacement

The IAT sensor is an integral component of the Mass Air Flow (MAF) sensor (see Section 10).

## 8  Knock sensor - replacement

♦ Refer to illustration 8.7

➥**Note: There are two knock sensors, one on each side of the block. Knock sensor 1 is on the left side of the block; knock sensor 2 is on the right side. Because they're located below the exhaust manifolds, you cannot access the knock sensors from above.**

1  On V8 models, loosen the left or right front wheel lug nuts.

2  On all models, raise the front of the vehicle and place it securely on jackstands.

3  On V8 models, remove the left or right front wheel.

4  If you're removing knock sensor 2 (the right side sensor) on a V8 model, remove the right wheel well splash shield (see Chapter 11).

5  If you're removing knock sensor 2 on a V8 model, remove the starter motor (see Chapter 5).

6  On V6 models, remove the knock sensor heat shield retaining bolt and remove the heat shield.

7  Disconnect the electrical connector from the knock sensor (see illustration).

8  Remove the knock sensor retaining bolt and remove the knock sensor.

9  Installation is the reverse of removal. Be sure to tighten the knock sensor retaining bolt to the torque listed in this Chapter's Specifications. Tighten the wheel lug nuts to the torque listed in the Chapter 1 Specifications.

**8.7  Depress the release tab and disconnect the electrical connector, then remove the knock sensor retaining bolt**

# EMISSIONS AND ENGINE CONTROL SYSTEMS

## 9  Manifold Absolute Pressure (MAP) sensor - replacement

### V6 MODELS

▶ Refer to illustration 9.1

➡ **Note:** *The MAP sensor is located on top of the intake manifold, on the right side, between the throttle body and the fuel meter body.*

1  Disconnect the MAP sensor electrical connector (see illustration).
2  Remove the MAP sensor by pulling it straight up out of the intake manifold.
3  Remove the old sealing grommet from the MAP sensor and inspect it. If the grommet is cracked, torn or otherwise deteriorated, replace it.
4  When installing the MAP sensor, use a little clean engine oil on the grommet so that it doesn't tear when pushing the MAP sensor into its mounting hole.
5  Installation is otherwise the reverse of removal.

### V8 MODELS

▶ Refer to illustrations 9.8a and 9.8b

➡ **Note:** *The MAP sensor is located on top of the intake manifold.*

6  Remove the intake manifold cover.
7  Disconnect the MAP sensor electrical connector.
8  Remove the MAP sensor retainer and remove the sensor by pulling it straight up (see illustrations).
9  Remove the old sealing grommet (see illustration 9.8b) from the MAP sensor and inspect it. If the grommet is cracked, torn or otherwise deteriorated, replace it. When installing the MAP sensor, use a little clean engine oil on the grommet so that it doesn't tear when pushing the MAP sensor into its mounting hole.
10  When installing the retainer, make sure that it snaps into place. Installation is otherwise the reverse of removal.

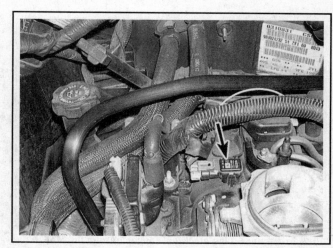

9.1  To remove the MAP sensor from the intake manifold on a V6 engine, disconnect the electrical connector, then pull the sensor straight up

9.8a  To detach the MAP sensor from the intake manifold on a V8 model, carefully pry each end of the retainer out and up and remove the retainer . . .

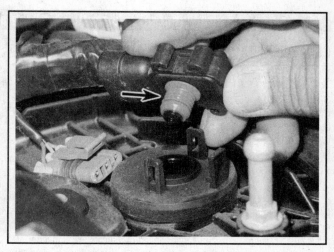

9.8b  . . . then grasp the sensor firmly and pull it straight up and out of the manifold. Be sure to inspect the sensor grommet

## 10 Mass Air Flow (MAF) sensor - replacement

◆ Refer to illustrations 10.2 and 10.3

➡ Note: The MAF sensor is located between the air filter housing and the air intake duct.

1  On V6 models, remove the air intake duct (see Chapter 4, Section 10).
2  Disconnect the electrical connector from the MAF sensor (see illustration).
3  Loosen the hose clamp screws that secure the MAF sensor to the air filter housing and, on V8 models, the air intake duct (see illustration).
4  On V6 models, remove the MAF sensor by pulling it off the air filter housing cover. On V8 models, remove the MAF sensor from between the air filter housing and the air intake duct by pulling off the air intake duct, then pulling the MAF sensor off the air filter housing.
5  Installation is the reverse of removal.

10.2  To disconnect the electrical connector from the MAF sensor, depress this release tab and pull off the connector

10.3  To detach the MAF sensor from the air filter housing and, on V8 models, from the air intake duct, loosen these two hose clamp screws

## 11 Oxygen sensors - general information and replacement

### GENERAL INFORMATION

1  An oxygen sensor is a galvanic battery that produces a very small voltage output in response to the amount of oxygen in the exhaust gases. This voltage signal is the input side of the feedback loop between the oxygen sensor and the Powertrain Control Module (PCM). Without it, the PCM would be unable to correct the injector on-time (which determines the air/fuel ratio) to maintain the perfect (known as stoichiometric) air/fuel ratio of 14.7:1 that the catalyst needs for optimal operation.

2  All vehicles covered by this manual have On-Board Diagnostics II (OBD-II) engine management systems, which means they have the ability to monitor the performance of the catalytic converter. They accomplish this by using an oxygen sensor ahead of the catalytic converter and another oxygen sensor behind the catalytic converter. By comparing the amount of oxygen in the post-catalyst exhaust gas to the oxygen content of the exhaust gas before it enters the catalyst, the PCM can determine the efficiency of the converter.

3  All models covered by this manual have four heated oxygen sensors: one upstream sensor (ahead of the catalytic converter) per cylinder bank and one downstream sensor (after the catalytic converter) per cylinder bank. The upstream sensors are located on the upper part of the front exhaust pipe assembly, above the catalytic converters. The downstream sensors are located under the vehicle, in the exhaust pipes, just behind the catalytic converters.

4  The upstream and downstream oxygen sensors are heated to speed up the warm-up time during which the sensors are unable to produce an accurate voltage signal. The circuit for each oxygen sensor heater is controlled by the PCM, which opens the ground side of the circuit to shut off the heater as soon as the sensor reaches its normal operating temperature.

5  Special care must be taken whenever a sensor is serviced.
  a) *Oxygen sensors have a permanently attached pigtail and an electrical connector that cannot be removed. Damaging or removing the pigtail or electrical connector will render the sensor useless.*
  b) *Keep grease, dirt and other contaminants away from the electrical connector and the louvered end of the sensor.*
  c) *Do not use cleaning solvents of any kind on an oxygen sensor.*
  d) *Oxygen sensors are extremely delicate. Do not drop a sensor, throw it around or handle it roughly.*
  e) *Make sure that the silicone boot on the sensor is installed in the correct position. Otherwise, the boot might melt and it might prevent the sensor from operating correctly.*

### REPLACEMENT

➡ Note: Because they're installed in the exhaust pipes, which contract as they cool down, the oxygen sensors can be very difficult to loosen when the engine is cold. Rather than risk damage to a sensor or its mounting threads, start and run the engine for a minute or two, then shut it off. Be careful not to burn yourself during the following procedure.

# EMISSIONS AND ENGINE CONTROL SYSTEMS

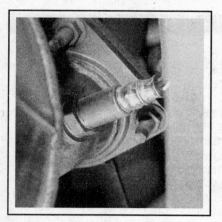

11.7a  Left upstream oxygen sensor location, viewed from underneath the vehicle, looking forward (4WD V8 model shown, V6 models similar)

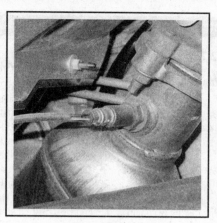

11.7b  Right upstream oxygen sensor location, viewed from the right wheel well, with the splash shield removed (V8 model shown, V6 models similar)

11.8  Using an oxygen sensor socket, unscrew the upstream oxygen sensor from the exhaust manifold (right upstream sensor shown)

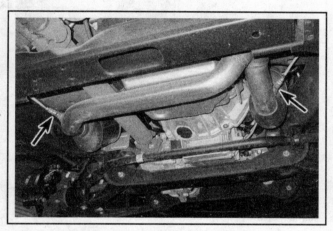

11.12a  Locate the left or right downstream oxygen sensor . . .

11.12b  . . . then trace the sensor electrical lead to the electrical connector, depress the release tab and disconnect the connector (left downstream oxygen sensor and connector shown)

## Upstream oxygen sensor

▸ Refer to illustrations 11.7a, 11.7b, and 11.8

➡Note: The upstream oxygen sensors are located on the upper part of the front exhaust pipe assembly, just above the catalytic converters.

6  Raise the vehicle and place it securely on jackstands.
7  Locate the oxygen sensor (see illustrations), then trace the sensor electrical lead to the electrical connector and disconnect the connector.
8  Using an oxygen sensor socket (available at most auto parts stores), unscrew the upstream oxygen sensor (see illustration). If the sensor is difficult to loosen, spray some penetrating oil onto the sensor threads and allow it to soak in for awhile.
9  If you're going to install the old sensor, apply anti-seize compound to the threads of the sensor to facilitate future removal. If you're going to install a new oxygen sensor, it's not necessary to apply anti-seize compound to the threads. The threads on new sensors already have anti-seize compound on them.
10  Installation is otherwise the reverse of removal. Be sure to tighten the sensor to the torque listed in this Chapter's Specifications.

## Downstream oxygen sensor

▸ Refer to illustrations 11.12a and 11.12b

➡Note: The downstream oxygen sensor is located on top of the exhaust pipe, just behind the catalytic converter.

11  Raise the vehicle and place it securely on jackstands.
12  Locate the downstream oxygen sensor (see illustration), then trace the sensor electrical lead to the connector (see illustration) and disconnect it.
13  Unscrew the downstream oxygen sensor with an oxygen sensor socket (see illustration 11.8). If the sensor is difficult to loosen, spray some penetrating oil onto the sensor threads and allow it to soak in for awhile.
14  If you're going to install the old sensor, apply anti-seize compound to the threads of the sensor to facilitate future removal. If you're going to install a new oxygen sensor, it's not necessary to apply anti-seize compound to the threads. The threads on new sensors already have anti-seize compound on them.
15  Installation is otherwise the reverse of removal. Be sure to tighten the sensor to the torque listed in this Chapter's Specifications.

## 6-22  EMISSIONS AND ENGINE CONTROL SYSTEMS

### 12  Transmission Range (TR) switch - replacement and adjustment

**Refer to illustrations 12.3, 12.4, 12.5 and 12.7**

**Note:** *The TR switch is located on the left side of the transmission.*

1  Apply the parking brake and put the shift lever in NEUTRAL.
2  Raise the vehicle and place it securely on jackstands.
3  Disconnect the electrical connector from the TR switch (see illustration).
4  Disconnect the shift cable from the manual lever (see illustration).
5  Remove the manual lever nut (see illustration) and remove the manual lever.
6  Remove the TR switch mounting bolts (see illustration 12.5) and remove the TR switch.
7  If you're installing a new switch, align the slots on the switch (where the shaft is inserted) with the notch on the switch body (see illustration). Then install the switch onto the shaft.
8  If you're installing the old unit, simply align the flats of the shift shaft with the flats of the TR switch and install the switch.
9  Installation is the reverse of removal. After installing the TR switch, verify that the engine will start only in PARK and NEUTRAL. If it starts in any other gear, readjust the switch.
10  To adjust the TR switch, loosen the switch mounting bolts and turn it slightly one way or the other until the engine starts only in PARK and NEUTRAL, then tighten the mounting bolts securely.

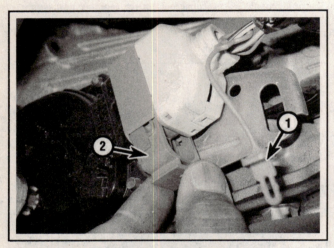

12.3  Pull out the retainer lock (1), slide the retainer (2) out and pull off the connector

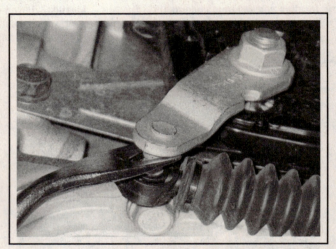

12.4  Pry the shift cable from the pin on the end of the manual lever

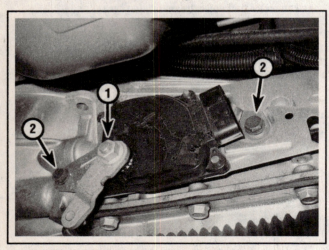

12.5  To remove the manual lever from the manual shaft, remove this nut (1). To detach the TR switch from the transmission, remove the switch mounting bolts (2)

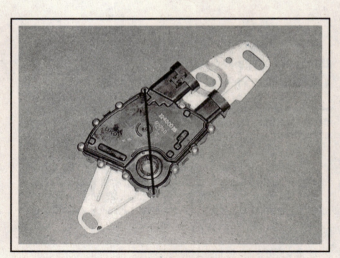

12.7  Before installing the TR switch, align the tabs on the switch with the notch on the switch body. The switch is in the NEUTRAL position in this photo

# EMISSIONS AND ENGINE CONTROL SYSTEMS    6-23

## 13  Vehicle Speed Sensor (VSS) - replacement

♦ Refer to illustrations 13.2a and 13.2b

➡ Note 1: On 2WD models, the VSS is located on the right side of the transmission extension housing. On 4WD models, it is located on the rear upper left part of the transfer case.

➡ Note 2: The photos accompanying this Section depict the Vehicle Speed Sensors used on 2WD and 4WD vehicles. However, some vehicles are equipped with 6L50, 6L80 or 6L90 automatic transmissions on which the speed sensors are located inside the transmission housing, on the valve body. If the VSS is not on the extension housing (2WD models) or on the transfer unit (4WD models), it's because your vehicle has one of these newer automatics.

1  Raise the vehicle and place it securely on jackstands.
2  Disconnect the electrical connector from the VSS (see illustrations).
3  On 2WD models, remove the VSS hold-down bolt and remove the VSS from the transmission. On 4WD models, unscrew the VSS with a deep socket.
4  Remove the old O-ring from the VSS and discard it. Be sure to use a new O-ring when installing the VSS (even if you're planning to reuse the old VSS sensor).
5  Installation is the reverse of removal. Be sure to tighten the VSS hold-down bolt (2WD models) or the VSS (4WD models) securely.

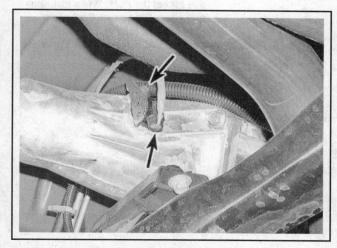

13.2a  To remove the VSS from the extension housing, disconnect the electrical connector, then remove the sensor hold-down bolt (2WD models)

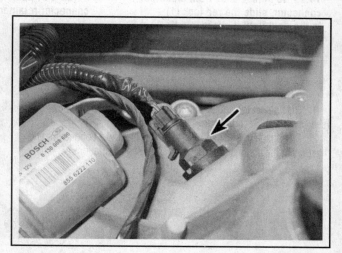

13.2b  Disconnect the electrical connector from the VSS, then unscrew the VSS from the transfer case with a deep socket (4WD models)

## 14  Powertrain Control Module (PCM) - removal and installation

♦ Refer to illustrations 14.2a 14.2b and 14.3

**✲✲ CAUTION:**

To avoid electrostatic discharge damage to the PCM, handle the PCM only by its case. Do not touch the electrical terminals during removal and installation. If available, ground yourself to the vehicle with an anti-static ground strap, available at computer supply stores.

➡ Note 1: The PCM is mounted in the left front corner of the engine compartment.

➡ Note 2: The procedures in this Section apply only to disconnecting, removing and installing the PCM that is already installed in your vehicle. If, however, you need to replace the PCM, it must be programmed with new software and calibrations. This procedure requires the use of GM's TECH-2 scan tool and GM's latest PCM-programming software, so you WILL NOT BE ABLE TO REPLACE THE PCM AT HOME.

# 6-24 EMISSIONS AND ENGINE CONTROL SYSTEMS

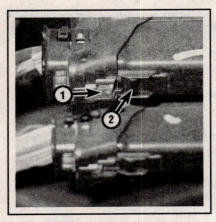

14.2a  To unlock each PCM electrical connector, slide the red lock (1) to the right until it clicks into its unlocked position, then depress the release tab (2) . . .

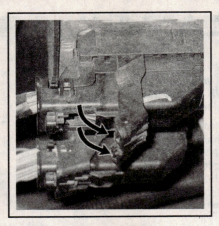

14.2b  . . . and flip open the connector retainer

14.3  Pry up the lock tabs at the top and lift out the PCM. When installing, make sure the lugs on the bottom fit behind the tabs of the mounting bracket, and make sure the upper end snaps into place

1  Disconnect the cable from the negative terminal of the battery (see Chapter 5, Section 1).
2  Disengage the retainers on the PCM electrical connectors and disconnect the connectors from the PCM (see illustrations).

3  Disengage the retainer tabs that secure the PCM to its mounting bracket (see illustration) and remove the PCM.
4  Installation is the reverse of removal.

## 15 Catalytic converters - description, check and replacement

➡ **Note:** *Because of a Federally-mandated extended warranty which covers emission-related components such as the catalytic converter, check with a dealer service department before replacing the converter at your own expense.*

### DESCRIPTION

1  A catalytic converter (or catalyst) is an emission control device in the exhaust system that reduces certain pollutants in the exhaust gas stream. There are two types of converters. An oxidation catalyst reduces hydrocarbons (HC) and carbon monoxide (CO). A reduction catalyst reduces oxides of nitrogen (NOx). Catalysts that can reduce all three pollutants are known as "three-way catalysts." The vehicles covered in this manual are equipped with two three-way catalysts.

### CHECK

2  The test equipment for a catalytic converter (a loaded-mode dynamometer and a five-gas analyzer) is expensive. If you suspect that the converter on your vehicle is malfunctioning, take it to a dealer or authorized emission inspection facility for diagnosis and repair.
3  Whenever you raise the vehicle to service underbody components, inspect the converter for leaks, corrosion, dents and other damage. Carefully inspect the welds and/or flange bolts and nuts that attach the front and rear ends of the converter to the exhaust system. If you note any damage, replace the converter.
4  Although catalytic converters don't break too often, they can become plugged up. The easiest way to check for a restricted converter is to use a vacuum gauge to diagnose the effect of a blocked exhaust on intake vacuum.

   a) *Connect a vacuum gauge to an intake manifold vacuum source (see Chapter 2).*
   b) *Warm the engine to operating temperature, place the transmission in Park (automatic models) or Neutral (manual models) and apply the parking brake.*
   c) *Note the vacuum reading at idle and write it down.*
   d) *Quickly open the throttle to near its wide-open position, then quickly get off the throttle and allow it to close. Note the vacuum reading and write it down.*
   e) *Do this test three more times, recording your measurement after each test.*
   f) *If your fourth reading is more than one in-Hg lower than the reading that you noted at idle, the exhaust system might be restricted (the catalytic converter could be plugged, OR an exhaust pipe or muffler could be restricted).*

# EMISSIONS AND ENGINE CONTROL SYSTEMS

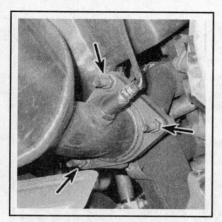

**15.7a  To detach the upper left end of the front exhaust pipe assembly from the left exhaust manifold, remove these nuts**

**15.7b  To detach the upper right end of the front exhaust pipe assembly from the right exhaust manifold, remove these nuts**

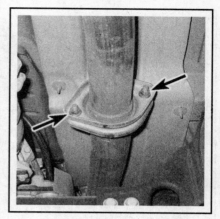

**15.8  To detach the rear end of the front exhaust pipe assembly from the rest of the exhaust system, remove these nuts**

## REPLACEMENT

♦ Refer to illustrations 15.7a, 15.7b and 15.8

➡ **Note:** *The catalytic converters are integral components of the front exhaust pipe assembly, which connects the exhaust manifolds to the rest of the exhaust pipe assembly. Both catalysts are welded into the front exhaust pipe assembly, which is a one-piece assembly.*

5  Raise the vehicle and place it securely on jackstands.

6  Disconnect the electrical connectors for the upstream and downstream oxygen sensors and remove all four sensors (see Section 11).

7  Remove the three nuts at each of the upper flanges (see illustrations) that secure the front exhaust pipe assembly to the exhaust manifold flanges.

8  Remove the two nuts (see illustration) that secure the rear flange of the front exhaust pipe assembly to the rest of the exhaust system and remove the front exhaust pipe assembly.

9  Remove and discard the old flange gaskets from all three mounting flanges. Be sure to use new gaskets at both mounting flanges. Also use new fasteners at both flanges.

10  Coat the threads of the nuts and bolts with anti-seize compound to facilitate future removal. Tighten the nuts that secure the front exhaust pipe assembly to the exhaust manifold and to the rest of the exhaust system securely. Installation is otherwise the reverse of removal

## 16  Evaporative emissions control (EVAP) system - description and component replacement

### DESCRIPTION

1  The Evaporative Emissions Control (EVAP) system prevents fuel system vapors (which contain unburned hydrocarbons) from escaping into the atmosphere. On warm days, vapors trapped inside the fuel tank expand until the pressure reaches a certain threshold, at which point the fuel vapors are routed from the fuel tank through the fuel vapor vent valve and the fuel vapor control valve to the EVAP canister, where they're stored temporarily, until they can be consumed by the engine during normal operation. When the conditions are right (engine warmed up, vehicle up to speed, moderate or heavy load on the engine, etc.) the Powertrain Control Module (PCM) opens the canister purge solenoid, which allows the fuel vapors to be drawn from the canister into the intake manifold, where they mix with the air/fuel mixture before being consumed in the combustion chambers. This system is complex and virtually impossible to troubleshoot without the right tools and training. However, the following description should give you a good idea of how it works:

2  The EVAP canister is located under the vehicle, in front of or to the side of the fuel tank, depending on the model. The EVAP canister, which contains activated charcoal, is the repository for storing the fuel vapors, and is designed to be maintenance-free (it should last the life of the vehicle).

3  The fuel tank pressure sensor, which is located on top of the mounting flange for the in-tank fuel pump/fuel level sending unit module, monitors the pressure inside the tank, and transmits its measurement to the PCM during an OBD-II leak test.

4  The EVAP canister vent solenoid is normally open. But it seals off the EVAP system for inspection and maintenance (I/M 240) testing and for OBD-II leak and pressure tests.

5  The EVAP canister purge solenoid, which is under the control of the Powertrain Control Module (PCM), regulates the flow of vapors being purged from the EVAP canister into the intake manifold. The canister purge solenoid is normally closed. It opens only when directed to do so by the PCM, which uses the availability of intake manifold vacuum and data from various information sensor inputs to determine when and how long to open the valve. The interval of time during which the purge valve is opened by the PCM is known as its duty cycle.

### General system checks

6  The most common symptom of a faulty EVAP system is a strong fuel odor (particularly during hot weather). If you smell fuel while driv-

# 6-26 EMISSIONS AND ENGINE CONTROL SYSTEMS

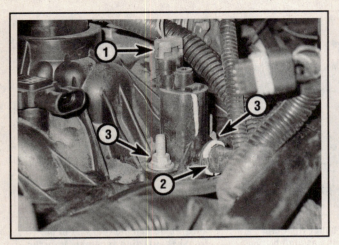

**16.10 Canister purge solenoid details - V6 engine**

1. Electrical connector
2. Purge hose
3. Mounting fasteners

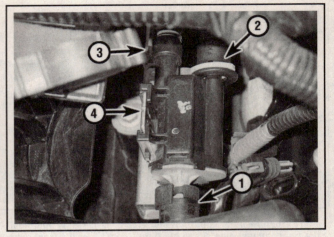

**16.15 Canister purge solenoid details - V8 engines**

1. Outlet purge line
2. Inlet purge line
3. Electrical connector
4. Mounting bracket

ing or (more likely) right after you park the vehicle and turn off the engine, check the fuel filler cap first. Make sure that it's screwed onto the fuel filler neck all the way. If the odor persists, inspect all EVAP hose connections, both in the engine compartment and under the vehicle. Be sure to inspect each hose attached to the canister for damage and leakage along its entire length. Repair or replace as necessary. Inspect the canister for damage and look for fuel leaking from the bottom. If fuel is leaking or the canister is otherwise damaged, replace it.

7  Poor idle, stalling, and poor driveability can be caused by a defective fuel vapor vent valve or canister purge solenoid, a damaged canister, cracked hoses, or hoses connected to the wrong tubes. Fuel loss or fuel odor can be caused by fuel leaking from fuel lines or hoses, a cracked or damaged canister, or a defective vapor valve.

8  To check for excessive fuel vapor pressure in the fuel tank, remove the gas cap and listen for the sound of pressure release. If the fuel tank emits a whooshing sound when you open the filler cap, fuel tank vapor pressure is excessive. Inspect the canister vapor hoses and the canister inlet port for blockage or collapsed hoses. Also inspect the vapor vent valve. A complete test can only be done with a professional-level OBD-II scan tool, which will run a series of checks to detect excessive pressure. You'll have to take the vehicle to a dealer service department or other qualified repair shop to have the EVAP system professionally diagnosed.

## COMPONENT REPLACEMENT

### EVAP canister purge solenoid

#### V6 engines

♦ Refer to illustration 16.10

➡ **Note: The EVAP canister purge solenoid valve is located on the right side of the intake manifold, next to the fuel meter body.**

9  Remove the air intake duct (see Chapter 4).
10  Disconnect the canister purge solenoid electrical connector (see illustration).
11  Disconnect the EVAP purge line quick-connect fitting from the canister purge solenoid. If you're unfamiliar with quick-connect fittings,

refer to Chapter 4.
12  Remove the canister purge solenoid mounting nut and remove the purge solenoid.
13  Installation is the reverse of removal.

#### V8 engines

♦ Refer to illustration 16.15

➡ **Note: The EVAP canister purge solenoid valve is located above the front end of the left fuel rail.**

14  Remove the air intake duct (see Chapter 4).
15  Disconnect the outlet purge line quick-connect fitting from the canister purge solenoid (see illustration).
16  Disconnect the electrical connector from the EVAP canister purge valve.
17  Disconnect the inlet purge line quick-connect fitting from the canister purge solenoid.
18  Cap both EVAP purge lines to prevent dirt, dust and moisture from entering the EVAP system while the line is open. If you're unfamiliar with quick-connect fittings, refer to Section 4 in Chapter 4.
19  Remove the canister purge solenoid from its mounting bracket.
20  Installation is the reverse of removal.

### EVAP canister vent solenoid

♦ Refer to illustrations 16.22, 16.23 and 16.24

➡ **Note: On some models, the EVAP canister vent solenoid is located in front of the EVAP canister. On other models, the vent solenoid is located at the back of the fuel tank.**

21  Raise the vehicle and place it securely on jackstands.
22  Disconnect the electrical connector from the vent solenoid (see illustration).
23  Follow the inlet and outlet EVAP lines from the vent solenoid to their quick-connect fittings, then disconnect the fittings (see illustration). Also disconnect the lines from any clips securing them.
24  Remove the vent solenoid from its mounting bracket (see illustration).
25  Installation is the reverse of removal.

# EMISSIONS AND ENGINE CONTROL SYSTEMS    6-27

16.22 To disconnect the electrical connector from the vent solenoid, carefully pry the locking retainer away from the locking lug on the solenoid with a small screwdriver and pull off the connector

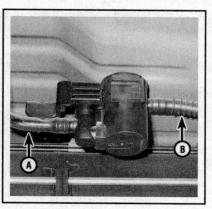

16.23 Follow the inlet (A) and outlet (B) EVAP lines to their quick-connect fittings, then disconnect them

16.24 To separate the canister vent solenoid from its mounting bracket, pry the lock tab on the bracket loose from the lug on the vent solenoid and pull the solenoid off the bracket

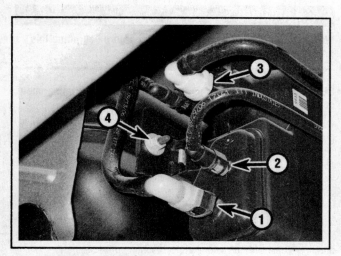

16.27 EVAP canister details:

1  EVAP inlet line (from fuel tank)
2  EVAP purge line (to canister purge solenoid)
3  EVAP vent line (to vent solenoid)
4  Mounting nut

## EVAP canister

♦ Refer to illustration 16.27

➥Note: On models with the EVAP canister mounted in front of the fuel tank, it's installed vertically, on its side. On models with the canister mounted next to the fuel tank, it's mounted horizontally. All vehicles use the same canister and the installations are similar.

26  Raise the vehicle and place it securely on jackstands.
27  Clearly label all EVAP hoses connected to the EVAP canister, then disconnect the quick-connect fittings from the canister (see illustration). For more information on quick-connect fittings, see Chapter 4.
28  Remove the canister mounting nut (see illustration 16.27a) and remove the canister from its mounting bracket.

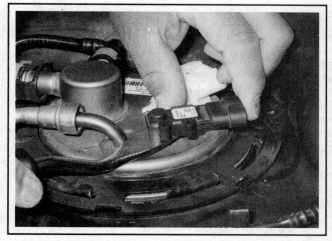

16.33 Pry the fuel tank pressure sensor out of the fuel pump/fuel level sending unit mounting flange

29  When installing the canister, make sure that the horizontal and vertical slots in the back end of the canister are aligned with and slide onto their corresponding tabs on the canister mounting bracket.
30  Installation is the reverse of removal. Make sure that all three EVAP line quick-connect fittings are fully engaged.

## Fuel tank pressure sensor

♦ Refer to illustration 16.33

➥Note: The fuel tank pressure sensor is located on top of the mounting flange for the fuel pump/fuel level sending unit module.

31  Raise the vehicle and place it securely on jackstands.
32  Remove the fuel tank (see Chapter 4).
33  Remove the fuel tank pressure sensor from the fuel pump/fuel level sending unit mounting flange (see illustration).
34  Installation is the reverse of removal.

## 6-28 EMISSIONS AND ENGINE CONTROL SYSTEMS

### 17 Positive Crankcase Ventilation (PCV) system - description

1 The Positive Crankcase Ventilation (PCV) system reduces hydrocarbon emissions by scavenging crankcase vapors and burning them along with the air-fuel mixture.

2 The PCV system uses a PCV orifice (V8 engines) or a PCV valve (V6 engine), which is located in the left valve cover. There are two hoses in this system. The fresh air inlet hose carries outside air from the air intake duct to a pipe on the right valve cover, then into the crankcase, where it mixes with blow-by gases and crankcase vapors. These vapors are drawn from the crankcase through the orifice or valve, through the crankcase ventilation hose (PCV hose), then into the intake manifold, where they mix with incoming air.

### 18 Camshaft Position (CMP) Actuator System - description and component replacement

#### DESCRIPTION

1 The Camshaft Position (CMP) Actuator System is an electro-hydraulic system that changes the angle, or timing, of the camshaft relative to the crankshaft position. By controlling this, the CMP actuator system reduces emissions by using exhaust gases to help dilute the intake charge, broadens the engine torque range and increases fuel mileage.

2 The CMP actuator system consists of the Powertrain Control Module (PCM), the CMP actuator solenoid and the camshaft position actuator. The PCM controls the amount of oil that flows through an oil passage to the CMP actuator by sending a pulse-width-modulated signal to the CMP actuator solenoid. The oil flows through two passages; one passage is for advancing the camshaft and one passage is for retarding the cam. The CMP actuator is on the front end of the camshaft. When oil is directed by the solenoid through one passage, the variable cam sprocket advances the camshaft timing; when oil is directed through the other passage, the sprocket retards cam timing.

3 Any of the following factors can affect the operation of the CMP Actuator System:
   Aftermarket oil additives
   Incorrect engine oil level
   Incorrect engine oil pressure
   Incorrect engine oil temperature
   Incorrect engine oil viscosity

#### COMPONENT REPLACEMENT

**Camshaft Position (CMP) actuator magnet**

**✳✳ WARNING:**

**Wait until the engine is completely cool before beginning this procedure.**

➡ **Note: The CMP actuator magnet is located on the timing chain cover.**

4 Drain the cooling system (see Chapter 1), then remove the water pump (see Chapter 3).

5 Disconnect the electrical connector from the CMP actuator magnet.

6 Remove the three CMP actuator magnet mounting bolts and remove the magnet.

7 Remove and discard the old CMP actuator magnet gasket.

8 Make sure that the gasket surface is clean and free of all debris.

9 Installation is the reverse of removal. Be sure to use a new gasket and tighten the CMP actuator magnet mounting bolts securely.

10 Refill the cooling system (see Chapter 1).

**Camshaft Position (CMP) actuator/solenoid valve**

11 To replace the CMP actuator or solenoid valve, refer to Chapter 2B, Section 12.

### 19 Cylinder Deactivation System - description and component replacement

➡ **Note: The cylinder deactivation system is used on 5.3L and 6.0L V8 engines only.**

#### DESCRIPTION

1 The Cylinder Deactivation System (or Active Fuel Management System) improves fuel economy and lowers emissions by deactivating four of the engine's eight cylinders. During starting, idling and medium or heavy throttle conditions, the engine operates normally. But during light-load cruising, the Powertrain Control Module (PCM) deactivates cylinders one and seven in the front cylinder bank and cylinders two and four in the rear cylinder bank, effectively turning the engine into a V4.

2 The system consists of the Valve Lifter Oil Manifold (VLOM) assembly and eight specially designed valve lifters (four intake and four exhaust) for cylinders 1, 4, 6 and 7. The VLOM assembly consists of four electrically operated solenoids. Each solenoid directs the flow of

# EMISSIONS AND ENGINE CONTROL SYSTEMS   6-29

19.7  To remove the VLOM oil filter and O-ring, disconnect the electrical connector from the VLOM oil pressure sensor . . .

19.8  . . . unscrew the sensor with a deep socket and remove the sensor and washer . . .

pressurized engine oil to the special intake and exhaust valve lifters. An oil pressure sensor, which is mounted on the left end of the VLOM assembly, monitors engine oil pressure and functions as an information sensor for the PCM. An oil filter, which is located directly below the oil pressure sensor, helps control contamination inside the hydraulic circuit of the Cylinder Deactivation System. An oil pressure relief valve, which is located at the left rear corner of the oil pan, regulates engine oil pressure to the lubrication system and to the VLOM assembly.

3  When operating in V8 mode, the special valve lifters function just like conventional lifters. The solenoids in the VLOM assembly are in their closed position and no pressurized oil is directed to the valve lifters. Spring-loaded locking pins in the lifters extend outward, mechanically locking the pin housings to the outer bodies of the valve lifters.

4  When the conditions are right, the PCM grounds the solenoid control circuits, which allows current to flow through the solenoid windings. When the solenoid windings are energized, the normally-closed solenoid valves open, which directs pressurized engine oil through the VLOM assembly into eight vertical oil passages in the engine block lifter valley. The eight vertical passages, two per cylinder, direct pressurized oil to the lifter bores of cylinders 1, 4, 6 and 7. The pressurized oil forces the locking pins inside the lifters inward, locking up the pushrods and preventing them from traveling up and down. The outer bodies of the lifters continue moving up and down independently of the pin housings. Additionally, the PCM turns off the fuel injectors to those cylinders.

5  When the PCM turns off the system, the solenoids in the VLOM assembly close, blocking the flow of pressurized oil to the valve lifters. The oil pressure within the lifters decreases and the locking pins again move out to mechanically lock up the pin housing with the outer lifter body.

## COMPONENT REPLACEMENT

### VLOM oil pressure sensor and oil filter

▶ Refer to illustrations 19.7, 19.8 and 19.9

6  Remove the engine cover.
7  Disconnect the electrical connector from the VLOM oil pressure sensor (see illustration).

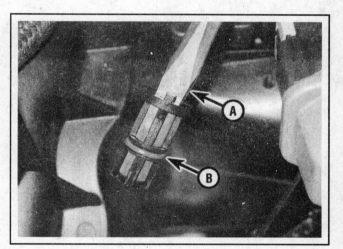

19.9  . . . then carefully wedge the tip of a screwdriver (A) into the top of the oil filter and pull out the filter. Inspect the condition of the screen and the condition of the O-ring (B). If either is damaged, replace the filter and O-ring as a set

8  Remove the VLOM oil pressure sensor (see illustration) and remove the sensor washer
9  Remove the oil filter and the filter O-ring (see illustration). If the filter is plugged or the O-ring is cracked, torn or otherwise deteriorated, replace the filter and O-ring as a set.
10  Installation is the reverse of removal.

### Valve Lifter Oil Manifold (VLOM) assembly

▶ Refer to illustrations 19.14, 19.15 and 19.16

11  Remove the engine cover.
12  Disconnect the electrical connector from the VLOM oil pressure sensor (see illustration 19.7).
13  Remove the fuel rail (see Chapter 4) and the intake manifold (see Chapter 2B).
14  Disconnect the electrical connector from the VLOM solenoid (see illustration).

# 6-30 EMISSIONS AND ENGINE CONTROL SYSTEMS

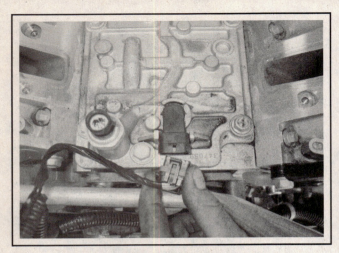

**19.14 Disconnect the electrical connector from the VLOM solenoid**

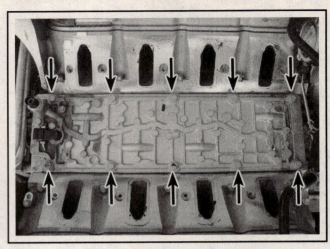

**19.15 To detach the VLOM assembly from the engine, remove these mounting bolts**

15  Remove the VLOM assembly mounting bolts (see illustration) and remove the VLOM assembly.

16  Inspect the VLOM assembly gasket (see illustration). If it's cracked, torn or otherwise deteriorated, replace it.

17  Installation is the reverse of removal. Be sure to tighten the VLOM assembly mounting bolts to the torque listed in this Chapter's Specifications.

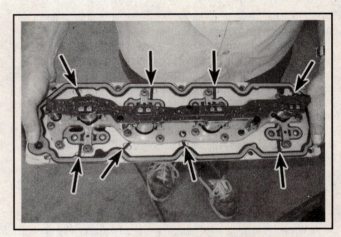

**19.16 To remove the old gasket from the VLOM assembly, carefully cut the eight retaining straps at the indicated points to separate the outer gasket from the rest of the old gasket. Replace only the outer gasket, leaving the rest in place**

## Torque specifications — Ft-lbs (unless otherwise indicated)

➡ Note: One foot-pound (ft-lb) of torque is equivalent to 12 inch-pounds (in-lbs) of torque. Torque values below approximately 15 ft-lbs are expressed in inch-pounds, since most foot-pound torque wrenches are not accurate at these smaller values.

| | |
|---|---|
| Engine coolant temperature sensor | 180 in-lbs |
| Knock sensor retaining bolt | 18 |
| Oxygen sensors | 31 |
| Valve Lifter Oil Manifold (VLOM) assembly mounting bolts | 18 |

**Section**

1 General information
2 Diagnosis - general
3 Shift cable - removal, installation and adjustment
4 Park/Lock system - description and component replacement
5 Extension housing oil seal (2WD) - replacement
6 Transmission mount - check and replacement
7 Automatic transmission - removal and installation

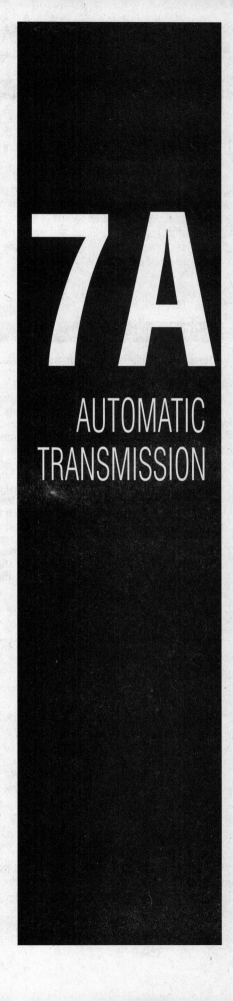

# 7A
## AUTOMATIC TRANSMISSION

# 7A-2 AUTOMATIC TRANSMISSION

## 1 General information

The models covered by this manual use either a 4L60-E, 4L65-E or a 4L70-E four-speed automatic transmission or a 6L50, 6L80 or 6L90 six-speed transmission. They are all electronically controlled and are equipped with a torque converter clutch (TCC), which provides a direct connection between the engine and the drive wheels for improved efficiency and economy during certain operating conditions. The TCC consists of a solenoid controlled by the Powertrain Control Module (PCM) that locks the converter in third or fourth gear (four-speed transmissions) or fifth or sixth gear (six-speed transmissions) when the vehicle is cruising on level ground and the engine is fully warmed up. Some models are also equipped with an auxiliary transmission cooler that is mounted in front of the radiator and air conditioning condenser.

Due to the complexity of the automatic transmissions covered in this manual and the need for specialized equipment to perform most service operations, this Chapter contains only general diagnosis, adjustment and removal and installation procedures.

If the transmission requires major repair work, it should be left to a dealer service department or an automotive or transmission repair shop. You can, however, remove and install the transmission yourself and save the expense, even if the repair work is done by a transmission shop.

## 2 Diagnosis - general

➙Note: Automatic transmission malfunctions may be caused by five general conditions: poor engine performance, improper adjustments, hydraulic malfunctions, mechanical malfunctions or malfunctions in the Powertrain Control Module or its signal network. Diagnosis of these problems should always begin with a check of the easily repaired items: fluid level and condition (see Chapter 1), and shift cable adjustment (see Section 3). Next, perform a road test to determine if the problem has been corrected or if more diagnosis is necessary. Because the transmission relies on many sensors in the engine control system, and since the transmission shift points are controlled by the Powertrain Control Module, you'll also want to check to see if any trouble codes have been stored in the PCM (see Chapter 6 for a list of trouble codes and how to extract them). If the problem persists after the preliminary tests and corrections are completed, additional diagnosis should be done by a dealer service department or transmission repair shop. Refer to the Troubleshooting Section at the front of this manual for transmission problem diagnosis.

### PRELIMINARY CHECKS

1 Drive the vehicle to warm the transmission to normal operating temperature.
2 Check the fluid level as described in Chapter 1:
   a) If the fluid level is unusually low, add enough fluid to bring the level within the designated area of the dipstick, then check for external leaks.
   b) If the fluid level is abnormally high, drain off the excess, then check the drained fluid for contamination by coolant. The presence of engine coolant in the automatic transmission fluid indicates that a failure has occurred in the internal radiator walls that separate the coolant from the transmission fluid (see Chapter 3).
   c) If the fluid is foaming, drain it and refill the transmission, then check for coolant in the fluid or a high fluid level.
3 Check the engine idle speed.

➙Note: If the engine is malfunctioning, do not proceed with the preliminary checks until it has been repaired and runs normally.

4 Inspect the shift cable (see Section 3). Make sure that it's properly adjusted and that it operates smoothly.
5 Check the Transmission Range (TR) switch adjustment (see Chapter 6).

### FLUID LEAK DIAGNOSIS

6 Most fluid leaks are easy to locate visually. Repair usually consists of replacing a seal or gasket. If a leak is difficult to find, the following procedure may help.
7 Identify the fluid. Make sure it's transmission fluid and not engine oil or brake fluid (automatic transmission fluid is a deep red color).
8 Try to pinpoint the source of the leak. Drive the vehicle several miles, then park it over a large sheet of cardboard. After a minute or two, you should be able to locate the leak by determining the source of the fluid dripping onto the cardboard.
9 Make a careful visual inspection of the suspected component and the area immediately around it. Pay particular attention to gasket mating surfaces. A mirror is often helpful for finding leaks in areas that are hard to see.
10 If the leak still cannot be found, clean the suspected area thoroughly with a degreaser or solvent, then dry it.
11 Drive the vehicle for several miles at normal operating temperature and varying speeds. After driving the vehicle, visually inspect the suspected component again.
12 Once the leak has been located, the cause must be determined before it can be properly repaired. If a gasket is replaced but the sealing flange is bent, the new gasket will not stop the leak. The bent flange must be straightened.
13 Before attempting to repair a leak, check to make sure that the following conditions are corrected or they may cause another leak.

➙Note: Some of the following conditions cannot be fixed without highly specialized tools and expertise. Such problems must be referred to a transmission shop or a dealer service department.

**Gasket leaks**

14 Check the pan periodically. Make sure the bolts are tight, no bolts are missing, the gasket is in good condition and the pan is flat (dents in the pan may indicate damage to the valve body inside).
15 If the pan gasket is leaking, the fluid level or the fluid pressure may be too high, the vent may be plugged, the pan bolts may be too tight, the pan sealing flange may be warped, the sealing surface of the transmission housing may be damaged, the gasket may be damaged or the transmission casting may be cracked or porous. If sealant instead of gasket material has been used to form a seal between the pan and the transmission housing, it may be the wrong sealant.

# AUTOMATIC TRANSMISSION 7A-3

### Seal leaks

16  If a transmission seal is leaking, the fluid level or pressure may be too high, the vent may be plugged, the seal bore may be damaged, the seal itself may be damaged or improperly installed, the surface of the shaft protruding through the seal may be damaged or a loose bearing may be causing excessive shaft movement.

17  Make sure the dipstick tube seal is in good condition and the tube is properly seated. Periodically check the area around the speedometer gear or vehicle speed sensor for leakage. If transmission fluid is evident, check the O-ring for damage. Also inspect the driveshaft oil seal for leakage.

### Case leaks

18  If the case itself appears to be leaking, the casting is porous and will have to be repaired or replaced.

19  Make sure the oil cooler hose fittings are tight and in good condition. The transmission oil cooler lines on these models are equipped with quick connect fittings - always inspect the O-rings if a leak is suspected.

### Fluid comes out vent pipe or fill tube

20  If this condition occurs, the transmission is overfilled, there is coolant in the fluid, the case is porous, the dipstick is incorrect, the vent is plugged or the drain back holes are plugged.

## 3  Shift cable - removal, installation and adjustment

### REMOVAL

→Note: The shift cable on these models is a two piece design, therefore one end of the cable can be replaced without replacing the other.

1  Disconnect the cable from the negative terminal of the battery (see Chapter 5, Section 1).
2  Place the transmission in PARK and apply the parking brake.
3  Block the rear wheels so the vehicle will not accidentally roll in either direction.

### Lower cable - transmission end

▶ Refer to illustrations 3.4, 3.5a, 3.5b, 3.6, 3.7 and 3.8

4  Disconnect the shift cable from the transmission shift lever (see illustration).
5  Disengage the shift cable from the cable bracket on the transmission (see illustration). Unclip the cable from its retainer (see illustration).

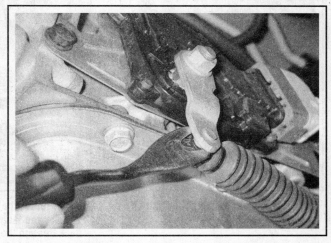

**3.4  Use a screwdriver to pry the shift cable off the shift lever at the transmission**

**3.5a  Remove the retaining clip and detach the shift cable from the bracket**

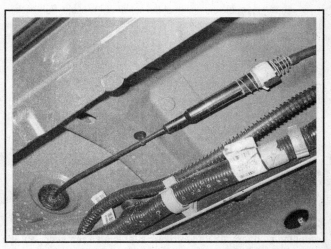

**3.5b  Disconnect the cable from the clip that secures it to the floor of the vehicle**

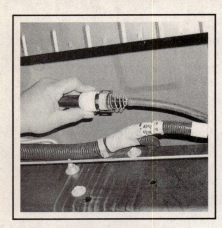
3.6 Pull back the white plastic collar to expose the cable connector retaining clip

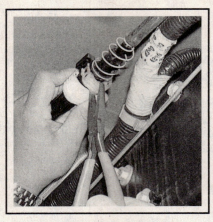
3.7 Pull downward on the center tabs of the connector retaining clip to remove it

3.8 Detach the cable retaining clip, separate the cable halves and remove the transmission end of the cable from the vehicle

3.11 Pry the end of the shift cable off the shift lever arm

3.12 Detach the cable retaining clip, depress the tangs and push the cable back through the hole in the steering column bracket

3.16 The cable grommet is located on the floorboard just in front of the drivers seat

6  Locate the cable connector at the center of the cable. Pull back the white plastic collar to expose the cable connector retaining clip (see illustration).

7  Remove the cable connector retaining clip (see illustration).

8  Remove the cable retaining clip (see illustration).

➡ **Note: Always replace the cable retaining clip (E-clip) with a new one upon installation.**

9  Separate the cable halves and remove the transmission end of the shift cable from the vehicle.

### Upper cable - steering column end

◆ Refer to illustrations 3.11, 3.12 and 3.16

10  Remove the steering column trim covers and the knee bolster from below the steering column (see Chapter 11).

➡ **Note: It will be easier to access the shift cable and the surrounding components if the drivers seat is also removed, but it's not absolutely necessary.**

11  Pry the shift cable ball socket from the shift lever ball pivot (see illustration).

12  Remove the retaining clip securing the cable to the steering column bracket. Depress the tangs and slide the cable out of the steering column bracket (see illustration).

13  Working below the steering column, remove the bolt and the cable support wire securing the cable to the lower half of the steering column.

14  Working under the vehicle, perform Steps 6 through 9 to separate the upper shift cable from the lower shift cable.

15  Working back inside the passenger compartment, remove the door sill plate and the left kick panel from the vehicle, then peel back the carpeting.

16  Trace the cable to the cable grommet (the point at which it goes through the floor board). Pry out the grommet and pull the upper cable up through the hole in the floor to remove it (see illustration).

# AUTOMATIC TRANSMISSION  7A-5

## INSTALLATION AND ADJUSTMENT

**Refer to illustration 3.20**

17  Installation of the either cable is the reverse of removal with the following exceptions:

18  Make sure the driver's shift lever and the transmission shift lever are in the Park position.

19  Before joining the cable halves together it will be necessary to insert a new cable retaining clip (E-clip) onto the female half of the cable at the cable connector (see illustration 3.8).

20  Using two hands, align the male half of the cable with the female half of the cable and push the connector halves together until the blue adjustment spring is fully compressed. This engages the male end of the cable into the E-clip on the female end of the cable and locks the cable together (see illustration).

21  Release the lower (transmission) end of the cable and allow the blue spring to self-adjust the cable. The blue spring must be free standing with no manual help to properly adjust the cable.

22  Pull back the white plastic collar on the upper (steering column) end of the cable connector and insert the cable connector retaining clip (see illustration 3.6).

23  Verify that the cable connector retaining clip is fully seated and that the white plastic collar slides back over the cable connector retaining clip. Test the vehicle for proper shift operation.

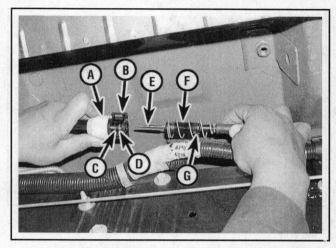

3.20  Shift cable (adjustment) connector installation details

| | | | |
|---|---|---|---|
| A | White plastic collar | D | Shift cable (female half) |
| B | Shift cable connector (steering column end) | E | Shift cable (male half) |
| | | F | Shift cable connector (transmission end) |
| C | Shift cable retaining clip (E-clip) | G | Adjustment spring |

## 4  Park/Lock system - description and component replacement

### DESCRIPTION

1  The Park/Lock system prevents the shift lever from being moved out of Park unless the brake pedal is depressed simultaneously. It also prevents the ignition key from being removed from the ignition switch unless the shift lever is in the Park position. When the car is started, a solenoid is energized, locking the shift lever in Park; when the brake pedal is depressed, the solenoid is de-energized, unlocking the shift lever so that it can be moved into some other gear.

### ACTUATOR REPLACEMENT

**Refer to illustrations 4.4, 4.5a, 4.5b, 4.8 and 4.11**

**WARNING:**

Models covered by this manual are equipped with a Supplemental Restraint System (SRS), more commonly known as airbags. Always disable the airbag system before working in the vicinity of any airbag system component to avoid the possibility of accidental deployment of the airbag, which could cause personal injury (see Chapter 12).

2  Park the vehicle with the shift lever in the Neutral position. Set the parking brake.

3  Disable the airbag system (see Chapter 12), then remove the driver's knee bolster (see Chapter 11, Section 26).

4  Disconnect the wiring from the shift lock actuator (see illustration).

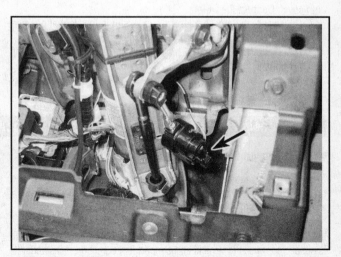

4.4  Disconnect the wiring from the brake/transmission shift interlock actuator

## 7A-6 AUTOMATIC TRANSMISSION

4.5a Pry the brake/transmission shift interlock actuator from the arm . . .

4.5b . . . and from the steering column bracket

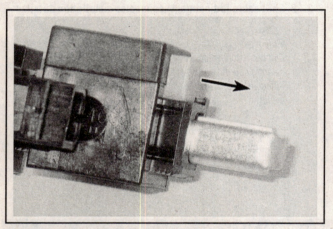

4.8 Pull out the white release tab to release the adjustment mechanism . . .

4.11 . . . then press the adjuster block down and slide it away from the actuator as far as possible; push the white lock tab back into place until it clicks to lock it

5  Pry the shift lock actuator off of the cable shift cam and the steering column (see illustrations).
6  Snap the actuator onto the cable shift cam and the steering column cover. Connect the wiring.
7  Make sure that the shift lever clevis is still in the Neutral position.
8  Pull out the tab on the actuator (see illustration).
9  Press the adjuster block. This will disengage the internal teeth of the mechanism.

10  Slide the adjuster block away from the actuator as far as you can.
11  Push in on the tab to lock it in place (see illustration).
12  Reconnect the battery, turn the ignition key to the On position, then check the operation of the actuator by making sure that the lever can only be moved out of Park when the brake pedal is depressed. Readjust if necessary.
13  Reinstall the knee bolster.

## 5  Extension housing oil seal (2WD) - replacement

▶ **Refer to illustrations 5.4 and 5.5**

1  Oil leaks frequently occur due to wear of the extension housing oil seal. Replacement of this seal is relatively easy, since it can be performed without removing the transmission from the vehicle.
2  The extension housing oil seal is located at the extreme rear of the transmission, where the driveshaft is attached. If leakage at the seal is suspected, raise the vehicle and support it securely on jackstands. If the seal is leaking, transmission lubricant will be built up on the front of the driveshaft and may be dripping from the rear of the transmission.
3  Remove the driveshaft (see Chapter 8).

4  Using a seal removal tool or a large screwdriver, carefully pry the oil seal out of the rear of the transmission (see illustration). Do not damage the splines on the transmission output shaft.
5  Using a seal driver or a very large deep socket as a drift, install the new oil seal (see illustration). Drive it into the bore squarely and make sure it's completely seated.
6  Lubricate the splines of the transmission output shaft and the outside of the driveshaft yoke with lightweight grease, then install the driveshaft (see Chapter 8). Be careful not to damage the lip of the new seal.

# AUTOMATIC TRANSMISSION 7A-7

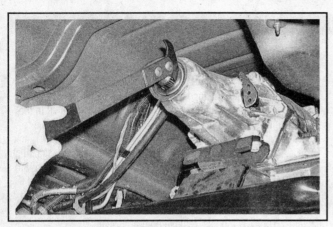

5.4 Carefully pry the old seal out of the extension housing - don't damage the splines on the output shaft

5.5 Drive the new seal into place with a seal driver or a large socket and hammer

## 6 Transmission mount - check and replacement

### CHECK

▸ **Refer to illustration 6.2**

1  Raise the vehicle and support it securely on jackstands.
2  Insert a large screwdriver or prybar into the space between the transmission extension housing and the crossmember and try to pry the transmission up slightly (see illustration).
3  The transmission should not move much at all - if the mount is cracked or torn, replace it.

### REPLACEMENT

▸ **Refer to illustrations 6.4a and 6.4b**

4  To replace the mount, remove the fasteners attaching the mount to the crossmember and the to the transmission (see illustrations).
5  Raise the transmission slightly with a jack and remove the mount.
6  Installation is the reverse of the removal procedure.

6.2 To check the transmission mount, insert a large screwdriver or prybar between the crossmember and the transmission and try to pry the transmission up - it should move very little (2WD model shown)

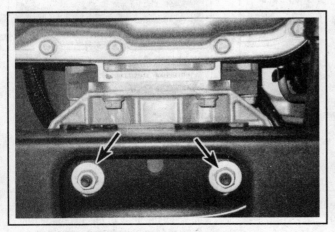

6.4a Transmission mount-to-crossmember nuts

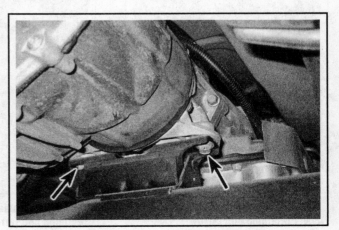

6.4b Transmission mount-to-transmission bolts

# 7A-8 AUTOMATIC TRANSMISSION

## 7 Automatic transmission - removal and installation

### REMOVAL

▸ Refer to illustrations 7.6, 7.8a, 7.8b, 7.9, 7.17, 7.18a and 7.18b

**✳✳ CAUTION:**

The transmission and torque converter must be removed as a single assembly. If you try to leave the torque converter attached to the driveplate, the converter driveplate, pump bushing and oil seal will be damaged. The driveplate is not designed to support the load, so none of the weight of the transmission should be allowed to rest on the plate during removal.

1  Disconnect the cable from the negative terminal of the battery (see Chapter 5, Section 1).
2  Raise the vehicle and support it securely on jackstands. Remove the skid plate and skid plate crossmember, if equipped.
3  Remove the transmission oil pan drain plug and drain the transmission fluid (see Chapter 1).
4  Remove all exhaust components that interfere with transmission removal (see Chapter 4).
5  Remove the torque converter inspection plug at the bottom of the bellhousing.
6  Working through the inspection plug hole, mark the relationship of the torque converter to the driveplate so they can be installed in the same position (see illustration).
7  Remove the starter motor (see Chapter 5).
8  Detach the covers on each side of the bellhousing (see illustrations).
9  Remove the torque converter-to-driveplate bolts (see illustration). Turn the crankshaft for access to each bolt. Turn the crankshaft in a clockwise direction only (as viewed from the front).
10  Mark the position of the yoke and remove the driveshaft (see Chapter 8). On 4WD models, remove both driveshafts.
11  Working on the left side of the transmission, disconnect the shift cable from the transmission (see Section 3) and unplug the electrical connectors from the Park/Neutral position switch. Also remove the bolt securing the wiring harness bracket to the left side of the transmission. Unbolt the fuel line support bracket from the transmission.
12  Working on the right side of the transmission, remove the heat shield. Unplug the electrical connectors from the transmission solenoid and the Vehicle Speed Sensor (see Chapter 6).
13  Disconnect the transmission cooler lines from the right side of the transmission and the engine. To disconnect the lines from the transmission, simply unsnap the plastic collar from the quick connect fitting, then pry off the quick connect fitting retaining clip and remove the lines. Plug the ends of the lines to prevent fluid from leaking out after you disconnect them. Always be sure to inspect the O-rings on the cooler lines before reinstallation.
14  On 4WD models, remove the transfer case (see Chapter 7B).

➥ **Note: If you are not planning to replace the transmission, but are removing it in order to gain access to other components such as the torque converter, it isn't really necessary to remove the transfer case. However, the transmission and transfer case are awkward and heavy when removed and installed as a single assembly; they're much easier to maneuver off and on as separate units.**

If you decide to leave the transfer case attached, disconnect the shift rod (manual shift models only) from the transfer case shift lever. Also disconnect the electrical connectors from the transfer case speed sensors and detach the transfer case vent tube (see Chapter 7B).

**7.6** Pry off the round inspection plug and mark the relationship between the torque converter and the driveplate

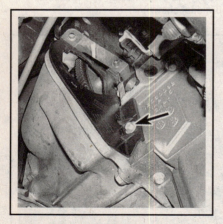

**7.8a** Passenger side bellhousing cover retaining bolt (arrow)

**7.8b** Driver's side bellhousing cover retaining bolt (arrow)

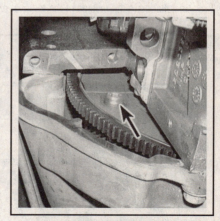

**7.9** With a large screwdriver wedged between the teeth of the driveplate ring gear and the bellhousing, remove the torque converter-to-driveplate bolts

# AUTOMATIC TRANSMISSION  7A-9

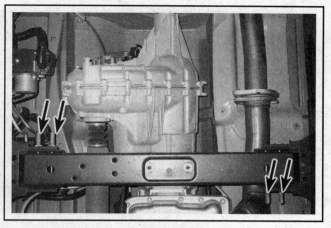

7.17  Transmission crossmember mounting bolts

7.18a  Driver's side bellhousing bolts

### ✱✱ WARNING:

If you decide to leave the transfer case attached to the transmission, be sure to use safety chains to help stabilize the transmission and transfer case assembly and to prevent it from falling off the jack head, which could cause serious damage to the transmission and/or transfer case and serious bodily injury to you.

15  Support the engine with a jack positioned under the projection on the rear of the engine block, near the transmission bellhousing.

16  Support the transmission with a jack - preferably a jack made for this purpose (available at most tool rental yards). Safety chains will help steady the transmission on the jack.

17  Remove the bolts securing the transmission mount to the crossmember (see Section 6). Raise the transmission slightly and remove the crossmember (see illustration).

18  Remove the bolts and/or studs securing the transmission to the engine (see illustrations). A long extension and a U-joint socket will greatly simplify this step.

➡ Note: The upper bolts are easier to remove after the transmission has been lowered (see the next Step).

19  Lower the engine and transmission slightly and remove the fill/dipstick tube bracket bolt and pull the tube out of the transmission. Don't lose the tube seal (it can be reused if it's still in good shape).

20  Clamp a small pair of locking pliers on the bellhousing case through the lower inspection hole. Clamp the pliers just in front of the torque converter, behind the driveplate. The pliers will prevent the torque converter from falling out while you're removing the transmission. Move the transmission to the rear to disengage it from the engine block dowel pins and make sure the torque converter is detached from the driveplate. Lower the transmission with the jack.

## INSTALLATION

21  Prior to installation, make sure the torque converter is securely engaged in the pump. If you've removed the converter, apply a small amount of transmission fluid on the torque converter rear hub, where the transmission front seal rides. Install the torque converter onto the front input shaft of the transmission while rotating the converter back and forth. It should engage into the transmission front pump in stages. Spin the converter while pushing it into place to make sure it is fully engaged in the transmission pump. Reinstall the locking pliers to hold

7.18b  Passenger side bellhousing bolts - there are three more bolts at the top of the bellhousing that are not visible in this photo

the converter in this position.

22  With the transmission secured to the jack, raise it into position.

23  Turn the torque converter to line up the holes with the holes in the driveplate. The marks on the torque converter and driveplate made in Step 6 must line up.

24  Move the transmission forward carefully until the dowel pins and the torque converter are engaged. Make sure the transmission mates with the engine with no gap. If there's a gap, make sure there are no wires or other objects pinched between the engine and transmission and also make sure the torque converter is completely engaged in the transmission front pump. Try to rotate the converter - if it doesn't rotate easily, it's probably not fully engaged in the pump. If necessary, lower the transmission and install the converter fully.

25  Install the transmission dipstick tube and seal into the transmission housing, then install the transmission-to-engine bolts and tighten them securely. As you're tightening the bolts, make sure that the engine and transmission mate completely at all points. If not, find out why. Never try to force the engine and transmission together with the bolts or you'll break the transmission case!

26  Raise the rear of the transmission and install the transmission crossmember.

27  Remove the jacks supporting the transmission and the engine.

28  Install the torque converter-to-driveplate bolts. Tighten them

# 7A-10 AUTOMATIC TRANSMISSION

to the torque listed in this Chapter's Specifications. Install the plastic bellhousing covers.

29 Install the starter motor (see Chapter 5).

30 Install new retaining rings onto the quick-connect fittings.

➡ **Note: Don't push the retaining rings onto the fittings. Instead, hook one of the ends of the clip into a slot in the fitting, then rotate the other end of the ring into the other slot. If the retaining ring isn't installed like this, it may become spread-out and won't be able to retain the cooler lines securely.**

Connect the transmission fluid cooler lines to the fittings, making sure they click into place, then push the plastic caps onto the fittings.

31 Plug in the transmission electrical connectors and install the heat shield.

32 Connect the shift cable (see Section 3).

33 Install the torque converter inspection cover.

34 On 4WD models, install the transfer case, if removed (see Chapter 7B).

35 Install the driveshaft(s) (see Chapter 8).

36 Adjust the shift cable (see Section 3).

37 Install any exhaust system components that were removed or disconnected (see Chapter 4).

38 Remove the jackstands and lower the vehicle.

39 Fill the transmission with the specified fluid (see Chapter 1), run the engine and check for fluid leaks.

## Specifications

### General

| | |
|---|---|
| Transmission fluid type | See Chapter 1 |

### Torque specifications — Ft-lbs (unless otherwise indicated)

➡ **Note: One foot-pound (ft-lb) of torque is equivalent to 12 inch-pounds (in-lbs) of torque. Torque values below approximately 15 foot-pounds are expressed in inch-pounds, because most foot-pound torque wrenches are not accurate at these smaller values.**

| | |
|---|---|
| Transmission fluid pan bolts | |
|   4L60-E, 4L65-E and 4L70-E transmissions | 97 in-lbs |
|   6L50, 6L80 and 6L90 transmissions | 80 in-lbs |
| Transmission-to-engine block bolts | 37 |
| Driveplate-to-torque converter bolts | 47 |
| Transmission mount bolts | |
|   2010 and earlier models | 37 |
|   2011 models | |
|     Transmission mount-to-crossmember nuts | 40 |
|     Transmission-to-transmission mount bolts | 50 |
|   2012 and later LD models | |
|     Transmission mount-to-crossmember nuts | 40 |
|     Transmission-to-transmission mount bolts | 50 |
|   2012 and later HD models | |
|     Transmission mount-to-crossmember nuts | 79 |
|     Transmission-to-transmission mount bolts (2WD) | 50 |
|     Transmission-to-transmission mount bolts (4WD) | |
|       Step 1 | 33 |
|       Step 2 | Tighten an addtional 50 degrees |

# 7B TRANSFER CASE

**Section**

1. General information
2. Shift lever (manual-shift models) - removal and installation
3. Shift linkage (manual-shift models) - adjustment
4. Transfer case control switch (electric-shift models) - replacement
5. Transfer case shift motor/actuator (electric-shift models) - replacement
6. Transfer case speed sensors (electric-shift models) - general information
7. Transfer case control module (electric-shift models) - replacement
8. Oil seal - replacement
9. Transfer case - removal and installation
10. Transfer case overhaul - general information

# 7B-2 TRANSFER CASE

## 1 General information

Four-wheel drive (4WD) models are equipped with a transfer case mounted on the rear of the transmission. Drive is transmitted from the engine, through the transmission and the transfer case, to the front and rear axles by driveshafts.

Here is a list of the transfer cases available on the vehicles covered by this manual:

**Manual transfer cases**

NVG 261 NP2
MP 1222, 1225, 1226-NQG

**Automatic transfer cases**

MP 1625, 1626-NQF
MP 3023, 3024-NQH
BW 4485-NR3

We don't recommend trying to rebuild any of these transfer cases at home. They're difficult to overhaul without special tools, and rebuilt units are available (on an exchange basis) for less than it would cost to rebuild your own.

## 2 Shift lever (manual-shift models) - removal and installation

1  Raise the front of the vehicle and support it securely on jackstands.
2  Remove the transfer case skid plate, if equipped.
3  Remove the front driveshaft (see Chapter 8).
4  Use a small screwdriver to release the control rod at each end. Remove the control rod.
5  Remove the knob from the lever by putting a wrench under the knob and tapping it off with a hammer.
6  Remove the three bezel screws and remove the bezel.
7  Remove the four lever bolts and lift out the lever assembly.
8  Installation is the reverse of removal.

## 3 Shift linkage (manual-shift models) - adjustment

1  Place the shift lever in the 2WD High position. Have an assistant hold the lever in the indicated position, or tape it in place.
2  Raise the vehicle and support it securely on jackstands.
3  Loosen the adjustment bolt at the center of the shift linkage.
4  Make sure the ball sockets at each end of the shift linkage are securely seated on the ball pivots at the transfer case and the shift lever.
5  Verify that the transfer case range lever is in the 2WD High position.

➡ **Note:** When the transfer case range lever is in the 2WD High position on the NVG 261 transfer case, it will be positioned in the third detent from the rear (two clicks from the rearmost position).

6  Tighten the shift linkage adjustment bolt to the torque listed in this Chapter's Specifications.
7  Lower the vehicle and check the operation of the transfer case.

## 4 Transfer case control switch (electric-shift models) - replacement

♦ **Refer to illustration 4.4**

1  Disconnect the cable from the negative terminal of the battery (see Chapter 5, Section 1).
2  Remove the center trim panel from the dashboard (see Chapter 11, Section 26).
3  Unplug the electrical connector from the rear of the switch.
4  Release the retaining tabs and remove the transfer case control switch (see illustration).
5  Installation is the reverse of removal.

**4.4 After removing the trim panel, release the retaining tabs to remove the switch**

# TRANSFER CASE 7B-3

## 5  Transfer case shift motor/actuator (electric-shift models) - replacement

♦ **Refer to illustration 5.3**

1  Raise the vehicle and place it securely on jackstands.
2  Remove the stone shields from below the transfer case.
3  Unplug the shift motor electrical connector (see illustration).
4  Unscrew the retaining bolts and remove the electric shift motor/actuator assembly.
5  Inspect the seal for damage and replace it if necessary.
6  Position the gasket and the shift motor in place on the transfer case and install the bolts. Tighten the shift motor retaining bolts to the torque listed in this Chapter's Specifications.
7  Installation is the reverse of removal.

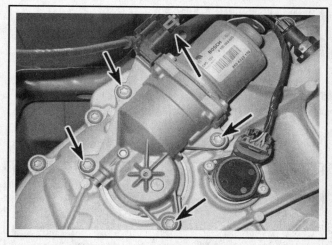

**5.3  Disconnect the electrical connector and remove the electric shift motor retaining bolts**

## 6  Transfer case speed sensors (electric shift models) - general information

There are a variety of electric-shift transfer cases on the models covered by this manual and several different combinations of speed sensors available. The Vehicle Speed Sensor (VSS) sends information to the PCM for the driveability program and is located at the rear output shaft. The procedure for checking and replacement of the output shaft speed sensor is essentially the same as the procedure for the Vehicle Speed Sensor (VSS). A scan tool must be used to access speed sensor data and trouble codes. Refer to the Vehicle Speed Sensor (VSS) replacement procedure in Chapter 6.

## 7  Transfer case control module (electric-shift models) - replacement

♦ **Refer to illustration 7.1**

The transfer case control module is located in the left end of the instrument panel (see illustration). If a problem arises with the transfer case control module, it is recommended that the job be left to a dealership service department or other qualified repair shop. A new module must be programmed with a factory scan tool by a dealership service department or other shop equipped with the proper tool and software, so even if you were to replace the old module with a new unit, it wouldn't work until the vehicle was towed to a dealer (or other properly equipped repair facility) for programming.

**7.1  The transfer case control module is located to the left of the steering column**

# 7B-4 TRANSFER CASE

## 8  Oil seal - replacement

▸ **Refer to illustrations 8.4 and 8.6**

➡ **Note:** *This procedure applies to both the front and rear output shaft seals.*

1  Raise the vehicle and support it securely on jackstands.
2  Remove the skid plate.
3  If you're replacing the front seal, remove the front driveshaft; if you're replacing the rear seal, remove the rear driveshaft (see Chapter 8).
4  To remove the either output shaft seal, simply pry the seal out with a screwdriver or a seal removal tool (see illustration). Don't damage the seal bore.
5  Lubricate the new seal lips with petroleum jelly. If the new seal has a weep hole, put it at the bottom and the notch at the top.
6  Drive the seal into place with a seal driver or a large socket (see illustration). The outside diameter of the socket should be slightly smaller than the outside diameter of the seal.
7  The remainder of installation is the reverse of removal.

**8.4  The output shaft seal can be removed with a conventional seal extractor or a large screwdriver**

**8.6  Drive the new transfer case seals into place with a seal driver or large socket**

## 9  Transfer case - removal and installation

▸ **Refer to illustrations 9.7, 9.11a and 9.11b**

1  Disconnect the cable from the negative terminal of the battery (see Chapter 5, Section 1).
2  On models with a manually shifted transfer case, put the transfer case in the 2WD High position.
3  Raise the vehicle and support it securely on jackstands.
4  Remove the stone shields (if equipped).
5  Drain the transfer case lubricant (see Chapter 1).
6  Remove the front and rear driveshafts (see Chapter 8).
7  Unplug all electrical connectors and detach the vent hose from the top of the transfer case (see illustration).
8  On manually shifted models, disconnect the shift linkage from the transfer case. This is done by simply pulling the shift linkage ball socket off the ball pivot on the transfer case shift lever.
9  Raise the transmission enough to remove the crossmember (see Chapter 7A). With the crossmember removed, support the transmission with a floor jack.

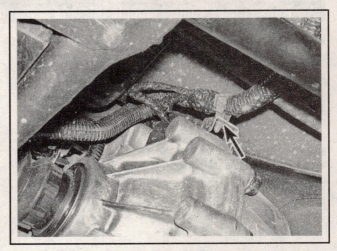

**9.7  Detach all wiring harness retaining clips and vent hoses from the transfer case**

# TRANSFER CASE    7B-5

**9.11a  Left side transfer case mounting nuts (crossmember removed for clarity)**

**9.11b  Right side transfer case mounting nuts**

10  Support the transfer case with a jack - preferably a special jack made for this purpose. Safety chains will help steady the transfer case on the jack.

11  Remove the transfer case mounting nuts (see illustrations).

12  Make a final check that all wires and hoses have been disconnected from the transfer case, then move the transfer case and jack toward the rear of the vehicle until it's clear of the transmission. Keep the transfer case level as this is done. Once the input shaft is clear, rotate the transfer case as necessary, lower it and remove it from under the vehicle.

13  Installation is the reverse of removal. Replace the gasket with a new one. Be sure to tighten the transmission-to-transfer case nuts to the torque listed in this Chapter's Specifications. On manually shifted models, be sure to adjust the transfer case shift linkage (see Section 3).

## 10  Transfer case overhaul - general information

Overhauling a transfer case is a difficult job for the do-it-yourselfer. It involves the disassembly and reassembly of many small parts. Numerous clearances must be precisely measured and, if necessary, changed with select-fit spacers and snap-rings. As a result, if transfer case problems arise, it can be removed and installed by a competent do-it-yourselfer, but overhaul should be left to a transmission repair shop. Rebuilt transfer cases may be available - check with your dealer parts department and auto parts stores. At any rate, the time and money involved in an overhaul is almost sure to exceed the cost of a rebuilt unit.

Nevertheless, it's not impossible for an inexperienced mechanic to rebuild a transfer case if the special tools are available and the job is done in a deliberate step-by-step manner so nothing is overlooked.

The tools necessary for an overhaul include internal and external snap-ring pliers, a bearing puller, a slide hammer, a set of pin punches, a dial indicator and possibly a hydraulic press. In addition, a large, sturdy workbench and a vise or transmission stand will be required.

During disassembly of the transfer case, make careful notes of how each piece comes off, where it fits in relation to other pieces and what holds it in place. Note how parts are installed when you remove them; this will make it much easier to get the transfer case back together.

## TRANSFER CASE

### Torque specifications — Ft-lbs (unless otherwise indicated)

➡ **Note:** One foot-pound (ft-lb) of torque is equivalent to 12 inch-pounds (in-lbs) of torque. Torque values below approximately 15 foot-pounds are expressed in inch-pounds, because most foot-pound torque wrenches are not accurate at these smaller values.

| | |
|---|---|
| Shift lever-to-floor pan bolts | 20 |
| Shift linkage adjustment bolt | 132 in-lbs |
| Shift motor/actuator mounting bolts | 156 in-lbs |
| Transfer case-to-transmission nuts | 37 |
| Transfer case-to-transmission support braces | 37 |

## Section

1. General information
2. Driveshaft and universal joints - general information and inspection
3. Driveshaft(s) - removal and installation
4. Driveshaft center support bearing - replacement
5. Universal joints - replacement
6. Axles - description and check
7. Axleshaft (rear) - removal and installation
8. Axleshaft oil seal (rear, semi-floating axle) - replacement
9. Axleshaft bearing (rear, semi-floating axle) - replacement
10. Rear hub, wheel bearing and seal (full-floating axle) - removal, bearing/seal replacement and installation
11. Pinion oil seal - replacement
12. Axle assembly (rear) - removal and installation
13. Driveaxles (4WD models) - general information and inspection
14. Driveaxle (4WD models) - removal and installation
15. Driveaxle boot (4WD models) - replacement
16. Front axle shift motor (4WD models) - replacement
17. Right axleshaft, tube, bearing and shift fork (4WD and AWD models) - removal, component replacement and installation
18. Axleshaft oil seals and bearings (front, 4WD models) - replacement
19. Front differential carrier - removal and installation

## Reference to other Chapters

Differential lubricant change - See Chapter 1
Differential lubricant level check - See Chapter 1

# 8
DRIVELINE

# 8-2 DRIVELINE

## 1 General information

The information in this Chapter deals with the components from the rear of the engine to the rear wheels, except for the transmission (and transfer case, if equipped), which is dealt with in the previous Chapter. For the purposes of this Chapter, these components are grouped into two categories: driveshaft and axles. Separate Sections within this Chapter offer general descriptions and checking procedures for components in each of the two groups.

Since nearly all the procedures covered in this Chapter involve working under the vehicle, make sure it's securely supported on sturdy jackstands or on a hoist where the vehicle can be easily raised and lowered.

## 2 Driveshaft and universal joints - general information and inspection

### GENERAL INFORMATION

1 A driveshaft is a tube, or a pair of tubes, that transmits power between the transmission (or transfer case on 4WD models) and the differential. Universal joints are located at either end of the driveshaft and in the center on two-piece driveshafts.

2 Single piece driveshafts employ a splined yoke at the front, which slips into the extension housing of the transmission. This arrangement allows the driveshaft to slide back-and-forth within the transmission during vehicle operation to compensate for changes in length due to suspension movement. An oil seal prevents leakage of fluid at this point and keeps dirt from entering the transmission. If leakage is evident at the front of the driveshaft, replace the oil seal (see Chapter 7, Part A).

3 If a two-piece driveshaft is used, a slip joint is employed on the front of the rear driveshaft section.

4 Two-piece driveshafts also have a center support bearing. The center bearing is a ball-type bearing mounted in a rubber cushion attached to a frame crossmember. The bearing is pre-lubricated and sealed at the factory.

5 On all models, the driveshaft assembly requires very little service. The universal joints are lubricated for life and must be replaced if problems develop. The driveshaft must be removed from the vehicle for this procedure.

6 Since the driveshaft is a balanced unit, it's important that no undercoating, mud, etc. be allowed to stay on it. When the vehicle is raised for service it's a good idea to clean the driveshaft and inspect it for any obvious damage. Also, make sure the small weights used to originally balance the driveshaft are in place and securely attached. Whenever the driveshaft is removed it must be reinstalled in the same relative position to preserve the balance.

7 Problems with the driveshaft are usually indicated by a noise or vibration while driving the vehicle. A road test should verify if the problem is the driveshaft or another component. Refer to the *Troubleshooting* Section at the front of this manual. If you suspect trouble, inspect the driveline.

### INSPECTION

8 Raise the rear of the vehicle and support it securely on jackstands. Block the front wheels to keep the vehicle from rolling off the stands.

9 Crawl under the vehicle and visually inspect the driveshaft. Look for any dents or cracks in the tubing. If any are found, the driveshaft must be replaced.

10 Check for oil leakage at the front and rear of the driveshaft. Leakage where the driveshaft enters the transmission or transfer case indicates a defective transmission/transfer case seal (see Chapter 7). Leakage where the driveshaft enters the differential indicates a defective pinion seal (see Section 11).

11 While under the vehicle, have an assistant rotate a rear wheel so the driveshaft will rotate. As it does, make sure the universal joints are operating properly without binding, noise or looseness. Listen for any noise from the center bearing (if equipped), indicating it's worn or damaged. Also check the rubber portion of the center bearing for cracking or separation, which will necessitate replacement.

12 The universal joint can also be checked with the driveshaft motionless, by gripping your hands on either side of the joint and attempting to twist the joint. Any movement at all in the joint is a sign of considerable wear. Lifting up on the shaft will also indicate movement in the universal joints.

13 Finally, check the driveshaft mounting bolts at the ends to make sure they're tight.

14 On 4WD models, the above driveshaft checks should be repeated on the front driveshaft as well. In addition, check for leakage around the sleeve yoke, indicating failure of the yoke seal.

15 Check for leakage where the driveshafts connect to the transfer case and front differential. Leakage indicates worn oil seals.

16 At the same time, check for looseness in the joints of the front driveaxles. Also check for grease or oil leakage from around the driveaxles by inspecting the rubber boots and both ends of each axle. Oil leakage around the axle flanges indicates a defective axleshaft oil seal. Grease leakage at the CV joint boots means a damaged rubber boot. For servicing of these components, see the appropriate Sections.

# DRIVELINE  8-3

## 3  Driveshaft(s) - removal and installation

### REAR DRIVESHAFT

#### Removal

▶ Refer to illustrations 3.3 and 3.4

1  Raise the rear of the vehicle and support it securely on jackstands. Block the front wheels to prevent the vehicle from rolling.

2  Place the transmission in Neutral with the parking brake off.

3  Make reference marks on the driveshaft and the pinion flange in line with each other (see illustration). This is to make sure the driveshaft is reinstalled in the same position to preserve the balance.

4  Remove the rear universal joint bolts and retainers. Turn the driveshaft (or wheels) as necessary to bring the bolts into the most accessible position. To prevent the driveshaft from turning when you loosen the bolts, insert a large screwdriver through the driveshaft yoke (see illustration).

5  On vehicles with a two-piece driveshaft, remove the fasteners from the center support bearing.

6  On all models, tape the bearing caps to the spider to prevent the caps from coming off during removal.

7  Lower the rear of the driveshaft. Slide the front of the driveshaft out of the transmission or transfer case.

8  Wrap a plastic bag over the transmission or transfer case housing and hold it in place with a rubber band. This will prevent loss of fluid and protect against contamination while the driveshaft is out.

#### Installation

9  Remove the plastic bag from the transmission or transfer case and wipe the area clean. Inspect the oil seal carefully. Procedures for replacement of this seal can be found in Chapter 7.

10  Slide the front of the driveshaft into the transmission or transfer case.

11  On models with a two-piece driveshaft, raise the center support bearing into position, install the fasteners and tighten them to the torque listed in this Chapter's Specifications.

12  Raise the rear of the driveshaft into position, checking to be sure the marks are in alignment. If not, turn the rear wheels to match the pinion flange and the driveshaft.

13  Remove the tape securing the bearing caps and install the clamps and bolts. Tighten all bolts to the torque listed in this Chapter's Specifications.

### FRONT DRIVESHAFT (4WD MODELS)

#### Removal

▶ Refer to illustrations 3.15 and 3.16

14  Raise the front of the vehicle and support it securely on jackstands. Remove the differential carrier splash shield, if equipped.

15  Pry open the clamp securing the boot at the transfer case output shaft (see illustration). Disengage the boot from the shaft and slide it forward.

16  Mark the relationship of the driveshaft to the front differential companion flange (see illustration).

**3.3  Mark the relationship of the rear driveshaft to the differential pinion flange**

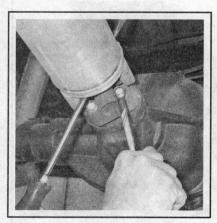

**3.4  Insert a screwdriver through the driveshaft yoke to prevent the shaft from turning when you loosen the bolts**

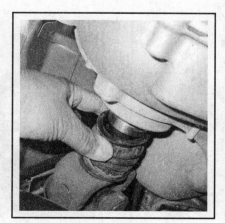

**3.15  Pry up the end of the boot clamp and dislodge the boot from the transfer case output shaft**

**3.16  Mark the relationship of the front driveshaft to the front differential companion flange, then remove the bolts and clamps**

## 8-4 DRIVELINE

17 Remove the bolts and clamps from the differential flange.
18 Push the driveshaft to the rear far enough to separate it from the differential flange, then lower it and pull the shaft out of the transfer case.

### Installation

19 Slide the rear of the driveshaft into the splines in the transfer case output shaft.
20 Attach the front end of the shaft to the differential companion flange (be sure to line up the marks), install the clamps and bolts and tighten all of the bolts to the torque listed in this Chapter's Specifications.
21 At the rear end of the shaft, push the boot up over the transfer case output shaft and seat it into its groove. Insert a small screwdriver between the boot and the shaft to equalize pressure inside the boot, then install a new clamp and crimp it into place with a pair of clamp-crimping pliers.
22 Install the front differential carrier splash shield (if equipped).

## 4  Driveshaft center support bearing - replacement

1 Remove the driveshaft (see Section 3).
2 Mark the relationship of the front portion of the driveshaft to the rear portion of the driveshaft (it's best to make the mark on the slip yoke). Slide the rear shaft off the front shaft.
3 If you have access to a hydraulic press (one tall enough to accommodate the shaft) and the necessary fixtures, press the shaft out of the center support bearing. Reverse this operation to install the new bearing.
4 If you do not have the necessary equipment, take the shaft to an automotive machine shop or other qualified repair facility to have the old bearing pressed off and the new one pressed on.

## 5  Universal joints - replacement

➡ **Note:** Always purchase a universal joint service kit for your model vehicle before beginning this procedure. Also, read through the entire procedure before beginning work.

1 Remove the driveshaft (see Section 3).

### OUTER SNAP-RING TYPE

▶ **Refer to illustrations 5.3, 5.4, 5.5 and 5.6**

2 Place the driveshaft on a bench equipped with a vise.
3 Remove the snap-rings with a small pair of pliers (see illustration).
4 Place a piece of pipe or a large socket, having an outside diameter slightly larger than the outside diameter of one of the bearing caps, over one of the bearing caps. Position a socket with an outside diameter slightly smaller than that of the opposite bearing cap against the cap (see illustration) and use the vise or press to force the bearing cap out (inside the pipe or large socket).
5 Press the U-joint through as far as possible, then grip the bearing cap with pliers and remove it (see illustration).

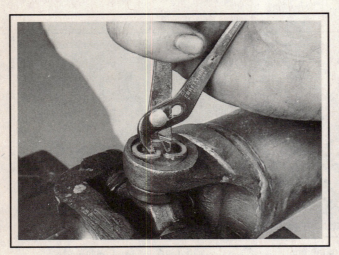

**5.3  Use a small pair of pliers to remove the snap-rings from the ends of the universal joint yokes**

**5.4  To remove the U-joint from the driveshaft, use a vise as a press - the small socket will push the U-joint and bearing cap into the large socket**

# DRIVELINE 8-5

5.5 Locking pliers can be used to remove the bearing caps from the yoke

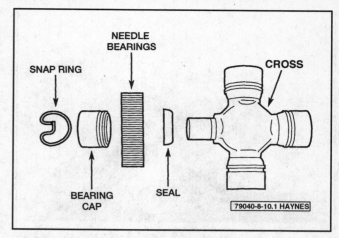

5.6 Outer snap-ring type U-joint

6  A universal joint repair kit will contain a new U-joint, seals, bearings, caps and snap-rings (see illustration).

7  Inspect the bearing cap bores in the yokes for wear and damage.

8  If the bearing cap bores in the yoke are so worn that the caps are a loose fit, the driveshaft will have to be replaced with a new one.

9  Make sure the dust seals are properly located on the U-joint.

10  Using a vise, press one bearing cap into the yoke approximately 1/4-inch.

11  Use chassis grease to hold the needle rollers in place in the caps.

12  Insert the U-joint into the partially installed bearing cap, taking care not to dislodge the needle rollers.

13  Hold the U-joint in correct alignment and press both caps into place by slowly and carefully closing the jaws of the vise.

14  Use a socket slightly smaller in diameter than the caps to press them into the yoke. Press in one side, install the snap-ring, then press the other side to shift the U-joint assembly tight against the installed snap-ring and install the other snap-ring.

15  Repeat the operations for the remaining two bearing caps. Proceed to Step 22.

## INJECTED PLASTIC (INNER SNAP-RING) TYPE

♦ **Refer to illustrations 5.16, 5.20 and 5.21**

16  If the joint has been previously rebuilt, remove the snap-rings (bearing retainers) located on the inner part of each bearing cap (see illustration).

17  If this is the first time the joint is being rebuilt, it will not be necessary to remove the snap-rings, since there aren't any; the pressing operation will shear the molded plastic retaining material.

➡ **Note: It may be necessary to heat the U-joint over 500-degrees (melting the plastic retaining material) before pressing the U-joint apart.**

18  Press out the bearing caps as described in Steps 4 and 5.

19  Remove the U-joint and clean all plastic material from the yoke. Use a small punch to remove the plastic from the injection holes.

20  Reassembly is the same as for the outer snap-ring joint described in Steps 9 through 15, except the snap-rings are on the inner part of each bearing cap (see illustration).

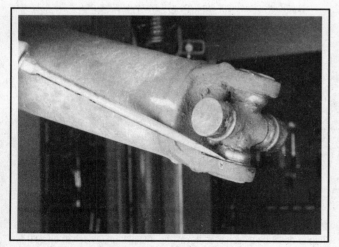

5.16 Remove the inner snap-rings from the U-joint by tapping them off with a screwdriver and hammer

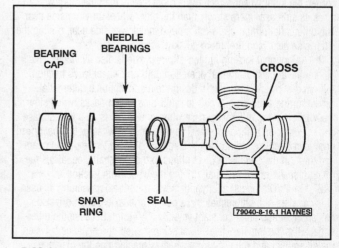

5.20 Inner snap-ring type U-joint

# 8-6 DRIVELINE

**5.21 Installing a snap-ring on an inner snap-ring type U-joint**

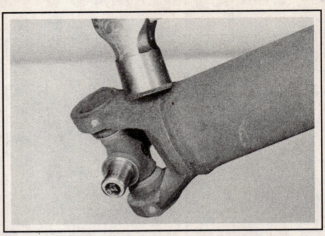

**5.22 Strike the yoke sharply with a hammer to spring the yoke ears, which will free-up the joint**

21 When installing the bearing cap, press it in until the snap-ring can be installed (see illustration).

## ALL MODELS

▸ **Refer to illustration 5.22**

22 If the joint is stiff after assembly, strike the yoke sharply with a hammer (see illustration). This will spring the yoke ears slightly and free up the joint.

## 6 Axles - description and check

### DESCRIPTION

1 The rear axle assembly is a hypoid (the centerline of the pinion gear is below the centerline of the ring gear), semi-floating type. When the vehicle goes around a corner, the differential allows the outer rear wheel to turn at a higher speed than the inner tire. The axleshafts are splined to the differential side gears, so when the vehicle goes around a corner, the inner wheel, which turns more slowly than the outer wheel, turns its side gear more slowly than the outer wheel turns its side gear. The differential pinion gears roll around the slower side gear, driving the outer side gear (and tire) more quickly.

2 An optional locking limited-slip rear axle is also available. This differential allows for normal operation until one wheel loses traction. A limited-slip unit is similar in design to a conventional differential, except for the addition of a pair of multi-disc clutch packs which slow the rotation of the differential case when one wheel is on a firm surface and the other on a slippery one. The difference in wheel rotational speed produced by this condition applies additional force to the pinion gears and through the cone, which is splined to the axleshafts, equalizes the rotation speed of the axleshaft driving the wheel with traction.

3 On 4WD models, a fully independent front axle assembly is used. This consists of a differential and a pair of driveaxles. Each driveaxle has an inner and outer constant velocity (CV) joint. Because the differential - like the transfer case - is offset to the left, the distance between the differential and the right front wheel is greater than the distance from the differential to the left wheel. In order to use two equal-length driveaxles, an extension axleshaft is employed on the right side to make up the difference.

### CHECK

4 Often, a suspected axle problem lies elsewhere. Do a thorough check of other possible causes before assuming the axle is the problem.

5 The following noises are those commonly associated with axle diagnosis procedures:

a) Road noise is often mistaken for mechanical faults. Driving the vehicle on different surfaces will show whether or not the road surface is the cause of the noise. Road noise will remain the same if the vehicle is under power or coasting.

b) Tire noise is sometimes mistaken for mechanical problems. Tires which are worn or low on pressure are particularly susceptible to emitting vibrations and noises. Tire noise will remain about the same during varying driving situations, where axle noise will change during coasting, acceleration, etc.

c) Engine and transmission noise can be deceiving because it will travel along the driveline. To isolate engine and transmission noises, make a note of the engine speed at which the noise is most pronounced. Stop the vehicle, place the transmission in Neutral and run the engine to the same speed. If the noise is the same, the axle is not at fault.

6 Because of the special tools needed, overhauling the differential isn't cost effective for a do-it-yourselfer. The procedures included in this Chapter describe axleshaft removal and installation, axleshaft oil seal replacement, axleshaft bearing replacement and removal of the entire unit for repair or replacement. Any further work should be left to a qualified repair shop.

# DRIVELINE 8-7

## 7  Axleshaft (rear) - removal and installation

### SEMI-FLOATING AXLESHAFT

### Removal

♦ **Refer to illustrations 7.3, 7.4, 7.5a and 7.5b**

1  Loosen the rear wheel lug nuts. Raise the rear of the vehicle, support it securely on jackstands and block the front wheels. Remove the wheel and brake disc or drum (see Chapter 9).

2  Drain the differential lubricant (see Chapter 1). Remove the differential cover.

➥**Note: Only the 8.6 inch differential requires the differential cover to be removed to drain the fluid. All other models are equipped with a drain plug.**

3  Remove the pinion shaft lock screw (see illustration).

4  On models with a conventional differential (non-locking), remove the pinion shaft. On models with a locking differential, withdraw the pinion shaft part way, then rotate the differential until the shaft touches the case, providing enough clearance for access to the C-locks (see illustration).

5  Push in on the outer flanged end of the axleshaft, then remove the C-lock from the groove on the inner end of the shaft (see illustration).

➥**Note: On models with a locking differential, use a screwdriver to rotate the C-lock until the open end points in (see illustration).**

6  With the C-lock removed, withdraw the axleshaft, taking care not to damage the oil seal (but note that it is a good idea to replace the seal whenever the axleshaft is removed - see Section 8). Some models have a thrust washer in the differential; make sure it doesn't fall out when the axleshaft is removed.

### ✸✸ CAUTION:

**Do NOT rotate the axleshaft (and don't rotate the other axleshaft, either). The pinion (or "spider") gears in the differential will become misaligned and it will be difficult to put them back in position during reassembly.**

7.3  Remove the pinion shaft lock screw

7.4  Withdraw the pinion shaft for access to the C-locks (don't turn the axleshafts or differential carrier after the shaft has been pulled out, or the spider gears may become misaligned)

7.5a  Push the axle flange in, then remove the C-lock from the inner end of the axleshaft

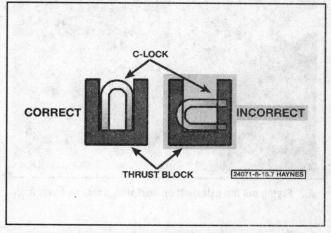

7.5b  On models with a locking differential, the C-lock must be positioned as shown before it can be removed

# 8-8 DRIVELINE

## Installation

7  To install, carefully insert the axleshaft into the housing until it engages with the differential side gear.

8  Install the C-lock in the axleshaft groove and pull out on the flange to lock it.

9  Insert the pinion shaft, align the hole in the shaft with the lock screw hole and install the lock screw.

➡ **Note:** *Apply a non-hardening, thread-locking compound to the threads of the lock screw before installing it.*

Tighten the lock screw to the torque listed in this Chapter's Specifications.

10  Install the cover and tighten the bolts to the torque listed in this Chapter's Specifications. Fill the differential with the lubricant specified in Chapter 1.

11  Install the brake disc or drum. On models with rear disc brakes, install the caliper mounting bracket and caliper and tighten the fasteners to the torque listed in the Chapter 9 Specifications. Install the wheel and lug nuts, then lower the vehicle. Tighten the lug nuts to the torque listed in the Chapter 1 Specifications.

## FULL-FLOATING AXLESHAFT

12  Remove the bolts that attach the axleshaft flange to the hub.

13  Tap the flange with a soft-face hammer to loosen the shaft, then grip the rib in the face of the flange with a pair of locking pliers. Twist the shaft slightly in both directions and withdraw it from the housing. Place a drip pan under the outer end of the axle to catch any lubricant which might leak out while the axle is removed.

14  Installation is the reverse of removal. Be sure to hold the axleshaft level to engage the splines at the inner end with those in the differential side gear. Always use a new gasket on the flange and keep both the flange and hub mating surfaces free of grease and oil. Tighten the axleshaft flange bolts to the torque listed in this Chapter's Specifications.

## 8  Axleshaft oil seal (rear, semi-floating axle) - replacement

▸ Refer to illustrations 8.2 and 8.3

1  Remove the axleshaft (see Section 7).

2  Pry the oil seal out of the end of the axle housing (see illustration).

3  Apply a film of multi-purpose grease to the oil seal recess and tap the new seal evenly into place with a hammer and seal installation tool (see illustration), large socket or piece of pipe so the lips are facing in and the metal face is visible from the end of the axle housing. When correctly installed, the face of the oil seal should be flush with the end of the axle housing.

4  Install the axleshaft (see Section 7).

**8.2 Prying out the axleshaft oil seal with a seal removal tool**

**8.3 Using a seal driver to install the axleshaft oil seal - drive the seal in until it's flush with the bore**

# DRIVELINE 8-9

## 9  Axleshaft bearing (rear, semi-floating axle) - replacement

▶ Refer to illustrations 9.2, 9.3 and 9.4

1  Remove the axleshaft (see Section 7) and the oil seal (see Section 8).

2  A bearing puller which grips the bearing from behind will be required for this job (see illustration).

3  Attach a slide hammer to the puller and extract the bearing from the axle housing (see illustration).

4  Clean out the bearing recess and drive in the new bearing with a bearing installer or a piece of pipe positioned against the outer bearing race (see illustration). Make sure the bearing is tapped in to the full depth of the recess.

5  Install a new oil seal (see Section 8), then install the axleshaft (see Section 7).

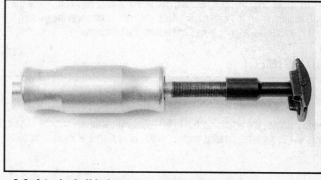

9.2  A typical slide hammer and axleshaft bearing remover attachment

9.3  Removing the axleshaft bearing with a slide hammer

9.4  Use a bearing driver or a large socket to tap the bearing evenly into the axle housing

## 10  Rear hub, wheel bearing and seal (full-floating axle) - removal, bearing/seal replacement and installation

### REMOVAL

1  Remove the axleshaft (see Section 7).

2  Loosen the rear wheel lug nuts, raise the rear of the vehicle and support it securely on jackstands. Block the front wheels and remove the rear wheels.

3  Remove the brake disc (see Chapter 9).

4  Remove the retaining ring and key (if equipped) from the end of the axle housing.

5  Remove the adjusting nut, using a special socket available at most auto parts stores.

6  Pull the hub assembly straight off the axle tube.

7  Remove and discard the oil seal from the back of the hub.

8  Use solvent to clean the bearings, hub and axle tube, then spray the bearings with brake system cleaner, which will remove the solvent and allow the bearings to dry much more rapidly.

9  Carefully inspect the bearings for cracks, wear and damage. Check the axle tube flange, studs and hub splines for damage and corrosion. Check the bearing cups (races) for pitting or scoring. Worn or damaged components must be replaced with new ones.

10  To further disassemble the hub, use a hammer and a long bar or drift punch to knock out the inner bearing cup (races).

11  Remove the outer retaining ring, then knock the outer bearing cup from the hub.

12  Clean the old sealing compound from the seal bore in the hub.

13  Inspect the brake disc (see Chapter 9).

14  Reassemble the hub by reversing the disassembly procedure. Use only the proper size bearing driver when installing the new bearing cups (races).

15  Lubricate the bearings and the axle tube contact areas with wheel bearing grease. Work the grease completely into the bearings, forcing it between the rollers, cone and cage.

# 8-10 DRIVELINE

## INSTALLATION

16 Make sure the axle housing oil deflector is in position. Place the hub assembly on the axle tube, taking care not to damage the oil seals.
17 Install the adjusting nut and adjust the bearings.

## ADJUSTMENT

18 Rotate the hub, making sure it turns freely.
19 While rotating the hub in the normal direction of rotation (forward), tighten the adjusting nut to 50 ft-lbs with a torque wrench. Again, this will require the special socket, which is available at most auto parts stores.

20 Back the nut off until it's loose, then tighten the nut hand-tight with the special socket.
21 If necessary, turn the nut counterclockwise to align the closest slot in the nut with the keyway in the spindle, then install the key.

### ✳✳ CAUTION:
**Don't turn the nut more than one slot to achieve alignment.**

22 Install the retaining ring.
23 Wiggle the hub assembly; you shouldn't be able to detect any play, but the hub should turn freely (there shouldn't be any preload on the bearings, but there shouldn't be any freeplay, either).
24 Install the axleshaft (see Section 7) and lower the vehicle.

## 11 Pinion oil seal - replacement

♦ **Refer to illustrations 11.3, 11.4, 11.8 and 11.9**

➡ **Note: This procedure applies to the front and rear pinion oil seals.**

1 Loosen the wheel lug nuts. Raise the front (for front differential) or rear (for rear differential) of the vehicle and support it securely on jackstands. Block the opposite set of wheels to keep the vehicle from rolling off the stands. Remove the wheels.
2 Disconnect the driveshaft from the differential pinion flange and fasten it out of the way (see Section 3).
3 Rotate the pinion a few times by hand. Use a beam-type or dial-type inch-pound torque wrench to check the torque required to rotate the pinion (see illustration). Record it for use later.
4 Mark the relationship of the pinion flange to the shaft (see illustration), then count and write down the number of exposed threads on the shaft.
5 A special tool, available at most auto parts stores, can be used to keep the companion flange from moving while the self-locking pinion nut is loosened. A chain wrench can also be used to immobilize the flange.
6 Remove the pinion nut.
7 Withdraw the flange. It may be necessary to use a two-jaw puller engaged behind the flange to draw it off. Do not attempt to pry or hammer behind the flange or hammer on the end of the pinion shaft.
8 Pry out the old seal and discard it (see illustration).
9 Lubricate the lips of the new seal, then tap it evenly into position with a seal installation tool or a large socket (see illustration). Make sure it enters the housing squarely and is tapped in to its full depth.
10 Install the pinion flange, lining up the marks made in Step 4. If necessary, tighten the pinion nut to draw the flange into place. Do not try to hammer the flange into position.
11 Apply a bead of RTV sealant to the ends of the splines visible in the center of the flange so oil will be sealed in.
12 Install the washer and a new pinion nut. Tighten the nut until the

**11.3 Use an inch-pound torque wrench to check the torque required to rotate the pinion shaft**

**11.4 Before removing the nut, mark the position of the flange to the shaft and count the number of exposed threads**

# DRIVELINE 8-11

11.8 Use a seal removal tool or a large screwdriver to remove the pinion seal (be careful not to disturb the pinion while doing this)

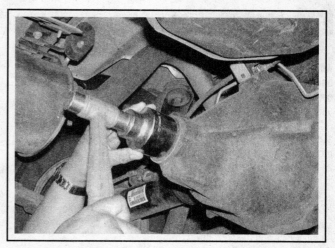

11.9 A large socket with a diameter the same as that of the new pinion seal can be used to drive the seal into the differential housing

number of threads recorded in Step 4 are exposed.

13 Measure the torque required to rotate the pinion and tighten the nut in small increments (no more than 5 ft-lbs) until it matches the figure recorded in Step 3. To compensate for the drag of the new oil seal, the nut should be tightened a little more until the rotational torque of the pinion exceeds the earlier recording by 5 in-lbs.

14 Reinstall all components removed previously by reversing the removal Steps, tightening all fasteners to their specified torque values. Tighten the wheel lug nuts to the torque listed in the Chapter 1 Specifications.

## 12 Axle assembly (rear) - removal and installation

### REMOVAL

1  Loosen the rear wheel lug nuts, raise the rear of the vehicle and support it securely on jackstands. Block the front wheels to keep the vehicle from rolling off the stands. Remove the rear wheels.

2  Position a jack under the rear axle differential housing.

3  Remove the driveshaft (see Section 3).

4  Disconnect the shock absorbers at their lower mounts (see Chapter 10).

5  Disconnect the vent hose from the fitting on the axle housing.

6  Disconnect the brake hose from the junction block on the axle housing, then plug the hose to prevent fluid leakage.

7  Remove the rear brake assemblies (including the parking brake shoes on disc brake models) (see Chapter 9).

8  Disconnect the parking brake cables from the brackets on the rear axle housing.

9  Remove the rear wheel speed sensors (see Chapter 9).

### Leaf spring models

10  Disconnect the spring U-bolts (see Chapter 10). Remove the spring plates.

11  Lower the jack under the differential, then remove the rear axle assembly from under the vehicle.

### Coil spring models

12  Detach the links for the electronic suspension control sensors, if equipped, from the trailing arms.

13  Detach the track bar from the rear axle (see Chapter 10).

14  Remove the stabilizer bar (see Chapter 10).

15  Remove the coil springs (see Chapter 10).

16  Disconnect the upper and lower trailing arms from the axle assembly (see Chapter 10).

17  Lower the jack under the differential, then remove the rear axle assembly from under the vehicle.

### INSTALLATION

18  Installation is the reverse of removal. Tighten the U-joint clamp bolts to the torque listed in this Chapter's Specifications. Tighten all suspension fasteners to the torque values listed in the Chapter 10 Specifications. Tighten the brake fasteners to the torque values listed in the Chapter 9 Specifications.

19  Bleed the brakes (see Chapter 9).

20  Install the wheels and lug nuts, then lower the vehicle and tighten the lug nuts to the torque listed in the Chapter 1 Specifications.

# 8-12 DRIVELINE

## 13 Driveaxles (4WD models) - general information and inspection

1 Power is transmitted from the front differential/axle to the front wheels through a pair of driveaxles. The inner end of each driveaxle is bolted to an axleshaft connected to the differential side gears; the outer end of each driveaxle has a stub shaft that is splined to the front hub and bearing assembly and locked in place with a large nut.

2 The inner ends of the driveaxles are equipped with sliding constant velocity (CV) joints, which are capable of both angular and axial motion. Each inner CV joint assembly consists of a tripot-type bearing and a housing in which the joint is free to slide in-and-out as the driveaxle moves up-and-down with the wheel.

3 The outer ends of the driveaxles are equipped with ball-and-cage type CV joints, which are capable of angular but not axial movement. Each outer CV joint consists of six caged ball bearings running between an inner race and the housing.

4 The boots should be inspected periodically for damage and leaking lubricant. Torn CV joint boots must be replaced immediately or the joints will be damaged. If either boot of a driveaxle is damaged, that driveaxle must be removed in order to replace the boot.

5 Should a boot be damaged, the CV joint can be disassembled and cleaned (see Section 15), but if any parts are damaged, the entire driveaxle assembly must be replaced as a unit.

6 The most common symptom of worn or damaged CV joints, besides lubricant leaks, is a clicking noise in turns, a clunk when accelerating after coasting and vibration at highway speeds. To check for wear in the CV joints and driveaxle shafts, grasp each axle (one at a time) and rotate it in both directions while holding the CV joint housings, feeling for play indicating worn splines or sloppy CV joints. Also check the driveaxle shafts for cracks, dents and distortion.

## 14 Driveaxle (4WD models) - removal and installation

### REMOVAL

◆ Refer to illustrations 14.2, 14.3 and 14.5

1 Loosen the wheel lug nuts, raise the front of the vehicle and support it securely on jackstands. Remove the wheel.
2 Pry off the hub cover (see illustration).
3 Remove the driveaxle/hub nut. To prevent the hub from rotating, brace a large prybar across two of the wheel studs (see illustration), or insert a long punch or screwdriver through the window in the brake caliper and into the disc cooling vanes.
4 Remove the differential carrier splash shield, if equipped.
5 Remove the driveaxle-to-axleshaft flange bolts (see illustration). Have an assistant apply the brake as you loosen the bolts to prevent the driveaxle from turning. Separate the driveaxle from the axleshaft flange.
6 Lower the inner end of the driveaxle, then pull the stub shaft out of the hub. Carefully guide the driveaxle out from under the vehicle.
➡ Note 1: It may be necessary to remove the stabilizer bar link to provide clearance for driveaxle removal.
➡ Note 2: If the stub shaft sticks in the hub splines, tap on the end of the shaft with a brass punch and a hammer. If that doesn't free the splines, push the driveaxle from the hub with a puller.

### INSTALLATION

7 Installation is the reverse of removal. Before installing the driveaxle, lubricate the splines on the stub shaft with multi-purpose grease.

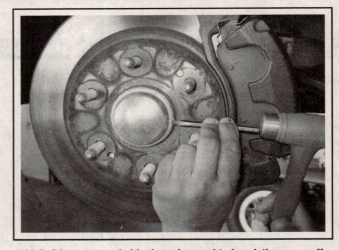

**14.2 A hammer and chisel can be used to knock the cover off the hub**

Be sure to tighten the driveaxle/hub nut (new) and the flange bolts to the torque values listed in this Chapter's Specifications. Tighten the wheel lug nuts to the torque listed in the Chapter 1 Specifications.

### ✳✳ WARNING:

**The hub nut should not be reused. Install a new hub nut when installing the shaft.**

# DRIVELINE 8-13

14.3 A large prybar can be used to immobilize the hub while loosening the nut, or a screwdriver can be inserted through the window in the brake caliper and into the disc cooling vanes

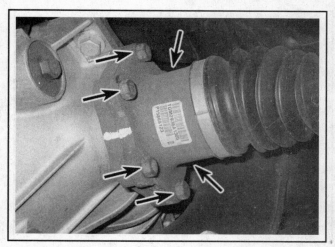

14.5 Remove the driveaxle-to-axleshaft flange bolts

## 15 Driveaxle boot (4WD models) - replacement

➡ **Note:** If the CV joint boots must be replaced, explore all options before beginning the job. Complete rebuilt driveaxles are available on an exchange basis, which eliminates much time and work. Whichever route you choose to take, check on the cost and availability of parts before disassembling the vehicle.

1  Remove the driveaxle (see Section 14).

2  Place the driveaxle in a vise lined with rags to avoid damage to the axleshaft. Check the CV joint for excessive play in the radial direction, which indicates worn parts. Check for smooth operation throughout the full range of motion for each CV joint. If a boot is torn, disassemble the joint, clean the components and inspect for damage due to loss of lubrication and possible contamination by foreign matter.

➡ **Note:** Some models are equipped with a protective cover that clamps around the larger diameter of each boot. Use diagonal cutting pliers to remove the clamps, then slide the cover off for access to the boots.

### INNER CV JOINT

◆ **Refer to illustrations 15.3a through 15.3t**

3  To replace the inner boot, refer to the accompanying illustrations (see illustrations 15.3a through 15.3t).

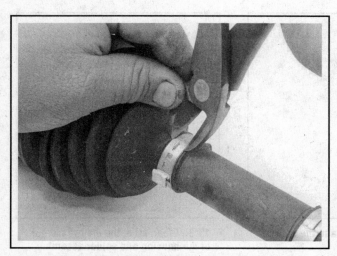

15.3a Cut off the old boot clamps with a pair of diagonal cutting pliers

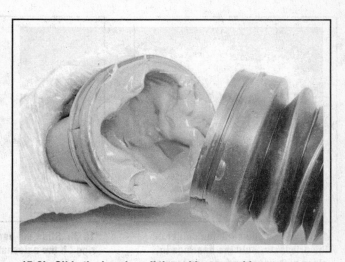

15.3b Slide the housing off the spider assembly

15.3c Slide the boot towards the center of the driveaxle

15.3d Spread the ends of the stop ring apart and slide it towards the center of the shaft

15.3e Slide the spider assembly back to expose the retaining ring, then pry off the ring

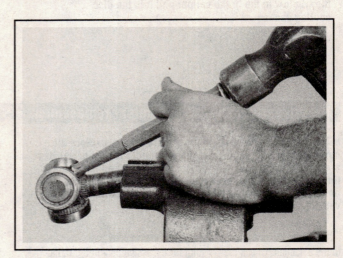

15.3f Carefully tap the spider off the axleshaft with a brass punch (but don't hit it so hard that it flies off, or you'll be picking up needle bearings!)

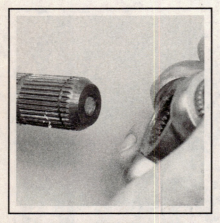

15.3g When you slide the spider off the driveaxle, hold the bearings in place with your hand, or use tape or a cloth wrapped around the spider bearing assembly to retain them

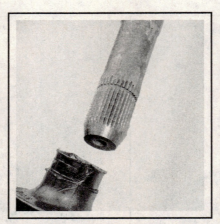

15.3h Slide the boot and the stop ring off the axleshaft

15.3i Clean all of the old grease out of the housing and spider assembly, then remove each bearing, one at time

DRIVELINE **8-15**

15.3j  Carefully disassemble each section of the spider assembly, clean the needle bearings with solvent and inspect the rollers, spider U-joint, bearings and housing for scoring, pitting and other signs of abnormal wear

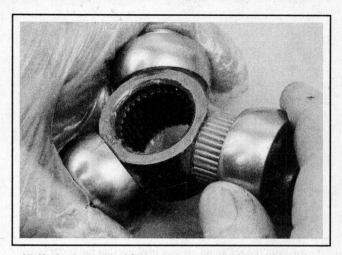

15.3k  Apply a coat of CV joint grease to the inner bearing surfaces to hold the needle bearings in place and slide the bearing over them

15.3l  Wrap the axleshaft splines with tape to avoid damaging the boot, then slide the small clamp and boot onto the axleshaft

15.3m  Slide the spider stop ring onto the axleshaft, past the groove in which it seats

15.3n  Install the spider bearing with the recess in the counterbore facing the end of the driveaxle

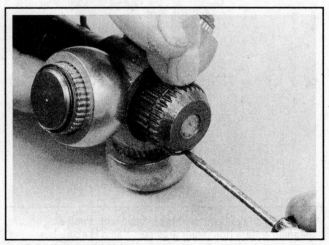

15.3o  Install the spider retaining ring, then slide the spider assembly against it and seat the stop ring in its groove

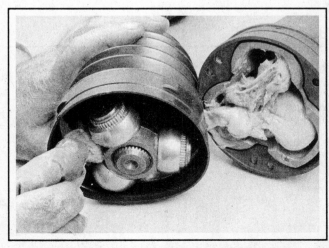

15.3p  Pack the housing with half of the grease furnished with the new boot and place the remainder in the boot

# 8-16 DRIVELINE

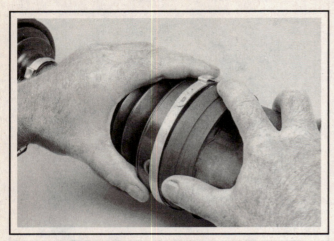

15.3q  With the retaining clamps in place (but not tightened), install the tripot housing

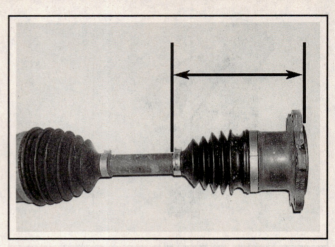

15.3r  Seat the boot in the housing and axle seal grooves, then adjust the length of the joint to the dimension listed in this Chapter's Specifications

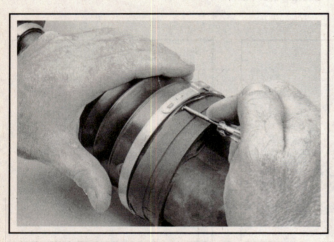

15.3s  With the joint set to the proper length, equalize the pressure in the boot by inserting a small screwdriver between the boot and the housing (make sure the boot isn't dimpled, stretched or out of shape) . . .

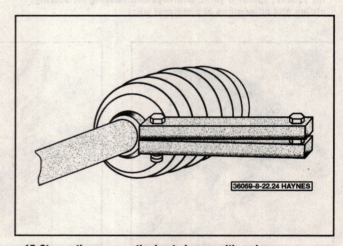

15.3t  . . . then secure the boot clamps with a clamp crimping tool (available at auto parts stores)

## OUTER CV JOINT

▶ **Refer to illustrations 15.4a through 15.4r**

4   Refer to the accompanying illustrations and perform the outer CV joint boot replacement procedure (see illustrations 15.4a through 15.4r).

15.4a  Cut off the boot retaining clamps with a pair of diagonal cutters (the larger diameter clamp is actually a swage ring; you may have to use a hand-held grinder to cut through it)

# DRIVELINE 8-17

**15.4b** Spread apart the ends of the internal snap-ring, then slide the CV joint off the shaft

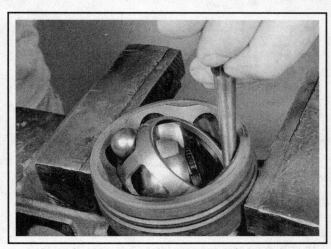

**15.4c** Press down on the inner race far enough to allow a ball bearing to be removed - if it's difficult to tilt, gently tap the cage and inner race with a brass punch and hammer

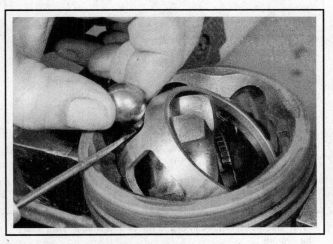

**15.4d** Pry the balls out of the cage, one at a time

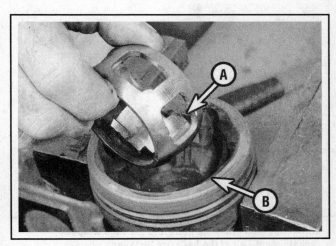

**15.4e** Tilt the inner race and cage 90-degrees, then align the windows in the cage (A) with the lands of the housing (B) and rotate the inner race and cage up and out of the housing

**15.4f** Align an inner race land with a cage window and rotate the inner race out of the cage

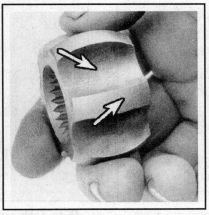

**15.4g** After cleaning the components with solvent, check the inner race lands and grooves for pitting and score marks

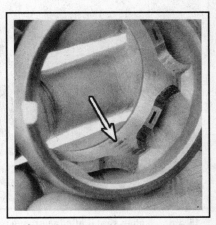

**15.4h** Check the cage for cracks, pitting and score marks - shiny spots are normal and don't affect operation

## 8-18 DRIVELINE

15.4i  With the race and cage tilted 90-degrees, lower the assembly into the housing

15.4j  Rotate the assembly by gently tapping with a hammer and brass punch . . .

15.4k  . . . then press the balls into the cage windows, repeating until all of the balls are installed

15.4l  Use needle-nose pliers to lower a new snap-ring into the groove . . .

15.4m  . . . then seat it into the groove with snap-ring pliers

15.4n  Apply grease through the splined hole, then insert a wooden dowel (with a diameter slightly less than that of the axle) through the splined hole and push down - the dowel will force the grease into the joint - repeat until the bearing is completely packed

# DRIVELINE 8-19

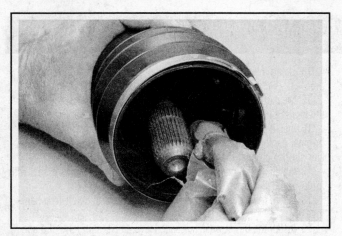

15.4o Install the small clamp and the boot on the driveaxle and apply grease to the inside of the axle boot . . .

15.4p . . . until the level is up to the end of axle

15.4q Position the CV joint assembly on the driveaxle, aligning the splines, then use a soft-face hammer to drive the joint onto the driveaxle until the snap-ring is seated in the groove (try to pull the CV joint off the shaft to make sure it's completely seated)

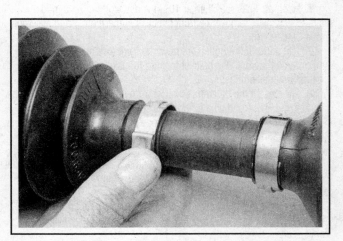

15.4r Seat the inner end of the boot in the groove and install the retaining clamp, then do the same on the other end of the boot - tighten the boot clamps with the special tool (see illustration 15.3t)

## 16 Front axle shift motor (4WD models) - replacement

♦ Refer to illustration 16.3

1  Raise the front of the vehicle and support it securely on jackstands. Remove the differential carrier splash shield, if equipped.
2  Disconnect the electrical connector from the shift motor.
3  Unscrew the motor from the axle tube (see illustration).
4  Before installing the motor, coat the threads with RTV sealant. Tighten the motor securely.

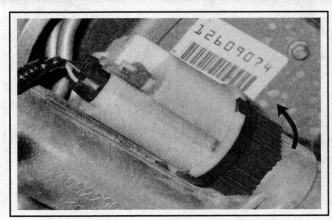

16.3 The front axle shift motor simply unscrews from the axle tube

# 8-20 DRIVELINE

## 17 Right axleshaft, tube, bearing and shift fork (4WD and AWD models) - removal, component replacement and installation

1  Loosen the wheel lug nuts, raise the vehicle, place it securely on jackstands and remove the right front wheel.
2  Remove the differential carrier splash shield, if equipped.
3  Drain the lubricant from the front differential (see Chapter 1).

### SELECTABLE 4WD SYSTEMS

◆ Refer to illustration 17.9

4  Unbolt the right driveaxle from the axleshaft flange (see Section 14). Support the driveaxle out of the way with a piece of wire - don't let it hang by the outer CV joint.
5  Remove the front axle shift motor (see Section 16).
6  If you're working on a model with a 9.25 inch axle, disconnect the front stabilizer link (see Chapter 10). If you're working on a model with a 8.25 inch axle, remove the front shock absorber/coil spring assembly (see Chapter 10).
7  On 2500/3500 models, remove the front axle bracket bolts and nuts. Slide the front axle bracket forward to make additional clearance for the inner axle housing.
8  To prevent the differential carrier from cocking when the axle tube is removed, support it with a floor jack and a block of wood.
9  Remove the bolts securing the axle tube to the differential carrier (see illustration).
10  Remove the nuts that attach the tube to its support bracket.
11  Carefully remove the output shaft tube. Make sure you don't allow any of the components to fall out of the tube. Remove the shift sleeve, thrust washer and seal.
12  Mount the tube in a vise, with the jaws clamping on the flange. Remove the damper spring, clip, fork, sleeve, gear, thrust washer and shift shaft.
13  Remove the output shaft from the tube by striking the inside of the flange with a soft-faced hammer while holding the tube.

14  Using a large screwdriver, pry the deflector and seal from the tube.
15  Measure the installed depth of the axleshaft bearing, then remove the bearing using a slide hammer and bearing remover attachment.
16  Install the bearing, with the square shoulder facing in, to the original depth using a bearing driver or a socket with an outside diameter slightly smaller than that of the bearing.
17  Install a new seal and deflector. Lubricate the lips of the seal with multi-purpose grease.
18  Install the axleshaft, carefully driving it in with a soft-face hammer.
19  Install the thrust washer, gear, sleeve, shaft, shift fork, clip and spring.
20  Apply a bead of RTV sealant to the tube-to-differential housing mating surface.
21  Apply some grease to the thrust washer and install the washer. Install the gear and shift sleeve.
22  Carefully install the output shaft tube assembly and install the bolts. Tighten the bolts to the torque listed in this Chapter's Specifications.
23  Install the two tube-to-chassis nuts and tighten them to the torque listed in this Chapter's Specifications.
24  Install the driveaxle (see Section 14).
25  Install the front axle shift motor (see Section 16).
26  Fill the differential with the proper lubricant (see Chapter 1).
27  Install the differential carrier splash shield.
28  Install the wheel and lug nuts. Lower the vehicle, tighten the lug nuts to the torque listed in the Chapter 1 Specifications, then check for proper operation.

### ALL-WHEEL DRIVE (AWD) SYSTEMS

29  Remove the stabilizer bar link (see Chapter 10).
30  If you're working on a model with an 8.25 inch axle, remove the front shock absorber.
31  Unbolt the right driveaxle from the axleshaft flange (see Section 14). Support the driveaxle out of the way with a piece of wire - don't let it hang by the outer CV joint.
32  To prevent the differential carrier from cocking when the axle tube is removed, support it with a floor jack and a block of wood.
33  Remove the nuts that attach the axle shaft to its support bracket.
34  Remove the bolts securing the axle shaft flange to the differential carrier.
35  Use a brass punch and hammer to tap on the axle shaft and separate it from the differential case side gear.
36  Remove the inner axle shaft assembly and the axle shaft from the vehicle.
37  Remove the inner seal and bearing. A special tool is required to pull the bearing from the clutch shaft. Install a new bearing and seal.
38  Apply RTV sealant to the differential sealing surface and Install the inner axle shaft housing assembly. Allow the RTV sealant to set-up, install the bolts and tighten the bolts to the torque listed in this Chapter's Specifications.
39  Install the axle shaft to the support bracket, install the nuts and tighten them to the torque listed in this Chapter's Specifications.
40  Align the splines of the inner axle shaft with the splines on the differential side gear by slowly rotating the inner axle shaft. Tap the

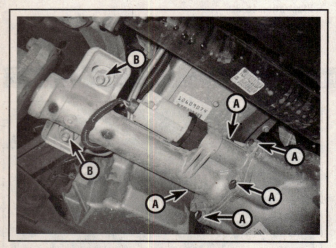

**17.9 Differential carrier and axle tube details on a 1500 Silverado model**

A  Axle tube-to-differential carrier bolts (two upper bolts hidden from view)
B  Axle tube mounting nuts

inner axle shaft with a mallet until the ring (clip) on the axle shaft is aligned with the groove in the differential case side gear.

41 Install the axle shaft to the inner flange. Install the axle shaft flange bolts and tighten them to the torque listed in this Chapter's Specifications.

42 Install the stabilizer link (see Chapter 10).

43 Fill the differential with the proper lubricant (see Chapter 1).

44 Install the differential carrier splash shield.

45 Install the wheel and lug nuts. Lower the vehicle, tighten the lug nuts to the torque listed in the Chapter 1 Specifications, then check for proper operation.

## 18 Axleshaft oil seals and bearings (front, 4WD models) - replacement

### RIGHT SIDE

1 Refer to Section 17 for the right-side axleshaft seal and bearing replacement procedure.

### LEFT SIDE

2 Loosen the wheel lug nuts, raise the vehicle, place it securely on jackstands and remove the left front wheel. Remove the differential carrier splash shield from under the front axle.

3 Remove the left driveaxle (see Section 14).

4 Drain the lubricant from the front differential (see Chapter 1).

5 Pull the left axleshaft out with a slide hammer and adapter.

6 Remove the deflector and seal from the differential.

7 Remove the bearing using a slide hammer and bearing remover attachment.

8 Install the bearing, with the square shoulder facing in, using a bearing driver or a socket with an outside diameter slightly smaller than that of the bearing.

9 Install a new seal and deflector. Lubricate the lips of the seal with multi-purpose grease.

10 Install the axleshaft, carefully driving it in with a soft-face hammer.

11 Fill the differential with the proper lubricant (see Chapter 1).

12 Install the driveaxle (see Section 14).

13 Install the differential carrier splash shield.

14 Install the wheel and lug nuts. Lower the vehicle, tighten the lug nuts to the torque listed in the Chapter 1 Specifications, then check for proper operation.

## 19 Front differential carrier - removal and installation

1 Disconnect the cable from the negative terminal of the battery (see Chapter 5, Section 1).

### 1500 AND 2500 LD MODELS

▶ Refer to illustrations 19.12a and 19.12b

2 Loosen the front wheel lug nuts, raise the front of the vehicle and support it securely on jackstands placed under the frame rails. Remove the wheels.

3 Remove the differential carrier splash shield, if equipped.

4 Drain the lubricant from the front differential (see Chapter 1).

5 Remove the front driveshaft (see Section 3).

6 Remove the lower control arm crossmember (see illustration 19.12a).

7 Detach the inner ends of the driveaxles from the axleshaft flanges (see Section 14). Suspend the driveaxles with lengths of wire - don't let them hang by the outer CV joints.

8 Disconnect the electrical connector from the shift motor (see illustration 16.3).

9 Detach the vent hose from the differential housing.

10 Support the differential carrier with a floor jack. If a transmission jack adapter is available, use it - it will hold the assembly more securely.

11 Remove the two nuts that attach the axle tube (differential carrier) to the chassis (see illustration 19.12a).

## 8-22 DRIVELINE

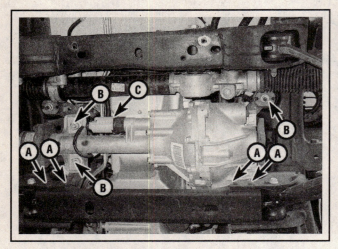

**19.12a Differential carrier details on a 1500 Silverado model**

A  Lower control arm crossmember bolts
B  Differential carrier mounting bolts/nuts
C  Front axle shift motor

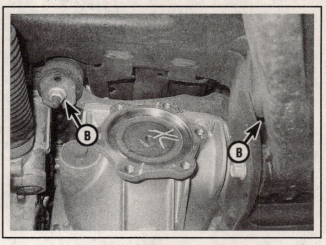

**19.12b The differential carrier bolts (B) are difficult to access - one bolt hidden from view**

12  Remove the differential carrier rear mounting bolts (see illustrations).
13  Slowly lower the jack and guide the carrier out from under the vehicle.
14  Check the bushings in the front and rear mounting bosses; if they are in need of replacement, take the carrier to an automotive machine shop or other qualified repair facility to have the old bushings pressed out and the new ones pressed in.
15  Installation is the reverse of removal. Tighten all fasteners to the proper torque values. Refill the differential with the proper lubricant (see Chapter 1).

### 2500 HD AND 3500 MODELS

16  Turn the steering wheel all the way to the left.
17  Loosen the wheel lug nuts, raise the front and rear of the vehicle and support it securely on jackstands placed under the frame rails. Remove the wheels.
18  Remove the differential carrier splash shield, if equipped.
19  Drain the lubricant from the front differential (see Chapter 1).
20  Remove the front driveshaft (see Section 3).
21  Disconnect the relay rod from the idler arm and the Pitman arm (see Chapter 10). Move the relay rod forward, out of the way.
22  Detach the inner ends of the driveaxles from the axleshaft flanges (see Section 14). Suspend the driveaxles with lengths of wire - don't let them hang by the outer CV joints.
23  Disconnect the electrical connector from the shift motor (see illustration 16.3).
24  Detach the vent hose from the differential housing.
25  Support the differential carrier with a floor jack. If a transmission jack adapter is available, use it - it will hold the assembly more securely.
26  Remove the two nuts that attach the axle tube to the chassis.
27  Remove the differential carrier rear upper mounting bolt and nut.
28  Pivot the differential carrier down and away from the vehicle to access the lower mounting bolt and nut.
29  Slowly lower the jack and remove the lower mounting bolt. Guide the carrier out from under the vehicle.
30  Check the bushings in the front and rear mounting bosses; if they are in need of replacement, take the carrier to an automotive machine shop or other qualified repair facility to have the old bushings pressed out and the new ones pressed in.
31  Installation is the reverse of removal. Tighten all fasteners to the proper torque values. Refill the differential with the proper lubricant (see Chapter 1).

# DRIVELINE 8-23

## Specifications

### General

Inner CV joint length (see illustration 15.3r)
    1500 models                                          6-11/16 inches
    2500 and 3500 models                   7 inches

### Torque specifications                      Ft-lbs (unless otherwise indicated)

### Driveshaft

Front driveshaft U-joint clamp bolts           18
Rear driveshaft
    U-joint clamp bolts                           18
    Center support bearing fasteners           30

### Rear axle

Pinion shaft lock screw
    8.6-inch axle                                 27
    9.5-inch axle                                 37
Full floating axle flange bolts
    10.5-inch axle                              115
    11.5-inch axle                              136
Differential cover bolts                           30

### Driveaxles (4WD models)

Driveaxle inner joint-to-axleshaft flange bolts     58
Driveaxle hub nut                              177

### Front differential carrier (4WD models)

Right axleshaft tube/shaft-to-chassis nuts       75
Right axleshaft tube/shaft-to-differential housing bolts
    8.25-inch axle                              41
    9.25-inch axle                              30
Differential housing-to-chassis bolts/nuts       75
Front suspension frame crossmember bolts
    1500 and 2500 LD
        Step 1                                   37
        Step 2                                   Tighten an additional 120 degrees
    2500 HD and 3500                     89
Shift motor                                       15

# 8-24 DRIVELINE

## Notes

**Section**

1 General information and precautions
2 Anti-lock Brake System (ABS), Traction Control (TCS) and vehicle stability control (StabiliTrack) systems - general information
3 Disc brake pads - replacement
4 Brake caliper - removal and installation
5 Brake disc - inspection, removal and installation
6a Drum brake shoes (2008 and earlier models) - replacement
6b Drum brake shoes (2009 and later models) - replacement
7 Wheel cylinder - removal and installation
8 Master cylinder - removal, installation and reservoir/O-ring replacement
9 Brake hoses and lines - check and replacement
10 Brake hydraulic system - bleeding
11 Power brake booster - check, removal and installation
12 Brake pedal travel - check
13 Parking brake - adjustment
14 Parking brake shoes - replacement
15 Brake light switch - check, adjustment and replacement

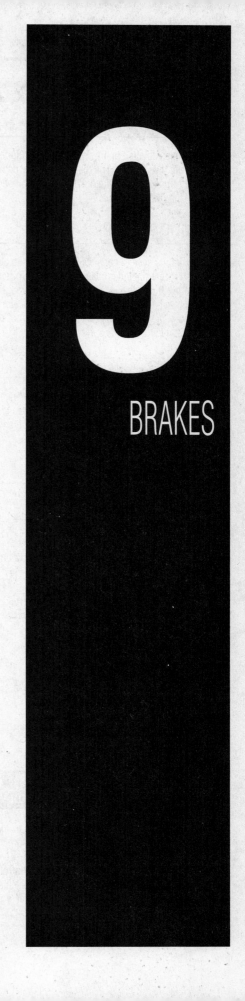

# 9

BRAKES

# 9-2 BRAKES

## 1 General information and precautions

### GENERAL

The vehicles covered by this manual are equipped with hydraulically operated front disc brakes. The rear brakes may be disc or drum type. Both types of brakes are self adjusting and automatically compensate for pad or shoe wear.

### HYDRAULIC SYSTEM

The hydraulic system consists of two separate circuits, split front-to-rear. The master cylinder has separate reservoirs for the two circuits, and, in the event of a leak or failure in one hydraulic circuit, the other circuit will remain operative and a warning indicator will light up on the instrument panel when a substantial amount of brake fluid is lost, showing that a failure has occurred.

### POWER BRAKE BOOSTER

The power brake booster uses either engine manifold vacuum or hydraulic pressure from the power steering pump to provide assistance to the brakes. It is mounted on the firewall in the engine compartment, directly behind the master cylinder.

### PARKING BRAKE

The parking brake operates the rear brakes only, through cable actuation. It's activated by a pedal mounted under the left end of the instrument panel. The parking brake cables actuate the rear brake shoes on drum type brakes. On rear disc type brakes, the cables actuate parking brake shoes mounted inside of the drum portion of each rear brake disc.

### SERVICE

After completing any operation involving disassembly of any part of the brake system, always test drive the vehicle to check for proper braking performance before resuming normal driving. When testing the brakes, perform the tests on a clean, dry, flat surface. Conditions other than these can lead to inaccurate test results.

Test the brakes at various speeds with both light and heavy pedal pressure. The vehicle should stop evenly without pulling to one side or the other.

Tires, vehicle load and wheel alignment are other factors which also affect braking performance.

### PRECAUTIONS

There are some general cautions and warnings involving the brake system on this vehicle:

a) *Use only brake fluid conforming to DOT 3 specifications.*
b) *The brake pads and linings contain fibers which are hazardous to your health if inhaled. Whenever you work on brake system components, clean all parts with brake system cleaner. Do not allow the fine dust to become airborne. Also, wear an approved filtering mask.*
c) *Safety should be paramount whenever any servicing of the brake components is performed. Do not use parts or fasteners which are not in perfect condition, and be sure that all clearances and torque specifications are adhered to. If you are at all unsure about a certain procedure, seek professional advice. Upon completion of any brake system work, test the brakes carefully in a controlled area before putting the vehicle into normal service. If a problem is suspected in the brake system, don't drive the vehicle until it's fixed.*
d) *Used brake fluid is considered a hazardous waste and it must be disposed of in accordance with federal, state and local laws. DO NOT pour it down the sink, into septic tanks or storm drains, or on the ground.*
e) *Clean up any spilled brake fluid immediately and then wash the area with large amounts of water. This is especially true for any finished or painted surfaces.*

## 2 Anti-lock Brake System (ABS), Traction Control (TCS) and vehicle stability control (StabiliTrak) systems - general information

1 The Anti-lock Brake System (ABS) helps to maintain vehicle maneuverability, directional stability, and optimum deceleration under severe braking conditions on most road surfaces. It does so by monitoring the rotational speed of the wheels and controlling the brake line pressure to the wheels during braking. This prevents the wheels from locking up on slippery roads or during hard braking.

### ELECTRO-HYDRAULIC CONTROL UNIT (EHCU)

▶ **Refer to illustration 2.2**

2 The Electro-Hydraulic Control Unit (EHCU), mounted on the left-side frame rail underneath the cab, controls hydraulic pressure to the brake calipers by modulating hydraulic pressure to prevent wheel

# BRAKES 9-3

lock-up (see illustration). It is made up of the Brake Pressure Modulator Valve (BPMV) and the Electronic Brake Control Module (EBCM). Basically, the BPMV reduces pressure in a brake line when the Electronic Brake Control Module (EBCM) detects an abnormal deceleration in the speed of a wheel (via a wheel speed sensor signal). When the speed of the wheel is restored to normal, the modulator once again allows full pressure to the brake. This cycle is repeated as many times as necessary, which results in a pulsing of the brake pedal.

3  In addition to sensing and processing information received from the brake switch and wheel speed sensors to control the hydraulic line pressure and avoid wheel lock up, the EBCM also continually monitors the system and stores fault codes which indicate specific problems.

## TRACTION CONTROL SYSTEM

4  Some models are equipped with a Traction Control System (TCS). When wheel slip is detected, the Electronic Brake Control Module (EBCM) will activate the traction control mode. A signal is sent from the EBCM to the PCM (Powertrain Control Module) commanding less torque to the drive wheels. Torque is reduced by retarding the ignition timing and by controlling the throttle control actuator (TCA).

5  The Traction Control System (TCS) is deactivated when the transmission shift lever is selected into the LOW position, the driver manually selects OFF on the TCS switch on the dash or the EBCM automatically shuts off the TCS during cruising or non-hazard conditions.

## STABILITY CONTROL SYSTEM

6  Some models may be equipped with StabiliTrak, a stability control system. This system is designed to assist in correcting over/under steering. This system is integrated with the ABS and TCS systems but uses additional sensors.

## WHEEL SPEED SENSORS

7  Generally, there is a wheel speed sensor designated for each wheel. Each sensor generates a signal in the form of a low-voltage electrical current or a frequency when the wheel is turning. A variable signal is generated as a result of a square-toothed ring (tone-ring, exciter-ring, reluctor, etc.) that rotates very close to the sensor. The signal is directly proportional to the wheel speed and is interpreted by an electronic module (computer).

8  The front sensors are mounted to the hub and wheel bearing assemblies and the tone-rings are integrated within the assemblies.

9  The rear sensors are mounted to the rear axle housing. The tone-rings are integrated with the rear axle shafts.

## WARNING LIGHTS

10  The ABS system has self-diagnostic capabilities. Each time the vehicle is started, the EBCM runs a self-test. There are two warning lights on the instrument panel, a red BRAKE light and an amber ABS light, each with their own functions. During starting, these lights should come on briefly then go out. If the red BRAKE light stays on, it indicates a problem with the main braking system, such as low fluid level detected or the parking brake is still on. If the light stays on after the parking brake is released, check the brake fluid level in the master cylinder reservoir (see Chapter 1).

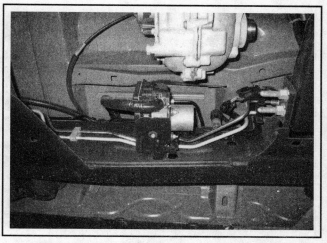

**2.2 The ABS Electro-Hydraulic Control Unit (EHCU) is located along the left frame rail, underneath the driver**

11  The amber ABS light indicates a problem with the ABS system, not the main or basic brake system. If the light stays on, it indicates that there is a problem with the ABS system, but the main system is still working. Take the vehicle to a dealer service department or other qualified repair shop for diagnosis and repair.

## DIAGNOSIS AND REPAIR

12  If a dashboard warning light comes on and stays on while the vehicle is in operation, the ABS, or other related systems require attention. Information will also be displayed in the Driver Information Center (DIC), as equipped. Although special diagnostic testing and tools are necessary to properly diagnose the system, you can perform a few preliminary checks before taking the vehicle to a dealer service department.
  a) *Check the brake fluid level in the reservoir.*
  b) *Check the electrical connectors at the EBCM and the hydraulic modulator/motor assembly.*
  c) *Check the fuses.*
  d) *Follow the wiring harness to each wheel and verify that all connections are secure and that the wiring is undamaged.*

13  If the above preliminary checks do not rectify the problem, the vehicle should be diagnosed by a dealer service department or other qualified repair shop. Due to the complexity of this system, all actual repair work must be done by a qualified automotive technician.

**❄❄ WARNING:**

**Do NOT try to repair a wheel speed sensor wiring harness. These systems are sensitive to even the smallest changes in resistance. Repairing the harness could alter resistance values and cause the system to malfunction. If the wiring harness is damaged in any way, it must be replaced.**

**❄❄ CAUTION:**

**Make sure the ignition is turned off before unplugging or reattaching any electrical connections.**

# 9-4 BRAKES

**2.17a Front wheel speed sensor location**

**2.17b Rear wheel speed sensor location**

## WHEEL SPEED SENSOR - REMOVAL AND INSTALLATION

▸ Refer to illustrations 2.17a and 2.17b

### ✳✳ WARNING:

**The dust created by the brake system is harmful to your health. Never blow it out with compressed air and don't inhale any of it. An approved filtering mask should be worn when working on the brakes. Do not, under any circumstances, use petroleum-based solvents to clean brake parts. Use brake system cleaner only!**

14  Loosen the wheel lug nuts, raise the vehicle and support it securely on jackstands. Remove the wheel.

15  Make sure the ignition key is turned to the OFF position.

16  On front sensors, remove the front brake disc (see Section 5).

17  Remove the mounting fastener and carefully pull the sensor out from the front hub assembly or the rear axle housing (see illustrations).

18  Follow the wiring harness to the electrical connector and disconnect it. Remove the harness from any brackets that may secure it to other components.

19  Installation is the reverse of the removal procedure. Tighten the mounting fastener to the torque listed in this Chapter's Specifications.

20  Install the wheel and lug nuts, lower the vehicle and tighten the lug nuts to the torque listed in the Chapter 1 Specifications.

## 3  Disc brake pads - replacement

▸ Refer to illustrations 3.5a through 3.5m

### ✳✳ WARNING:

**Disc brake pads must be replaced on both front or both rear wheels at the same time - never replace the pads on only one wheel. Also, the dust created by the brake system is harmful to your health. Never blow it out with compressed air and don't inhale any of it. An approved filtering mask should be worn when working on the brakes. Do not, under any circumstances, use petroleum-based solvents to clean brake parts. Use brake system cleaner only!**

➡ **Note:** This procedure applies to the front and rear brake pads.

1  Remove the cap from the brake fluid reservoir. Remove about two-thirds of the fluid from the reservoir, then reinstall the cap.

### ✳✳ CAUTION:

**Brake fluid will damage paint. If any fluid is spilled, wash it off immediately with plenty of clean, cold water.**

2  Loosen the front or rear wheel lug nuts, raise the front or rear of the vehicle and support it securely on jackstands. Block the wheels at the opposite end.

3  Remove the wheels. Work on one brake assembly at a time, using the assembled brake for reference if necessary.

4  Inspect the brake disc carefully as outlined in Section 5. If machining is necessary, follow the information in that Section to remove the disc.

5  Follow the accompanying photo sequence for the actual pad replacement procedure (see illustrations 3.5a through 3.5m). Be sure to stay in order and read the caption under each illustration.

# BRAKES 9-5

3.5a Before disassembling the brake, wash it thoroughly with brake system cleaner and allow it to dry - position a drain pan under the brake to catch the residue - DO NOT use compressed air to blow off brake dust!

3.5b To make room for the new pads, use a C-clamp to depress the piston(s) into the caliper before removing the caliper and pads - do this a little at a time, keeping an eye on the fluid level in the master cylinder to make sure it doesn't overflow. ➡Note: It's a good idea to loosen the bleed screw and allow the brake fluid to exit the caliper while compressing the caliper pistons (see illustration 10.8). This step will keep any debris and sediment that may be in the calipers from contaminating the ABS Electro-Hydraulic Control Unit (EHCU) and keep the master cylinder from overflowing

3.5c Front caliper mounting details

A  Caliper mounting bolts
B  Brake hose inlet fitting (banjo) bolt
C  Caliper mounting bracket bolts

3.5d If you're replacing the front brake pads (or the rear pads on 2500 and 3500 models), remove the lower mounting bolt and pivot the caliper up, supporting it in this position with a piece of wire

3.5e If you're replacing the front or rear pads on a 1500 model, hold the caliper slide pin with an open-end wrench, loosen the lower mounting bolt with another wrench, then pivot the caliper up and support it in this position for access to the brake pads

6  When reinstalling the caliper, be sure to tighten the mounting bolts to the torque listed in this Chapter's Specifications. Tighten the wheel lug nuts to the torque listed in the Chapter 1 Specifications.

7  After the job has been completed, slowly depress the brake pedal about two-thirds of its travel a few times until the pedal is firm. This will bring the pads into contact with the disc. Check the level of the brake fluid, adding some if necessary (see Chapter 1).

8  Break in the new pads by making several low speed stops (from approximately 30 mph). Allow time in between each stop for the discs to cool. About twenty stops should be sufficient to break in the new pads.

9  Confirm that the brakes are fully operational before resuming normal driving.

## 9-6 BRAKES

3.5f  Remove the inner brake pad

3.5g  Remove the outer brake pad

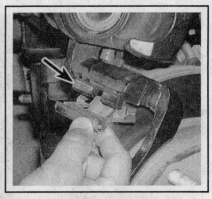

3.5h  Remove the upper and lower pad retainers and clean them. If they are cracked or distorted, replace them. Clean the bracket-to-pad retainer mating surface with a stiff wire brush, then wipe it clean with a rag. Apply a thin film of high-temperature brake grease to the mating surface only

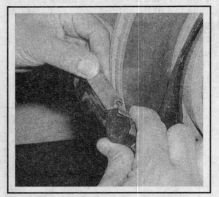

3.5i  Install the upper and lower pad retainers on the caliper mounting bracket

3.5j  Install the inner brake pad . . .

3.5k  . . . and the outer brake pad

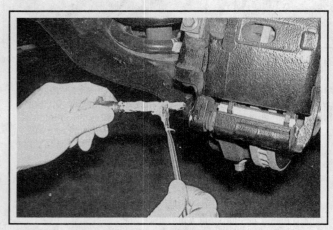

3.5l  On 2500 and 3500 models, inspect the caliper mounting bolt/slide pin for scoring and corrosion, then lubricate it with high-temperature brake grease. If it was dry, remove the caliper mounting bracket, remove the upper mounting bolt/slide pin and lubricate it too

3.5m  On 1500 models, before lowering the caliper over the pads, check the condition of the slide pin and the rubber boot, then lubricate the pin with high-temperature brake grease (if it was dry, pivot the caliper up again, pull the upper slide pin and boot out of the bracket and lubricate it, too)

# BRAKES 9-7

## 4  Brake caliper - removal and installation

### ✴✴ WARNING:

The dust created by the brake system is harmful to your health. Never blow it out with compressed air and don't inhale any of it. An approved filtering mask should be worn when working on the brakes. Do not, under any circumstances, use petroleum-based solvents to clean brake parts. Use brake system cleaner only!

### REMOVAL

◆ Refer to illustration 4.2

1  Loosen the front or rear wheel lug nuts, raise the front or rear of the vehicle and place it securely on jackstands. Block the wheels at the opposite end. Remove the front or rear wheel.

2  Remove the inlet fitting bolt and disconnect the brake hose from the caliper. Discard the old sealing washers (see illustration). Plug the brake hose immediately to keep contaminants and air out of the brake system and to prevent losing any more brake fluid than is necessary.

➡ Note: If you are simply removing the caliper for access to other components, leave the brake hose connected and suspend the caliper with a length of wire - don't let it hang by the hose (see illustration 5.2).

3  Remove the caliper mounting bolts and detach the caliper from the mounting bracket (see illustration 3.5c). When removing a rear caliper on a 1500 model, hold the slide pins with an open-end wrench to prevent them from turning when the mounting bolts are unscrewed (see illustration 3.5e).

➡ Note: On 2500 and 3500 models, the rear caliper mounting bracket will have to be removed in order to remove the upper caliper mounting bolt/slide pin.

### INSTALLATION

4  Installation is the reverse of removal. Don't forget to use new sealing washers on each side of the brake hose inlet fitting and be sure to tighten the fitting bolt and the caliper mounting bolts to the torque listed in this Chapter's Specifications.

5  Bleed the brake system (see Section 10).

➡ Note: If the brake hose was not disconnected, bleeding won't be required.

Make sure there are no leaks from the hose connections. Confirm that the brakes are fully operational before resuming normal driving.

4.2  There is a sealing washer on either side of the brake hose inlet fitting; be sure to replace these with new ones when reconnecting the hose

## 5  Brake disc - inspection, removal and installation

### ✴✴ WARNING:

The dust created by the brake system is harmful to your health. Never blow it out with compressed air and don't inhale any of it. An approved filtering mask should be worn when working on the brakes. Do not, under any circumstances, use petroleum-based solvents to clean brake parts. Use brake system cleaner only!

### INSPECTION

◆ Refer to illustrations 5.2, 5.3, 5.4a, 5.4b, 5.5a and 5.5b

1  Loosen the wheel lug nuts, raise the vehicle and support it securely on jackstands. Remove the wheel and reinstall the lug nuts to hold the disc in place.

➡ Note: If the lug nuts don't contact the disc when screwed on all the way, install washers under them.

2  Remove the brake caliper. It isn't necessary to disconnect the brake hose. After removing the caliper bolts, suspend the caliper out of the way with a piece of wire (see illustration).

5.2  Hang the caliper out of the way with a piece of wire - don't let it hang by the brake hose!

## 9-8 BRAKES

**5.3  The brake pads on this vehicle were obviously neglected, as they wore down completely and cut deep grooves into the disc - wear this severe means the disc must be replaced**

**5.4a  To check disc runout, mount a dial indicator as shown and rotate the disc**

**5.4b  Using a swirling motion, remove the glaze from the disc with sandpaper or emery cloth**

**5.5a  The minimum thickness is cast into the disc (typical)**

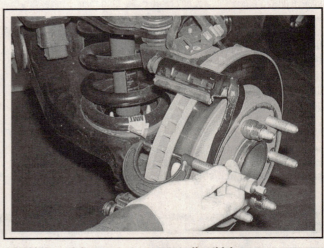

**5.5b  Use a micrometer to measure disc thickness**

3   Visually check the disc surface for score marks, cracks and other damage. Light scratches and shallow grooves are normal after use and may not always be detrimental to brake operation. Deep score marks or cracks may require disc refinishing by an automotive machine shop, or disc replacement (see illustration). Be sure to check both sides of the disc. If the brake pedal pulsates during brake application, suspect disc runout.

➡**Note: The most common symptoms of damaged or worn brake discs are pulsation in the brake pedal when the brakes are applied or loud grinding noises caused from severely worn brake pads. If these symptoms are extreme, it is very likely that the discs will need replacing.**

4   To check disc runout, place a dial indicator at a point about 1/2-inch from the outer edge of the disc (see illustration). Set the indicator to zero and turn the disc. Indicator readings that exceed 0.003 of an inch could cause pulsation upon brake application and will require disc refinishing by an automotive machine shop or disc replacement.

➡**Note: If disc refinishing or replacement is not necessary, you can deglaze the brake pad surface on the disc with emery cloth or sandpaper (use a swirling motion to ensure a non-directional finish) (see illustration).**

5   It's absolutely critical that the disc not be machined to a thickness under the specified minimum thickness. The minimum wear (or discard) thickness is cast into the underside of front discs (see illustration) and on the outside of rear discs. The disc thickness can be checked with a micrometer (see illustration).

### REMOVAL

♦ **Refer to illustrations 5.6a, 5.6b and 5.7**

➡**Note: On models equipped with dual rear wheels, the wheel hub must be removed (see Chapter 8) and the wheel studs pressed out of the hub/disc assembly.**

# BRAKES 9-9

5.6a Caliper mounting bracket bolts - front

5.6b Caliper mounting bracket bolts - rear

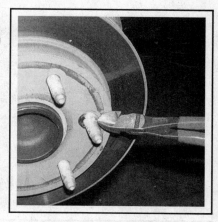

5.7 Cut off and discard the disc retaining washers, if present (it isn't necessary to reinstall them)

6  Remove the two caliper mounting bracket bolts and detach the mounting bracket (see illustrations).

7  Remove the lug nuts which you installed to hold the disc in place and slide the disc off the hub. If pressed-metal retaining clips are present on any of the wheel studs, cut them off (see illustration).

➥**Note: It isn't necessary to reinstall these pressed-metal retaining clips.**

## INSTALLATION

8  Place the disc in position over the threaded studs.

9  Install the mounting bracket and tighten the bolts to the torque listed in this Chapter's Specifications. Install the brake pads.

10  Install the caliper onto the mounting bracket, tightening the bolts to the torque listed in this Chapter's Specifications.

11  Install the wheel and lug nuts. Lower the vehicle and tighten the lug nuts to the torque listed in the Chapter 1 Specifications. Depress the brake pedal a few times to bring the brake pads into contact with the disc. Bleeding won't be necessary unless the brake hose was disconnected from the caliper. Confirm that the brakes are fully operational before resuming normal driving.

## 6a  Drum brake shoes (2008 and earlier models) - replacement

♦ Refer to illustrations 6a.2a, 6a.2b, 6a.3, 6a.4, 6a.5, 6a.9, 6a.11, 6a.16a and 6a.16b

### ✹✹ WARNING:

Drum brake shoes must be replaced on both wheels at the same time - never replace the shoes on only one wheel. Also, the dust created by the brake system is harmful to your health. Never blow it out with compressed air and don't inhale any of it. An approved filtering mask should be worn when working on the brakes. Do not, under any circumstances, use petroleum-based solvents to clean brake parts. Use brake system cleaner only!

1  Loosen the wheel lug nuts, raise the rear of the vehicle and support it securely on jackstands. Block the front wheels to keep the vehicle from rolling. Remove the wheels.

2  Release the parking brake and remove the brake drums (see illustration). If the shoes have worn into the drum, preventing drum removal, remove the access plug from the backing plate, push the lever

6a.2a  If the drum is retained by pressed-metal washers, cut them off and discard them (there is no need to reinstall them)

## 9-10 BRAKES

**6a.2b** If the shoes have worn into the drum, preventing its removal, remove the plug from the backing plate and push the parking brake lever off its stop like this, then insert a brake adjusting tool or another screwdriver and turn the adjuster star wheel upward to retract the shoes

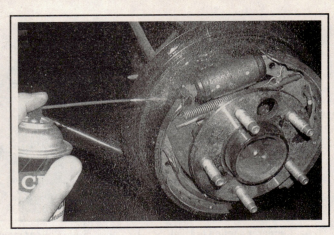

**6a.3** Before removing any internal drum brake components, wash them off with brake system cleaner and allow them to dry - position a drain pan under the brake to catch the residue - DO NOT USE COMPRESSED AIR TO BLOW THE DUST FROM THE PARTS

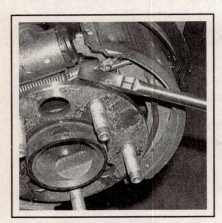

**6a.4** Grip the adjuster spring with a pair of pliers and detach its end from the adjuster lever

**6a.5** Pull the end of the retractor spring out of the hole in each shoe

**6a.9** Lubricate the brake shoe contact areas on the backing plate with high-temperature grease

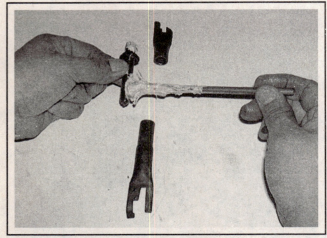

**6a.11** Lubricate the threads and socket end of the adjuster screw assembly with high-temperature grease

off the adjuster star wheel with one narrow screwdriver while turning the adjuster star wheel with another narrow screwdriver (or a brake adjusting tool) (see illustration).

3  Before disassembling anything, clean off the brake assembly with brake system cleaner (see illustration).

4  Detach the spring from the adjuster lever (see illustration).

5  Pull the retractor spring out of the hole in each shoe (see illustration).

6  Remove the trailing shoe and adjuster lever, then the adjuster screw assembly.

7  Pull the retractor spring out of the way, then remove the leading shoe.

8  Detach the parking brake lever from the trailing shoe.

9  Clean the backing plate, then lubricate the shoe contact areas with a thin film of high-temperature grease (see illustration).

10  Connect the parking brake lever to the trailing shoe, position the shoe on the backing plate and install the end of the retractor spring in its hole.

11  Clean the adjuster screw assembly, then lubricate the threads and socket end with high-temperature grease (see illustration).

# BRAKES 9-11

**6a.16a  The maximum permissible diameter is cast into the drum (typical)**

**6a.16b  Remove the glaze from the drum surface with sandpaper or emery cloth**

12  Install the adjuster screw assembly, making sure it engages properly with the leading shoe.

13  Lightly lubricate the adjuster lever and install it on the trailing shoe.

14  Position the trailing shoe on the backing plate, making sure it (and the parking brake lever and adjuster lever) engage properly with the adjuster screw assembly, then insert the retractor spring into its hole in the shoe.

15  Insert the actuator spring into its hole in the leading shoe, then stretch it across and connect it to the adjuster lever.

16  Before reinstalling the drum it should be checked for cracks, score marks, deep scratches and hard spots, which will appear as small, discolored areas. If the hard spots cannot be removed with emery cloth or if any of the other conditions listed above exist, the drum must be taken to an automotive machine shop to have it resurfaced.

→Note: Professionals recommend resurfacing the drums whenever a brake job is done. Resurfacing will eliminate the possibility of out-of-round or tapered drums. If the drums are worn so much that they can't be resurfaced without exceeding the maximum allowable diameter (stamped into the drum) (see illustration), then new ones will be required. At the very least, if you elect not to have the drums resurfaced, remove the glazing from the surface with emery cloth or sandpaper using a swirling motion (see illustration).

17  When installing the drum, adjust the brake shoes by turning the star wheel on the adjuster until the drum just slips over the shoes. When turning the drum, the shoes should not rub; if they do, remove the drum and back off the star wheel a little bit so they don't. This adjustment is just to get the shoes close to the drum; the brake shoes will self-adjust after making several stops while driving in reverse.

18  Confirm that the brakes are fully operational before resuming normal driving.

## 6b  Drum brake shoes (2009 and later models) - replacement

▶ Refer to illustrations 6b.4a through 6b.4w

### ※※ WARNING:

**Drum brake shoes must be replaced on both wheels at the same time - never replace the shoes on only one wheel. Also, the dust created by the brake system is harmful to your health. Never blow it out with compressed air and don't inhale any of it. An approved filtering mask should be worn when working on the brakes. Do not, under any circumstances, use petroleum-based solvents to clean brake parts. Use brake system cleaner only!**

1  Loosen the wheel lug nuts, raise the rear of the vehicle and support it securely on jackstands. Block the front wheels to keep the vehicle from rolling.

2  Release the parking brake.

3  Remove the wheel.

→Note: All four rear brake shoes must be replaced at the same time, but to avoid mixing up parts, work on only one brake assembly at a time.

4  Follow the accompanying illustrations for the brake shoe replacement procedure (see illustrations 6b.4a through 6b.4w). Be sure to stay in order and read the caption under each illustration.

**6b.4a  If the brake drum has never been removed, you'll have to pull off the drum retainer clip - it can be discarded (it's for assembly line purposes - there's no need to install a new one)**

## 9-12  BRAKES

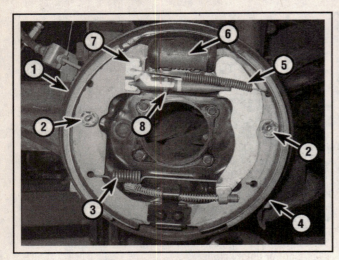

**6b.4b  Rear brake shoe details (left side shown)**

1. Leading shoe
2. Shoe retainer and spring
3. Lower return spring
4. Trailing shoe
5. Upper return spring
6. Wheel cylinder
7. Adjuster lever
8. Star-wheel assembly

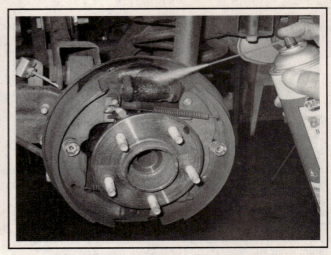

**6b.4c  Before removing anything, wash the brake assembly with brake system cleaner and allow it to drip dry into a pan - DO NOT USE COMPRESSED AIR TO BLOW BRAKE DUST FROM PARTS!**

➥Note: *If the brake drum cannot be easily removed, make sure the parking brake is completely released. If the drum still cannot be pulled off, the brake shoes will have to be retracted. This is done by first removing the plug from the backing plate. With the plug removed, push the lever off the adjuster star wheel with a narrow screwdriver while turning the adjuster wheel with another screwdriver, moving the shoes away from the drum (see illustration 6b.4w). The drum should now come off.*

5  Before reinstalling the drum, it should be checked for cracks, score marks, deep scratches and hard spots, which will appear as small discolored areas. If the hard spots cannot be removed with fine emery cloth or if any of the other conditions listed above exist, the drum must be taken to an automotive machine shop to have it resurfaced.

➥Note: *Professionals recommend resurfacing the drums each time a brake job is done. Resurfacing will eliminate the possibility of out-of-round drums. If the drums are worn so much that they can't be resurfaced without exceeding the maximum allowable diameter (stamped into the drum), then new drums will be required. At the very least, if you elect not to have the drums resurfaced, remove the glaze from the surface with emery cloth using a swirling motion.*

6  Install the brake drum on the axle flange. Pump the brake pedal a couple of times, then turn the drum and listen for the sound of the shoes rubbing. If they are, turn the adjuster star wheel until the shoes stop rubbing.

7  Install the wheel and lug nuts, then lower the vehicle. Tighten the lug nuts to the torque listed in the Chapter 1 Specifications.

8  Make a number of forward and reverse stops to seat the shoes to the drum and to actuate the auto-adjusting mechanism.

9  Check the operation of the parking brake and adjust, if necessary.

10  Check the operation of the brakes carefully before driving the vehicle.

**6b.4d  Unhook the lower spring and remove it**

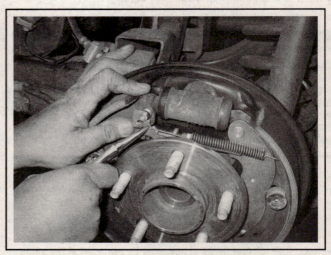

**6b.4e  The upper spring can now be removed from the adjuster lever - note how it engages**

# Brakes 9-13

6b.4f  Remove the adjuster lever

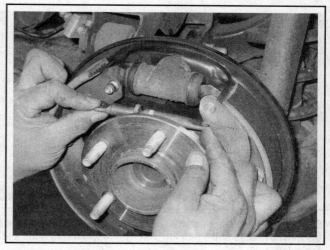
6b.4g  Remove the star-wheel assembly

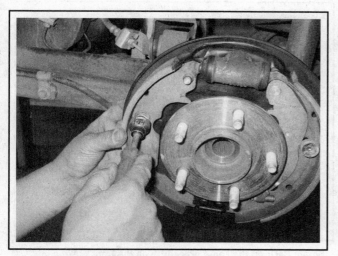
6b.4h  Depress the front shoe retainer and turn it to remove it

6b.4i  Remove the front brake shoe

6b.4j  Remove the rear shoe retainer and allow the shoe to come free

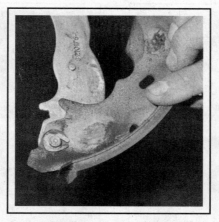
6b.4k  Pry open the clip to disconnect the parking brake lever from the rear shoe

6b.4l  Lubricate the backing plate contact points after thoroughly cleaning it

## 9-14 BRAKES

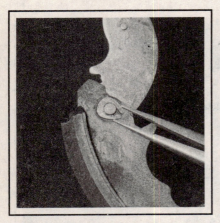

6b.4m Crimp the clip to attach the new rear shoe to the parking brake lever

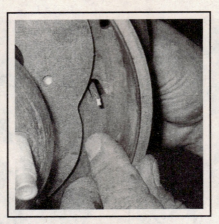

6b.4n Install the pin through the backing plate and the rear shoe

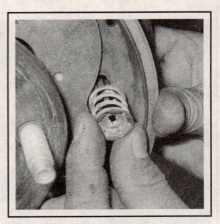

6b.4o Install the spring and the retainer to the pin

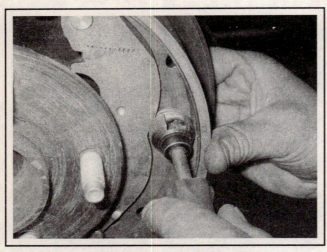

6b.4p Use the special tool to compress the spring and engage the retainer

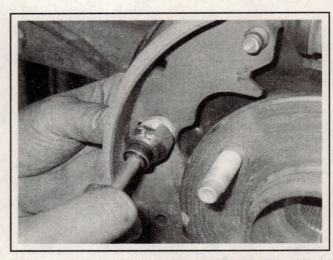

6b.4q Perform the same procedure on the front brake shoe

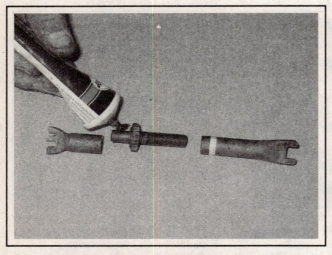

6b.4r Clean the star-wheel adjuster and lubricate it lightly

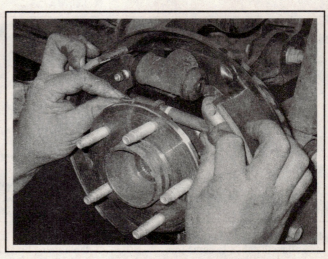

6b.4s Install the star-wheel assembly with the star to the front

# Brakes  9-15

6b.4t  Install the adjuster lever

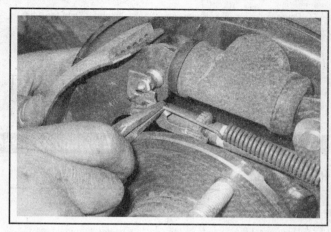

6b.4u  Install the upper spring, hooking it into the adjuster

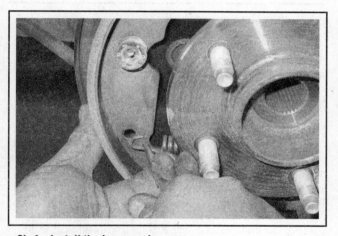

6b.4v  Install the lower spring

6b.4w  Adjust the star-wheel to give a very light drag when the brake drum is turned

## 7  Wheel cylinder - removal and installation

### ※ WARNING:

The dust created by the brake system is harmful to your health. Never blow it out with compressed air and don't inhale any of it. An approved filtering mask should be worn when working on the brakes. Do not, under any circumstances, use petroleum-based solvents to clean brake parts. Use brake system cleaner only!

### REMOVAL

♦ Refer to illustration 7.4

1  Raise the rear of the vehicle and support it securely on jackstands. Block the front wheels to keep the vehicle from rolling.
2  Remove the brake shoe assembly (see Section 6a or 6b).
3  Remove all dirt and foreign material from around the wheel cylinder.
4  Using a flare-nut wrench, unscrew the brake line fitting from the wheel cylinder. Don't pull the line away from the wheel cylinder (it could get bent) (see illustration).

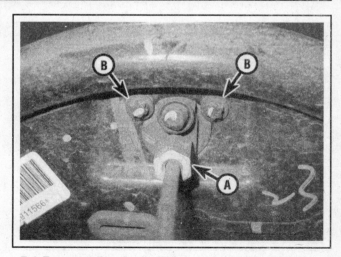

7.4  To remove the wheel cylinder, unscrew the brake line fitting (A) with a flare-nut wrench, then remove the wheel cylinder mounting bolts (B)

# 9-16 BRAKES

5  Remove the wheel cylinder mounting bolts and detach the wheel cylinder from the brake backing plate.

## INSTALLATION

6  Place the wheel cylinder in position and connect the brake line fitting, being careful not to cross-thread it. Install the mounting bolts, tightening them to the torque listed in this Chapter's Specifications, then tighten the brake line fitting securely.

7  Install and adjust the brake shoes (see Section 6a or 6b).
8  Bleed the brakes (see Section 10).
9  Confirm that the brakes are fully operational before resuming normal driving.

## 8  Master cylinder - removal, installation and reservoir/O-ring replacement

### REMOVAL

♦ **Refer to illustration 8.2**

1  Disconnect the cable from the negative battery terminal (see Chapter 5, Section 1).
2  Unplug the electrical connector for the fluid level warning switch (see illustration).
3  Remove as much fluid as possible from the reservoir with a suction gun, large syringe or a poultry baster.

### ※※ WARNING:

**If a poultry baster is used, never again use it for the preparation of food.**

4  Place rags under the line fittings and prepare caps or plastic bags to cover the ends of the lines once they're disconnected.

### ※※ CAUTION:

**Brake fluid will damage paint. Cover all body parts and be careful not to spill fluid during this procedure.**

Loosen the fittings at the ends of the brake lines where they enter the master cylinder. To prevent rounding off the flats, use a flare-nut wrench, which wraps around the fitting hex.

5  Pull the brake lines away from the master cylinder and plug the ends to prevent contamination.
6  Remove the nuts attaching the master cylinder to the power booster. Pull the master cylinder off the studs to remove it. Again, be careful not to spill fluid as this is done.

### INSTALLATION

♦ **Refer to illustrations 8.8, 8.13 and 8.16**

7  Bench bleed the new master cylinder before installing it. Mount the master cylinder in a vise, with the jaws of the vise clamping on the mounting flange.
8  Attach a pair of master cylinder bleeder tubes to the outlet ports of the master cylinder (see illustration).
9  Fill the reservoir with brake fluid of the recommended type (see Chapter 1).
10  Slowly push the pistons into the master cylinder (a large Phillips screwdriver can be used for this) - air will be expelled from the pressure chambers and into the reservoir. Because the tubes are submerged in fluid, air can't be drawn back into the master cylinder when you release the pistons.
11  Repeat the procedure until no more air bubbles are present.
12  Remove the bleed tubes, one at a time, and install plugs in the open ports to prevent fluid leakage and air from entering. Install the reservoir cap.
13  Replace the O-ring seal on the end of the master cylinder (see illustration), then install it to the booster, leaving the mounting nuts finger tight for now.
14  Thread the brake line fittings into the master cylinder. Since

**8.2  Master cylinder mounting details:**

1  Fluid level sensor electrical connector
2  Brake line fittings
3  Mounting nuts

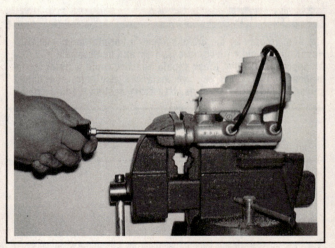

**8.8  The best way to bleed air from the master cylinder before installing it on the vehicle is with a pair of bleeder tubes that direct brake fluid into the reservoir during bleeding**

# BRAKES 9-17

8.13 Install a new O-ring on the master cylinder

8.16 Have an assistant depress the brake pedal and hold it down, then loosen the fitting nut, allowing air and fluid to escape; repeat this procedure on both fittings until the fluid is clear of air bubbles

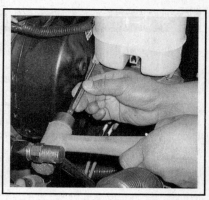

8.19 Driving out the roll pins that retain the master cylinder fluid reservoir on 2500 and 3500 models

the master cylinder is still a bit loose, it can be moved slightly so the fittings thread in easily. Don't strip the threads as the fittings are tightened.

15 Tighten the mounting nuts to the torque listed in this Chapter's Specifications. Tighten the brake line fittings securely.

16 Fill the master cylinder reservoir with fluid, then bleed the lines at the master cylinder, followed by bleeding the remainder of the brake system (see Section 10). To bleed the lines at the master cylinder, have an assistant depress the brake pedal and hold it down. Loosen the fitting to allow air and fluid to escape (see illustration). Tighten the fitting, then allow your assistant to return the pedal to its rest position. Repeat this procedure on both fittings until the fluid is free of air bubbles, then bleed the rest of the system. Confirm that the brakes are fully operational before resuming normal driving.

### ❋❋ WARNING:

**If you do not have a firm brake pedal at the end of the bleeding procedure, or have any doubts as to the effectiveness of the brake system, DO NOT drive the vehicle. Have it towed to a dealer service department or other qualified repair shop for diagnosis.**

## RESERVOIR/O-RING REPLACEMENT

▸ Refer to illustrations 8.19 and 8.21

➡Note: The brake fluid reservoir can be replaced separately from the master cylinder body if it becomes damaged. If there is leakage between the reservoir and the master cylinder body, the O-rings on the reservoir can be replaced.

17 Remove as much fluid as possible from the reservoir with a suction gun, large syringe or a poultry baster.

### ❋❋ WARNING:

**If a poultry baster is used, never again use it for the preparation of food.**

18 Place rags under the master cylinder to absorb any fluid that may spill out once the reservoir is detached from the master cylinder.

### ❋❋ CAUTION:

**Brake fluid will damage paint. Cover all body parts and be careful not to spill fluid during this procedure.**

19 On 2500 and 3500 models, use a hammer and a small punch to drive out the roll pins that retain the reservoir to the master cylinder (see illustration). On 1500 models, carefully pry the retaining tabs outwards while pulling the reservoir up to release it.

20 Pull the reservoir out of the master cylinder body.

21 If you are simply replacing the O-rings, carefully remove the old O-rings and install new ones (see illustration).

22 Lubricate the reservoir O-rings with clean brake fluid, then press the reservoir into place on the master cylinder body and secure it. Use new roll pins if applicable.

23 Refill the reservoir with the recommended brake fluid (see Chapter 1) and check for leaks.

24 Bleed the master cylinder (see illustration 8.16).

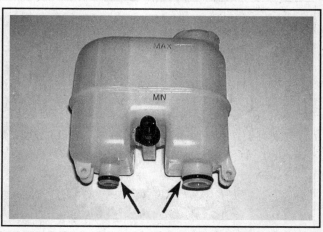

8.21 Location of the reservoir O-rings (2500/3500 model shown - 1500 similar)

## 9  Brake hoses and lines - check and replacement

1  About every six months, with the vehicle raised and placed securely on jackstands, the flexible hoses which connect the steel brake lines with the front and rear brake assemblies should be inspected for cracks, chafing of the outer cover, leaks, blisters and other damage. These are important and vulnerable parts of the brake system and inspection should be complete. A light and mirror will be needed for a thorough check. If a hose exhibits any of the above defects, replace it with a new one.

### FLEXIBLE HOSES

▶ Refer to illustrations 9.2a, 9.2b and 9.3

➡ Note: The replacement hose must be an exact duplicate of the original and have all of the appropriate brackets, if equipped.

2  Clean all dirt away from the ends of the hose (see illustrations).
3  Disconnect the brake line from the hose fitting (see illustration). Be careful not to bend the frame bracket or line. If necessary, soak the connections with penetrating oil.

4  Remove the U-clip from the female fitting at the bracket and remove the hose from the bracket.
5  Follow the hose's path and remove any mounting fasteners for brackets that secure it to other components.
6  On models where the hose is connected directly to the caliper, disconnect the hose fitting from the caliper by removing the banjo bolt. Discard the copper washers on both sides of the fitting. Use new copper washers and attach the new brake hose to the caliper.
7  Route the hose exactly as the original one was and mount any hose brackets that secure the hose to other components, as equipped.
8  Pass the female fitting through the frame or frame bracket. With the least amount of twist in the hose, install the fitting in position.
9  Install the U-clip in the female fitting at the frame bracket.
10  Using a back-up wrench on the fitting, attach the brake line to the hose fitting. Tighten the tube nut securely.
11  Confirm that the replacement hose follows the same path as the original one. This will avoid contact with moving suspension or steering components.
12  Bleed the brake lines as described in Section 10.

### METAL BRAKE LINES

13  When replacing brake lines, be sure to use the correct parts. Don't use copper tubing for any brake system components. Purchase steel brake lines from a dealer parts department or auto parts store.
14  Prefabricated brake lines, with the ends already flared and fittings installed, are available at auto parts stores and dealer service departments. If necessary, carefully bend the line to the proper shape. A tube bender is necessary for this.

**⁕⁕⁕ CAUTION:**

**Do not crimp or damage the line.**

**9.2a  A front brake hose on a 4WD model (other models similar)**

**9.2b  Location of the rear brake hoses above the rear axle housing**

**9.3  Brake hose fitting details:**

1  Metal tube nut (unscrew first - use a flare-nut wrench here)
2  Retaining clip (pull straight out with pliers to release the hose fitting from the bracket)
3  Brake hose fitting (hold with an open-end wrench while loosening the tube nut)

# BRAKES   9-19

15 When installing the new line, make sure it's well supported in the brackets and has plenty of clearance between moving or hot components.

16 After installation, check the brake fluid reservoir level and add fluid as necessary (see Chapter 1). Bleed the brake system as outlined in Section 10 and confirm that the brakes are fully operational before resuming normal driving.

## 10  Brake hydraulic system - bleeding

♦ Refer to illustration 10.8

**WARNING:**

Wear eye protection when bleeding the brake system. If the fluid comes in contact with your eyes, immediately rinse them with water and seek medical attention.

→ Note: Bleeding the hydraulic system is necessary to remove any air that manages to find its way into the system when it's been opened during removal and installation of a hose, line, caliper or master cylinder.

1  You'll probably have to bleed the system at all four brakes if air has entered it due to low fluid level, or if the brake lines have been disconnected at the master cylinder.

2  If a brake line was disconnected only at a wheel, then only that caliper must be bled.

3  If a brake line is disconnected at a fitting located between the master cylinder and any of the brakes, that part of the system served by the disconnected line must be bled.

4  Remove any residual vacuum (vacuum booster) or pressure (hydraulic booster) from the power brake booster by applying the brake several times with the engine off.

5  Remove the master cylinder reservoir cap and fill the reservoir with brake fluid. Reinstall the cap.

→ Note: Check the fluid level often during the bleeding operation and add fluid as necessary to prevent the fluid level from falling low enough to allow air bubbles into the master cylinder.

6  Have an assistant on hand, as well as a supply of new brake fluid, a clear container partially filled with clean brake fluid, a length of clear tubing to fit over the bleeder valve and a wrench to open and close the bleeder valve.

7  Beginning at the right rear wheel, loosen the bleeder valve slightly, then tighten it to a point where it's snug but can still be loosened quickly and easily.

8  Place one end of the tubing over the bleeder valve and submerge the other end in brake fluid in the container (see illustration).

9  Have the assistant depress the brake pedal slowly and hold it in the depressed position.

10  While the pedal is held down, open the bleeder valve just enough to allow a flow of fluid to leave the valve. Watch for air bubbles to exit the submerged end of the tube. When the fluid flow slows after a couple of seconds, close the valve and have your assistant release the pedal.

11  Repeat Steps 9 and 10 until no more air is seen leaving the tube, then tighten the bleeder valve and proceed to the left rear wheel, the right front wheel and the left front wheel, in that order, and perform the same procedure. Be sure to check the fluid in the master cylinder reservoir frequently.

12  Never use old brake fluid. It contains moisture which can cause the fluid to boil, rendering the brake system inoperative.

13  Refill the master cylinder with fluid at the end of the operation.

14  Check the operation of the brakes. The pedal should feel solid when depressed, with no sponginess. If necessary, repeat the entire process.

15  Before driving the vehicle, sit in the driver's seat and perform the following test:

  a) Take your foot off the brake pedal
  b) Start the engine and let it run for a minimum of 10 seconds. Watch the amber ABS light on the dash.
  c) If the light comes on and does not turn off after 10 seconds, have the vehicle towed to a dealer service department or other qualified repair shop. A scan tool will have to be used to diagnose the ABS system.
  d) If the ABS light goes off after three seconds or so, turn off the ignition.
  e) Repeat steps a) through d) one more time. If the amber ABS light turns off, test drive the vehicle in an isolated area before returning the vehicle to normal service.

**WARNING:**

Do not operate the vehicle if you're in doubt about the effectiveness of the brake system.

10.8  When bleeding the brakes, a hose is connected to the bleed screw at the caliper and submerged in brake fluid - air will be seen as bubbles in the tube and container (all air must be expelled before moving to the next wheel)

# 9-20 BRAKES

## 11 Power brake booster - check, removal and installation

### CHECK

**Vacuum booster**

**Operating check**

1  Depress the pedal and start the engine. If the pedal goes down slightly, operation is normal.
2  Depress the brake pedal several times with the engine running and make sure that there is no change in the pedal reserve distance.

**Airtightness check**

3  Start the engine and turn it off after one or two minutes. Depress the brake pedal several times slowly. If the pedal goes down farther the first time but gradually rises after the second or third depression, the booster is airtight.
4  Depress the brake pedal while the engine is running, then stop the engine with the pedal depressed. If there is no change in the pedal reserve travel after holding the pedal for 30 seconds, the booster is airtight.

**Hydraulic booster**

5  Turn the engine off, then depress the brake pedal several times to deplete the pressure in the accumulator.
6  Push down on the brake pedal, exerting approximately 40 pounds of force, then start the engine. If the booster is working properly, the brake pedal will sink towards the floor then rise back up against your foot.
7  If the booster does not work as described, check the fluid level in the power steering fluid reservoir, adding as necessary. Also check the hoses from the power steering pump to the booster for kinks. If everything checks out OK, the booster or power steering pump is defective. Have the power steering pump output pressure checked. If the pump is developing sufficient pressure, replace the booster.

### REMOVAL

▶ **Refer to illustrations 11.10, 11.14 and 11.15**

8  Disconnect the cable from the negative battery terminal (see Chapter 5, Section 1).
9  Depress the brake pedal several times to deplete the pressure in the accumulator (hydraulic booster) or a few times to remove vacuum (vacuum booster).
10  If you're working on a model with a vacuum booster, detach the vacuum hose from the booster (see illustration).
11  If you're working on a model with a hydraulic booster, detach the pressure and return lines from the booster. Cap the lines to prevent fluid leakage.
12  Remove the master cylinder without detaching the brake lines. Pull it forward and position it aside. Be careful not to bend or kink the brake lines.
13  Disconnect any electrical connectors from the components mounted to the booster, as equipped.
14  Remove the pushrod retaining clip (see illustration) and slip the brake light switch and the pushrod off the large pin on the brake pedal arm (or assembly on adjustable pedals).

➡ **Note: Some models may be equipped with a bolt in addition to the retaining clip.**

15  Remove the four nuts holding the brake booster to the firewall (see illustration).
16  Slide the booster straight out from the firewall until the studs clear the holes and pull the booster and gasket from the engine compartment.

### INSTALLATION

17  Installation is the reverse of removal. Be sure to use a new

**11.10  Pull the vacuum hose fitting straight out of the grommet in the booster**

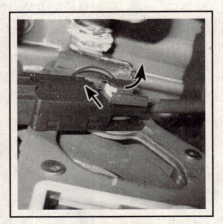

**11.14  Pry the center of the retaining clip up while moving the entire clip upwards off the end of the large pin on the brake pedal, then slide the brake light switch and pushrod off of the pin**

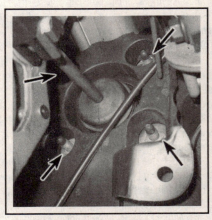

**11.15  Remove the booster mounting nuts (not all nuts are visible in this photo - vicinity given)**

gasket, and tighten the booster mounting nuts and the master cylinder mounting nuts to the torque values listed in this Chapter's Specifications.

18  If you're working on a model with a hydraulic booster, bleed the power steering system as described in Chapter 10. Check the power steering fluid level and add some, if necessary, to bring it up to the appropriate level.

## 12  Brake pedal travel - check

➡ **Note: On models with electrically adjustable pedals, perform the following Steps with the pedals set to their lowest position (closest to the floor).**

1  The brake pedal travel should be checked if the pedal seems to go too low. You'll need a tape measure, yardstick or ruler for this procedure.

2  Depress the brake pedal several times to deplete the pressure in the accumulator (hydraulic booster) or a few times to remove vacuum (vacuum booster).

3  Measure the position of the pedal at rest. You can either measure from the floor to the pedal or from the pedal to the steering wheel. Record your reading.

4  Now, depress the pedal (exerting approximately 70 lbs. of force) and measure how far the pedal has traveled. Compare your findings with the measurement listed in this Chapter's Specifications.

5  If the pedal travel is excessive, check for air in the system (bleed the brakes - see Section 10). A failed seal in the master cylinder could also cause excessive pedal travel.

## 13  Parking brake - adjustment

The parking brake cable is self-adjusting, but if the parking brake pedal travel is excessive or won't hold the vehicle on an incline, the parking brake shoes may need to be replaced or adjusted (see Section 14).

## 14  Parking brake shoes - replacement

❋❋ **WARNING:**
The dust created by the brake system is harmful to your health. Never blow it out with compressed air and don't inhale any of it. An approved filtering mask should be worn when working on the brakes. Do not, under any circumstances, use petroleum-based solvents to clean brake parts. Use brake system cleaner only!

➡ **Note: This procedure only applies to models with rear disc brakes.**

1  Loosen the rear wheel lug nuts, raise the rear of the vehicle and support it securely on jackstands. Release the parking brake. Block the front wheels to prevent the vehicle from rolling, then remove the rear wheels.

2  Remove the brake caliper (see Section 4), mounting bracket and the brake disc (see Section 5).

### 1500 MODELS

▶ Refer to illustrations 14.3, 14.4 and 14.6

3  Wash the brake assembly with brake system cleaner. Remove the clip securing the bottom of the shoe (see illustration), then slide the shoe up and off of the actuator.

➡ **Note: Detach the parking brake cable from the actuator if necessary.**

14.3  On 1500 models, the parking brake shoe is secured by clips held with bolts

## 9-22 BRAKES

**14.4 Lift one end of the parking brake shoe over the axle flange, then wind the rest of the shoe over the flange (1500 models)**

**14.6 Make sure the ends of the shoe seat in the adjuster screw slot (A) and the tappet slot (B); C is the adjuster screw star wheel (1500 models)**

4  Lift one end of the shoe over the axle flange, then work the shoe over the flange and remove it (see illustration).

5  Before installing the new shoe, turn the adjuster screw star wheel in, then make sure the slots in the adjusting screw and the tappet are parallel with the backing plate.

6  To install the shoe, reverse the removal procedure. Make sure the ends of the shoe seat properly in the slots in the adjuster screw and tappet (see illustration)

7  When installing the new shoe and lining assembly, turn the adjuster screw until the shoe lining just drags on the braking surface inside the disc. Then remove the disc and back-off the adjuster screw until the shoe lining doesn't drag when the disc is installed and turned. The actual clearance between the lining surface of the shoe and the braking surface inside the disc should be 0.026-inch.

8  Installation is otherwise the reverse of the removal procedure. Be sure to tighten the caliper bracket bolts and the caliper mounting bolts to the torque listed in this Chapter's Specifications, and the wheel lug nuts to the torque listed in the Chapter 1 Specifications.

### 2500/3500 MODELS

▶ Refer to illustration 14.9

9  Wash the brake assembly with brake system cleaner, then remove the hold-down clip from each parking brake shoe (see illustration).

10  Unhook the return springs from the top and bottom of each shoe, then remove the shoes from the backing plate.

11  Installation of the new shoes is the reverse of the removal procedure. When installing the new shoes, turn the adjuster screw until

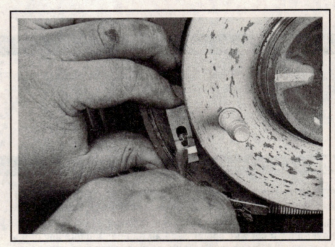

**14.9 Push in on the hold-down clip and turn the pin 90-degrees, then remove the clip (2500 and 3500 models)**

the shoe lining just drags on the braking surface inside the disc. Then remove the disc and back-off the adjuster screw until the shoe lining doesn't drag when the disc is installed and turned. The actual clearance between the lining surface of the shoe and the braking surface inside the disc should be 0.026-inch.

12  Be sure to tighten the caliper bracket bolts and the caliper mounting bolts to the torque listed in this Chapter's Specifications, and the wheel lug nuts to the torque listed in the Chapter 1 Specifications.

## 15 Brake light switch - check, adjustment and replacement

### CHECK

◆ **Refer to illustration 15.1**

1  The brake light switch (see illustration) is located on the side of the brake pedal and is retained by the same clip that retains the booster pushrod. The switch activates the brake lights at the rear of the vehicle when the pedal is depressed.

2  If the brake lights are inoperative, check the fuse first (see Chapter 12).

3  If the fuse is good, check for voltage to the switch on the feed wire (refer to the wiring diagrams at the end of this manual for the proper color wire to check). If no voltage is present, repair the wire between the switch and the fuse box.

4  If voltage is present, depress the brake pedal and check for voltage at the output wire terminal (again, refer to the wiring diagrams). If no voltage is present, replace the switch.

5  If voltage is present, check for power on the brake light wires at the tail light housings (with the brake pedal depressed). If voltage is not present, repair the circuit between the switch and the brake lights.

6  If voltage is present, check for a bad ground; using a jumper wire connected to a good ground, probe the ground wire terminal at the tail light connector. If the brake lights go on, repair the ground circuit (follow the ground wire from the tail light housing).

7  Keep in mind that the brake light bulbs could be burned out, but the likelihood of all the bulbs being burned out is very unlikely.

### ADJUSTMENT

8  The brake light switch on these vehicles is not adjustable. If it doesn't work as described above, replace it.

### REPLACEMENT

9  Unplug the electrical connector from the switch (see illustrations 15.1).

10  Remove the clip that retains the switch and pushrod to the pin on the brake pedal arm (or assembly on adjustable pedals) (see illustrations 11.14 and 15.1) and slip the brake light switch off the pin.

➡**Note: Some models may be equipped with a bolt in addition to the retaining clip.**

11  To install the new switch, reverse the removal procedure. Make sure the retaining clip is properly installed.

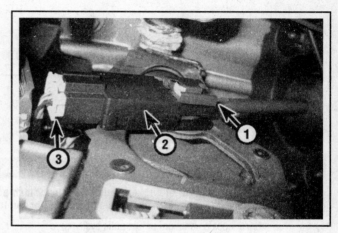

**15.1 Brake light switch mounting details:**

1  Pushrod/brake light switch retaining clip
2  Brake light switch
3  Brake light switch electrical connector

# 9-24 BRAKES

## Specifications

### General
| | |
|---|---|
| Brake fluid type | See Chapter 1 |
| Brake pedal travel (maximum) | |
|     With vacuum power booster | 2.36 inches |
|     With hydraulic power booster | 3.54 inches |

### Disc brakes
| | |
|---|---|
| Minimum pad thickness | See Chapter 1 |
| Brake disc minimum thickness | Cast into disc |
| Maximum disc runout | |
|     1500 models | 0.002 inch |
|     All others | 0.005 inch |
| Maximum disc thickness variation | |
|     1500 model rear discs | 0.0004 inch |
|     All others | 0.001 inch |

### Drum brakes
| | |
|---|---|
| Minimum shoe lining thickness | See Chapter 1 |
| Maximum radial runout | 0.0024 inch |
| Maximum drum diameter | Cast into drum |

## Torque specifications      Ft-lbs (unless otherwise indicated)

➥ **Note:** One foot-pound (ft-lb) of torque is equivalent to 12 inch-pounds (in-lbs) of torque. Torque values below approximately 15 foot-pounds are expressed in inch-pounds, because most foot-pound torque wrenches are not accurate at these smaller values.

| | |
|---|---|
| Brake booster (vacuum or hydraulic systems) | |
|     Mounting nuts | 24 |
|     Pushrod clip bolt | 89 in-lbs |
| Brake caliper | |
|     Caliper mounting bolts (slide pin bolts) | |
|         Front | |
|             1500 models | 74 |
|             All others | 80 |
|         Rear | |
|             1500 models | 38 |
|             All others | 80 |
|     Caliper mounting bracket bolts | |
|         Front | |
|             1500 models | 129 |
|             All others | 221 |
|         Rear | |
|             1500 models | 148 |
|             All others | |
|                 Models with single rear wheels | 148 |
|                 Models with dual rear wheels | 221 |
| Brake hose-to-caliper banjo bolt | 30 |
| Master cylinder-to-brake booster mounting nuts | 24 |
| Wheel cylinder mounting bolts | 156 in-lbs |
| Wheel speed sensor mounting bolts | |
|     Front | 156 in-lbs |
|     Rear | 80 in-lbs |
| Wheel lug nuts | See Chapter 1 |

# 10
# SUSPENSION AND STEERING SYSTEMS

**Section**

1 General information
2 Shock absorber (front) - removal and installation
3 Stabilizer bar and bushings (front) - removal and installation
4 Torsion bar - removal and installation
5 Upper control arm - removal and installation
6 Lower control arm - removal and installation
7 Balljoints - check and replacement
8 Hub and bearing assembly (front) - removal and installation
9 Wheel studs - replacement
10 Steering knuckle - removal and installation
11 Shock absorber (rear) - removal and installation
12 Stabilizer bar and bushings (rear) - removal and installation
13 Leaf spring - removal and installation
14 Coil spring (rear) - removal and installation
15 Suspension arms (rear) - removal and installation
16 Steering wheel - removal and installation
17 Steering column - removal and installation
18 Intermediate shaft and coupler - removal and installation
19 Tie-rod ends - removal and installation
20 Steering gear boots - replacement
21 Steering linkage - inspection, removal and installation
22 Steering gear - removal and installation
23 Power steering pump - removal and installation
24 Power steering system - bleeding
25 Wheels and tires - general information
26 Front end alignment - general information

**Reference to other Chapters**

Power steering fluid level check - See Chapter 1
Suspension and steering check - See Chapter 1
Tire and tire pressure checks - See Chapter 1
Tire rotation - See Chapter 1

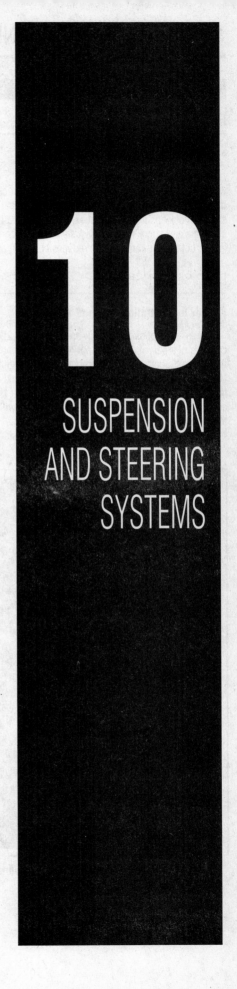

# 10-2  SUSPENSION AND STEERING SYSTEMS

## 1 General information

### FRONT SUSPENSION

♦ Refer to illustrations 1.1a and 1.1b

The front suspension (see illustrations) is fully independent. Each wheel is connected to the frame by a steering knuckle, upper and lower balljoints and upper and lower control arms. Some models use coil spring/shock absorber assemblies, while others are equipped with torsion bars and shock absorbers. A stabilizer bar connected to the frame and to both lower control arms reduces body roll during cornering.

### REAR SUSPENSION

♦ Refer to illustrations 1.2 and 1.3

The rear suspension on pick-up models and 2500 SUV models consists of a pair of multi-leaf springs and two shock absorbers (see illustration). The rear axle assembly is attached to the leaf springs by U-bolts. The front ends of the springs are attached to the frame at the front hangers, through rubber bushings. The rear ends of the springs are attached to the frame by shackles which allow the springs to alter their length as they compress and rebound.

The rear suspension on 1500 SUV models is a five-link design, using coil springs, upper and lower control arms, a lateral link, two shock absorbers and a stabilizer bar (see illustration).

### STEERING SYSTEM

There are two types of steering systems used on these models: rack-and-pinion steering gear and recirculating-ball steering gearbox. The recirculating-ball steering gearbox system is typically used on heavy duty applications.

The rack-and-pinion steering gear has two adjustable tie-rods and is power assisted.

The recirculating-ball steering gearbox has a Pitman arm, idler arm, relay rod, two adjustable tie-rod assemblies (each consisting of an inner tie-rod, adjuster tube and outer tie-rod) and, on some models, a steering damper. When the steering wheel is turned, the gear rotates the Pitman arm, which forces the relay rod to one side. The tie-rods, which are connected to the relay rod, transfer steering force to the steering knuckles. The tie-rods are adjustable and are used for toe-in adjustments. The relay rod is supported by the Pitman arm and idler arm. The idler arm pivots on a support attached to the right frame rail. The steering damper, if equipped, is attached to a bracket on the frame and to the relay rod.

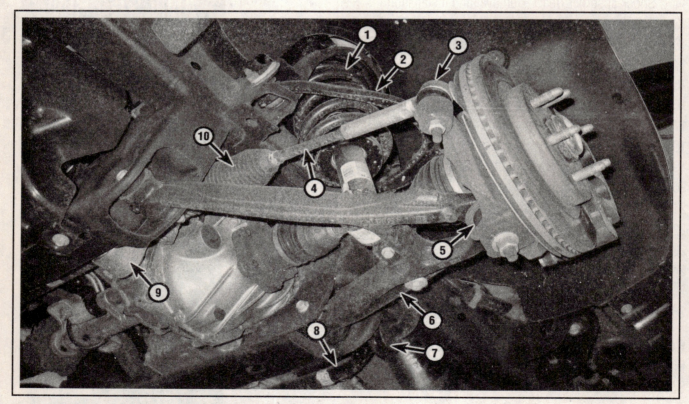

**1.1a Front suspension and steering components (4WD 1500 model)**

| | | |
|---|---|---|
| 1 Shock absorber/coil spring assembly | 5 Lower balljoint | 8 Stabilizer bar |
| 2 Upper control arm | 6 Lower control arm | 9 Steering gear |
| 3 Tie-rod end | 7 Stabilizer bar bracket | 10 Steering gear boot |
| 4 Tie-rod | | |

# SUSPENSION AND STEERING SYSTEMS    10-3

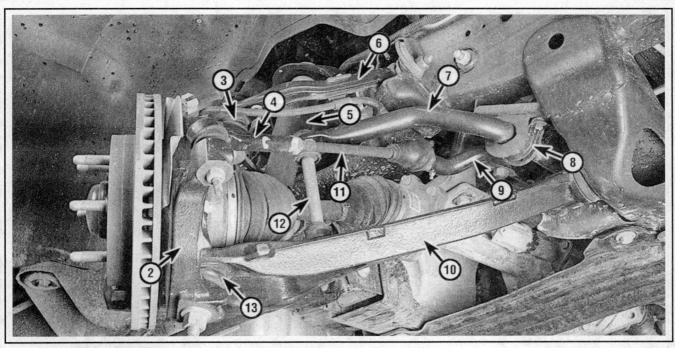

**1.1b Front suspension and steering components (4WD 2500 model)**

1  Torsion bar
2  Steering knuckle
3  Upper balljoint
4  Tie-rod end
5  Shock absorber
6  Upper control arm
7  Stabilizer bar
8  Stabilizer bar clamp
9  Relay rod
10  Lower control arm
11  Tie-rod
12  Stabilizer bar link
13  Lower balljoint

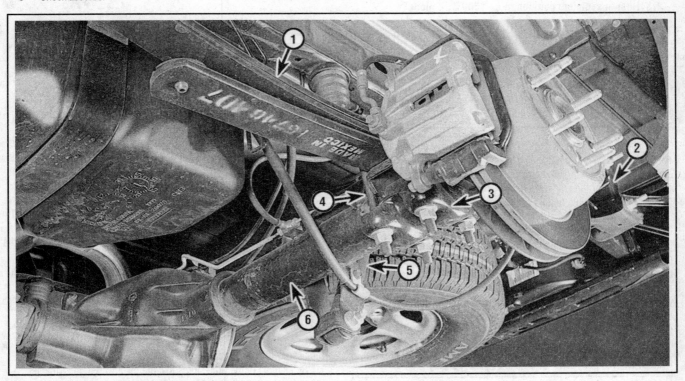

**1.2 Rear suspension components - 1500 pick-up**

1  Leaf spring
2  Shackle
3  Spring plate
4  U-bolt
5  Shock absorber
6  Rear axle

# 10-4 SUSPENSION AND STEERING SYSTEMS

**1.3 Rear suspension components - 1500 SUV model**

1. Track bar
2. Stabilizer bar link
3. Shock absorber
4. Trailing arm (lower)
5. Stabilizer bar clamp
6. Stabilizer bar
7. Rear axle
8. Coil spring

## RIDE CONTROL SYSTEMS

Light duty (1500) truck and SUV models may have an optional ride control system called Autoride®. This system electronically controls shock absorber damping and also incorporates an automatic leveling control system for the rear suspension. These computer controlled and fully automatic systems require no input from the driver and work in conjunction with the tow/haul mode. Autoride® is a system designed to optimize vehicle ride and handling and adjusts based on road conditions. The automatic leveling control keeps the rear of the vehicle at the proper ride height based on payload weight. It utilizes an electronically controlled air-shock system. It is likely that removing and installing any related suspension components on vehicles equipped with this option will require calibration with the use of a specialized scan tool; this will require service at a dealer service department or a qualified repair location.

## PRECAUTIONS

Frequently, when working on the suspension or steering system components, you may come across fasteners which seem impossible to loosen. These fasteners on the underside of the vehicle are continually subjected to water, road grime, mud, etc., and can become rusted or frozen, making them extremely difficult to remove. In order to unscrew these stubborn fasteners without damaging them (or other components), be sure to use lots of penetrating oil and allow it to soak in for a while. Using a wire brush to clean exposed threads will also ease removal of the nut or bolt and prevent damage to the threads. Sometimes a sharp blow with a hammer and punch is effective in breaking the bond between a nut and bolt threads, but care must be taken to prevent the punch from slipping off the fastener and ruining the threads. Heating the stuck fastener and surrounding area with a torch sometimes helps too, but isn't recommended because of the obvious dangers associated with fire. Long breaker bars and extension, or cheater, pipes will increase leverage, but never use an extension pipe on a ratchet - the ratcheting mechanism could be damaged. Sometimes, turning the nut or bolt in the tightening (clockwise) direction first will help to break it loose. Fasteners that require drastic measures to unscrew should always be replaced with new ones.

Since most of the procedures that are dealt with in this Chapter involve jacking up the vehicle and working underneath it, a good pair of jackstands will be needed. A hydraulic floor jack is the preferred type of jack to lift the vehicle, and it can also be used to support certain com-

# SUSPENSION AND STEERING SYSTEMS

ponents during various operations.

> ✳✳ **WARNING:**
> Never, under any circumstances, rely on a jack to support the vehicle while working on it. Also, whenever any of the suspension or steering fasteners are loosened or removed they must be inspected and, if necessary, replaced with new ones of the same part number or of original equipment quality and design. Torque specifications must be followed for proper reassembly and component retention. Never attempt to heat or straighten suspension or steering components. Instead, replace bent or damaged parts with new ones.

## 2  Shock absorber (front) - removal and installation

➡ **Note:** Electronically controlled shock absorbers that are part of the Autoride® system must be replaced with compatible replacement shock absorbers.

1  Loosen the front wheel lug nuts. Raise the front of the vehicle and support it securely on jackstands, then remove the wheel.

2  Support the outer end of the lower control arm with a floor jack (the shock absorber serves as the down-stop for the suspension). The jack must remain in this position throughout the entire procedure.

3  If equipped with a ride control system, unlock and unplug the electrical connector from the top of the shock absorber.

### COIL-OVER SHOCK ABSORBER MODELS

◆ Refer to illustrations 2.5 and 2.6

➡ **Note:** It is possible to replace the shocks or springs individually, but the unit will have to be disassembled by a qualified repair shop with the proper equipment, and this could add considerable cost to the project. Compare the cost of replacing the complete assemblies yourself to the cost of replacing individual components (with the help of a shop).

4  Detach the tie-rod end from the steering knuckle (see Section 19).

5  Remove the shock absorber assembly's upper mounting nuts (see illustration).

6  Working underneath the vehicle, remove the two bolts that attach the lower end of the shock absorber to the lower control arm (see illustration) and pull the shock out from below.

7  Installation is the reverse of removal. Be sure to tighten the mounting fasteners to the torque listed in this Chapter's Specifications.

### TORSION BAR MODELS

8  Using a back-up wrench on the stem, remove the shock absorber upper mounting nut and related hardware.

9  Remove the shock absorber lower mounting nut and bolt. Note the direction in which the bolt points.

10  Remove the shock absorber.

11  Installation is the reverse of removal. Be sure to install the lower mounting bolt so it's pointing in the same direction as it was prior to removal. Replace all of the related hardware and tighten all fasteners to the torque listed in this Chapter's Specifications.

➡ **Note:** The upper insulator is the larger of the two insulators.

### ALL MODELS

12  If equipped with a suspension control system, reconnect the electrical connector to the top of the shock absorber.

13  Install the wheels and lug nuts, lower the vehicle and tighten the lug nuts to the torque listed in the Chapter 1 Specifications.

2.5  Shock absorber upper mounting nuts (1500 models)

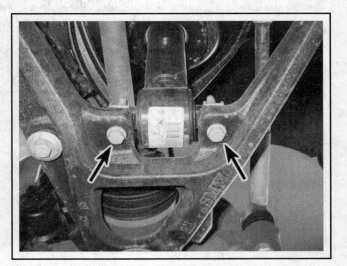

2.6  Shock absorber lower mounting bolts (1500 models)

## 3  Stabilizer bar and bushings (front) - removal and installation

**Refer to illustrations 3.2 and 3.3**

1  Raise the vehicle and support it securely on jackstands. On models so equipped, remove the oil pan skid plate or engine shield.
2  Remove the nuts from the link bolts and remove the link bolts (see illustration).
3  Remove the stabilizer bar bracket bolts (see illustration).
4  Remove the stabilizer bar.
5  Remove the rubber bushings.
6  Inspect all parts for wear and damage.
7  When you install the rubber bushings on the stabilizer bar, be sure to position them so the slits are facing the proper direction, as follows:

   1500 models - slits face forward
   2500/3500 models - slits face rearward

8  Installation is otherwise the reverse of removal. Be sure to tighten all fasteners to the torque listed in this Chapter's Specifications and to use a thread-locking compound (blue) on the threads of the bolts.

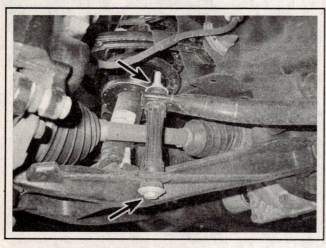

3.2 Stabilizer bar link bolt and nut

3.3 Remove the stabilizer bar bracket bolts

## 4  Torsion bar - removal and installation

**Refer to illustrations 4.2, 4.3, 4.4, 4.5, 4.6, 4.8a, 4.8b and 4.9**

**Note:** *The torsion bars must be removed as a pair, since the torsion bar crossmember must be removed to provide clearance to slide the bars to the rear.*

1  With the vehicle unloaded, on level ground and with the wheels pointing forward, measure the ride height at the front by measuring between a point on the frame and the ground. Mark the spot on the frame from where the measurement was made so another measurement can be taken using the exact same spot. This step is crucial to establish the vehicle's ride height so that it can be restored after the torsion bars have been reinstalled.
2  Raise the front of the vehicle and support it securely on jackstands. Count the number of threads showing on the torsion bar adjuster bolt and mark the relationship of the bolt to the torsion bar adjuster nut (see illustration).
3  In the torsion bar adjuster arm there's a small dimple. Install a puller with its bolt centered on this dimple (see illustration).
4  Turn the puller bolt until all tension is removed from the torsion bar adjuster bolt, then unscrew the torsion bar adjuster bolt and remove the nut (see illustration). Slowly unscrew the puller bolt until the torsion bar unwinds completely and no more tension is on the adjuster arm. Remove the puller.
5  Mark the relationship of the forward end of the torsion bar to the lower control arm (see illustration). Also mark the relationship of the rear end of the torsion bar to the adjuster arm.
6  Push the torsion bar forward, through the lower control arm, until the rear end of the bar clears the crossmember, then remove the torsion bar adjuster arm (see illustration).
7  Repeat Steps 1 through 6 on the other torsion bar.
8  Unbolt the torsion bar crossmember and remove it (see illustration). Inspect the bushings in the frame bosses where the crossmember mounts. If they're worn or otherwise deteriorated they can be replaced. They are staked in place from behind; once unstaked, they can be driven out of their bores toward the front of the vehicle (see illustration). On models equipped with a link between the crossmember and the frame, check the links for wear. If they're not in good shape, replace them.

# SUSPENSION AND STEERING SYSTEMS

4.2 To ensure proper adjustment of the torsion bar upon reassembly, count the number of threads showing on the torsion bar adjuster bolt and mark the relationship of the bolt to the torsion bar adjuster nut as insurance

4.3 Install a two-jaw puller as shown, with the fingers hooked around the flange running along each side of the crossmember; make sure the puller bolt is centered on the dimple in the torsion bar adjuster arm. Tighten the puller bolt until all tension is removed from the adjuster bolt

4.4 With tension removed from the adjuster bolt, unscrew the bolt and remove the adjuster nut

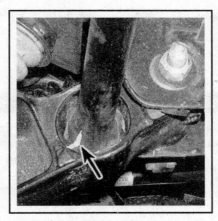

4.5 Mark the relationship of the torsion bar to the lower control arm

4.6 Slide the torsion bar forward through the lower control arm far enough to pull the rear end of the bar out of the crossmember, then remove the torsion bar adjuster arm

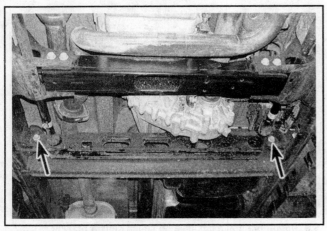

4.8a Remove the torsion bar crossmember mounting bolts from both sides, then remove the crossmember

4.8b On some models, the crossmember mounting bushings in the frame are replaceable; they must first be unstaked, then driven out towards the front of the vehicle

# 10-8 SUSPENSION AND STEERING SYSTEMS

9   Slide the torsion bar to the rear, lower the rear end of the bar down and guide it out through the hole in the frame crossmember (see illustration). If the front end of the bar hangs up in the lower control arm, drive it out of the control arm with a brass drift.

10   Installation is the reverse of removal, but if both torsion bars were removed, be sure to install them on the proper side of the vehicle (they are marked L and R on the ends). Clean out the hexagonal hole in the lower control arm and lube it with multi-purpose grease before inserting the torsion bar into the arm. Also apply some grease to the hex ends of the torsion bar, to the top of the adjuster arm and to the adjuster bolt. Make sure that the alignment marks you made on the torsion bar and the control arm and adjuster arm line up, and that the torsion bar is completely engaged with the adjuster arm.

11   Tighten the crossmember mounting fasteners to the torque values listed in this Chapter's Specifications.

12   Tighten the torsion bar adjuster bolt until the same number of threads are showing and the marks you made on the adjuster bolt and nut are lined up.

13   Lower the vehicle, jounce the front suspension a couple of times, then roll the vehicle back-and-forth a few feet to settle the suspension.

14   Repeat Steps 12 and 13 until both sides of the vehicle are set to the original ride height.

15   Have the front end alignment checked and, if necessary, adjusted.

**4.9  Guide the torsion bar to the rear and down, being careful not to nick it in the process**

Be sure to inform the repair facility that the ride height must be inspected before any alignment procedures are performed.

## 5   Upper control arm - removal and installation

### REMOVAL

♦ Refer to illustrations 5.2, 5.3, 5.5a and 5.5b

1   Loosen the wheel lug nuts, raise the front of the vehicle and support it securely on jackstands. Remove the wheel. Position a floor jack under the lower control arm in the area underneath the balljoint. Raise the jack slightly to take the spring pressure off the upper control arm.

### ※※ WARNING:

**The jack must remain in this position throughout the entire procedure.**

2   Mark the relationship of the adjusting cams to the brackets on the frame (see illustration).

3   Unbolt the brake hose/wheel speed sensor bracket from the upper control arm (see illustration). Also remove the wheel speed sensor (see Chapter 9) and detach the wiring harness bracket from the steering

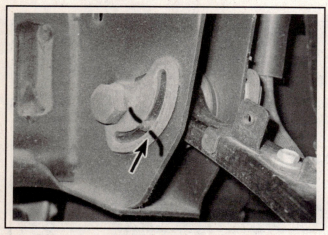

**5.2  Mark the relationship of all adjusting cams to the frame brackets**

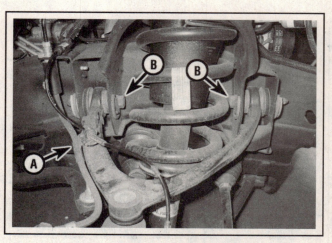

**5.3  Unbolt the brake hose/wheel speed sensor bracket (A) from the upper control arm and note the direction of the pivot bolts (B)**

# SUSPENSION AND STEERING SYSTEMS

5.5a Separating the upper control arm balljoint from the steering knuckle using a puller

5.5b Separating the upper control arm balljoint from the steering knuckle using a picklefork type separator

knuckle and move it aside.

4  If you're working on a 4WD model, remove the driveaxle (see Chapter 8).

→Note: If you use a picklefork type balljoint separator, it isn't necessary to perform this step.

5  To disconnect the upper control arm from the steering knuckle, loosen the upper balljoint nut a few turns (don't remove it), install a puller or balljoint separator and break the balljoint loose from the knuckle (see illustration). Now remove the nut.

→Note 1: If you don't have the proper balljoint removal tool, a hammer and a drift can sometimes be used to break the ballstud loose from the knuckle (see illustration 6.7). A picklefork-type balljoint separator can also be used, but keep in mind that this type of tool will probably damage the balljoint boot (see illustration).

→Note 2: The manufacturer recommends replacing the balljoint nut with a new one whenever it has been removed.

6  Remove the upper control arm pivot bolts and nuts, noting which way the bolts are installed (see illustration 5.3). Remove the control arm.

## INSTALLATION

7  Position the arm in the frame brackets and install the bolts and nuts, but don't tighten them yet.

8  Attach the balljoint to the steering knuckle, install a new nut and tighten it to the torque listed in this Chapter's Specifications.

9  The remainder of installation is the reverse of removal. Make sure the marks you made prior to disassembly are aligned, then tighten the nuts to the torque listed in this Chapter's Specifications.

→Note: The pivot bolt nuts should be tightened with the vehicle at normal ride height. This can be done after the vehicle has been lowered to the ground (on vehicles with adequate clearance), or it can be simulated by raising the lower control arm with a floor jack.

Tighten the wheel lug nuts to the torque listed in the Chapter 1 Specifications.

10  Have the front end alignment checked and, if necessary, adjusted.

## 6  Lower control arm - removal and installation

### REMOVAL

♦ Refer to illustrations 6.7 and 6.8

1  Loosen the wheel lug nuts, raise the vehicle and support it securely on jackstands placed under the frame rails. Remove the wheel.

2  Disconnect the stabilizer bar link from the control arm (see Section 3).

3  Remove the shock absorber lower mounting fasteners (see Section 2).

4  On 4WD models, remove the front driveaxle (see Chapter 8).

5  If you're working on a model with torsion bar front suspension, remove the torsion bar (see Section 4).

6  Using rope or wire, support the upper control arm and steering knuckle from the top of the shock absorber.

# 10-10 SUSPENSION AND STEERING SYSTEMS

7  To disconnect the lower control arm from the steering knuckle, loosen the balljoint nut a few turns (don't remove it), install a balljoint separator and break the balljoint loose from the knuckle. Now remove the nut.

➡ **Note 1: If you don't have the proper balljoint removal tool, a hammer and a drift can sometimes be used to break the ballstud loose from the knuckle (see illustration). A picklefork-type balljoint separator can also be used, but keep in mind that this type of tool will probably damage the balljoint boot.**

➡ **Note 2: The manufacturer recommends replacing the balljoint nut with a new one whenever it has been removed.**

8  Remove the lower control arm pivot bolts and nuts, noting which way the bolts are installed and, on 4WD models, how the washers are installed (see illustration). Pull the lower arm from its frame brackets.

## INSTALLATION

9  Position the arm in the frame brackets and install the bolts and nuts, but don't tighten them yet.

10  Attach the balljoint to the steering knuckle, install a new nut and tighten it to the torque listed in this Chapter's Specifications.

11  The remainder of installation is the reverse of removal. Be sure to tighten all fasteners to the torque values listed in this Chapter's Specifications.

➡ **Note: The pivot bolt nuts should be tightened with the vehicle at normal ride height. This can be done after the vehicle has been lowered to the ground (on vehicles with adequate clearance), or it can be simulated by raising the lower control arm with a floor jack.**

12  Install the wheel and lug nuts. Lower the vehicle and tighten the lug nuts to the torque listed in the Chapter 1 Specifications.

13  If you're working on a model with torsion bar front suspension, be sure to check and adjust the ride height (see Section 4).

14  Have the front end alignment checked and, if necessary, adjusted.

**6.7  A drift and a hammer can be used to loosen the ballstud from the steering knuckle if a balljoint separator isn't readily available**

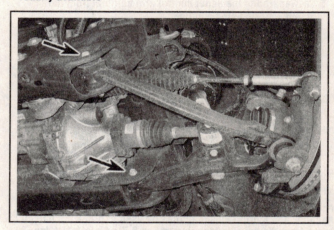

**6.8  The location of the lower control arm pivot bolts (1500 4WD model)**

## 7  Balljoints - check and replacement

### CHECK

1  Inspect the control arm balljoints for looseness anytime either of them is separated from the steering knuckle. See if you can turn the ballstud in its socket with your fingers. If the balljoint is loose, or if the ballstud can be turned, replace the balljoint. You can also check the balljoints with the suspension assembled as follows.

### Upper balljoints

2  Raise the front of the vehicle and support it securely on jackstands placed under the frame rails. Place a floor jack under the lower control arm and raise it slightly.

3  Attach a dial indicator to the upper suspension arm, with the plunger of the indicator touching the steering knuckle, in line with the axle.

4  Carefully pry between the two components while observing movement from the dial indicator. There should be no more than 0.079-inch deflection. If the indicated reading exceeds this figure, replace the upper arm.

➡ **Note: It may be possible to have the balljoint pressed out of the arm by an automotive machine shop or a repair shop that specializes in suspension work. A replacement part may be** available as an aftermarket (non-OEM) part, but this could not be confirmed at the time of this manual's writing.

### Lower balljoints

5  Raise the front of the vehicle and support it securely on jackstands placed under the frame rails. Place a floor jack under the lower control arm and raise it slightly.

6  Attach a dial indicator to the lower control arm and position the indicator plunger against the steering knuckle, in-line with the axle.

7  Carefully pry between the two components while observing movement from the dial indicator. There should be no more than 0.079-inch deflection. If the indicated reading exceeds this figure, replace the balljoint.

### REPLACEMENT

### Lower balljoint

➡ **Note 1: Before removing the lower control arm to replace the balljoint, check on the availability of parts. At the time of writing, the manufacturer indicated that only models with coil-over shock absorber-type front suspension have replaceable balljoints.**

# SUSPENSION AND STEERING SYSTEMS

➡ **Note 2:** If the lower control arm is made of aluminum, the manufacturer states that the balljoints on these control arms are not replaceable. The control arm and balljoint must be replaced as an assembly.

➡ **Note 3:** If you don't have access to the necessary equipment (see Step 10), you can remove the control arm and take it to an automotive machine shop to have the balljoint replaced.

8  Remove the lower control arm (see Section 6).

9  Mount the control arm in a vise, then use a hammer and chisel to knock back the crimped areas on the balljoint.

10  Press the balljoint from the control arm. A balljoint press (they're available at equipment rental yards and many auto parts stores), or a hydraulic press with the proper adapters, can be used.

11  Press the new balljoint into the control arm. Make sure the press adapter only applies force to the flange of the balljoint. Once pressed in fully, use a hammer and punch to crimp or stake the balljoint in place.

12  Install the control arm (see Section 6).

13  Have the front end alignment checked and, if necessary, adjusted.

### Upper balljoint

14  The upper balljoints are not replaceable separately, necessitating replacement of the entire upper control arm.

## 8  Hub and bearing assembly (front) - removal and installation

### ✳✳ WARNING:

The dust created by the brake system is harmful to your health. Never blow it out with compressed air and don't inhale any of it. Do not, under any circumstances, use petroleum-based solvents to clean brake parts. Use brake system cleaner only.

➡ **Note:** The hub and bearing assembly is sealed-for-life. If worn or damaged, it must be replaced as a unit.

### REMOVAL

▸ Refer to illustrations 8.5a, 8.5b and 8.6

1  Loosen the front wheel lug nuts, raise the vehicle and support it securely on jackstands. Remove the wheel.

2  If you're working on a 4WD model, remove the hub cover, then unscrew the driveaxle/hub nut with a socket and large breaker bar (see Chapter 8). Brace a large prybar across two of the wheel studs or insert a large screwdriver through the center of the brake caliper and into the disc cooling vanes to prevent the hub from turning as the nut is loosened.

3  Remove the brake disc (see Chapter 9).

4  Remove the wheel speed sensor from the hub (see Chapter 9).

5  Working from the back side of the steering knuckle, remove the hub retaining bolts from the steering knuckle (see illustrations). Remove the disc shield.

6  Remove the hub from the steering knuckle. If you're working on a 4WD model, pull the assembly off the driveaxle splines (see illustration).

### ✳✳ CAUTION:

Be careful not to pull outward on the driveaxle, as this could separate the inner CV joint components.

If the driveaxle splines stick in the hub, attach a two-jaw puller to the hub flange and push the stub axle out of the hub. The hub assembly should come right out of the steering knuckle, but if it doesn't, tap it from side-to-side to free it.

8.5a  The hub and bearing assembly on 1500 models is retained by three bolts - on other models it's retained by four bolts

8.5b  On 4WD models, there's not much room to get a wrench onto the hub bolt heads, especially when they are backed-out a few turns. As the bolts are unscrewed, push in on the driveaxle and pull the hub and bearing out of the knuckle to provide clearance

8.6  When removing the hub assembly on a 4WD model, don't pull out on the driveaxle; the inner CV joint could become separated

# 10-12 SUSPENSION AND STEERING SYSTEMS

## INSTALLATION

7  Clean the mating surfaces on the steering knuckle, bearing flange and knuckle bore. Make sure the O-ring came out with the hub assembly, and be sure to install a new O-ring on the back of the hub before fitting the hub to the steering knuckle.

8  Insert the hub and bearing assembly into the steering knuckle and, on 4WD models, onto the end of the driveaxle.

➡ **Note:** *On 4WD models, lubricate the splines of the driveaxle with multi-purpose grease before installing the hub. Position the disc shield and install the bolts, tightening them to the torque listed in this Chapter's Specifications.*

9  Install the wheel speed sensor (see Chapter 9).

10  Install the brake disc, caliper mounting bracket and caliper (see Chapter 9).

11  On 4WD models, install the driveaxle/hub nut and tighten it to the torque listed in the Chapter 8 Specifications.

12  Install the wheel, lower the vehicle and tighten the lug nuts to the torque listed in the Chapter 1 Specifications.

## 9  Wheel studs - replacement

♦ Refer to illustration 9.3

➡ **Note:** *This procedure applies to both the front and rear wheel studs.*

1  Loosen the wheel lug nuts, raise the vehicle and support it securely on jackstands. Remove the wheel.

2  Remove the brake disc or drum (see Chapter 9).

3  Push the stud out of the hub flange with a press tool (see illustration).

4  Insert the new stud into the hub flange from the back side and install some flat washers and a lug nut on the stud.

5  Tighten the lug nut until the stud is seated in the flange.

6  Reinstall the disc and caliper, or brake drum (see Chapter 9). Install the wheel and lug nuts. Lower the vehicle and tighten the lug nuts to the torque listed in the Chapter 1 Specifications.

**9.3  Use a small press tool such as this to push the stud out of the flange**

## 10  Steering knuckle - removal and installation

1  Loosen the wheel lug nuts, raise the vehicle and support it securely on jackstands. Remove the wheel.

2  If you're working on a 4WD model, remove the driveaxle/hub nut (see Chapter 8). Brace a large prybar across two of the wheel studs or insert a large screwdriver through the center of the brake caliper and into the disc cooling vanes to prevent the hub from turning as the nut is loosened.

3  On models equipped with a torsion bar suspension, support the lower control arm with a floor jack. Raise the jack slightly.

❋❋ **WARNING:**

*The jack must remain in this position throughout the entire procedure.*

4  Remove the brake caliper and brake disc (see Chapter 9). Hang the caliper out of the way on a piece of wire (don't disconnect the brake hose).

5  Remove the hub and bearing assembly (see Section 8).

6  Remove the disc splash shield from the steering knuckle.

7  Unbolt the brake hose bracket from the top of the steering knuckle.

8  Disconnect the tie-rod end from the steering knuckle (see Section 19).

9  Disconnect the upper and lower arms from the steering knuckle (see Sections 5 and 6).

10  Remove the steering knuckle.

11  Installation is the reverse of removal. Be sure to tighten the hub and bearing assembly, balljoint and tie-rod end fasteners to the torque values listed in this Chapter's Specifications. Tighten the caliper mounting bolts to the torque values listed in the Chapter 9 Specifications. Tighten the driveaxle/hub nut to the torque listed in the Chapter 8 Specifications (4WD models). Tighten the wheel lug nuts to the torque listed in the Chapter 1 Specifications.

# SUSPENSION AND STEERING SYSTEMS

## 11 Shock absorber (rear) - removal and installation

▶ Refer to illustration 11.4

➡ **Note:** Electronically controlled shock absorbers that are part of the Autoride® system must be replaced with compatible replacement shock absorbers.

1  Raise the rear of the vehicle and support it securely on jackstands placed underneath the frame rails. Block the front wheels so the vehicle doesn't roll off the stands.

➡ **Note:** It isn't necessary to remove the rear wheels, but doing so will improve access to the shock absorbers.

2  If the vehicle is equipped with a ride control system, disconnect the electrical connector and the air line from the shock absorber.

### ❊❊ WARNING:

**Relieve the air pressure in the ride control system by loosening the line fitting to the shock a few turns and letting the air leak out over a period of time. Wear safety glasses as a precaution when removing the line completely.**

3  Support the rear axle with a floor jack placed under the axle tube closest to the shock absorber being removed.

4  Remove the shock absorber upper and lower mounting fasteners (see illustration).

11.4  Rear shock absorber mounting fasteners

5  Remove the shock absorber.

6  Installation is the reverse of removal. Tighten the fasteners to the torque values listed in this Chapter's Specifications.

## 12 Stabilizer bar and bushings (rear) - removal and installation

▶ Refer to illustrations 12.2 and 12.4

1  Loosen the rear wheel lug nuts, raise the rear of the vehicle and support it securely on jackstands. Block the front wheels to keep the vehicle from rolling off the stands. Remove the rear wheels.

2  Remove the stabilizer bar link-to-frame nuts/bolts (see illustration).

3  Remove the nuts from the lower ends of the links, then separate the link from the bar.

4  Remove the stabilizer bar clamp bolts (see illustration) and remove the stabilizer bar assembly.

➡ **Note: On 3500 models, remove the driveshaft to remove the bar. However, the links and bushings can be replaced with the driveshaft in place.**

5  Inspect the stabilizer bar bushings and link bushings for cracks, tears and other signs of deterioration. Replace as necessary.

6  Installation is the reverse of removal. Be sure to tighten all fasteners to the torque listed in this Chapter's Specifications.

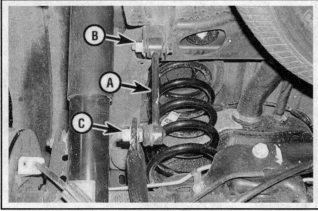

12.2  Stabilizer bar link mounting details (1500 model shown - other models similar)

A  Link
B  Link-to-frame bolt/nut
C  Link-to-stabilizer bar nut

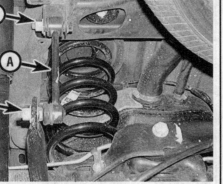

12.4  Remove the stabilizer bar clamp bolts from the axle housing

# 10-14 SUSPENSION AND STEERING SYSTEMS

## 13 Leaf spring - removal and installation

### REMOVAL

▶ **Refer to illustrations 13.7, 13.8 and 13.9**

1  If you're removing the left leaf spring or either spring on a 3500 model, relieve the fuel system pressure (see Chapter 4).

2  Loosen the rear wheel lug nuts, raise the rear of the vehicle and support it securely on jackstands placed underneath the frame rails. Block the front wheels to keep the vehicle from rolling off the stands. Remove the rear wheels.

3  Remove the trailer hitch, if so equipped.

4  If you're removing the left leaf spring, remove the fuel tank (see Chapter 4).

➡ **Note: On 3500 models, the auxiliary fuel tank must be removed to remove the rear shackle bolts on either side.**

5  If you're removing the right leaf spring, unbolt the rear portion of the exhaust system from the flange forward of the muffler, then detach the system from its rubber hangers to position it toward the center of the vehicle (this is to provide clearance for the front mounting bolt removal).

6  Support the axle with a floor jack placed under the axle tube and raise it slightly to take the weight of the axle.

7  Remove the four U-bolt nuts and washers (see illustration), the spring plate and the two U-bolts. Discard the U-bolts and nuts (the manufacturer recommends using new ones during installation).

8  At the front end of the spring, remove the nut and bolt from the spring-to-front bracket (see illustration).

9  At the rear end of the spring, remove the spring shackle-to-frame nut and bolt (see illustration).

10  Lower the jack slightly and remove the spring assembly. If necessary, remove the shackle from the spring. Remove the U-bolt top plate and spring spacer (if equipped) as necessary.

11  If the bushings at the ends of the spring are worn or deteriorated, an automotive machine shop or dealer service department can usually press the old ones out and press new ones in.

### INSTALLATION

12  If removed, install the shackle to the spring, but don't tighten the nut yet.

13  Place the spring in position and install the forward mounting bolt and nut, but don't tighten the nut yet.

14  Raise the axle on the jack until it mates properly with the spring. Install the spring plate and new U-bolts, then install the nuts and washers. Tighten the U-bolt nuts to the torque listed in this Chapter's Specifications.

15  Tighten the front mounting bolt/nut, the shackle-to-frame bolt/nut and the spring-to-shackle bolt/nut to the torque values listed in this Chapter's Specifications.

16  The remainder of installation is the reverse of the removal procedure.

17  Install the wheel and lug nuts. Lower the vehicle and tighten the lug nuts to the torque listed in the Chapter 1 Specifications.

**13.7 Remove the four U-bolt nuts and washers, then remove the spring plate and the two U-bolts**

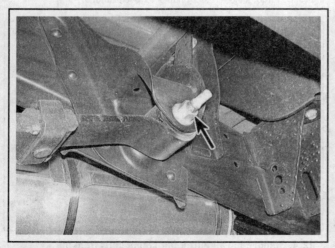

**13.8 At the front end of the leaf spring, remove the nut and bolt that attach the spring to the frame bracket**

**13.9 At the rear end of the spring, remove the spring shackle-to-frame nut and bolt (A); the shackle-to-spring nut and bolt (B) can't be removed until the spring and shackle are removed from the vehicle**

# SUSPENSION AND STEERING SYSTEMS 10-15

## 14 Coil spring (rear) - removal and installation

### REMOVAL

1  Loosen the rear wheel lug nuts. Raise the rear of the vehicle and support it securely on jackstands placed underneath the frame rails. Block the front wheels to prevent the vehicle from rolling. Remove the rear wheels.

2  If your vehicle is equipped with Autoride®, unbolt the height sensor linkage from the bracket on the trailing arm.

➡ **Note: Do not change the height sensor linkage adjustment or pry on it to detach it. Pivot the linkage up and support it to protect it from damage.**

3  Support the rear axle housing with a floor jack placed underneath the differential.

4  Detach the rear axle vent hose.

5  Unbolt the track bar from the bracket on the rear axle housing (see Section 15).

6  Disconnect the lower ends of the shock absorbers from the axle housing (see Section 11).

7  Slowly lower the floor jack in order to extend the coil springs. Do not exceed the slack in the flexible brake lines or any wire harnesses. Detach the hoses and wire harnesses if necessary.

8  When the coil springs are fully extended, remove the springs and insulators.

9  Check the condition of the insulators. If they're cracked, hardened or otherwise deteriorated, replace them.

### INSTALLATION

10  Place the springs and insulators in position on the axle and raise the axle until the ends of the springs engage properly with their upper mounts (an assistant would be helpful).

11  Continue to raise the axle until the shock absorbers can be connected to the axle housing. Install the bolts and nuts, tightening them to the torque listed in this Chapter's Specifications.

12  Connect the track bar to the axle, tightening the bolt to the torque listed in this Chapter's Specifications.

13  Reattach the vent hose.

14  Reconnect the brake hoses and wire harnesses if necessary, then bleed the rear brakes (see Chapter 9).

15  If equipped with Autoride®, reattach the height sensor linkage.

16  Install the wheels and lug nuts. Lower the vehicle and tighten the lug nuts to the torque listed in the Chapter 1 Specifications.

## 15 Suspension arms (rear) - removal and installation

➡ **Note: This procedure applies to models with coil spring rear suspension only.**

### TRAILING ARMS

▸ Refer to illustrations 15.3a and 15.3b

※※ **WARNING:**

**Remove and install only one arm at a time. This will prevent the axle housing from shifting on the jack.**

1  Loosen the rear wheel lug nuts. Raise the rear of the vehicle and support it securely on jackstands placed underneath the frame rails. Block the front wheels to prevent the vehicle from rolling. Remove the wheel(s).

2  If equipped with Autoride®, unbolt the height sensor linkage from the bracket on the trailing arm.

➡ **Note: Do not change the height sensor linkage adjustment or pry on it to detach it. Pivot the linkage up and support it to protect it from damage.**

3  Support the rear axle with a floor jack, then remove the nuts, washers and bolts from each end of the trailing arm (see illustrations)

4  Remove the arm. Check the bushings in the arm for cracking, hardness or other signs of deterioration. If the bushings are in need

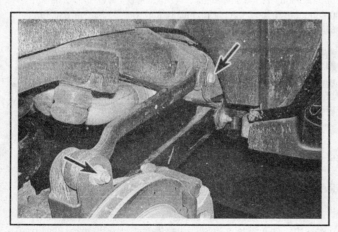

15.3a  Upper trailing arm mounting fastener locations

15.3b  Lower trailing arm mounting fastener locations

# 10-16 SUSPENSION AND STEERING SYSTEMS

of replacement, check with your local auto parts store or dealer parts department regarding the availability of replacement bushings. If replacement bushings are available, take the arm and the bushings to an automotive machine shop or other qualified repair facility to have the old ones pressed out and the new ones pressed in.

5  Installation is the reverse of removal. Before tightening the bolts/nuts, raise the rear axle with a floor jack to simulate normal ride height, then tighten the fasteners to the torque listed in this Chapter's Specifications.

## TRACK BAR

▶ Refer to illustration 15.8

6  Raise the rear of the vehicle and support it securely on jackstands placed underneath the frame rails. Block the front wheels to prevent the vehicle from rolling.

7  Detach the parking brake cable from the clips on the arm.

8  Remove the nuts/bolts from each end of the bar (see illustration).

9  Remove the bar. Check the bushings in the bar for cracking, hardness or other signs of deterioration. If the bushings are in need of replacement, check with your local auto parts store or dealer parts department regarding the availability of replacement bushings. If replacement bushings are available, take the bar and the bushings to an automotive machine shop or other qualified repair facility to have the old ones pressed out and the new ones pressed in.

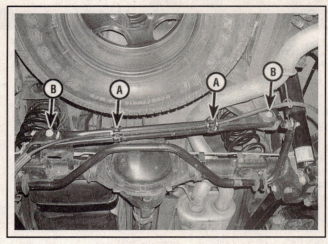

**15.8 Track bar mounting details**

A  Parking brake cable clip    B  Mounting bolt

10  Installation is the reverse of removal. Before tightening the bolts/nuts, raise the rear axle with a floor jack to simulate normal ride height, then tighten the fasteners to the torque listed in this Chapter's Specifications.

## 16  Steering wheel - removal and installation

### ※※ WARNING:

**These models are equipped with airbags. Always disable the airbag system before working in the vicinity of any airbag system component to avoid the possibility of accidental deployment of the airbag, which could cause personal injury (see Chapter 12).**

### REMOVAL

▶ Refer to illustrations 16.3a, 16.3b, 16.4, 16.5 and 16.7

1  Park the vehicle with the wheels pointing straight ahead. Disconnect the cable from the negative terminal of the battery (see Chapter 5).

2  Disable the airbag system (see Chapter 12).

3  Insert a small diameter rod into the hole on one side of the steering wheel to release the airbag module on that side. Repeat the procedure for the opposite side and carefully lift the module from the steering wheel (see illustrations).

4  To disconnect the electrical connectors, pry the connector lock up, then squeeze the small tabs on each side of the connector while pulling it from the module (see illustration). Set the module aside in a safe, isolated area, with the airbag side of the module facing UP.

### ※※ WARNING:

**When carrying the airbag module, keep the driver's (trim) side of it away from your body, and when you set it down, make sure the driver's side is facing up.**

5  Unplug the electrical connector from the clockspring (see illustration).

6  Remove the steering wheel retaining nut and mark the position of the steering wheel to the shaft, if marks don't already exist or don't line up (see illustration 16.5).

7  Wiggle the steering wheel away from the shaft to remove it. If necessary, use a puller to release the steering wheel from the shaft (see illustration).

# SUSPENSION AND STEERING SYSTEMS    10-17

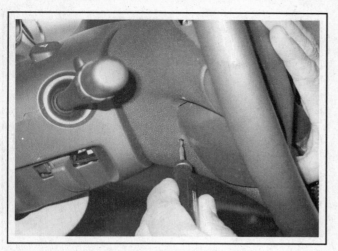

16.3a  Insert a tool into the hole to release the airbag module from the steering wheel, one side at a time

16.3b  Lift the module from the steering wheel to locate the two electrical connectors

16.4  Pry the connector lock up, then squeeze the small tabs on each side of the connector to pull it out of the airbag module

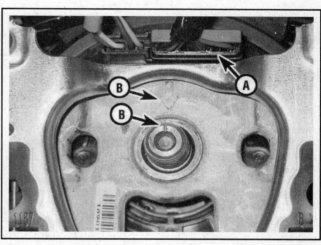

16.5  Remove the electrical connector from the clockspring (A) and note the alignment marks for the steering wheel hub and the steering shaft (B)

**✱✱ CAUTION:**

Do not use a hammer for removal.

8  Lift the steering wheel from the shaft while carefully guiding the airbag module harnesses through the steering wheel hub.

**✱✱ CAUTION:**

Changing the direction of the front wheels with the steering wheel removed will rotate the steering shaft. The steering shaft must be centered (with the front wheels pointing straight) before the steering wheel is installed or damage to the clockspring can occur. Also, other systems will be disabled as a result. It's best not to move the wheels with the steering wheel removed.

16.7  A puller can be used to remove the steering wheel from the steering shaft

# 10-18 SUSPENSION AND STEERING SYSTEMS

16.9 Steering shaft index mark (A) and corresponding gap on the shaft splines (B)

16.10 Clockspring mounting details:

1. Snap ring
2. Centering window
3. Index marks
4. Electrical connectors

## INSTALLATION

♦ Refer to illustrations 16.9 and 16.10

9  Before installing the steering wheel, make sure that the steering shaft is centered.

➥Note: *The front wheels should be pointing straight-forward and the index mark on the steering shaft (and gap on shaft splines) should be pointing directly up (see illustration). The clockspring is synchronized with the steering shaft. In the event that the steering shaft has been disconnected from the steering gear and rotated or the clockspring has been removed, the clockspring will require centering.*

10  If the airbag system clockspring is not centered, remove the steering column covers (see Chapter 11). Disconnect the electrical connectors from the bottom, remove the snap-ring from the steering shaft and carefully release any retaining tabs while lifting the clockspring off the steering column (see illustration).

11  Center the clockspring.

➥Note: *There is a centering window and index marks on the clockspring. New clocksprings are equipped with a centering pin which is removed after it is installed.*

Rotate the hub in one direction until it stops (don't apply too much force). Now turn the hub 2-1/2 turns in the opposite direction until the index marks are aligned and the centering window displays color (see illustration 16.10).

12  Install the clockspring and snap-ring. Connect the electrical connectors while making sure that the harness is in its original position. Install the steering column covers. Install the steering wheel onto the steering shaft while aligning the index marks.

### ✶✶ CAUTION:

**Do not force the steering wheel onto the shaft if there is resistance. Check for a blocked tooth on the steering wheel hub splines and match it to the splines on the steering shaft.**

13  Install the steering wheel nut and tighten it to the torque listed in this Chapter's Specifications.
14  Reconnect the steering wheel wiring harness to the clockspring
15  Connect the airbag connector to the back of the airbag module. Make sure the connector lock is securely engaged.
16  Position the airbag module on the steering wheel and push it in until the pins on the module engage with the spring retainers.
17  Refer to Chapter 12 for the procedure to enable the airbag system.

## 17 Steering column - removal and installation

### ✶✶ WARNING:

**These models are equipped with airbags. Always disable the airbag system before working in the vicinity of any airbag system component to avoid the possibility of accidental deployment of the airbag, which could cause personal injury (see Chapter 12).**

### REMOVAL

♦ Refer to illustrations 17.6, 17.7 and 17.8

1  Park the vehicle with the wheels pointing straight ahead. Disconnect the cable from the negative terminal of the battery (see Chapter 5, Section 1). Disable the airbag system (see Chapter 12).

# SUSPENSION AND STEERING SYSTEMS          10-19

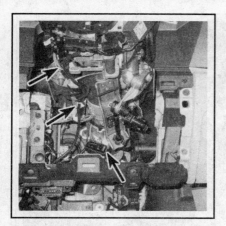

17.6 Follow the main harness to disconnect all related electrical connectors, then detach the harness from the steering column

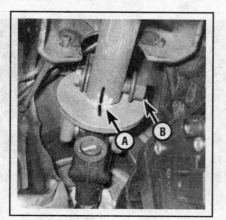

17.7 Mark the relationship of the upper intermediate shaft to the steering column (A), then remove the nut (B). The bolt is staked to an anti-rotation bracket (remove it after removing the nut first)

17.8 Remove the steering column mounting nuts (left side shown)

2   Remove the steering wheel (see Section 16), then turn the ignition key to the LOCK position to prevent the steering shaft from turning.

### ❈❈ CAUTION:

**If this is not done, the airbag clockspring could be damaged.**

3   Remove the steering column covers (see Chapter 11).
4   Remove the knee bolster and the steel reinforcement behind it (see Chapter 11).
5   Detach the shift cable from the column (see Chapter 7A).
6   Disconnect all electrical connectors from the steering column and detach the related wiring harnesses also (see illustration).
7   Mark the relationship of the intermediate shaft to the steering shaft from the column, then remove the fasteners securing the steering shaft to the upper intermediate shaft (see illustration).

➡ Note: Unscrew the nut; the bolt has a metal piece swaged to the head.

8   Before removing the steering column mounting nuts, note that there are two washers (spacers) beneath the rear mounting bracket for the column that must be placed back into position during installation (see illustration). Lower the column and pull it to the rear while making sure nothing is still connected. Separate the intermediate shaft from the steering shaft and remove the column.

### INSTALLATION

➡ Note: If a new column is being installed, check to see if a shipping lock pin is present. If so, remove it.

9   Guide the steering column into position, connect the intermediate shaft, then install the mounting nuts, but don't tighten them yet.

➡ Note: Make sure to place the two spacers for the column's rear mounting bracket back into position.

10  Install the fasteners for the steering shafts, then tighten the nut to the torque listed in this Chapter's Specifications.
11  Tighten the column mounting nuts to the torque listed in this Chapter's Specifications.
12  The remainder of installation is the reverse of removal. Adjust the shift cable following the procedures described in Chapter 7.

## 18  Intermediate shaft and coupler - removal and installation

◆ Refer to illustrations 18.3 and 18.4

1   When disconnecting any of the steering shaft components, the main thing to keep in mind is that they need to be reinstalled in the same orientation to one another so that synchronization with other components is maintained. Park the vehicle with the wheels pointing straight ahead. Disconnect the cable from the negative terminal of the battery (see Chapter 5, Section 1).

2   Turn the ignition key to the LOCK position to prevent the steering wheel (shaft) from turning.

### ❈❈ CAUTION:

**If this is not done, the airbag clockspring could be damaged.**

# 10-20  SUSPENSION AND STEERING SYSTEMS

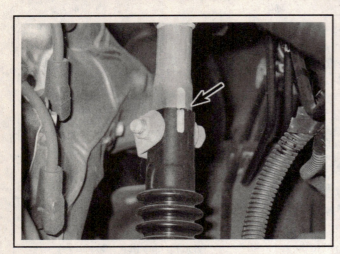

**18.3  Mark the relationship of the upper and lower halves of the intermediate shaft, then remove the fasteners to separate them**

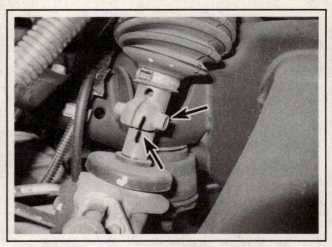

**18.4  Mark the relationship of the intermediate shaft coupler to the steering gear input shaft, then remove the pinch bolt from the bottom of the coupler**

3  Working under the hood, mark the relationship of the upper intermediate shaft to the lower intermediate shaft (see illustration). Remove the fasteners securing the lower intermediate shaft to the upper intermediate shaft.

4  Mark the relationship of the intermediate shaft coupler to the steering gear input shaft and remove the pinch bolt (see illustration).

➡ **Note: The lower intermediate shaft and the intermediate shaft coupler will be removed from the steering gear as an assembly. They can be separated and replaced individually as necessary.**

5  Slide the lower portion of the intermediate shaft downwards and off the upper intermediate shaft, the remove the lower intermediate shaft and steering shaft coupler from the steering gear.

6  The upper intermediate shaft can easily be removed from the shaft coming from the steering column if necessary (see Section 18).

7  Installation is the reverse of removal. Be sure to align the matchmarks and tighten the coupler bolt/nut and the shaft-to-steering gear pinch bolt to the torque listed in this Chapter's Specifications.

## 19  Tie-rod ends - removal and installation

▶ **Refer to illustrations 19.2a, 19.2b and 19.3**

1  Loosen the wheel lug nuts, raise the vehicle and place it securely on jackstands. Remove the wheel.

2  Loosen the tie-rod end locknut and mark the position of the tie-rod end on the threaded portion of the tie-rod (see illustrations).

3  Loosen the tie-rod end-to-steering knuckle nut about 1/4 inch from the tie-rod end balljoint stud, then install a balljoint separator or puller and separate the tie-rod end from the steering knuckle (see illustration). Remove the nut and detach the tie-rod end from the steering knuckle arm.

4  Unscrew the old tie-rod end and install the new one. Make sure the new tie-rod end is aligned with the mark you made on the threads of the tie-rod.

5  Installation is the reverse of removal. Be sure to tighten the tie-rod end-to-steering knuckle nut to the torque listed in this Chapter's Specifications. Tighten the locknut securely.

6  Tighten the lug nuts to the torque listed in the Chapter 1 Specifications.

7  Have the front end alignment checked and, if necessary, adjusted.

**19.2a  Loosen the tie-rod end locknut . . .**

# SUSPENSION AND STEERING SYSTEMS

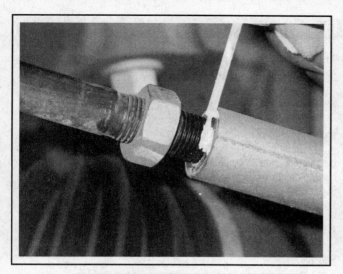

19.2b  ... then mark the position of the tie-rod end on the threaded portion of the tie-rod

19.3  Loosen (but don't remove) the castle nut from the tie-rod end ballstud, then install a balljoint separator or puller and separate the tie-rod end from the steering knuckle

## 20  Steering gear boots - replacement

♦ Refer to illustration 20.4

➡ Note: This procedure applies only to models with rack-and-pinion steering.

1  Loosen the wheel lug nuts, raise the front of the vehicle and support it securely on jackstands. Remove the wheels.
2  Remove the tie-rod ends from the tie-rods (see Section 19).
3  Remove the tie-rod end locknuts.
4  Remove the outer boot clamps with a pair of pliers, then cut off the inner boot clamps and discard them (see illustration).
5  Mark the location of the breather tube (if used) in relation to the rack assembly, then remove the boots and the tube. Apply some grease to the exposed ends of the toothed rack before installing new boots.
6  Install a new clamp on the inner end of the boot.
7  Apply multi-purpose grease to the groove on the tie-rod (where the outer end of the boot will ride) and the mounting grooves on the steering gear (where the inner boot end will be clamped).
8  Line up the breather tube with the marks made during removal (if applicable) and slide the new boot onto the steering gear housing.
9  Make sure the boot isn't twisted, then tighten the new inner clamp.
10  Install the outer clamps and tie-rod end locknuts.
11  Install the tie-rod ends (see Section 19).
12  Have the front-end alignment checked and, if necessary, adjusted.

20.4  The outer clamp (A) on the steering gear boot can be squeezed and removed with a pair of pliers; the inner clamp (B) must be cut off

## 10-22 SUSPENSION AND STEERING SYSTEMS

### 21 Steering linkage - inspection, removal and installation

**Note:** This Section applies only to models with a recirculating-ball steering gear.

## INSPECTION

**Refer to illustration 21.4**

1  The steering linkage connects the steering gear to the front wheels and keeps the wheels in proper relation to each other (see illustration 1.1b). The linkage consists of the Pitman arm, the idler arm, the relay rod, two adjustable tie-rods and, on some models, a steering damper. The Pitman arm, which is fastened to the steering gear shaft, moves the relay rod back-and-forth. The relay rod is supported on the other end by a frame-mounted idler arm. The back-and-forth motion of the relay rod is transmitted to the steering knuckles through a pair of tie-rod assemblies.

2  Unlock the steering wheel.

3  Raise the front of the vehicle and support it securely with jackstands placed underneath the frame rails.

4  Grasp the relay rod and the tie-rod and try to push them together and pull them apart (see illustration). If there is more than 0.039-inch of play, replace the relay rod/tie-rod assembly. Now push up and pull down on the tie-rod end; if it moves more than 0.039-inch, replace the tie-rod end.

5  Push up, then pull down on the relay rod end of the idler arm, exerting a force of approximately 25 pounds each way. Measure the total distance the end of the arm travels. If the play is greater than 0.078-inch, replace the idler arm.

6  Check for torn ballstud boots and bent or damaged linkage components.

## REMOVAL AND INSTALLATION

### Tie-rod end

7  Refer to Section 19 for the tie-rod end replacement procedure.

### Tie-rod (inner)

8  Refer to Section 19 for the tie-rod end replacement procedure.

9  Remove the under-vehicle splash shield. Also remove the skid plate bolts and the skid plate, if equipped.

10  Using a crows-foot wrench (or equivalent), detach the inner tie-rod from the relay rod (see illustration 21.4).

11  Thoroughly clean the area and the mounting threads with parts cleaner and allow everything to dry. Apply a heavy-duty thread-locker (red) to the threads, then install the inner tie-rod to the relay rod. Tighten the inner tie-rod to the torque listed in this Chapter's Specifications.

### Relay rod

12  Raise the vehicle and support it securely on jackstands. Apply the parking brake.

13  Detach the tie-rod ends from the steering knuckle arms, then remove the tie-rod ends from the tie-rods (see Section 19).

14  If equipped with a steering damper, separate the damper from the relay rod.

15  Separate the relay rod from the Pitman arm.

16  Separate the relay rod from the idler arm.

17  Installation is the reverse of the removal procedure. Use new nuts on all of the ballstuds. If the ballstuds spin when attempting to tighten the nuts, force them into the tapered holes with a large pair of pliers. Be sure to tighten all of the nuts to the torque listed in this Chapter's Specifications.

### Idler arm

**Refer to illustration 21.19**

18  Raise the vehicle and support it securely on jackstands. Apply the parking brake.

19  Loosen but do not remove the idler arm-to-relay rod nut (see illustration).

20  Separate the idler arm from the relay rod with a two jaw puller. Remove and discard the nut - don't reuse it.

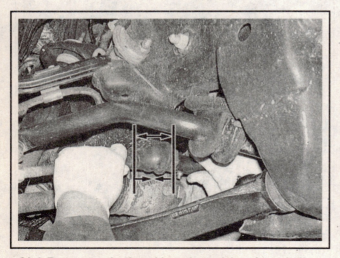

**21.4**  Try to move the tie-rod in-and-out of the relay rod - if there is more than 0.039-inch (1.0 mm) of play between the tie-rod and relay rod, replace the relay rod assembly

**21.19**  Remove the idler arm ballstud nut (A) and separate the ballstud from the relay rod with a puller, then remove the idler arm mounting nuts and bolts (B)

# SUSPENSION AND STEERING SYSTEMS       10-23

**21.28  With the steering gear secured in a vise, use a prybar to hold the Pitman arm and break loose the Pitman arm nut. Be sure to mark the relationship of the Pitman arm to the steering gear shaft before pulling it off**

**21.29  Remove the Pitman arm with a Pitman arm puller (shown here) or a heavy-duty two-jaw puller**

21   Remove the idler arm-to-frame bolts (see illustration 21.19).

22   To install the idler arm, position it on the frame and install the bolts, tightening them to the torque listed in this Chapter's Specifications.

23   Insert the idler arm ballstud into the relay rod and install a new nut. Tighten the nut to the torque listed in this Chapter's Specifications. If the ballstud spins when attempting to tighten the nut, force it into the tapered hole with a large pair of pliers.

### Pitman arm

♦ **Refer to illustrations 21.28 and 21.29**

24   Raise the vehicle and support it securely on jackstands.

25   Remove the relay rod nut from the Pitman arm ballstud. Discard the nut - don't reuse it.

26   Using a puller, separate the relay rod from the Pitman arm ballstud.

27   Remove the steering gear (see Section 22).

28   Remove the Pitman arm nut and washer and discard the nut - use a new one during installation (see illustration). Mark the Pitman arm and the steering gear shaft to ensure proper alignment at reassembly time (only if the same Pitman arm is going to be used).

29   Remove the Pitman arm with a Pitman arm puller or a two-jaw puller (see illustration).

30   Inspect the ballstud threads for damage. Inspect the ballstud seal for excessive wear. Clean the threads on the ballstud.

31   Installation is the reverse of removal. Make sure the marks you made on the Pitman arm and Pitman shaft are aligned, and be sure to use a new nut.

### Steering damper

32   Inspect the steering damper for fluid leakage. A slight film of fluid near the shaft seal is normal, but if there's excessive fluid present and it's obviously coming from the steering damper, replace the damper.

33   Inspect the steering damper bushing for excessive wear. If it's in bad shape, replace the damper.

34   To test the damper itself, disconnect it from the relay rod. Using as much travel as possible, extend and compress the damper. The resistance should be smooth and constant for each stroke. If any binding, dead spots or unusual noises are present, replace the damper.

35   Remove the damper ballstud-to-relay rod nut. Separate the damper from the relay rod using a small puller.

36   Remove the steering damper mounting bolt and nut, then remove the damper.

37   Installation is the reverse of removal.

**※※ WARNING:**

**The manufacturer recommends using new nuts on the ballstuds during installation.**

Tighten the fasteners to the torque values listed in this Chapter's Specifications.

## 22  Steering gear - removal and installation

**※※ WARNING:**

**DO NOT allow the steering column shaft to rotate with the steering gear removed or damage to the airbag system could occur. As a method of preventing the shaft from turning, wrap the seat belt around the rim of the steering wheel and buckle the belt in place.**

### MODELS WITH RACK-AND-PINION STEERING GEAR

♦ **Refer to illustrations 22.5 and 22.6**

1   Loosen the front wheel lug nuts, raise the front of the vehicle and support it securely on jackstands. Apply the parking brake. Remove the wheels.

# 10-24 SUSPENSION AND STEERING SYSTEMS

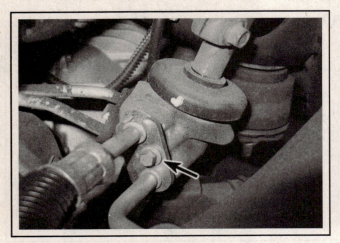

**22.5 The location of the bolt that secures the lines to the steering gear**

**22.6 Rack-and-pinion steering gear mounting fasteners**

2  Remove the under-vehicle splash shield. Also remove the skid plate bolts and the skid plate, if equipped. Refer to Section 3 and remove the stabilizer bar.

3  Mark the relationship of the intermediate shaft coupler to the steering gear input shaft and remove the pinch bolt (see illustration 18.4).

4  Detach the tie-rod ends from the steering knuckles (see Section 19).

5  Position a drain pan under the steering gear. Detach the power steering pressure and return lines from the steering gear by removing the bolt for the plate that secures them to the gear (see illustration). Cap or plug the lines to prevent leakage.

6  Unscrew the mounting nuts, remove the washers and slide the bolts forward into the frame. Lower the steering gear from the vehicle (see illustration).

7  Installation is the reverse of removal. Be sure to tighten all fasteners to the torque values listed in this Chapter's Specifications. Tighten the wheel lug nuts to the torque listed in the Chapter 1 Specifications. Check the power steering fluid level and add some, if necessary (see Chapter 1), then bleed the system as described in Section 24.

8  Have the front end alignment checked and, if necessary, adjusted.

## MODELS WITH RECIRCULATING-BALL STEERING GEAR

9  Raise the front of the vehicle and support it securely on jackstands. Apply the parking brake.

10  Remove the under-vehicle splash shield. Also remove the skid plate bolts and the skid plate, if equipped.

11  Mark the relationship of the intermediate shaft coupler to the steering gear input (stub) shaft, then remove the pinch bolt from the coupler.

12  Position a drain pan under the steering gear. Using a flare-nut wrench, unscrew the power steering lines from the steering gear.

13  Separate the relay rod from the Pitman arm (see Section 21).

14  Remove the steering gear retaining bolts from the frame rail, then detach the steering gear from the frame and remove it.

15  If you're installing a new steering gear or a new Pitman arm, remove the Pitman arm from the steering gear sector shaft (see Section 21).

16  Installation is the reverse of removal. Be sure to tighten all fasteners to the torque values listed in this Chapter's Specifications. Check the power steering fluid level and add some, if necessary (see Chapter 1), then bleed the system as described in Section 24.

## 23 Power steering pump - removal and installation

**⁕⁕ WARNING:**

DO NOT allow the steering wheel (shaft) to rotate when the intermediate shaft coupler is detached from the steering gear or damage to the airbag system could occur. As a method of preventing the shaft from turning, wrap the seat belt around the rim of the steering wheel and buckle the belt in place.

### REMOVAL

▸ Refer to illustrations 23.2, 23.6 and 23.7

1  Remove the drivebelt (see Chapter 1).

2  Using a power steering pump pulley remover, remove the pulley from the pump (see illustration).

➥ **Note: The pulley removal tool is available at most automotive parts or tool supply stores.**

# SUSPENSION AND STEERING SYSTEMS  10-25

23.2 Remove the pulley from the power steering pump with a pulley removal tool

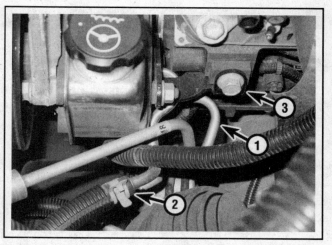

23.6 Power steering pump mounting details (V8 engine model shown - V6 model similar)

1  Pressure line
2  Feed hose
3  Rear bracket mounting bolt

3  Remove the oil pan skid plate or engine shield, if equipped.

4  Remove the intermediate shaft coupler from the steering gear input shaft and the lower intermediate shaft (see Section 18).

5  Unclip the wiring harness that's routed on the side of the pump and move it aside for clearance, if necessary.

6  Position a drain pan under the power steering pump. Disconnect the feed hose and the pressure line from the rear of the pump (see illustration). Disconnect the oil cooler hose from the pump, if equipped. Plug or cap all openings to prevent contaminants from entering any part of the system.

➡ **Note:** *The manufacturer suggests siphoning as much fluid from the reservoir as possible before disconnecting the line or hose(s) from the pump.*

7  Remove the pump mounting fasteners (see illustration 23.6 and the accompanying illustration) and lift the pump from the vehicle, taking care not to spill fluid on the painted surfaces.

➡ **Note:** *On V6 engine models, remove the rear bracket nuts (one on the lower front of the pump and one at the rear of the pump).*

## INSTALLATION

♦ Refer to illustration 23.10

8  Position the pump in the mounting bracket and install the mounting fasteners; tighten them to the torque listed in this Chapter's Specifications.

9  Connect the pressure line and feed hose to the pump, making sure that all connections are secure.

10  Press the pulley onto the shaft using a pulley installer tool (see illustration).

11  Install the drivebelt and engine shield (if equipped).

12  Fill the power steering reservoir with the recommended fluid (see Chapter 1) and bleed the system following the procedure described in the next Section.

23.7 The location of the power steering pump mounting bolts (V8 engine model shown - V6 model similar)

23.10 Press the pulley onto the shaft using a pulley installation tool - don't attempt to drive it on with a hammer or push it on with a traditional press!

## 24 Power steering system - bleeding

1  Following any operation in which the power steering fluid lines have been disconnected, the power steering system must be bled to remove all air and obtain proper steering performance.

2  With the front wheels in the straight ahead position, check the power steering fluid level and, if low, add fluid until it reaches the COLD mark on the dipstick or in the middle of the cross-hatched marks.

3  Raise the front of the vehicle just enough for the wheels to clear the ground and support it securely on jackstands. Apply the parking brake.

4  Turn the key to the ON position with the engine off, then turn the steering wheel fully in each direction (stop-to-stop) 12 times.

5  Check the fluid level and add more if necessary to reach the COLD fill mark.

6  Start the engine and allow it to warm up.

7  Turn the steering wheel from side-to-side, without hitting the stops and confirm that there is no noise coming from the system due to aeration of the fluid.

8  Road test the vehicle to be sure the steering system is functioning normally and noise free.

9  Recheck the fluid level to be sure it's up to the HOT mark on the dipstick while the engine is at normal operating temperature. Add fluid if necessary (see Chapter 1).

## 25 Wheels and tires - general information

▶ Refer to illustration 25.1

Most vehicles covered by this manual are equipped with metric-size fiberglass or steel belted radial tires (see illustration), or inch-pattern light truck tires. Use of other size or type of tires may affect the ride and handling of the vehicle. Don't mix different types of tires, such as radials and bias belted, on the same vehicle as handling may be seriously affected. It's recommended that tires be replaced in pairs on the same axle, but if only one tire is being replaced, be sure it's the same size, structure and tread design as the other.

Because tire pressure has a substantial effect on handling and wear, the pressure on all tires should be checked at least once a month or before any extended trips (see Chapter 1).

Wheels must be replaced if they're bent, dented, leak air, have elongated bolt holes, are heavily rusted, out of vertical symmetry or if the lug nuts won't stay tight. Wheel repairs that use welding or peening are not recommended.

Tire and wheel balance is important to the overall handling, braking and performance of the vehicle. Unbalanced wheels can adversely affect handling and ride characteristics as well as tire life. Whenever a tire is installed on a wheel, the tire and wheel should be balanced by a shop with the proper equipment.

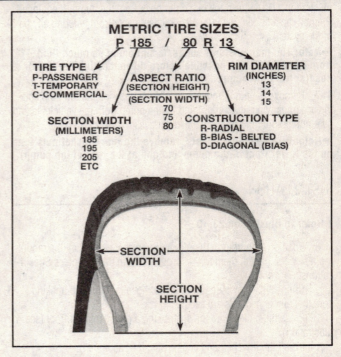

25.1 Metric tire size code

## 26 Front end alignment - general information

**Refer to illustration 26.1**

A front end alignment refers to the adjustments made to the front wheels so they're in proper angular relationship to the suspension and the ground (see illustration). Front wheels that are out of proper alignment not only affect steering control, but also increase tire wear.

Getting the proper front wheel alignment is a very exacting process, one in which complicated and expensive machines are necessary to perform the job properly. Because of this, you should have a technician with the proper equipment perform these tasks. We will, however, use this space to give you a basic idea of what is involved with front end alignment so you can better understand the process and deal intelligently with the shop that does the work.

Toe-in is the turning in of the front wheels. The purpose of a toe specification is to ensure parallel rolling of the front wheels. In a vehicle with zero toe-in, the distance between the front edges of the wheels will be the same as the distance between the rear edges of the wheels. The actual amount of toe-in is normally only a fraction of an inch. Toe-in is adjusted by turning the tie-rod in the tie-rod end to lengthen or shorten the tie-rod. Incorrect toe-in will cause the tires to wear improperly by making them scrub against the road surface.

Camber is the tilting of the front wheels from vertical when viewed from the front of the vehicle. When the wheels tilt out at the top, the camber is said to be positive (+). When the wheels tilt in at the top the camber is negative (-). The amount of tilt is measured in degrees from the vertical and this measurement is called the camber angle. This angle affects the amount of tire tread which contacts the road and compensates for changes in the suspension geometry when the vehicle is cornering or traveling over an undulating surface. Camber is adjusted by turning the upper control arm pivot bolts, one way or the other, in equal amounts.

Caster is the tilting of the top of the front steering axis from vertical. A tilt toward the rear is positive caster and a tilt toward the front is negative caster. Caster is adjusted by turning the upper control arm pivot bolts, one way or the other, in opposite directions.

When making adjustments to the front end alignment, the caster is set first, then the camber, then the toe-in.

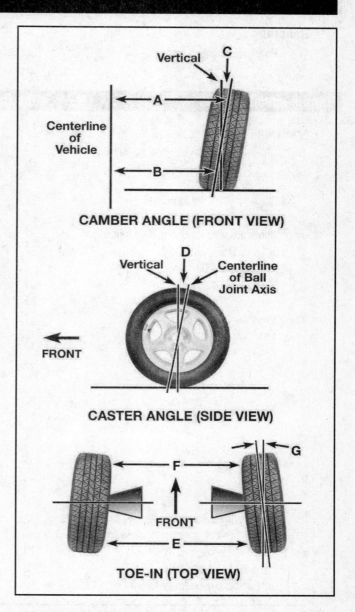

**26.1 Front end alignment details**

A minus B = C (degrees camber)
D = degrees caster
E minus F = toe-in (measured in inches)
G = toe-in (expressed in degrees)

# 10-28 SUSPENSION AND STEERING SYSTEMS

## Specifications

### General
Power steering fluid type — See Chapter 1

### Torque specifications — Ft-lbs (unless otherwise indicated)

#### Front suspension
Shock absorber upper mounting nut(s)
    1500 models — 37
    2500 models
        2007 through 2010 — 37
        2011 and later — 48
Shock absorber lower mounting fasteners
    1500 models — 37
    2500 models
        2007 through 2010 — 59
        2011 and later — 89
Upper control arm pivot bolt nuts — 140
Lower control arm pivot bolt nuts — 129
Balljoint-to-steering knuckle nut *
    Upper — 37
    Lower
        2007 through 2010 — 92
        2011 and later
            Step 1 — 37
            Step 2 — Tighten an additional 95-degrees
Stabilizer bar
    Link nut — 17
    Clamp bolts — 37
Hub/bearing assembly-to-steering knuckle bolts — 133
Torsion bar crossmember-to-frame (bushing) bolt
    2007 through 2010 — 70
    2011 and later — 92
Torsion bar crossmember-to-link bolt — 70

#### Rear suspension
Shock absorber mounting nuts/bolts — 85
Stabilizer bar
    1500 models
        Link nuts — 55
        Clamp bolts — 24
    3500 models
        Link to frame bolt — 70
        Link to bar nut — 39
        Clamp bolts — 24

* New nut must be used

# SUSPENSION AND STEERING SYSTEMS

Leaf spring
    U-bolts/nuts **
        1500 models        74
        2500 and 3500 models
            2008 and earlier        118
            2009 and later
            Step 1
                2009 and 2010        103
                2011 and later        74
            Step 2        Tighten an additional 180 degrees
    Spring-to-frame bolt/nut (front)
        2007 through 2010        148
        2011 and later        162
    Spring-to-hanger nut (front - 2500 and 3500 models)
        Step 1        125
        Step 2        Tighten an additional 48 degrees
    Rear shackle-to-spring bolt/nut        70
    Rear shackle-to-hanger bolt/nut        70
Trailing arm bolts/nuts
    Upper arm mounting nut/bolt
        Step 1        74
        Step 2        Tighten an additional 55 degrees
    Lower arm
        Arm-to-frame bolt
            Step 1        63
            Step 2        Tighten an additional 55 degrees
        Arm-to-axle bolt
            Step 1        74
            Step 2        Tighten an additional 95 degrees
Track bar bolts/nuts
    Bar-to-frame bolt
        Step 1        63
        Step 2        Tighten an additional 125 degrees
    Bar-to-axle bolt
        Step 1        74
        Step 2        Tighten an additional 100 degrees
** New U-bolt(s) and nut(s) must be used

## Steering

Steering gear-to-frame bolts/nuts (rack-and-pinion)
    Left side        148
    Right side        74
Steering gear-to-frame bolts (recirculating ball)        110
Intermediate shaft pinch bolt        33
Tie-rod end-to-steering knuckle nut        44
Inner tie-rod-to-steering gear        74

# 10-30 SUSPENSION AND STEERING SYSTEMS

**Torque specifications (continued)** — Ft-lbs (unless otherwise indicated)

### Steering (continued)

| | |
|---|---|
| Steering linkage (non-rack-and-pinion) | |
|   Pitman arm-to-input shaft | |
|     2007 through 2010 | 184 |
|     2011 and later | 273 |
|   Pitman arm-to-relay rod nut | |
|     2007 through 2010 | 46 |
|     2011 and later | 92 |
|   Idler arm mounting bolts/nuts | |
|     2007 through 2010 | 78 |
|     2011 and later | 122 |
|   Idler arm ballstud nut | |
|     2007 through 2010 | 46 |
|     2011 and later | 83 |
| Steering damper-to-relay rod nut | 30 |
| Steering damper-to-frame bolt/nut | 30 |

### Steering column

| | |
|---|---|
| Steering wheel nut | 30 |
| Steering column bracket nuts/bolts | 20 |
| Upper intermediate shaft-to-steering column shaft nut | 46 |
| Upper intermediate shaft-to-lower intermediate shaft bolt | 37 |
| Intermediate shaft coupler-to-steering gear pinch bolt | 35 |

### Power steering pump

| | |
|---|---|
| Mounting nuts (V6 engines) | 37 |
| Mounting bolts (V8 engines) | 37 |

### Wheel lug nuts

See Chapter 1

# 11 BODY

**Section**
1 General information
2 Body - maintenance
3 Vinyl trim - maintenance
4 Upholstery and carpets - maintenance
5 Body repair - minor damage
6 Body repair - major damage
7 Hinges and locks - maintenance
8 Windshield and fixed glass - replacement
9 Radiator grille - removal and installation
10 Hood - removal, installation and adjustment
11 Hood latch and release cable - removal and installation
12 Bumpers - removal and installation
13 Front fender - removal and installation
14 Cowl cover - removal and installation
15 Door trim panels - removal and installation
16 Door - removal and installation
17 Door latch, lock cylinder and handles - removal and installation
18 Door window glass - removal and installation
19 Door window glass regulator - removal and installation
20 Mirrors - removal and installation
21 Tailgate - removal and installation
22 Tailgate latch and handle - removal and installation
23 Liftgate and liftgate glass - removal and installation
24 Liftgate panels, latch and support struts - removal and installation
25 Center console - removal and installation
26 Dashboard trim panels - removal and installation
27 Steering column covers - removal and installation
28 Instrument panel - service positioning
29 Seats - removal and installation

# 11-2 BODY

## 1 General information

### GENERAL INFORMATION

The vehicles covered by this manual are built with body-on-frame construction. The frame is a hydro-formed ladder-type, consisting of C-shaped center rails welded to boxed front and rear sections, with both welded-in and bolt-in crossmembers.

The Pick-up body is in two separate sections, the cab and the bed, while the SUV body incorporates the cab, back-seat area and cargo compartment in one unitized structure.

Certain components are particularly vulnerable to accident damage and can be unbolted and repaired or replaced. Among these parts are the hood, doors, seats, tailgate, liftgate, bumpers and front fenders.

Only general body maintenance practices and body panel repair procedures within the scope of the do-it-yourselfer are included in this Chapter.

### FASTENERS AND TRIM REMOVAL

♦ Refer to illustrations 1.7a through 1.7h and 1.8

There is a variety of plastic fasteners used to hold trim panels, splash shields and other parts in place in addition to typical screws, nuts and bolts. Once you are familiar with them, they can usually be removed without too much difficulty.

The proper tools and approach can prevent added time and expense to a project by minimizing the number of broken fasteners and/or parts.

The following illustrations show various types of fasteners that are typically used on most vehicles and how to remove and install them (see illustrations).

Trim panels are typically made of plastic and their flexibility can help during removal. The key to their removal is to use a tool to pry

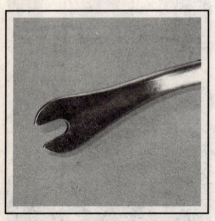

**1.7a** This tool is designed to remove special fasteners. A small pry tool used for removing nails will also work well in place of this tool

**1.7b** This fastener is used for interior panels. A Phillips head screwdriver can be used to release the center portion, but light pressure must be used because the plastic is easily damaged. Once the center is up, the fastener can be pried from its hole

**1.7c** Here is a view with the center portion fully released. Install the fastener as shown, then press the center in to set it

**1.7d** This fastener is used for interior panels. The center portion is pressed in to release the fastener

**1.7e** Here is a view of the fastener in the released position

**1.7f** Here is a view of the fastener in a reset position for installation

# BODY 11-3

**1.7g** This fastener is used for exterior panels and shields. The center portion must be pried up fully before the entire fastener can be removed or installed. Also, the center portion remains out for installation

**1.7h** This fastener is used for exterior and interior panels. It has no moving parts. Simply pry the fastener from its hole like the claw of a hammer removes a nail. Without a tool that can get under the top of the fastener, it can be very difficult to remove

the panel near its retainers (or fasteners) to release it without damaging surrounding areas or breaking off any retainers. The retainers or fasteners will usually snap out of their designated slot or hole after pressure is applied to them. Stiff plastic tools designed for prying on trim panels can be found at specialty automotive or tool supply stores and are ideal for ongoing projects (see illustration). Tools that are tapered and wrapped in protective tape, such as a screwdriver or small pry tool, are also very effective when used with care.

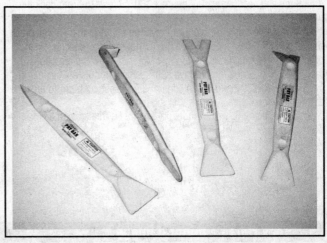

**1.8** These small plastic pry tools are the best choice of tool to use when prying off trim panels

## 2  Body - maintenance

1  The condition of your vehicle's body is very important, because the resale value depends a great deal on it. It's much more difficult to repair a neglected or damaged body than it is to repair mechanical components. The hidden areas of the body, such as the wheel wells, the frame and the engine compartment, are equally important, although they don't require as frequent attention as the rest of the body.

2  Once a year, or every 12,000 miles, it's a good idea to have the underside of the body steam cleaned. All traces of dirt and oil will be removed and the area can then be inspected carefully for rust, damaged brake lines, frayed electrical wires, damaged cables and other problems. The front suspension components should be greased after completion of this job.

3  At the same time, clean the engine and the engine compartment with a steam cleaner or water-soluble degreaser.

4  The wheel wells should be given close attention, since undercoating can peel away and stones and dirt thrown up by the tires can cause the paint to chip and flake, allowing rust to set in. If rust is found, clean down to the bare metal and apply an anti-rust paint.

5  The body should be washed about once a week. Wet the vehicle thoroughly to soften the dirt, then wash it down with a soft sponge and plenty of clean, soapy water. If the surplus dirt is not washed off very carefully, it can wear down the paint.

6  Spots of tar or asphalt thrown up from the road should be removed with a cloth soaked in tar remover or kerosene lamp oil.

7  Once every six months, wax the body and chrome trim. If a chrome cleaner is used to remove rust from any of the vehicle's plated parts, remember that the cleaner also removes part of the chrome, so use it sparingly.

# 11-4 BODY

## 3  Vinyl trim - maintenance

Don't clean vinyl trim with detergents, caustic soap or petroleum-based cleaners. Plain soap and water works just fine, with a soft brush to clean dirt that may be ingrained. Wash the vinyl as frequently as the rest of the vehicle. After cleaning, application of a high-quality rubber and vinyl protectant will help prevent oxidation and cracks. The protectant can also be applied to weather-stripping, vacuum lines and rubber hoses, which often fail as a result of chemical degradation, and to the tires.

## 4  Upholstery and carpets - maintenance

1  Every three months remove the floormats and clean the interior of the vehicle (more frequently if necessary). Use a stiff whiskbroom to brush the carpeting and loosen dirt and dust, then vacuum the upholstery and carpets thoroughly, especially along seams and crevices.

2  Dirt and stains can be removed from carpeting with basic household or automotive carpet shampoos available in spray cans. Follow the directions and vacuum again, then use a stiff brush to bring back the nap of the carpet.

3  Most interiors have cloth or vinyl upholstery, either of which can be cleaned and maintained with a number of material-specific cleaners or shampoos available in auto supply stores. Follow the directions on the product for usage, and always spot-test any upholstery cleaner on an inconspicuous area (bottom edge of a backseat cushion) to ensure that it doesn't cause a color shift in the material.

4  After cleaning, vinyl upholstery should be treated with a protectant.

➡ **Note:** Make sure the protectant container indicates the product can be used on seats - some products may make a seat too slippery.

❊❊ **CAUTION:**

**Do not use protectant on vinyl-covered steering wheels.**

5  Leather upholstery requires special care. It should be cleaned regularly with saddlesoap or leather cleaner. Never use alcohol, gasoline, nail polish remover or thinner to clean leather upholstery.

6  After cleaning, regularly treat leather upholstery with a leather conditioner, rubbed in with a soft cotton cloth. Never use car wax on leather upholstery.

7  In areas where the interior of the vehicle is subject to bright sunlight, cover leather seating areas of the seats with a sheet if the vehicle is to be left out for any length of time.

## 5  Body repair - minor damage

### PLASTIC BODY PANELS

The following repair procedures are for minor scratches and gouges. Repair of more serious damage should be left to a dealer service department or qualified auto body shop. Below is a list of the equipment and materials necessary to perform the following repair procedures on plastic body panels.

*Wax, grease and silicone removing solvent*
*Cloth-backed body tape*
*Sanding discs*
*Drill motor with three-inch disc holder*
*Hand sanding block*
*Rubber squeegees*
*Sandpaper*
*Non-porous mixing palette*
*Wood paddle or putty knife*
*Curved-tooth body file*
*Flexible parts repair material*

### Flexible panels (front and rear bumper trim)

1  Remove the damaged panel, if necessary or desirable. In most cases, repairs can be carried out with the panel installed.

2  Clean the area(s) to be repaired with a wax, grease and silicone removing solvent applied with a water-dampened cloth.

3  If the damage is structural, that is, if it extends through the panel, clean the backside of the panel area to be repaired as well. Wipe dry.

4  Sand the rear surface about 1-1/2 inches beyond the break.

5  Cut two pieces of fiberglass cloth large enough to overlap the break by about 1-1/2 inches. Cut only to the required length.

6  Mix the adhesive from the repair kit according to the instructions included with the kit, and apply a layer of the mixture approximately 1/8-inch thick on the backside of the panel. Overlap the break by at least 1-1/2 inches.

7  Apply one piece of fiberglass cloth to the adhesive and cover the cloth with additional adhesive. Apply a second piece of fiberglass cloth to the adhesive and immediately cover the cloth with additional adhesive in sufficient quantity to fill the weave.

8  Allow the repair to cure for 20 to 30 minutes at 60-degrees to 80-degrees F.

9  If necessary, trim the excess repair material at the edge.

10  Remove all of the paint film over and around the area(s) to be repaired. The repair material should not overlap the painted surface.

11  With a drill motor and a sanding disc (or a rotary file), cut a "V" along the break line approximately 1/2-inch wide. Remove all dust and loose particles from the repair area.

# BODY 11-5

12 Mix and apply the repair material. Apply a light coat first over the damaged area; then continue applying material until it reaches a level slightly higher than the surrounding finish.

13 Cure the mixture for 20 to 30 minutes at 60-degrees to 80-degrees F.

14 Roughly establish the contour of the area being repaired with a body file. If low areas or pits remain, mix and apply additional adhesive.

15 Block sand the damaged area with sandpaper to establish the actual contour of the surrounding surface.

16 If desired, the repaired area can be temporarily protected with several light coats of primer. Because of the special paints and techniques required for flexible body panels, it is recommended that the vehicle be taken to a paint shop for completion of the body repair.

## STEEL BODY PANELS

**See photo sequence**

### Repair of minor scratches

17 If the scratch is superficial and does not penetrate to the metal of the body, repair is very simple. Lightly rub the scratched area with a fine rubbing compound to remove loose paint, then use a wax-and-grease remover (available at auto parts stores) to clean the area. Rinse the area with clean water.

18 Apply touch-up paint to the scratch, using a small brush. Continue to apply thin layers of paint until the surface of the paint in the scratch is level with the surrounding paint. Allow the new paint at least two weeks to harden, then blend it into the surrounding paint by rubbing with a very fine rubbing compound. Finally, apply a coat of wax to the scratch area.

19 If the scratch has penetrated the paint and exposed the metal of the body, causing the metal to rust, a different repair technique is required. Remove all loose rust from the bottom of the scratch with a pocketknife, then apply rust inhibiting paint to prevent the formation of rust in the future. Using a rubber or nylon applicator, coat the scratched area with glaze-type filler. If required, the filler can be mixed with thinner to provide a very thin paste, which is ideal for filling narrow scratches. Before the glaze filler in the scratch hardens, wrap a piece of smooth cotton cloth around the tip of a finger. Dip the cloth in thinner and then quickly wipe it along the surface of the scratch. This will ensure that the surface of the filler is slightly hollow. The scratch can now be painted over as described earlier in this Section.

### Repair of dents

20 When repairing dents, the first job is to pull the dent out until the affected area is as close as possible to its original shape. There is no point in trying to restore the original shape completely as the metal in the damaged area will have stretched on impact and cannot be restored to its original contours. It is better to bring the level of the dent up to a point that is about 1/8-inch below the level of the surrounding metal. In cases where the dent is very shallow, it is not worth trying to pull it out at all.

21 If the backside of the dent is accessible, it can be hammered out gently from behind using a soft-face hammer. While doing this, hold a block of wood firmly against the opposite side of the metal to absorb the hammer blows and prevent the metal from being stretched.

22 If the dent is in a section of the body which has double layers, or some other factor makes it inaccessible from behind, a different technique is required. Drill several small holes through the metal inside the damaged area, particularly in the deeper sections. Screw long, self-tapping screws into the holes just enough for them to get a good grip in the metal. Now pulling on the protruding heads of the screws with locking pliers can pull out the dent.

23 The next stage of repair is the removal of paint from the damaged area and from an inch or so of the surrounding metal. This is easily done with a wire brush or sanding disk in a drill motor, although it can be done just as effectively by hand with sandpaper. To complete the preparation for filling, score the surface of the bare metal with a screwdriver or the tang of a file or drill small holes in the affected area. This will provide a good grip for the filler material. To complete the repair, see the subsection on filling and painting.

### Repair of rust holes or gashes

24 Remove all paint from the affected area and from an inch or so of the surrounding metal using a sanding disk or wire brush mounted in a drill motor. If these are not available, a few sheets of sandpaper will do the job just as effectively.

25 With the paint removed, you will be able to determine the severity of the corrosion and decide whether to replace the whole panel, if possible, or repair the affected area. New body panels are not as expensive as most people think and it is often quicker to install a new panel than to repair large areas of rust.

26 Remove all trim pieces from the affected area except those which will act as a guide to the original shape of the damaged body, such as headlight shells, etc. Using metal snips or a hacksaw blade, remove all loose metal and any other metal that is badly affected by rust. Hammer the edges of the hole in to create a slight depression for the filler material.

27 Wire-brush the affected area to remove the powdery rust from the surface of the metal. If the back of the rusted area is accessible, treat it with rust inhibiting paint.

28 Before filling is done, block the hole in some way. This can be done with sheet metal riveted or screwed into place, or by stuffing the hole with wire mesh.

29 Once the hole is blocked off, the affected area can be filled and painted. See the following subsection on filling and painting.

### Filling and painting

30 Many types of body fillers are available, but generally speaking, body repair kits which contain filler paste and a tube of resin hardener are best for this type of repair work. A wide, flexible plastic or nylon applicator will be necessary for imparting a smooth and contoured finish to the surface of the filler material. Mix up a small amount of filler on a clean piece of wood or cardboard (use the hardener sparingly). Follow the manufacturer's instructions on the package, otherwise the filler will set incorrectly.

31 Using the applicator, apply the filler paste to the prepared area. Draw the applicator across the surface of the filler to achieve the desired contour and to level the filler surface. As soon as a contour that approximates the original one is achieved, stop working the paste. If you continue, the paste will begin to stick to the applicator. Continue to add thin layers of paste at 20-minute intervals until the level of the filler is just above the surrounding metal.

32 Once the filler has hardened, the excess can be removed with a body file. From then on, progressively finer grades of sandpaper should be used, starting with a 180-grit paper and finishing with 600-grit wet-or-dry paper. Always wrap the sandpaper around a flat rubber or wooden block, otherwise the surface of the filler will not be completely flat. During the sanding of the filler surface, the wet-or-dry paper should be periodically rinsed in water. This will ensure that a very smooth finish is produced in the final stage.

These photos illustrate a method of repairing simple dents. They are intended to supplement Body repair - minor damage in this Chapter and should not be used as the sole instructions for body repair on these vehicles.

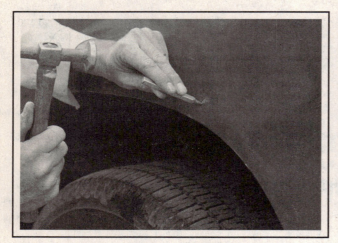

1  If you can't access the backside of the body panel to hammer out the dent, pull it out with a slide-hammer-type dent puller. In the deepest portion of the dent or along the crease line, drill or punch hole(s) at least one inch apart . . .

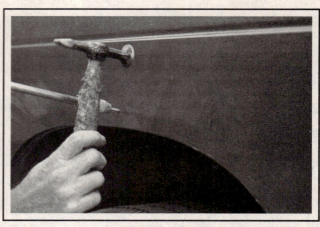

2  . . . then screw the slide-hammer into the hole and operate it. Tap with a hammer near the edge of the dent to help 'pop' the metal back to its original shape. When you're finished, the dent area should be close to its original contour and about 1/8-inch below the surface of the surrounding metal

3  Using coarse-grit sandpaper, remove the paint down to the bare metal. Hand sanding works fine, but the disc sander shown here makes the job faster. Use finer (about 320-grit) sandpaper to feather-edge the paint at least one inch around the dent area

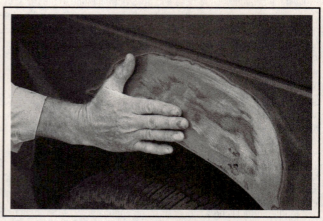

4  When the paint is removed, touch will probably be more helpful than sight for telling if the metal is straight. Hammer down the high spots or raise the low spots as necessary. Clean the repair area with wax/silicone remover

5  Following label instructions, mix up a batch of plastic filler and hardener. The ratio of filler to hardener is critical, and, if you mix it incorrectly, it will either not cure properly or cure too quickly (you won't have time to file and sand it into shape)

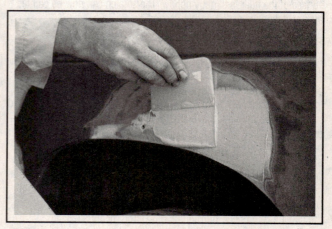

6  Working quickly so the filler doesn't harden, use a plastic applicator to press the body filler firmly into the metal, assuring it bonds completely. Work the filler until it matches the original contour and is slightly above the surrounding metal

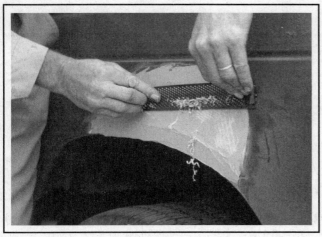

7 Let the filler harden until you can just dent it with your fingernail. Use a body file or Surform tool (shown here) to rough-shape the filler

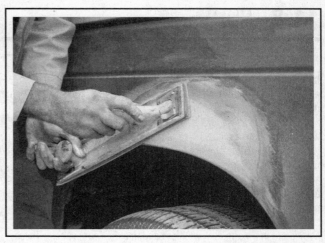

8 Use coarse-grit sandpaper and a sanding board or block to work the filler down until it's smooth and even. Work down to finer grits of sandpaper - always using a board or block - ending up with 360 or 400 grit

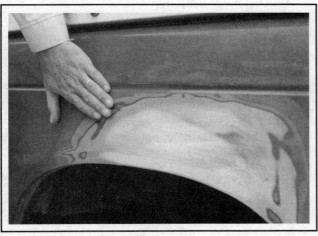

9 You shouldn't be able to feel any ridge at the transition from the filler to the bare metal or from the bare metal to the old paint. As soon as the repair is flat and uniform, remove the dust and mask off the adjacent panels or trim pieces

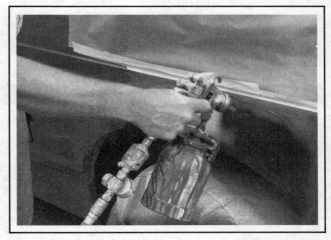

10 Apply several layers of primer to the area. Don't spray the primer on too heavy, so it sags or runs, and make sure each coat is dry before you spray on the next one. A professional-type spray gun is being used here, but aerosol spray primer is available inexpensively from auto parts stores

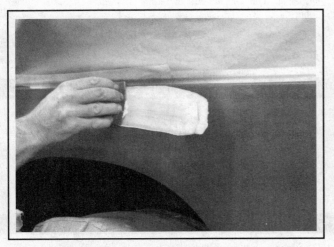

11 The primer will help reveal imperfections or scratches. Fill these with glazing compound. Follow the label instructions and sand it with 360 or 400-grit sandpaper until it's smooth. Repeat the glazing, sanding and respraying until the primer reveals a perfectly smooth surface

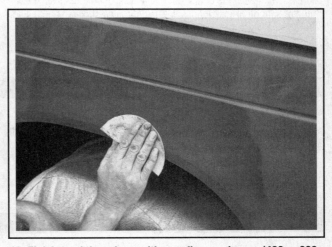

12 Finish sand the primer with very fine sandpaper (400 or 600-grit) to remove the primer overspray. Clean the area with water and allow it to dry. Use a tack rag to remove any dust, then apply the finish coat. Don't attempt to rub out or wax the repair area until the paint has dried completely (at least two weeks)

# 11-8 BODY

33 At this point, the repair area should be surrounded by a ring of bare metal, which in turn should be encircled by the finely feathered edge of good paint. Rinse the repair area with clean water until all of the dust produced by the sanding operation is gone.

34 Spray the entire area with a light coat of primer. This will reveal any imperfections in the surface of the filler. Repair the imperfections with fresh filler paste or glaze filler and once more smooth the surface with sandpaper. Repeat this spray-and-repair procedure until you are satisfied that the surface of the filler and the feathered edge of the paint are perfect. Rinse the area with clean water and allow it to dry completely.

35 The repair area is now ready for painting. Spray painting must be carried out in a warm, dry, windless and dust free atmosphere. These conditions can be created if you have access to a large indoor work area, but if you are forced to work in the open, you will have to pick the day very carefully. If you are working indoors, dousing the floor in the work area with water will help settle the dust that would otherwise be in the air. If the repair area is confined to one body panel, mask off the surrounding panels. This will help minimize the effects of a slight mismatch in paint color. Trim pieces such as chrome strips, door handles, etc., will also need to be masked off or removed. Use masking tape and several thickness of newspaper for the masking operations.

36 Before spraying, shake the paint can thoroughly, then spray a test area until the spray painting technique is mastered. Cover the repair area with a thick coat of primer. The thickness should be built up using several thin layers of primer rather than one thick one. Using 600-grit wet-or-dry sandpaper, rub down the surface of the primer until it is very smooth. While doing this, the work area should be thoroughly rinsed with water and the wet-or-dry sandpaper periodically rinsed as well. Allow the primer to dry before spraying additional coats.

37 Spray on the top coat, again building up the thickness by using several thin layers of paint. Begin spraying in the center of the repair area and then, using a circular motion, work out until the whole repair area and about two inches of the surrounding original paint is covered. Remove all masking material 10 to 15 minutes after spraying on the final coat of paint. Allow the new paint at least two weeks to harden, then use a very fine rubbing compound to blend the edges of the new paint into the existing paint. Finally, apply a coat of wax.

## 6 Body repair - major damage

1 Major damage must be repaired by an auto body shop specifically equipped to perform body and frame repairs. These shops have the specialized equipment required to do the job properly.

2 If the damage is extensive, the body must be checked for proper alignment or the vehicle's handling characteristics may be adversely affected and other components may wear at an accelerated rate.

3 Due to the fact that all of the major body components (hood, fenders, etc.) are separate and replaceable units, any seriously damaged components should be replaced rather than repaired. Sometimes the components can be found in a wrecking yard that specializes in used vehicle components, often at considerable savings over the cost of new parts.

## 7 Hinges and locks - maintenance

Once every 3000 miles, or every three months, the hinges and latch assemblies on the doors, hood and trunk should be given a few drops of light oil or lock lubricant. The door latch strikers should also be lubricated with a thin coat of grease to reduce wear and ensure free movement. Lubricate the door and trunk locks with spray-on graphite lubricant.

## 8 Windshield and fixed glass - replacement

Replacement of the windshield and fixed glass requires the use of special fast-setting adhesive/caulk materials and some specialized tools and techniques. These operations should be left to a dealer service department or a shop specializing in glasswork. On SUV models, the fixed glass also includes the quarter windows.

## 9 Radiator grille - removal and installation

▸ Refer to illustrations 9.1, 9.2 and 9.4

➡ Note: Refer to Section 1 for fastener types and removal techniques.

1 Open the hood and remove the plastic panel over the radiator, as equipped (see illustration).

2 Remove the mounting fasteners at the top of the grille that attach it to the radiator support (see illustration).

# BODY 11-9

3 Remove the bumper cover (see Section 12).

➥Note: *This Step is not necessary on Silverado models. On Sierra models, only the top trim of the bumper cover requires removal. However, the inner fenderwell liners will have to be removed.*

4 On Silverado models, disengage the grille's retaining tabs at the bottom of each corner by pulling each corner outward directly (see illustration). Remove the grille.

5 On Sierra models, remove any remaining lower mounting fasteners.

6 On all other models, remove the grille from the bumper cover by detaching the mounting screws.

7 Installation is the reverse of removal.

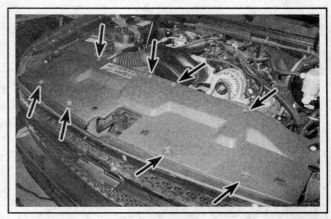

9.1 Remove the plastic fasteners, then remove the radiator panel (2007 Silverado model shown)

9.2 Remove the grille mounting fasteners (2007 Silverado model shown)

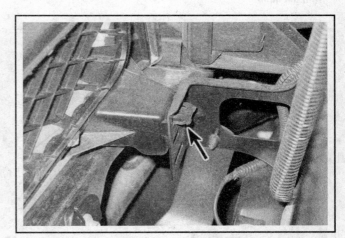

9.4 Location of a lower grille retaining tab on a Silverado

## 10 Hood - removal, installation and adjustment

### REMOVAL AND INSTALLATION

♦ Refer to illustrations 10.2, 10.4 and 10.5

➥Note: *The hood is heavy and somewhat awkward to remove and install - at least two people should perform this procedure.*

1 Use blankets or pads to cover the cowl area of the body and the fenders. This will protect the body and paint as the hood is lifted off.

2 If a hood hinge is going to be replaced, scribe alignment marks around the hinge flanges to insure proper hood alignment during installation (paint or a permanent-type felt-tip marker also will work for this) (see illustration). Skip this Step if the hood is going to be replaced.

3 Disconnect any electrical connectors attached to the hood (such as the under-hood light, ground wire, etc.).

10.2 Mark the hood around the hinge plates to aid alignment during reinstallation

# 11-10 BODY

**10.4 Pry the small clip off the end of the strut to detach it from the hood**

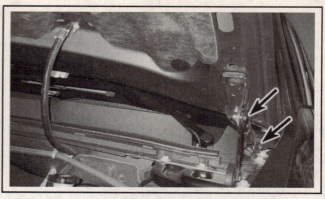

**10.5 Remove these bolts to detach the hood if the original hood and hood hinges are going to be reinstalled**

4  Some models may be equipped with a hood strut. Detach the strut from the hood by removing the small clip on the end, then pull the strut end off of the ball mount (see illustration).

### ✴ WARNING:

**Have an assistant support the weight of the hood while the strut is being removed.**

5  Have an assistant support the weight of the hood. Remove the hinge-to-hood bolts and remove the hood.

➡ **Note:** Alternatively, remove the pivot bolts and the hinge-to-spring bolts if the original hood and hinges are going to be reused. Remove the ends of the cowl cover to access the pivot bolts (see Section 14) (see illustration). Hood alignment will be preserved for reinstallation using this method of removal.

6  Installation is the reverse of removal.

## ADJUSTMENT

▸ **Refer to illustrations 10.10 and 10.11**

7  Fore-and-aft and side-to-side adjustment of the hood is done by moving the hood in relation to the hinge flanges after loosening the bolts.

8  Scribe or trace a line around the entire hinge plate so you can judge the amount of movement (paint or a permanent-type felt-tip marker also will work for this) (see illustration 10.2).

9  Loosen the nuts and move the hood into correct alignment. Move it only a little at a time. Tighten the hinge nuts and carefully lower the hood to check the alignment.

10  Adjust the hood latch so the hood closes securely. Detach the latch from the hood latch support bracket (see Section 11) in order to flatten the alignment tabs to allow movement of the latch (see illustration).

➡ **Note:** This type of adjustment is mostly necessary when structural damage has occurred and moved components out of alignment. It's best to see if the next Step will remedy the hood alignment before performing this Step.

11  Adjust the hood bumpers so that the hood is flush with the fenders when closed (see illustration).

12  The hood latch assembly, as well as the hinges, should be periodically lubricated with white lithium-base grease to prevent sticking and wear.

**10.10 Remove the hood latch mounting bolts (A) and detach the latch from the support bracket. Flatten the alignment tabs (B), then reattach the latch. The mounting bolts will have to be loosened and retightened during the adjustment process. Position the latch so that the hood is aligned when it's fully closed**

**10.11 Twist the rubber bumpers in or out to make fine adjustments to the hood's closed height**

# BODY 11-11

## 11 Hood latch and release cable - removal and installation

### ⚠ WARNING:

The models covered by this manual are equipped with Supplemental Restraint Systems (SRS), more commonly known as airbags. Always disable the airbag system before working in the vicinity of any airbag system components to avoid the possibility of accidental deployment of the airbags, which could cause personal injury (see Chapter 12).

### LATCH

▶ **Refer to illustration 11.4**

1  Remove the plastic panel over the radiator, as equipped (see Section 9).

2  On models with a plastic grille support mounted in front of the hood latch (not the grille itself), the support will have to be removed. Both headlight housings will require removal before the support can be removed (see Chapter 12).

3  Follow the wiring harness from the latch to the electrical connector mounted nearby, then disconnect it.

4  Remove the mounting bolts and detach the latch assembly (see illustration 10.10). Remove the cable from the latch (see illustration).

➡ **Note: If the alignment tabs on the hood latch support bracket have been removed, mark the relation of the latch to the hood latch support bracket before removing the latch.**

5  Installation is the reverse of removal. If necessary, adjust the latch so the hood engages securely when closed and the hood bumpers are slightly compressed (see Section 10).

### RELEASE CABLE

▶ **Refer to illustrations 11.7a, 11.7b, 11.8, 11.9a and 11.9b**

6  Remove the latch and disconnect the release cable from the hood latch assembly (see Step 1).

7  Working in the passenger compartment, remove the driver's kick panel to expose the hood latch release cable and handle (see illustrations).

8  Separate the cable end from the handle and use pliers to release the cable housing end from the tab on the handle (see illustration).

➡ **Note: If the handle assembly must be replaced, remove its mounting bolt and detach it.**

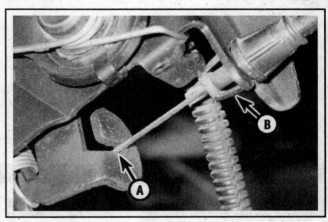

11.4  To disconnect the cable from the hood latch assembly, guide the cable end (A) off the lever and use pliers to disengage the cable housing end (B) from the assembly

11.7a  Firmly grasp the door sill trim and pull it up

11.7b  Pull out the driver's kick panel while guiding the release handle through the opening

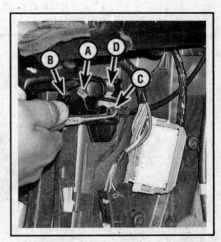

11.8  Work the cable housing end (A) out of the notch (B), then use pliers to twist the housing end (C) out of the bracket - to detach the hood latch release cable handle, remove the retaining nut (D)

## 11-12 BODY

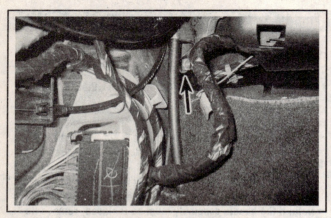

**11.9a Location of the parking brake pedal assembly lower mounting fastener**

**11.9b Location of the parking brake pedal assembly upper mounting fasteners**

9 On truck models, the cable is routed behind the parking brake pedal assembly. Loosen the mounting fasteners to the parking brake pedal assembly enough to allow the cable to pull through behind it (see illustrations).

10 Working in the engine compartment, locate the cable grommet at the firewall and pull it away from the firewall along with the cable.

11 Detach the cable from the mounting points within the engine compartment (as necessary), then pull the cable completely out of the hole in the firewall.

12 Guide the replacement cable through the firewall from the passenger compartment. Continue to guide the cable through the engine compartment until the grommet meets the firewall. Push the grommet into the hole in the firewall and seat it. Pull the grommet from the engine compartment side as necessary.

13 The remainder of installation is the reverse of removal.

## 12 Bumpers - removal and installation

### ✶✶ WARNING:

**The models covered by this manual are equipped with Supplemental Restraint Systems (SRS), more commonly known as airbags. Always disable the airbag system before working in the vicinity of any airbag system components to avoid the possibility of accidental deployment of the airbags, which could cause personal injury (see Chapter 12).**

1 Bumpers on truck models are chrome-plated steel, with black or color-matched plastic trim. Bumpers on SUV models are typically flexible plastic covers reinforced by metal framework.

### TRUCK MODELS

#### Front bumper

▸ Refer to illustrations 12.4a and 12.4b

2 Support the bumper with a jack or jackstand. Disconnect the electrical connectors at the fog lights, if equipped.

3 Refer to Section 9 and remove the radiator grille. Remove the fasteners securing the fender liners to the front bumper (see Section 13).

4 With an assistant supporting the bumper, remove the bolts/nuts retaining the bumper to the frame (see illustrations).

5 Disconnect any wiring harnesses or other components that would interfere with bumper removal and detach the bumper.

6 Installation is the reverse of removal.

**12.4a With the grille removed, there is access to the upper bumper mounting bolts (other components removed for clarity)**

**12.4b Remove the bolts for the bumper brace (left side shown - right similar)**

# BODY  11-13

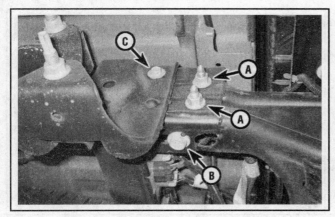

**12.9 Remove these bolts and nuts (left side shown) to remove the rear bumper with its brackets attached - (A) are the plate nuts, (B) is the lower frame rail bolt and (C) is the side frame rail bolt**

**12.14a Remove the bumper cover lower mounting fasteners (Tahoe model shown - other models similar)**

### Rear bumper

▶ Refer to illustration 12.9

7  Support the bumper with a jack or jackstand.

8  Disconnect the electrical connectors for the license plate lights and any other wiring harnesses or components that would interfere with the bumper removal.

9  With an assistant supporting the bumper, remove the bolts, nuts and stud plates retaining the bumper to the frame (see illustration).

➡ **Note:** *On models with a locking access hole for the spare-tire-lowering handle, pull the bumper straight back until the plastic tube on the back of the bumper clears the metal tube.*

10  Installation is the reverse of removal.

## SUV MODELS

### Front bumper cover

▶ Refer to illustrations 12.14a and 12.14b

11  Remove both headlight housings (see Chapter 12).

12  Remove the fasteners securing the fender liners to the front bumper (see Section 13).

13  Disconnect the electrical connectors at the fog lights, if equipped (see Chapter 12).

14  Detach the bumper cover at the top of each fenderwell and inside at the bottom on each side (see illustration). Remove the fasteners above the grill attaching the bumper cover to the radiator support (see illustration).

15  Disconnect any wiring harnesses or any other components that would interfere with the bumper cover removal and remove the cover.

16  Installation is the reverse of removal.

**12.14b Remove the bumper cover upper mounting fasteners (Tahoe model shown - other models similar)**

### Rear bumper cover

▶ Refer to illustrations 12.18, 12.19 and 12.21

➡ **Note:** *Refer to Section 1 for general information on removing body fasteners.*

17  Remove both rear fender liners (see illustration 13.7).

18  Remove the two lower fasteners at the bottom of the bumper cover (see illustration).

**12.18 Remove the lower bumper cover fasteners (Tahoe model shown - other models similar)**

**12.19 Remove the three fasteners from the inside of the bumper cover at the top edge (vicinity shown) (Tahoe model shown - other models similar)**

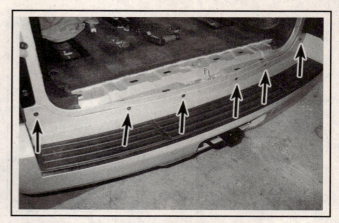

**12.21 Remove the fasteners from the top of the bumper cover**

19  Remove the three fasteners on the top and inside on each side of the bumper cover (see illustration).
20  Disconnect the electrical connectors for the rear back-up sensors, if equipped.
21  Remove the fasteners from the top of the bumper cover (see illustration).

22  Disconnect any wiring harnesses or any other components that would interfere with the bumper cover removal and remove the cover.
23  Installation is the reverse of removal.

## 13 Front fender - removal and installation

◆ Refer to illustration 13.7

➜ Note: Refer to Section 1 for fastener types and removal techniques.

1  Disconnect the cable from the negative battery terminal (see Chapter 5, Section 1). Loosen the wheel lug nuts, raise the vehicle, support it securely on jackstands and remove the front wheel.
2  Remove the headlight housing (see Chapter 12).
3  Remove the radiator grille (see Section 9).
4  Remove the end of the cowl (see Section 14).
5  Remove the elbow-shaped brace at the rear of the fender in the engine compartment.

6  Remove the hood, then remove all hinge components from the fender (see Section 10).
7  Remove the plastic fasteners and screws for the inner fenderwell liner, then remove it (see illustration). Refer to Section 1 for fastener types and removal techniques.

### LEFT FENDER

◆ Refer to illustration 13.8

8  Remove the auxiliary battery tray mounting fasteners (see illustration).

**13.7 Remove the plastic fasteners and the fenderwell liner (Silverado model shown - other models similar)**

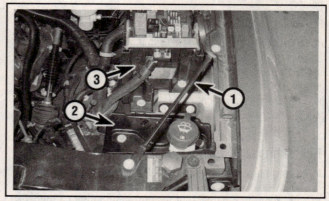

**13.8 Front fender (left) details (Silverado model shown - other models similar)**

1  Rod brace
2  Auxiliary battery tray
3  Fuse and relay box

# BODY    11-15

**13.15** Remove the lower fender mounting bolt at the rocker panel

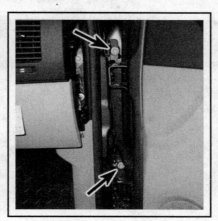

**13.16** Remove these fender bolts (arrows) in the door jamb

**13.17** Remove the front fender mounting bolts

9  Detach the fuse and relay box mounting fasteners but leave it in position.

10  Remove the rod brace between the fender and the radiator support.

11  Detach the hood release cable from the clips on the engine compartment side of the fender near the top.

## RIGHT FENDER

12  Remove the air filter housing (see Chapter 4), then unbolt its mounting tray and remove it from the engine compartment.

13  Remove the battery and battery tray (see Chapter 5).

14  Unbolt the surge tank from the fender (see Chapter 3).

## BOTH FENDERS

♦ Refer to illustrations 13.15, 13.16 and 13.17

15  Remove the lower fender mounting bolt at the rocker panel (see illustration).

16  Open the door and remove the fender mounting bolts from inside the door jamb (see illustration).

17  Remove the front fender mounting bolts (see illustration).

18  Unbolt the upper brace at the rear of the fender.

19  Detach the fender. It's a good idea to have an assistant support the fender while it's being moved away from the vehicle to prevent damage to the surrounding body panels.

20  Installation is the reverse removal. Tighten or install all fasteners securely.

## 14  Cowl cover - removal and installation

♦ Refer to illustrations 14.4, 14.5a and 14.5b

1  Mark the position of both windshield wiper blades on the windshield using tape or a grease pencil.

2  Remove the wiper arms and the radio antenna on the right side, if equipped (see Chapter 12).

3  Remove the retaining clips at the front of the cowl and the plastic fastener in the center.

4  At each side of the cowl, remove the side covers by pulling them up and off (see illustration).

5  Remove the cowl grille fasteners, remove the windshield washer hose and coupler, then detach the cowl grille from the vehicle (see illustrations).

6  Installation is the reverse of removal. Make sure to align the wiper blades with the marks made during removal.

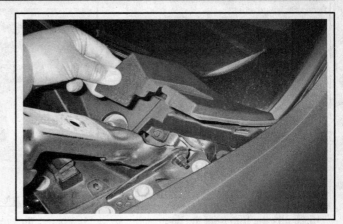

**14.4** The end sections of the cowl cover snap in place without fasteners

## 11-16 BODY

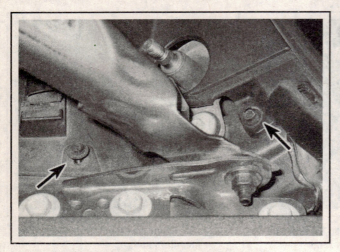

14.5a Remove the cowl fasteners (left side shown)

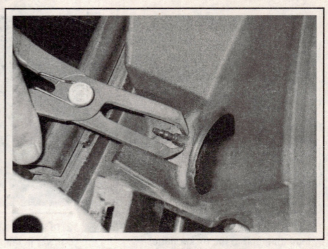

14.5b Once the cowl cover is raised, release the hose coupler

## 15 Door trim panels - removal and installation

➡ Note: Refer to Section 1 for general information on removing trim panels.

### CONVENTIONAL DOORS (PICK-UP AND SUV)

◆ Refer to illustrations 15.2a, 15.2b, 15.3a, 15.3b, 15.3c, 15.4a, 15.4b, 15.4c, 15.5, 15.6 and 15.7

1  On models with power door locks and/or power windows, disconnect the cable from the negative battery terminal (see Chapter 5, Section 1).

2  On models equipped with an additional door grab handle, pry the handle trim off to expose the mounting fasteners, then remove them and the handle (see illustrations)

3  On manual window models, remove the window crank (see illustration). On power window models, pry the switch plate from the door panel and disconnect the electrical connectors (see illustrations). Pull the door lock button up to the unlocked position, then use a small screwdriver to release the tab and pull the button up and off.

4  Pry the screw cover caps off and remove all door trim panel retaining screws (see illustrations).

➡ Note: There are two screws at the main handle area of the armrest and one behind the door release lever.

5  Remove the trim cover on the opposite side of the outside mirror (see illustration).

6  Pry the bottom and sides of the trim panel away from the door while pulling outward to release the door panel fasteners from the door (see illustration).

7  With the fasteners disengaged, raise the trim panel up to release the top of the panel from the door. Carefully rotate the trim to access the door release lever cable, then detach it from the lever (see illustration). Proceed to Step 13.

15.2a Pry the trim from the door grab handle (Silverado model shown - other models similar) . . .

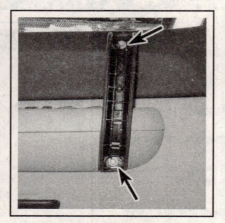

15.2b . . . then remove the door grab handle mounting fasteners

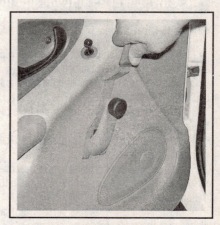

15.3a Use a hooked tool or a window crank removal tool like this one to remove the retaining clip, then detach the window crank handle

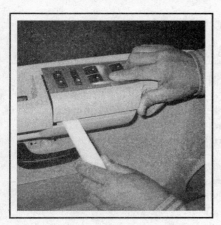

15.3b  Pry the switch plate up . . .

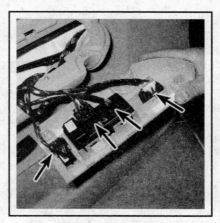

15.3c  . . . and disconnect the electrical connectors from the switches

15.4a  Pry the small trim covers to expose the mounting screws

15.4b  This model has a release handle above the armrest with a mounting screw beneath the cover

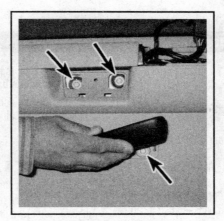

15.4c  Mounting screw locations

15.5  Pry the trim panel to expose the outside mirror mounting fasteners

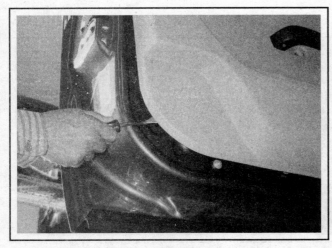

15.6  Pry the trim panel away from the bottom and side of the door, then lift the panel up to release it

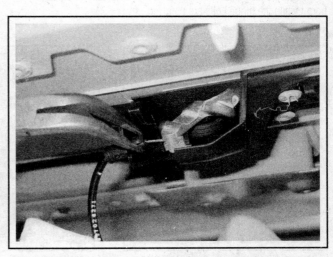

15.7  Detach the door lever release cable at the bottom of the assembly to free the panel from the door completely

## ACCESS DOORS (EXTENDED-CAB PICK-UPS)

8  Carefully pry the power window switch out of the armrest, disconnect the electrical connector, then remove the switch, if equipped.

9  Pry the trim panels from each side of the window starting at the bottom of each panel.

10  Pry the screw cover caps off and remove all door trim panel retaining screws (see illustrations 15.4a, 15.4b and 15.4c).

➡ **Note: There are two screws at the main handle area of the armrest and one behind the door release lever in front of the armrest.**

11  Pry the bottom and sides of the trim panel away from the door while pulling outward to release the door panel fasteners from the door.

12  With the fasteners disengaged, raise the trim panel up to release the top of the panel from the door. Detach the door release lever cable from the assembly.

## ALL DOORS

13  Disconnect any other wiring harness connectors, then remove the trim panel from the vehicle.

14  Prior to installing the door panel, be sure that all of the fasteners are intact and in the correct position on the panel and not the door; some fasteners may have been left in the door or broken during removal.

15  Installation is the reverse of removal. Don't forget to reconnect any electrical connectors and reattach the door release lever cable. Carefully align the door trim panel and fasteners with the door, then press the panel firmly to seat the fasteners into the holes in the door.

## 16  Door - removal, installation and adjustment

◆ Refer to illustrations 16.3, 16.4, 16.7a, 16.7b, 16.9a and 16.9b

➡ **Note: Refer to Section 1 for general information on removing trim panels.**

1  Remove the trim panel/fuse box cover from the left side of the instrument panel (see Chapter 12).

2  On front doors, remove the appropriate door sill trim and kick panel (see illustrations 11.7a and 11.7b).

3  Disconnect the electrical connectors for the door wiring harness (see illustration), then detach it from the instrument panel so that it can be removed with the door.

4  Detach the rubber conduit from the lower pillar area by releasing its retaining tabs, then guide the wiring harness out through the hole (see illustration).

5  Place a jack under the door or have an assistant on hand to support it when the hinge bolts are removed.

➡ **Note: If a jack is used, place a rag between it and the door to protect the door's painted surfaces.**

6  Remove the door check strap retaining fastner

7  Remove the hinge fasteners and carefully lift off the door (see illustrations).

8  Installation is the reverse of removal. All components can be transferred to a replacement door if necessary.

9  Rear doors are removed and installed as described above for front doors. The door is attached to the body "B" pillar; the electrical connector is just inside the pillar and can be disconnected after removing the rubber conduit (see illustrations).

10  On access type doors for extended cab models, open the door fully and pry the plastic trim from around the hinges rearward, then remove it. Detach the rubber conduit and disconnect any electrical connectors, if equipped. Remove the hinge-to-door bolts leaving the hinges attached to the "C" pillar.

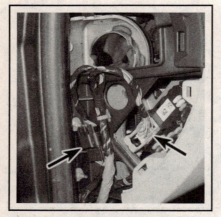

**16.3  Door wiring harness connectors at the side of the instrument panel**

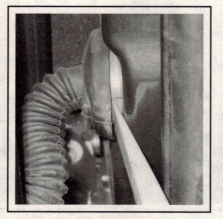

**16.4  Release the conduit retainer from the lower pillar by pressing the small retaining tabs**

**16.7a  The upper door hinge bolt**

# BODY  11-19

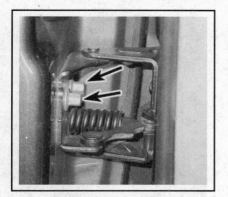

**16.7b** The lower door hinge bolts

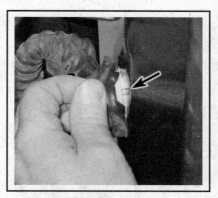

**16.9a** Release the conduit retainer from the lower pillar by pressing the small retaining tabs

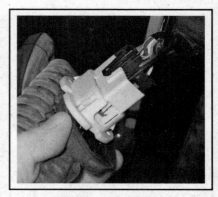

**16.9b** Pull the connector from the lower pillar and disconnect it

## 17 Door latch, lock cylinder and handles - removal and installation

♦ Refer to illustration 17.1

➡ **Note:** Refer to Section 1 for general information on removing trim panels.

1  Raise the window completely, then remove the door trim panel (see Section 15) and watershield (see illustration), where applicable.

### LATCH

#### Standard doors (pick-up and SUV)

♦ Refer to illustrations 17.2 and 17.3

2  Detach the latch rods from the lock cylinder, inside lock and outside handle (see illustration).

➡ **Note:** It may be easier to remove the latch first before detaching the latch rods or also detaching the rods from the other components first.

3  Remove the three Torx-head mounting screws, then remove the latch from the door (see illustration). Disconnect the rods (if not done previously) and any electrical connectors, detach the inside release lever cable, then separate the latch completely.

4  Installation is the reverse of the removal procedure.

### Access doors (extended-cab pick-ups)

5  The access doors use two latches, one at the top and one at the bottom. The door release handle has two cables to actuate the latch levers at the same time.

6  For upper latches, remove the trim from around the latch by prying it up from the bottom. The main door trim can stay installed.

7  Remove the mounting fasteners, remove the latch, then detach the door release handle cable from the latch.

8  Installation is the reverse of the removal procedure.

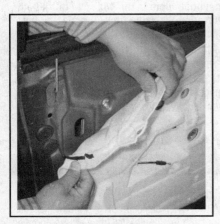

**17.1** Carefully remove the watershield by gently pulling it away from the door. Cut the adhesive with a sharp blade if necessary

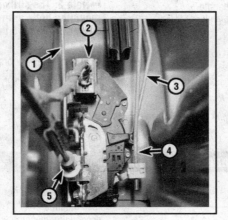

**17.2** Latch mounting details

1  Inside lock rod
2  Electrical connector
3  Lock cylinder rod
4  Outside handle rod
5  Inside release lever cable

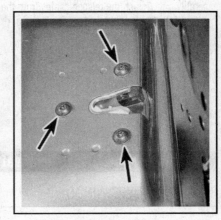

**17.3** Door latch mounting fasteners (Torx)

## 11-20 BODY

**17.10 Pry off the lock cylinder retaining clip towards the outside edge of the handle, then slip the cylinder out of the back**

**17.12 Location of the outside handle mounting bolts**

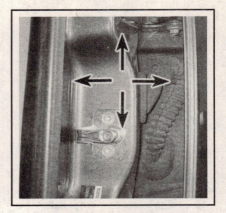

**17.17 The latch striker on the door jamb can be adjusted slightly up/down or in/out**

### LOCK CYLINDER

▶ Refer to illustration 17.10

9  Remove the outside door handle (see Step 12).
10  Pry the retaining clip off the lock cylinder and withdraw it from the door handle (see illustration).
11  Installation is the reverse of the removal procedure.

### OUTSIDE HANDLE

▶ Refer to illustration 17.12

12  Remove the hole plug in the door that provides access to one of the handle mounting bolts, then remove the bolts and detach the handle from the door (see illustration). Detach the rods for the lock cylinder and the handle itself.
13  On access doors (extended-cab pick-ups), remove the handle mounting screws in the front jamb of the door and pull the handle assembly out. Detach the release cables from the handle.
14  Installation is the reverse of the removal procedure.

### INSIDE RELEASE HANDLE

15  On standard doors, release the cable from the handle assembly (see illustration 15.7), then remove the mounting fasteners.
16  On access doors (extended-cab pick-ups), remove the mounting fasteners, then remove the release handle from the door trim panel. Detach the release cables from the lever.

### LATCH STRIKERS

▶ Refer to illustration 17.17

17  To make minor door adjustments for latch alignment, the bolts on the latch striker (which is mounted opposite the latch) can be loosened and the striker moved slightly (see illustration). Retighten the striker mounting bolts and check for proper latch operation.

➡ **Note 1: The striker can be moved up-and-down, or left-and-right to make adjustments by loosening the bolts.**

➡ **Note 2: The striker mounting plate in the body can fall if both mounting bolts are removed at the same time. If replacing the striker, make sure that one mounting bolt is in the mounting hole at all times.**

18  On access doors (extended-cab pick-ups), there are two latches and strikers for each door.
19  The striker for the upper latch is attached to the vehicle's roof. To access it for removal or adjustment, remove the trim piece around it.

## 18 Door window glass - removal and installation

### REGULATOR-TYPE WINDOWS

▶ Refer to illustrations 18.2 and 18.5

> ⚠ **CAUTION:**
> **The edges around the door frame openings are very sharp. Wear gloves for protection against cuts.**

1  Remove the door trim panel (see Section 15) and watershield (see illustration 17.1). Removal and installation of the window glass is essentially the same for front and rear doors; including access doors on extended-cab pick-ups with power windows.

➡ **Note: Temporarily reconnect the window switch or switch plate electrical connectors or attach the window regulator handle (without the retaining clip) to move the window up or down.**

2  Raise the regulator until the glass track bolts are visible in the door openings (see illustration). Tape the glass securely to the window

# BODY  11-21

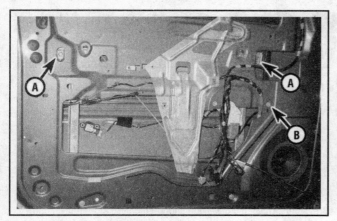

**18.2 Raise the window regulator so that the bolts (A) at the glass track can be loosened, then remove the run channel bolt (B)**

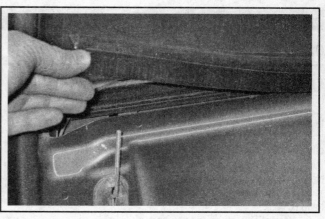

**18.5 Pull the inner seal off of the pinch-weld at the bottom of the window opening to remove it**

frame to support it in the raised position.

3  Remove the bolt from the front window glass channel, to allow the channel to move away from the glass.

4  Loosen the two bolts retaining the glass to the window regulator track (see illustration 18.2).

5  Remove the inner seal from the bottom of the window opening (see illustration).

6  Lower the regulator to a midway position while the glass is supported in the raised position.

7  Remove the tape and tilt the front of the glass down to maneuver it out of the window opening.

8  Installation is the reverse of the removal procedure.

## HINGE-TYPE WINDOWS

9  Remove the upper door trim panels from each side of the glass starting from the bottom of each panel.

➡ **Note: Refer to Section 1 for general information on removing trim panels.**

10  Tape the glass to the body to keep it in place.

11  Remove the mounting fasteners for the window latch, then remove the latch.

12  Remove the mounting fasteners for the hinges then remove them.

13  Carefully remove the tape and the window glass.

14  Installation is the reverse of the removal procedure.

## 19  Door window glass regulator - removal and installation

▶ Refer to illustration 19.3

### ※ CAUTION:

**The edges around the door frame openings are very sharp. Wear gloves for protection against cuts.**

➡ **Note: This procedure applies to front and rear door window regulators.**

1  Remove the door trim panel (see Section 15) and watershield.

2  Remove the door window glass (see Section 18).

3  Remove the large door trim bracket covering the door opening, if equipped (see illustration).

4  The window regulator assembly consists of two guides and cables. Remove the window regulator mounting fasteners (see illustration 19.3). When withdrawing the assembly from the door, push the two guides toward each other to collapse the assembly for easier removal. For access type rear doors on extended cab pick-up models, the regulator consists of one guide but removal is basically the same.

5  On power window equipped models, unplug the electrical connector for the regulator motor.

6  Remove the regulator from the door.

7  Installation is the reverse of removal.

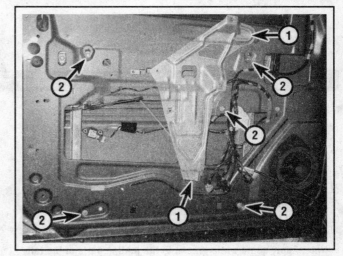

**19.3 Window regulator and door trim bracket mounting details (Silverado front door shown - other doors and models similar)**

1  Door trim bracket mounting fasteners
2  Window regulator mounting fasteners

# 11-22 BODY

## 20 Mirrors - removal and installation

### OUTSIDE MIRRORS

▶ **Refer to illustration 20.4**

1   On models with power mirrors, remove the door trim panel (see Section 15).
2   Remove the upper corner interior trim panel by pulling it straight off until the clips release from the door (see illustration 15.5).
3   On models with power mirrors, follow the wiring harness to the electrical connector and disconnect it.
4   Remove the nuts and detach the mirror from the door (see illustration).
5   Installation is the reverse of removal.

### INSIDE MIRROR

▶ **Refer to illustration 20.6**

6   On mirrors with electronic components, disconnect the electrical connector (see illustration).
7   Remove the setscrew, then slide the mirror up off the support base on the windshield.
8   Installation is the reverse of removal.
9   If the support base for the mirror has come off the windshield, it can be reattached with a special mirror adhesive kit available at auto parts stores. Clean the glass and support the base thoroughly. Follow the directions on the adhesive package, allowing the base to bond overnight before attaching the mirror.

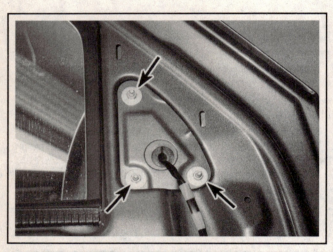

**20.4  Remove the three outside mirror mounting nuts**

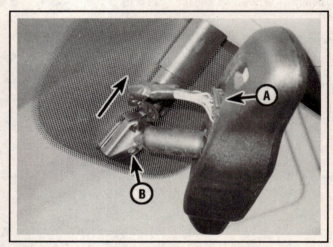

**20.6  Disconnect the electrical connector (A), then remove the setscrew (B) and slide the mirror towards the headliner and off of the support base**

## 21 Tailgate - removal and installation

▶ **Refer to illustrations 21.1, 21.2 and 21.4**

1   Open the tailgate and detach the retaining cables (see illustration).
2   Lower the tailgate until the flat on the right side hinge-pin aligns with the slot in the hinge pocket (about 45-degrees). Lift the tailgate out of the pocket (see illustration). With the help of an assistant to support the weight, withdraw the left hinge pin from the body and remove the tailgate from the vehicle.
3   Installation is the reverse of removal.

➥ **Note:** Apply some white grease to the mating parts of the tailgate hinge assembly before installation.

4   If alignment is necessary, the latches (one on each side) can be loosened and moved (see Section 22), and/or the latch striker pins can be moved slightly for the best tailgate fit (see illustration).

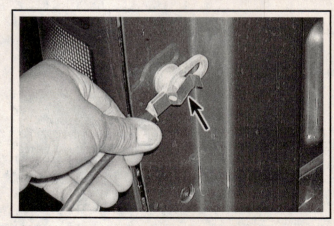

**21.1  Lift the spring retainer up and slide the cable end off the pin**

# BODY    11-23

**21.2 Lower the tailgate enough to align the open area on the right side hinge pin pocket with the straight sides of the pin, then lift the tailgate and slide it to the right and off the vehicle**

**21.4 The tailgate latch strikers (right side shown) can be loosened with a Torx bit to adjust their position slightly**

## 22 Tailgate latch and handle - removal and installation

▸ **Refer to illustrations 22.1, 22.2, 22.3 and 22.4**

1  From the bed side of the tailgate, remove the handle mounting bolts (see illustration).

2  Carefully pry or pull the handle trim cover off (see illustration).

➥**Note: Refer to Section 1 for general information on removing trim panels.**

3  Rotate the plastic retaining clips off the control rods and detach the rods from the handle (see illustration).

4  Remove the latch mounting bolts, then withdraw the latch from the end of the tailgate (see illustration). Detach any linkage to the lock cylinder, if equipped.

5  Installation is the reverse of removal. Tighten all fasteners securely.

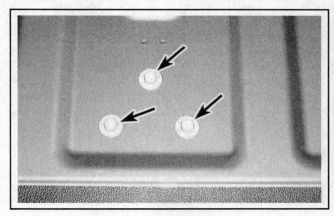

**22.1 Tailgate handle mounting bolts**

**22.2 Pry the handle trim off of the tailgate (the tailgate handle mounting bolts must be removed first - typical shown)**

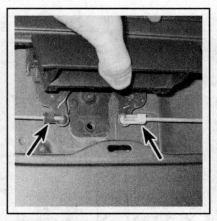

**22.3 Rotate the plastic retaining clips off of the rods to detach them from the handle levers**

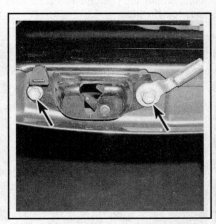

**22.4 Remove the tailgate latch mounting bolts and withdraw it with the control rod**

## 23 Liftgate and liftgate glass - removal and installation

### LIFTGATE

▸ Refer to illustrations 23.1a, 23.1b and 23.4

> **※ WARNING:**
>
> **The liftgate is heavy and awkward to hold. At least two people should perform this procedure.**

1  Open the liftgate and support it fully in this position. Pull the weatherstripping away from the door opening and remove the interior trim above the liftgate door opening (see illustrations).

➡ **Note:** Refer to Section 1 for general information on removing trim panels.

2  Detach the rubber conduit from the body by releasing its retaining tabs (see illustration 16.8a), then guide the wiring harness out through the hole and disconnect the electrical connector for the liftgate and disconnect the washer fluid hose.

3  Detach the support struts at the liftgate (see Section 24). On models equipped with a power liftgate, detach the actuator rod attached to the left side of the liftgate; it is removed in the same manner as the support struts.

4  Remove the liftgate hinge-to-body mounting bolts (see illustration). Remove the bolts while at least one assistant, preferably two, helps you hold the liftgate.

5  Installation is the reverse of the removal procedure.

➡ **Note:** On models equipped with a power liftgate, calibration may be necessary; this will have to be performed at a dealership service department or other repair shop equipped with a TECH 2 scan tool.

### LIFTGATE GLASS

▸ Refer to illustrations 23.7, 23.8 and 23.10

6  Disconnect the electrical connectors for the rear window defogger.

7  Open the liftgate glass and remove the mounting fasteners for the spoiler (see illustration).

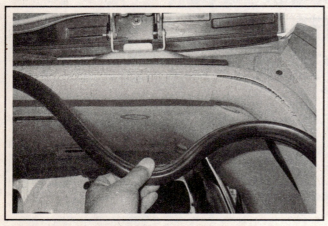

**23.1a** Carefully pull the weatherstripping away from the seam around the door opening to remove it. The adhesive may remain sticky

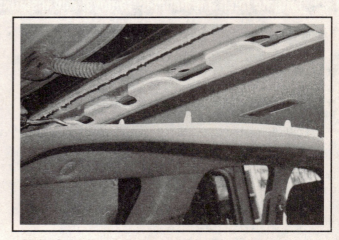

**23.1b** Carefully pry the trim panel directly away from the body to remove it

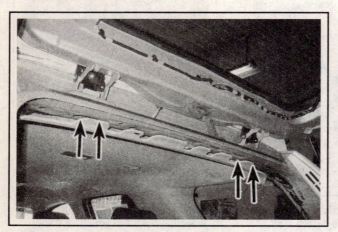

**23.4** Move the headliner aside enough to remove the liftgate hinge-to-body mounting nuts

**23.7** Locations of the spoiler mounting fasteners

# BODY  11-25

8  Ease the glass back down, then lift the spoiler off the glass (see illustration). Detach the electrical connector for the high-mount brake light, then remove the spoiler.

9  Open and support the liftgate glass, then remove both glass struts (see illustration 24.18).

10  With the glass up, remove the E-clip on the left side of each hinge (see illustration).

11  With the help of an assistant, slide the glass towards the left until the glass hinge slides off the pins on the liftgate hinges.

12  Installation is the reverse of the removal procedure. Make sure the E-clips snap firmly into the grooves on the hinge pins.

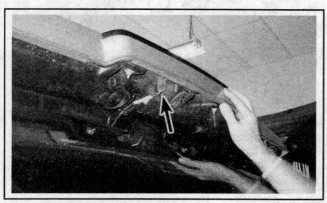

23.8  Lift the spoiler up, then disconnect the electrical connector for the high-mount brake light

23.10  Remove the E-clip at the glass hinge

## 24  Liftgate panels, latch and support struts - removal and installation

### INTERIOR TRIM PANELS

→Note: Refer to Section 1 for general information on removing trim panels.

#### Upper panel

▸ Refer to illustration 24.1

→Note: Remove the liftgate struts, if necessary.

1  Pull (or pry) each side of the panel away from the liftgate to disengage the retainers for the sides only (see illustration).

2  With the sides of the upper panel free, pull the panel towards the bottom of the liftgate to disengage the top retainers and remove it.

3  Installation is the reverse of the removal procedure.

#### Lower panel

▸ Refer to illustrations 24.5, 24.6 and 24.7

4  Remove the upper panel (see Steps 1 and 2).

5  Open the liftgate-glass, then remove the two fasteners securing the trim panel to the liftgate-glass latch (see illustration).

6  Remove the screws at the assist strap and at the liftgate grab handle (see illustration).

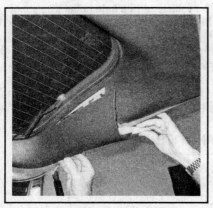

24.1  Pry the sides of the upper liftgate trim panel away from the liftgate, then pull the trim panel towards the bottom of the liftgate to disengage the retainers on the top of the panel

24.5  Remove the fasteners at the liftgate-glass latch

24.6  Remove the screws for the assist strap and the liftgate grab handle

# 11-26 BODY

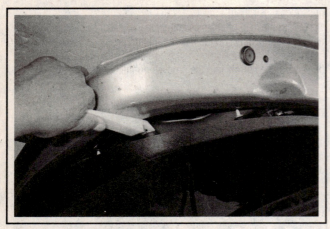

**24.7 Pry the trim panel away from the liftgate to disengage the retainers**

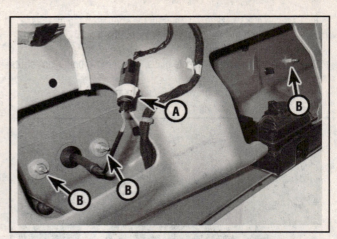

**24.11 The electrical connector for the liftgate release switch (A) and the handle mounting fasteners (B) (one fastener not visible in this photo)**

**24.14a Drill a hole through the trim panel at the index mark**

**24.14b Location of the release lever for the latch**

**24.15 Liftgate latch mounting fasteners**

7  With all external fasteners removed, pull (or pry) the bottom and sides of the panel outward and away from the liftgate (see illustration).

8  Lift the panel upwards and off of the liftgate-glass latch, then remove it.

9  Installation is the reverse of the removal procedure.

## OUTSIDE HANDLE

▶ **Refer to illustration 24.11**

10  Remove the trim panels (see Steps 1 through 9).

11  Remove the handle-to-liftgate mounting fasteners (see illustration).

12  Disconnect the electrical connector for the liftgate release switch, then remove the handle.

13  Installation is the reverse of the removal procedure.

## LATCHES

▶ **Refer to illustration 24.14a, 24.14b and 24.15**

14  Remove the interior liftgate trim panels (see Steps 1 through 9). There is an electrically operated latch mounted at the bottom edge of the liftgate where it meets the body sill. The latch is actuated by a switch mounted in the outside handle.

➡ **Note:** *If the liftgate cannot be opened due to a failed latch or switch, there is an over-ride procedure to operate the latch manually. From inside the vehicle, drill a hole in the trim panel where its marked (see illustration). Drill through the panel only. Insert a rod into the hole to move the lever near the side of the latch to open the liftgate (see illustration).*

15  To remove a latch, remove the latch mounting bolts from the bottom of the liftgate, detach the electrical connector, then remove the latch (see illustration).

16  Installation is the reverse of the removal procedure.

## SUPPORT STRUTS AND LIFTGATE ACTUATING ROD

▶ **Refer to illustration 24.18**

17  There are two support struts for the liftgate, and two for the glass. On models equipped with a power liftgate, there is an actuating rod mounted to the liftgate and linkage to a motor that is used to prop up the liftgate. Open the liftgate (or liftgate glass) and support it securely in its fully open position.

18  Release the small clip at each end of the strut, then pull the strut

# BODY 11-27

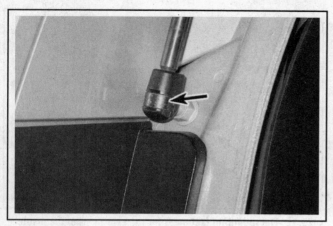

**24.18 Pry the small clip from the end of the strut, then pull the end off the ball mount**

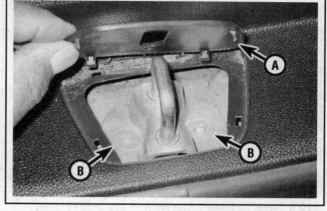

**24.21 Pry the small trim (A) from around the latch to access the mounting bolts (B)**

from the ball mount (see illustration). On models equipped with a power liftgate, remove the actuating rod in the same way.

19  Installation is the reverse of the removal procedure.

## STRIKER

▶ Refer to illustration 24.21

➡ Note: On models equipped with a power liftgate, calibration may be necessary if any adjustment is made to the striker; this will have to be performed at a dealership service department or other repair shop equipped with a TECH 2 scan tool.

20  The closing position of the liftgate can be adjusted slightly by moving either or both of the liftgate strikers.

21  Pry the small trim from around the striker and remove it (see illustration).

22  Loosen the striker mounting bolts, move the striker slightly, retighten the bolts and check the liftgate closing.

## 25  Center console - removal and installation

➡ Note: Refer to Section 1 for general information on removing trim panels.

### CENTER FLOOR CONSOLE

▶ Refer to illustrations 25.2, 25.3 and 25.4

1  Remove the front seats (see Section 29).

2  Pull or pry the front tray up and out of the console (see illustration).

3  Disconnect the electrical connectors for any of the wiring harnesses that lead to the main instrument panel (see illustration).

4  Remove the forward trim from each side of the console by pulling them directly away from the console (see illustration).

5  Carefully guide the console out of the vehicle.

6  Installation is the reverse of removal.

**25.2  Pull or pry the tray up from the console**

**25.3  Disconnect the electrical connector for the center console wiring harness**

**25.4  Remove the front trim by pulling it away from the console**

# 11-28 BODY

25.8 Mounting fasteners for the front overhead console

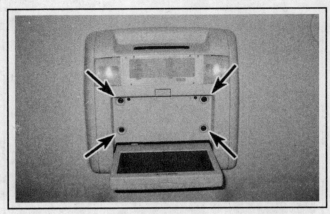

25.12 The mounting fasteners for the rear overhead console (DVD-equipped model shown)

## OVERHEAD CONSOLE

### Front

▶ Refer to illustration 25.8

7  Remove the map light lens (see Chapter 12).
8  Remove the mounting screws (see illustration).
9  Pull or pry the console directly down to release its retainers.
10  Unplug the electrical connectors and remove the console.

11  Installation is the reverse of removal.

### Rear

▶ Refer to illustration 25.12

12  If equipped with a DVD player, remove the mounting screws (see illustration).
13  Pull or pry the console directly down to release its retainers.
14  Unplug the electrical connectors and remove the console.
15  Installation is the reverse of removal.

## 26 Dashboard trim panels - removal and installation

**※※ WARNING:**

**The models covered by this manual are equipped with Supplemental Restraint Systems (SRS), more commonly known as airbags. Always disable the airbag system before working in the vicinity of any airbag system components to avoid the possibility of accidental deployment of the airbags, which could cause personal injury (see Chapter 12).**

➡ Note: Refer to Section 1 for general information on removing trim panels.

1  There are two dashboard configurations for models covered in this manual. For the sake of clarity, they will be referred to as the "Deluxe" trim or the "Standard" trim. They are easily distinguished because the Standard trim dashboard has an additional storage compartment above the glove box, while the Deluxe trim does not. Also, the Deluxe trim dashboard trim panels generally have color accents/faux woodgrain on them.

## DELUXE DASHBOARD TRIM

### Instrument cluster bezel and panel

▶ Refer to illustration 26.4

2  On models equipped with tilt steering columns, tilt the steering wheel down completely.
3  Remove the left and center air outlet panels (see Steps 9 through 11).
4  Remove the mounting fasteners for the bezel, then guide it away from the cluster (see illustration).

26.4 The bezel mounting fasteners

# BODY  11-29

**26.7 Carefully pry the center trim panel out to release its retainers**

**26.9 Pry the center air outlet panel out starting by its edge**

**26.11 Pry the left air outlet panel out starting by its edge**

**26.13 Pry the headlamp switch panel out starting by its edge**

**26.17 The instrument panel trim mounting screws beneath the bezel**

**26.18 Remove the instrument cluster trim by pulling it directly away from the top of the dash**

5  Grasp the cluster panel securely and pull it directly away from the instrument panel to detach its retainers.

6  Installation is the reverse of removal.

### Center trim panel

♦ Refer to illustration 26.7

7  Carefully pry the panel from the dashboard to release its retainers, then remove the panel (see illustration).

8  Installation is the reverse of removal.

### Center air outlet panel

♦ Refer to illustration 26.9

9  Starting at the right side of the panel, carefully pry it from the dashboard to release its retainers, then remove the panel (see illustration).

10  Installation is the reverse of removal.

### Left air outlet panel

♦ Refer to illustration 26.11

11  Carefully pry the air outlet panel directly out from the dashboard to release its retainers, then remove the panel (see illustration).

➥Note: Don't pry or pull on the center of the air outlet - pry only on the panel's edge.

12  Installation is the reverse of removal.

### Headlamp switch panel

♦ Refer to illustration 26.13

13  Carefully pry the headlamp switch panel directly out from the dashboard to release its retainers (see illustration).

14  With the panel detached from the dashboard, disconnect the electrical connectors and remove the panel.

15  Installation is the reverse of removal.

## STANDARD DASHBOARD TRIM

### Instrument cluster bezel and trim

♦ Refer to illustrations 26.17, 26.18 and 26.19

16  On models equipped with tilt steering columns, tilt the steering wheel down as far as possible.

17  Remove the trim mounting screws beneath the bezel (see illustration).

18  Grasp the trim securely and pull it directly away from the instrument panel to detach its retaining clips (see illustration).

# 11-30 BODY

19 Remove the bezel mounting screws and pull the bezel directly away from the instrument panel to detach its retaining clips (see illustration).

20 Installation is the reverse of removal.

### Center trim panel

▸ Refer to illustration 26.21

21 Carefully pry the panel directly out from the dashboard to release its retainers (see illustration).

22 With the panel detached from the dashboard, disconnect the electrical connectors and remove the panel.

23 Installation is the reverse of removal.

### Lower center trim panel

▸ Refer to illustration 26.25

24 Remove the knee bolster (see Steps 33 through 37).

**26.19** The bezel mounting screws

**26.21** Carefully pry the panel out to release its retainers

**26.25** Lower center trim panel mounting screws

**26.27** Pry the air outlet panel out by its edges

25 Remove the mounting screws, then pry the panel directly out from the dashboard to release its retainers and remove it (see illustration).

26 Installation is the reverse of removal.

### Left air outlet panel

▸ Refer to illustration 26.27

27 Carefully pry the air outlet panel directly out from the dashboard to release its retainers (see illustration).

28 With the panel detached from the dashboard, disconnect the electrical connectors and remove the panel.

29 Installation is the reverse of removal.

### Upper storage box

▸ Refer to illustration 26.30

30 Open the storage box door, then remove the mounting screws (see illustration).

31 Grip the door and the latch on the box, then pull it directly out to release its retainers.

32 Installation is the reverse of removal.

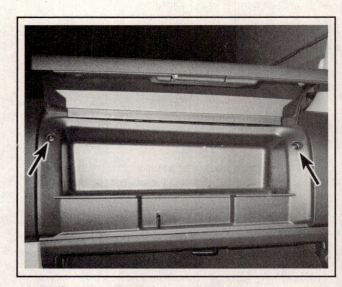

**26.30** Storage box mounting screws

# BODY  11-31

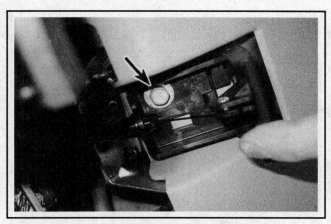

26.34 Parking brake release lever mounting bolt

26.35 Knee bolster mounting screws

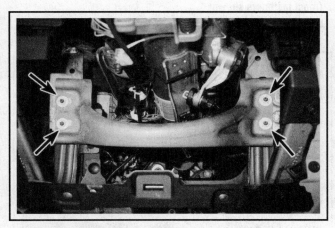

26.37 Steel knee bolster reinforcement mounting nuts

26.39 Pull the cover away from the instrument panel to release the retainers (left side shown - right side similar)

## BOTH DASHBOARD TRIMS

### Knee bolster

▸ Refer to illustrations 26.34, 26.35 and 26.37

33 Remove the side/fuse box cover panel from the left side of the instrument panel (see Step 39).

34 Remove the parking brake release lever mounting bolt (see illustration).

35 Remove the screws securing the driver's knee bolster, then pull it directly away from the dash (see illustration).

36 Detach the parking brake lever from the knee bolster by moving the lever off of the bolster to disengage its retaining tabs.

37 The steel knee bolster reinforcement can be removed by removing its mounting nuts, if necessary (see illustration).

38 Installation is the reverse of removal.

### Side cover panel

▸ Refer to illustration 26.39

39 Grasp the trim and pull it directly away from the instrument panel to detach its retaining clips (see illustration).

40 Installation is the reverse of removal.

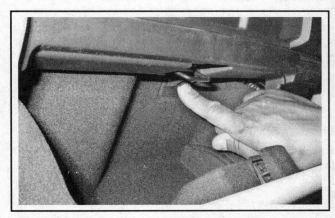

26.41 Push the stop up to allow the glove box to open completely

### Glove box

▸ Refer to illustrations 26.41 and 26.42

41 Open the glove box and push the stop up (see illustration), then guide the box downward completely.

26.42 Glove box hinge mounting screws

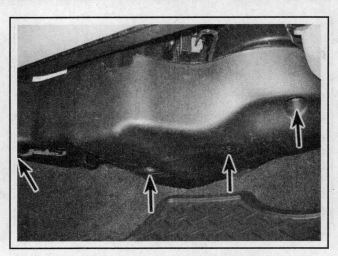

26.44 Vicinity of the insulator panel mounting fasteners (most fasteners shown)

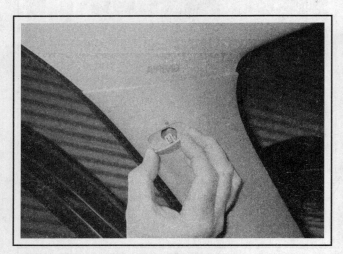

26.45a Front pillar trim mounting screw cover

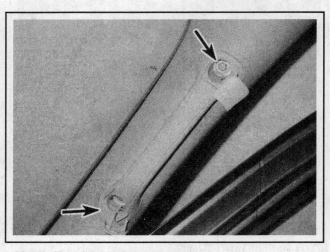

26.45b The grab handle screw covers

42 Remove the screws along the hinge of the glove box, then detach it (see illustration).
43 Installation is the reverse of removal.

### Insulator panel

▶ Refer to illustration 26.44

44 To access components under the right side of the instrument panel, remove the mounting fasteners, then remove the lower insulator panel from the vehicle (see illustration).

### Front pillar trim

▶ Refer to illustrations 26.45a and 26.45b

45 On the left (driver's) side, remove the cover over the mounting screw near the top, then remove the mounting screw (see illustration). On the right (passenger's) side, remove the covers over the grab handle mounting screws, then remove the screws and the handle (see illustration).

46 Pull the trim out then up to release the retainers on the inside and at the bottom to detach it.
47 With the trim detached from the pillar, disconnect the electrical connector for the speaker, if equipped.
48 Installation is the reverse of removal.

### Top trim panel

▶ Refer to illustrations 26.50 and 26.51

49 Remove the front pillar trim on both sides (see Steps 45 through 47).
50 Pry the trim panel up on the front edge to release those retainers, then pull the panel towards the rear to disengage the front tabs (see illustration).
51 With the panel detached from the dashboard, disconnect the electrical connector (see illustration), then remove the panel.

➡ **Note: Place tape over the light sensor so that it does not fall out accidentally.**

52 Installation is the reverse of removal

# BODY  11-33

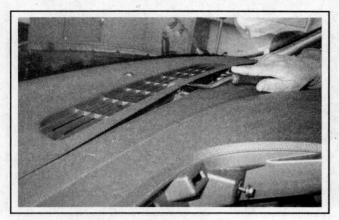

26.50  Pry the trim panel up at the front edge

26.51  Disconnect the electrical connector attached to the top trim panel

## 27  Steering column covers - removal and installation

▸ Refer to illustration 27.2

**※※ WARNING:**

**The models covered by this manual are equipped with Supplemental Restraint Systems (SRS), more commonly known as airbags. Always disable the airbag system before working in the vicinity of any airbag system components to avoid the possibility of accidental deployment of the airbags, which could cause personal injury (see Chapter 12).**

1  If equipped with a tilt-column, remove the tilt lever knob by pulling it straight out from the column. It's held to a lever by prongs.

2  Remove the mounting screws from underneath the lower half of the cover, if equipped, then separate both halves of the steering column cover (see illustration).

➡ **Note:** Some models may have screws that hold the cover halves together while others are held by clips only. Always check for the presence of screws underneath before pulling the cover apart.

3  Installation is the reverse of removal.

27.2  With the tilt lever knob and mounting screws removed (if equipped), separate the cover halves by pressing near the seam and pulling them apart (steering wheel removed for clarity)

## 28  Instrument panel - service positioning

▸ Refer to illustrations 28.11, 28.12a, 28.12b, 28.12c, 28.13a, 28.13b, 28.14a, 28.14b and 28.17

**※※ WARNING:**

**The models covered by this manual are equipped with Supplemental Restraint Systems (SRS), more commonly known as airbags. Always disable the airbag system before working in the vicinity of any airbag system components to avoid the possibility of accidental deployment of the airbags, which could cause personal injury (see Chapter 12).**

➡ **Note:** Service positioning for the instrument panel involves pivoting the instrument panel away from the firewall on the right (passenger's) side for service on the heating and air conditioning unit.

1  Turn the front wheels to the straight-ahead position and secure the steering column with the ignition lock.

2  Disconnect the cable from the negative battery terminal (see Chapter 5, Section 1).

3  Wait at least two minutes to allow the airbag system back-up power supply to become depleted (see Chapter 12).

4  Remove the front seats (see Section 29).

## 11-34 BODY

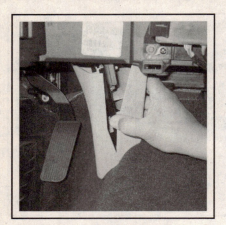

**28.11** After pulling the trim panels away from the support bracket, remove its fasteners and detach the bracket

**28.12a** Disconnect the electrical connectors above the instrument panel

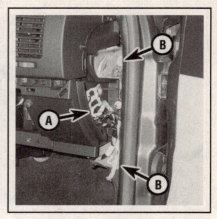

**28.12b** The location of the electrical connectors (A) on the right side of the instrument panel and the mounting bolts (B)

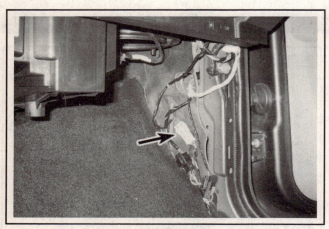

**28.12c** Disconnect the electrical connector below the instrument panel

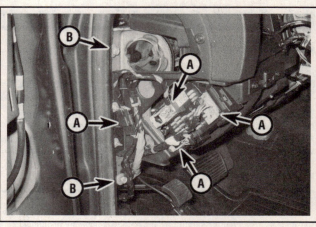

**28.13a** The location of the electrical connectors (A) on the left side of the instrument panel and the mounting bolts (B)

5  Remove the center console, if equipped (see Section 25).
6  Remove the knee bolster (see Section 26) and detach the shift cable from the steering column (see Chapter 7).
7  Disconnect the steering shaft (at the end of the steering column) from the intermediate shaft (see Chapter 10).
8  Remove the door sill trim and kick panels from each side of the vehicle (see illustrations 11.7a and 11.7b).
9  Remove the front pillar trim from each side of the vehicle (see Section 26).
10  Remove the top trim panel (see Section 26).
11  Pull the trim panels away from the instrument panel center support bracket then remove the bracket, if equipped (see illustration).
12  Working on the right (passenger's) side of the instrument panel, disconnect the electrical connectors to allow the panel to be moved away from the firewall (see illustrations).
13  Working on the left (driver's) side of the instrument panel, disconnect the electrical connectors to allow the panel to be pivoted away from the firewall (see illustrations).

➡ **Note:** It is possible that many of the electrical connectors can stay connected so long as the wiring harnesses have enough slack to allow the instrument panel to move

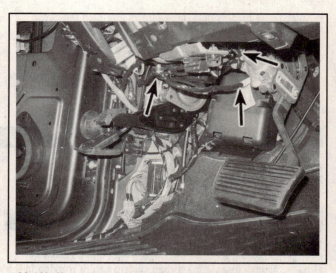

**28.13b** If necessary, detach the wiring harnesses and electrical connectors when pivoting the instrument panel

# BODY    11-35

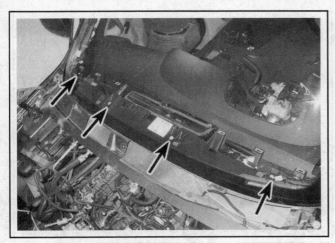

**28.14a  The location for the upper mounting bolts for the instrument panel**

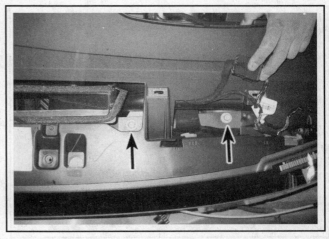

**28.14b  Additional mounting bolts for the support beneath the instrument panel**

14  Remove the upper mounting bolts for the instrument panel (see illustrations).

15  With the help of an assistant, remove the instrument panel mounting bolts on the right (passenger's) side and have the assistant hold it forward; the long guide pin will support the weight (see illustration 28.12b).

16  Loosen the instrument panel mounting bolts on the left (driver's side) (see illustration 28.13a).

17  With your assistant's help, slowly and carefully pivot the right side of the instrument panel away from the firewall. Make certain that any electrical connectors or components that hinder removal are detached to avoid damage to them while moving the instrument panel outwards. Support the instrument panel with a jackstand or equivalent (see illustration).

18  Installation is the reverse of removal. Be certain to reconnect all electrical connectors and secure their wiring harnesses during installation.

**28.17  The instrument panel in the service position supported with a jackstand**

## 29  Seats - removal and installation

### ✶✶ WARNING 1:

The models covered by this manual are equipped with Supplemental Restraint Systems (SRS), more commonly known as airbags. Always disable the airbag system before working in the vicinity of any airbag system components to avoid the possibility of accidental deployment of the airbags, which could cause personal injury (see Chapter 12).

### ✶✶ WARNING 2:

Some models are equipped with seat belt pre-tensioners, which are pyrotechnic (explosive) devices that tighten the seat belts during an impact of sufficient force. Always disable the airbag system before working in the vicinity of any restraint system component to avoid the possibility of accidental deployment of the seat belt pre-tensioners, which could cause personal injury (see Chapter 12).

➡ Note 1: The removal of seats can be very awkward due to their size and weight. It is best to have an assistant help you with these procedures.

➡ Note 2: Refer to Section 1 for general information on removing trim panels.

1  Individual front seats are offered on all models. An center front seat can be found on some models when a floor console is not installed. On truck models, a standard or 60/40 bench-type seat is used. On SUV models, the second row of seats may be a 60/40 bench-type seat or a pair of individual seats with armrests.

# 11-36 BODY

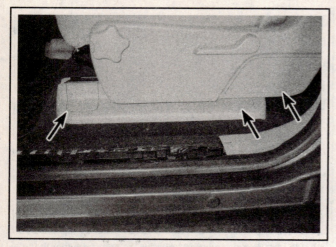

29.2 Remove the trim covers for access to the seat and seat belt mounting fasteners

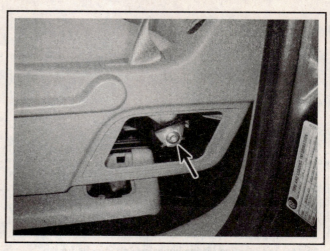

29.3 Location of the seat belt mounting bolt at the seat

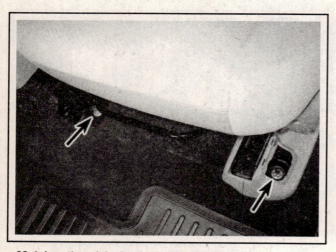

29.4 Location of the front mounting fasteners

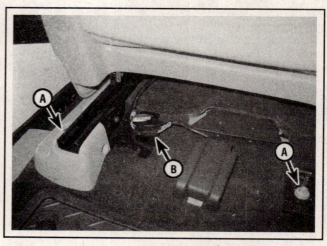

29.5 Location of the rear mounting fasteners (A) and the electrical connector (B)

29.7 The location of the electrical connector for the center seat

## FRONT SEATS (ALL MODELS)

▶ Refer to illustrations 29.2, 29.3, 29.4, 29.5 and 29.7

2  Remove the plastic trim covers to access the seat and seat belt mounting bolts (see illustration).

3  Detach the seat belt from the seat (see illustration).

4  Move the seat rearward, pull the plastic trim covering the inboard mounting fastener and remove the front mounting fasteners (see illustration).

5  Move the seat forward, pull the plastic trim covering the inboard mounting fastener and remove the rear mounting fasteners (see illustration).

6  Disconnect the electrical connector then lift the seat out of the vehicle.

7  The center seat can only be removed when the seats on each side are removed first. Detach the electrical connector before lifting the center seat out of the vehicle (see illustration).

8  Installation is the reverse of the removal procedure.

# BODY    11-37

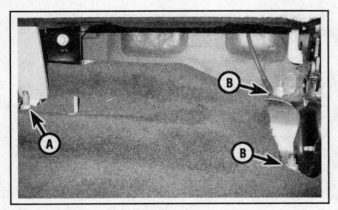

**29.9  Seat belt anchor fastener (A) and the left side seat mounting fasteners (B)**

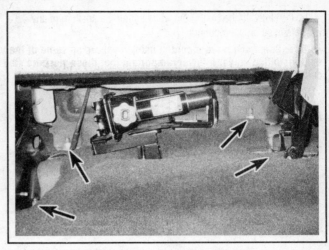

**29.10  The right side seat mounting fasteners**

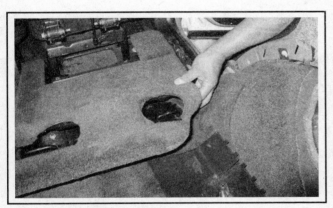

**29.13  Pull up the large floor trim from behind the seat**

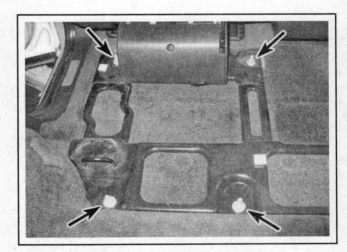

**29.14a  Rear seat mounting fasteners on the left side**

## REAR BENCH SEAT (PICK-UP MODELS)

▶ Refer to illustration 29.9 and 29.10

9  The rear seat is removed as an assembly. Remove the seat belt anchor bolt (see illustration).

10  Remove the seat bracket-to-floor fasteners and lift the rear seat assembly up and out (see illustration 29.9 and the accompanying illustration).

➡ **Note:** *Standard bench seats will have fewer mounting fasteners than the split type.*

11  With the floor mounting bolts removed, lift the assembly up to disengage the seat hooks from the body, then remove it.

➡ **Note:** *The 60/40 bench-type seats can be divided if necessary.*

12  Installation is the reverse of the removal procedure.

## SECOND ROW SEATS (SUV MODELS)

### Individual and typical bench seats

▶ Refer to illustrations 29.13, 29.14a and 29.14b

13  Fold the seatback forward and remove the large floor trim from behind the seat by pulling it directly up (see illustration).

14  Remove the mounting fasteners and disconnect any electrical connectors (see illustrations).

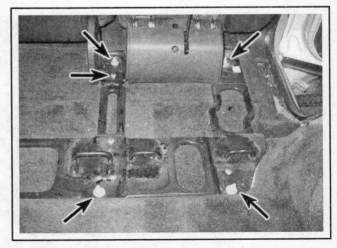

**29.14b  Rear seat mounting fasteners on the right side**

# 11-38 BODY

15 Carefully lift the seat(s) out while noting the position of any washers used on the mounting studs.

➡ **Note: Some seats are mounted using washers on some of the seat mounting studs. It is very important that these washers stay configured in their original positions.**

16 Installation is the reverse of removal.

## Bench seats on some Tahoe models

17 Fold the seatback forward and remove the trim covering on the sides and front of the seat by pulling them away from the seat.

18 Remove the hinge bolts that connect the seat to the brackets coming from the floor of the vehicle and disconnect any electrical connectors.

19 Carefully lift the seat assemblies out of the vehicle.

20 Installation is the reverse of removal.

## THIRD ROW SEATS (SUV MODELS)

21 Removal of the third row seat(s) is a standard feature which does not require any specialized tools. Simply fold the seatback and seat forward and pull up the release lever in the bottom rear, then remove the seat rearward.

# 12
# CHASSIS ELECTRICAL SYSTEM

**Section**

1. General information
2. Electrical troubleshooting - general information
3. Fuses - general information
4. Circuit breakers - general information
5. Relays - general information and testing
6. Keyless entry remote transmitter - battery replacement and programming
7. Turn signals and hazard flashers - general information
8. Turn signal/hazard flasher/multi-function switch - replacement
9. Key lock cylinder and ignition switch - replacement
10. Dashboard switches - replacement
11. Instrument cluster - removal and installation
12. Radio and speakers - removal and installation
13. Antenna - removal and installation
14. Headlight housing - replacement
15. Headlights - adjustment
16. Headlight bulb - replacement
17. Bulb replacement
18. Wiper motors - replacement
19. Horn - replacement
20. Rear window defogger - check and repair
21. Electric side view mirrors - general information
22. Cruise control system - general information
23. Power window system - general information
24. Power door lock system - general information
25. Daytime Running Lights (DRL) - general information
26. Airbag system - general information and precautions
27. Wiring diagrams - general information

# 12-2 CHASSIS ELECTRICAL SYSTEM

## 1 General information

The electrical system is a 12-volt, negative ground type. Power for the lights and all electrical accessories is supplied by a lead/acid-type battery, which is charged by the alternator.

This Chapter covers repair and service procedures for the various electrical components not associated with the engine. Information on the battery, alternator and starter motor can be found in Chapter 5.

It should be noted that when portions of the electrical system are serviced, the negative battery cable should be disconnected from the battery to prevent electrical shorts and/or fires.

## 2 Electrical troubleshooting - general information

▶ **Refer to illustrations 2.5a and 2.5b**

1  A typical electrical circuit consists of an electrical component, any switches, relays, motors, fuses, fusible links or circuit breakers related to that component and the wiring and connectors that link the component to both the battery and the chassis. To help you pinpoint an electrical circuit problem, wiring diagrams are included at the end of this Chapter.

2  Before tackling any troublesome electrical circuit, first study the appropriate wiring diagrams to get a complete understanding of what makes up that individual circuit. Noting whether other components related to the circuit are operating correctly, for instance, can often narrow down the location of potential trouble spots. If several components or circuits fail at one time, chances are the problem is in a fuse or ground connection, because several circuits are often routed through the same fuse and ground connections.

3  Electrical problems usually stem from simple causes, such as loose or corroded connections, a blown fuse, a melted fusible link or a failed relay. Visually inspect the condition of all fuses, wires and connections in a problem circuit before troubleshooting the circuit.

4  If test equipment and instruments are going to be utilized, use the diagrams to plan ahead of time where you will make the necessary connections in order to accurately pinpoint the trouble spot.

5  For electrical troubleshooting, you'll need a circuit tester, voltmeter or a 12-volt bulb with a set of test leads, a continuity tester and a jumper wire, preferably with a circuit breaker incorporated, which can be used to bypass electrical components (see illustrations). Before attempting to locate a problem with test instruments, use the wiring diagram(s) to decide where to make the connections.

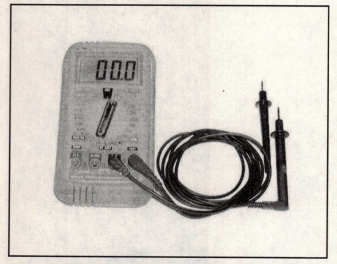

2.5a  The most useful tool for electrical troubleshooting is a digital multimeter that can check volts, amps, and test continuity

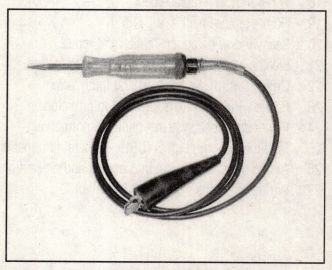

2.5b  A simple test light is a very handy tool for testing voltage

# CHASSIS ELECTRICAL SYSTEM    12-3

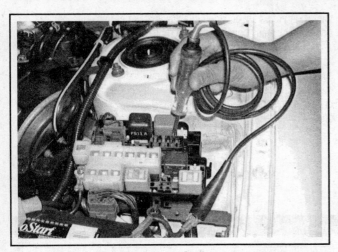

2.6  To use a test light, clip the lead to a known good ground, then use the pointed probe to test connectors, wires or electrical sockets. If the bulb lights, the circuit being tested has battery voltage

2.9  With a multimeter set to the ohm scale, you can check resistance across two terminals. When checking for continuity, a low reading indicates continuity, a high reading indicates high resistance and an infinite reading indicates lack of continuity

## VOLTAGE CHECKS

▶ Refer to illustration 2.6

6  Voltage checks should be performed if a circuit is not functioning properly. Connect one lead of a circuit tester to either the negative battery terminal or a known good ground. Connect the other lead to a connector in the circuit being tested, preferably nearest to the battery or fuse (see illustration). If the tester bulb lights, voltage is present, which means that the part of the circuit between the connector and the battery is problem free. Continue checking the rest of the circuit in the same fashion. When you reach a point at which no voltage is present, the problem lies between that point and the last test point with voltage. The problem can usually be traced to a loose connection.

➥Note: **Keep in mind that some circuits receive voltage only when the ignition key is in the ACC or RUN position.**

## FINDING A SHORT

7  One method of finding shorts in a live circuit is to remove the fuse and connect a test light in place of the fuse terminals (fabricate two jumper wires with small spade terminals, plug the jumper wires into the fuse box and connect the test light). There should be no voltage present in the circuit. Move the suspected wiring harness from side-to-side while watching the test light. If the bulb lights, there is a short to ground somewhere in that area, probably where the insulation has rubbed through.

## GROUND CHECK

8  Perform a ground test to check whether a component is properly grounded. Disconnect the battery and connect one lead of a continuity tester or multimeter (set to the ohm scale), to a known good ground. Connect the other lead to the wire or ground connection being tested. If the resistance is low (less than 5 ohms), the ground is good. If the bulb on a self-powered test light does not light, the ground is not good.

## CONTINUITY CHECK

▶ Refer to illustration 2.9

9  A continuity check determines whether there are any breaks in a circuit (whether it can no longer carry current from the voltage source to ground). With the circuit off (no power in the circuit), a self-powered continuity tester or multimeter can be used to check the circuit. Connect the test leads to both ends of the circuit (or to the power end and a good ground), and if the test light comes on the circuit is passing current properly (see illustration). If the resistance is low (less than 5 ohms), there is continuity; if the reading is 10,000 ohms or higher, there is a break somewhere in the circuit. The same procedure can be used to test a switch, by connecting the continuity tester to the switch terminals. With the switch turned to ON, the test light should come on (or low resistance should be indicated on a meter).

## FINDING AN OPEN CIRCUIT

10  When diagnosing for possible open circuits, it is often difficult to locate them by sight because the connectors hide oxidation or terminal misalignment. Merely wiggling a connector on a sensor or in the wiring harness may correct the open circuit condition. Remember this when an open circuit is indicated when troubleshooting a circuit. Intermittent problems may also be caused by oxidized or loose connections.

11  Electrical troubleshooting is simple if you keep in mind that all electrical circuits are basically electricity running from the battery, through the wires, switches, relays, fuses and fusible links to each electrical component (light bulb, motor, etc.) and to ground, from which it is passed back to the battery. Any electrical problem is an interruption in the flow of electricity to and from the battery.

## CONNECTORS

12  Most electrical connections on these vehicles are made with multi-wire plastic connectors. The mating halves of many connectors

# 12-4 CHASSIS ELECTRICAL SYSTEM

are secured with locking clips molded into the plastic connector shells. The mating halves of large connectors, such as some of those under the instrument panel, are held together by a bolt through the center of the connector.

13 To separate a connector with locking clips, use a small screwdriver to pry the clips apart carefully, then separate the connector halves. Pull only on the shell; never pull on the wiring harness as you may damage the individual wires and terminals inside the connectors. Look at the connector closely before trying to separate the halves. Often the locking clips are engaged in a way that is not immediately clear. Additionally, many connectors have more than one set of clips.

14 Each pair of connector terminals has a male half and a female half. When you look at the end view of a connector in a diagram, be sure to understand whether the view shows the harness side or the component side of the connector. Connector halves are mirror images of each other, and a terminal that is shown on the right side end-view of one half will be on the left side end view of the other half.

## 3  Fuses - general information

♦ **Refer to illustrations 3.1a, 3.1b and 3.2**

The electrical circuits of the vehicle are protected by a combination of fuses, circuit breakers and fusible links. Fuse and relay boxes are located in the engine compartment and inside the vehicle, in the left end of the instrument panel (see illustrations). Each of the fuses is designed to protect a specific circuit, and the various circuits are identified on the fuse panel cover. If the fuse panel cover is difficult to read, or missing, you can also refer to your owner's manual, which includes a complete guide to all fuses and relays in both fuse/relay boxes.

Miniaturized fuses are employed in the fuse blocks. If an electrical component fails, always check the fuse first. The best way to check a fuse is with a test light. Check for power at the exposed terminal tips of each fuse. If power is present on one side of a fuse but not the other, the fuse is blown. A blown fuse can also be confirmed by visually inspecting it (see illustration).

Be sure to replace blown fuses with the correct type. Fuses of different ratings are physically interchangeable, but only fuses of the proper rating should be used. Replacing a fuse with one of a higher or lower value than specified is not recommended. Each electrical circuit needs a specific amount of protection. The amperage value of each fuse is molded into the fuse body.

If the replacement fuse immediately fails, don't replace it again until the cause of the problem is isolated and corrected. In most cases, this will be a short circuit in the wiring caused by a broken or deteriorated wire.

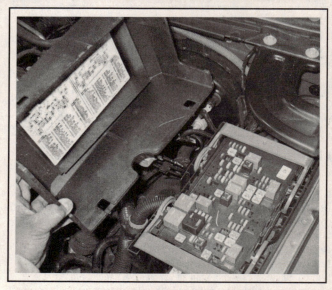

**3.1a  The engine compartment fuse and relay box is mounted on the left side of the engine compartment. The fuses and relays are identified on the backside of the cover**

**3.1b  The fuse and relay box inside the vehicle is located in the left end of the instrument panel**

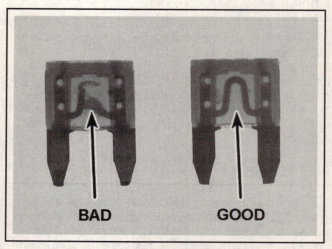

**3.2  When a fuse blows, the element between the terminals melts**

## CHASSIS ELECTRICAL SYSTEM    12-5

### 4  Circuit breakers - general information

Circuit breakers protect certain circuits, such as the power windows or heated seats. Depending on the vehicle's accessories, there may be one or two circuit breakers, located in the fuse/relay box in the engine compartment.

Because the circuit breakers reset automatically, an electrical overload in a circuit-breaker-protected system will cause the circuit to fail momentarily, then come back on. If the circuit does not come back on, check it immediately.

For a basic check, pull the circuit breaker up out of its socket on the fuse panel, but just far enough to probe with a voltmeter. The breaker should still contact the sockets.

With the voltmeter negative lead on a good chassis ground, touch each end prong of the circuit breaker with the positive meter probe. There should be battery voltage at each end. If there is battery voltage only at one end, the circuit breaker must be replaced.

Some circuit breakers must be reset manually.

### 5  Relays - general information and testing

#### GENERAL INFORMATION

1  Several electrical circuits in the vehicle use relays to transmit the electrical signal to the component. Relays use a low-current circuit (the control circuit) to open and close a high-current circuit (the power circuit). If the relay is defective, that component will not operate properly. Most relays are mounted in the engine compartment fuse/relay box, with some specialized relays located above the interior fuse box in the dash (see illustrations 3.1a and 3.1b). If a faulty relay is suspected, it can be removed and tested using the procedure below or by a dealer service department or a repair shop. Defective relays must be replaced as a unit.

#### TESTING

▶ Refer to illustrations 5.2a and 5.2b

2  Most of the relays used in these vehicles are of a type often called "ISO" relays, which refers to the International Standards Organization. The terminals of ISO relays are numbered to indicate their usual circuit connections and functions. There are two basic layouts of terminals on the relays used in these vehicles (see illustrations).

3  Refer to the wiring diagram for the circuit to determine the proper connections for the relay you're testing. If you can't determine the correct connection from the wiring diagrams, however, you may be able to determine the test connections from the information that follows.

4  Two of the terminals are the relay control circuit and connect to the relay coil. The other relay terminals are the power circuit. When the relay is energized, the coil creates a magnetic field that closes the larger contacts of the power circuit to provide power to the circuit loads.

5  Terminals 85 and 86 are normally the control circuit. If the relay contains a diode, terminal 86 must be connected to battery positive (B+) voltage and terminal 85 to ground. If the relay contains a resistor, terminals 85 and 86 can be connected in either direction with respect to B+ and ground.

6  Terminal 30 is normally connected to the battery voltage (B+) source for the circuit loads. Terminal 87 is connected to the circuit leading to the component being powered. If the relay has several alternate terminals for load or ground connections, they usually are numbered 87A, 87B, 87C, and so on.

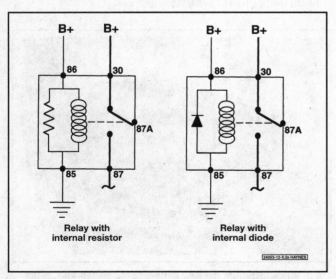

5.2a  Typical ISO relay designs, terminal numbering and circuit connections

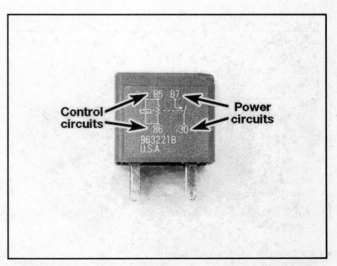

5.2b  Most relays are marked on the outside to easily identify the control circuits and the power circuits (four-terminal type shown)

## 12-6 CHASSIS ELECTRICAL SYSTEM

7  Use an ohmmeter to check continuity through the relay control coil.
   a) *Connect the meter according to the polarity shown in illustration 5.2a for one check; then reverse the ohmmeter leads and check continuity in the other direction.*
   b) *If the relay contains a resistor, resistance will be indicated on the meter, and should be the same value with the ohmmeter in either direction.*
   c) *If the relay contains a diode, resistance should be higher with the ohmmeter in the forward polarity direction than with the meter leads reversed.*
   d) *If the ohmmeter shows infinite resistance in both directions, replace the relay.*

8  Remove the relay from the vehicle and use the ohmmeter to check for continuity between the relay power circuit terminals. There should be no continuity between terminal 30 and 87 with the relay de-energized.
9  Connect a fused jumper wire to terminal 86 and the positive battery terminal. Connect another jumper wire between terminal 85 and ground. When the connections are made, the relay should click.
10 With the jumper wires connected, check for continuity between the power circuit terminals. Now, there should be continuity between terminals 30 and 87.
11 If the relay fails any of the above tests, replace it.

## 6  Keyless entry remote transmitter - battery replacement and programming

### BATTERY REPLACEMENT

▶ **Refer to illustrations 6.1, 6.2 and 6.3**

1  Carefully pry open the transmitter by inserting a screwdriver into the slot near the upper end of the unit and twisting it (see illustration).
2  Carefully pry out the old battery with a small screwdriver (see illustration).
3  Make sure that the positive side of the new battery faces down (see illustration).
4  Also make sure that the two halves of the cover snap together tightly.

**6.1  Find the small slot near the upper end of the remote transmitter, insert a small screwdriver into the slot and twist it until the two halves separate**

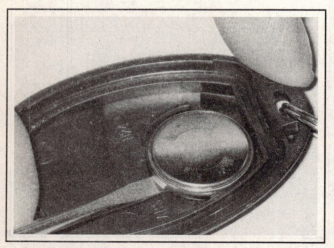

**6.2  Use a small screwdriver to carefully pry the old battery out of the transmitter**

**6.3  When installing the new battery in the transmitter, make sure that the positive side of the battery faces down (toward the transmitter body)**

# CHASSIS ELECTRICAL SYSTEM  12-7

## PROGRAMMING

➡ **Note:** You will need to use the Driver Information Center (DIC) to program the remote transmitter. The DIC buttons are located to the left or right of the instrument cluster, depending on the make and model (see your owner's manual if you're unfamiliar with the DIC). The display screen for the DIC is located on the instrument cluster. The DIC does a lot of things, but here we're interested only in how it's used to match the transmitter to the vehicle. If your vehicle is not equipped with a DIC, the transmitter must be programmed with a GM factory scan tool, so you will have to take the vehicle to a dealer for this procedure.

5  To use the DIC, turn the ignition key to ON. Press the VEHICLE INFORMATION button until PRESS TO RELEARN REMOTE KEY appears on the DIC display.
6  Press the SET/RESET button. The message REMOTE KEY LEARNING ACTIVE will appear on the DIC display.
7  Simultaneously press and hold the LOCK and UNLOCK buttons on the first transmitter for about 15 seconds. A chime will sound, indicating that the transmitter is matched to the vehicle.
8  To match other transmitters at this time, repeat Step 7. You can program up to eight transmitters.
9  To exit the programming mode, turn the ignition key to OFF.

## 7  Turn signals and hazard flashers - general information

1  There is no turn signal and hazard flasher relay on these vehicles. This function is handled by the Body Control Module (BCM).
2  If a bulb on one side of the vehicle flashes much faster than normal but the bulb at the other end of the vehicle (on the same side) doesn't light at all, the bulb that doesn't flash is probably faulty. Replace the bulb (see Section 17).

3  If both the left and right front or rear turn signal bulbs are not flashing, the turn signal and hazard flasher relay function in the BCM is probably defective. Have the BCM replaced by a dealer service department or other repair facility equipped with the proper tool. This is not a job that you can do at home because the BCM must be programmed with a factory scan tool when it's replaced.

## 8  Turn signal/hazard flasher/multi-function switch - replacement

▶ **Refer to illustrations 8.3 and 8.4**

✳✳ **WARNING:**
The models covered by this manual are equipped with a Supplemental Restraint System (SRS), more commonly known as airbags. Always disarm the airbag system before working in the vicinity of any airbag system component to avoid the possibility of accidental deployment of the airbag, which could cause personal injury (see Section 26).

1  Remove the steering wheel (see Chapter 10).
2  Remove the upper and lower steering column covers (see Chapter 11).
3  Disconnect the electrical connectors from the turn signal/hazard flasher/multi-function switch (see illustration).
4  Remove the turn signal/hazard flasher/multi-function switch mounting screws (see illustration) and remove the switch.
5  Installation is the reverse of removal.

8.3  To disconnect the electrical connectors from the turn signal multi-function switch, use a small screwdriver to disengage the locking tabs on the terminal from the lugs on the connectors

8.4  To detach the turn signal multi-function switch from the steering column, remove the two mounting screws

# 12-8  CHASSIS ELECTRICAL SYSTEM

## 9  Key lock cylinder and ignition switch - replacement

### ✷✷ WARNING:

The models covered by this manual are equipped with a Supplemental Restraint System (SRS), more commonly known as airbags. Always disarm the airbag system before working in the vicinity of any airbag system component to avoid the possibility of accidental deployment of the airbag, which could cause personal injury (see Section 26).

### KEY LOCK CYLINDER

▸ Refer to illustration 9.2

### ✷✷ CAUTION:

The following procedure is included only as a prelude to removing the ignition switch (see below). DO NOT try to replace the key lock cylinder at home. If you replace the key lock cylinder, you must have the transponder in the new key programmed by a dealer service department before the engine will start. Even if you were to use the old key in a new key lock cylinder you would still have to have it re-programmed at the dealer before the engine would start.

1  Disconnect the cable from the negative terminal of the battery (see Chapter 5, Section 1). Remove the upper and lower steering column covers (see Chapter 11).
2  Insert the ignition key in the key lock cylinder and turn the key to the START position. Insert a 1/8-inch Allen wrench, awl, punch or a small screwdriver through the hole in the casting that houses the key lock cylinder, then depress the lock cylinder release tab and pull out the lock cylinder (see illustration).
3  Insert a screwdriver into the lock cylinder housing and turn the lock cylinder gear clockwise to the START position, then let it spring back to the RUN position.

➡ **Note:** Push the electric park lock solenoid in while turning the gear. Slide the lock cylinder into the housing until it clicks into place.

4  The remainder of installation is the reverse of removal.

### IGNITION SWITCH

▸ Refer to illustrations 9.6, 9.7, 9.8, 9.9 and 9.10

5  Remove the key lock cylinder (see Steps 1 and 2).
6  Remove the theft deterrent control module from the key lock cylinder housing (see illustration), then trace the electrical harness from the module down to its electrical connectors and disconnect the main connector and the separate fused jumper connector.

**9.2  Turn the key to the Start position, insert a thin tool through this hole, depress the lock cylinder release tab and pull out the lock cylinder**

**9.6  To detach the theft deterrent control module from the key lock cylinder housing, carefully pry the lock tabs loose from the housing and pull off the module**

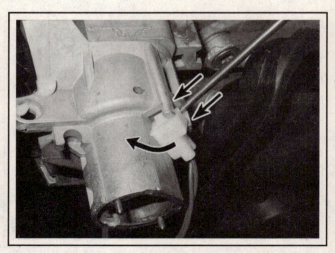

**9.7  Insert a small screwdriver between the alarm buzzer electrical connector and the housing to disengage the release tab, then turn the connector clockwise to free it from the lugs on the housing**

# CHASSIS ELECTRICAL SYSTEM  12-9

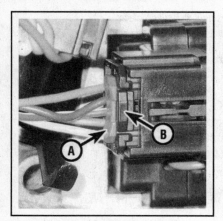

9.8 To disconnect the electrical connector from the ignition switch, slide out the connector lock (A), then depress the release tab (B) and pull off the connector

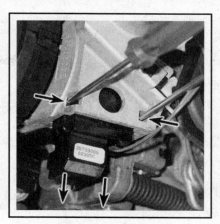

9.9 Insert a pair of screwdrivers into these two holes to free the locking tabs, then pull the switch straight down

9.10 When installing the ignition switch, make sure that the gears on top of the switch are positioned exactly as shown, or the engine might not start or the switch could cause a current drain

7  Disconnect the key alarm buzzer electrical connector from the key lock cylinder housing (see illustration).
8  Disconnect the electrical connector from the ignition switch (see illustration).
9  Remove the ignition switch (see illustration).
10  When installing the switch, make sure that the gears on top are positioned correctly (see illustration). Failure to do so will cause a misalignment between the gears on the ignition switch and the gears inside the lock cylinder housing, which could make it impossible to start the engine or could result in a battery current drain. Installation is otherwise the reverse of removal.

## 10  Dashboard switches - replacement

### ※ WARNING:

The models covered by this manual are equipped with a Supplemental Restraint System (SRS), more commonly known as airbags. Always disarm the airbag system before working in the vicinity of any airbag system component to avoid the possibility of accidental deployment of the airbag, which could cause personal injury (see Section 26).

### HEADLIGHT SWITCH/TRANSFER CASE CONTROL SWITCH

#### Headlight switch and dome light override knob (standard instrument panel)

▸ Refer to illustration 10.2

1  Remove the left air outlet plate (see Chapter 11).
2  Using three screwdrivers, disengage the three locking tabs from the three lugs on the switch (see illustration) and push the switch out of the left air outlet plate.

10.2  Use screwdrivers to disengage the locking tabs from the three lugs on the headlight switch housing (standard instrument panel)

# 12-10 CHASSIS ELECTRICAL SYSTEM

3  When installing the headlight switch, make sure that it snaps back into place and all three lugs on the switch are secured by the three locking tabs.

4  Installation is otherwise the reverse of removal.

### Headlight switch and transfer case control switch (deluxe instrument panel)

▸ **Refer to illustrations 10.5 and 10.6**

5  Carefully pry the headlight switch/transfer case control switch housing out of the air outlet plate with a trim panel removal tool (see illustration).

6  Disconnect the electrical connector(s) from the switch housing (see illustration).

7  Three lugs on the headlight switch unit are secured to the transfer case control switch housing by three locking tabs on the control switch housing. Disengage the three locking tabs from the three lugs on the headlight switch and pull the switch out of the control switch housing.

8  When installing the headlight switch into the transfer case control switch assembly, make sure it snaps into place.

9  Reconnect the electrical connector(s) to the headlight switch and, if equipped, the transfer case control switch and install the headlight switch/transfer case control switch in the air outlet plate until it snaps into place.

### HAZARD FLASHER SWITCH

10  The hazard flasher switch, which is located on top of the steering column, is an integral component of the turn signal/hazard flasher/multi-function switch (see Section 8).

### DRIVER INFORMATION CENTER (DIC) SWITCH ASSEMBLY

11  Remove the air outlet plate (standard instrument panel) or the center air outlet panel (deluxe instrument panel) as applicable (see Chapter 11).

12  Using screwdrivers, disengage the locking tabs from the lugs on the switch assembly and push the DIC switch out of the trim plate.

13  When installing the DIC switch assembly, make sure that the switch snaps into place.

14  Installation is otherwise the reverse of removal.

### AUTOMATIC TRANSFER CASE CONTROL SWITCH (STANDARD INSTRUMENT PANEL)

▸ **Refer to illustration 10.16**

➡ **Note: This procedure applies only to vehicles with a standard instrument panel, on which the transfer case control switch is located to the right of the instrument cluster, on the center trim panel. If the transfer case control switch is located on the air outlet plate to the left of the instrument cluster, refer to Step 5.**

15  Remove the center trim panel (see Chapter 11).

16  Using screwdrivers, disengage the locking tabs from the lugs on the switch assembly (see illustration) and push the switch out of the trim plate.

**10.5  To remove the switch housing for the headlight switch and, if equipped, the transfer case control switch, from the air outlet plate, carefully pry the switch housing out of the plate with a trim removal tool (deluxe instrument panel)**

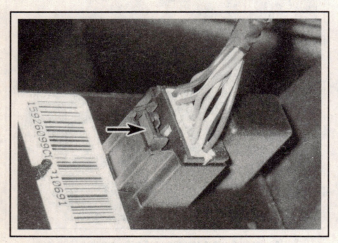

**10.6  To disconnect the electrical connector from the headlight switch (shown) or from the transfer case control switch, depress this release tab and pull off the connector**

**10.16  To remove the automatic transfer case control switch from the center trim panel, use three screwdrivers to disengage the locking tabs from the lugs on the switch, then push the switch out of the trim panel**

# CHASSIS ELECTRICAL SYSTEM  12-11

17  When installing the automatic transfer case control switch, make sure that it snaps into place. Installation is otherwise the reverse of removal.

## ACCESSORY SWITCHES

♦ Refer to illustration 10.19

18  Remove the center trim panel (see Chapter 11).
19  A pair of lugs on each end of the accessory switch unit are secured to the center trim panel by a corresponding pair of slots in the center trim panel. To disengage the slots on the center trim panel from the lugs on the switch unit, carefully pry loose the slots with screwdrivers, then pull out the switch (see illustration).
20  When installing the accessory switch unit, make sure that it snaps into place. Installation is otherwise the reverse of removal.

**10.19  Use screwdriver to carefully pry loose the slots in the center trim panel from the lugs on the accessory switch, then pull out the accessory switch unit**

## 11  Instrument cluster - removal and installation

♦ Refer to illustrations 11.2 and 11.3

**※※ WARNING:**

**The models covered by this manual are equipped with a Supplemental Restraint System (SRS), more commonly known as airbags. Always disarm the airbag system before working in the vicinity of any airbag system component to avoid the possibility of accidental deployment of the airbag, which could cause personal injury (see Section 26).**

1  Remove the instrument cluster trim panel (see Chapter 11).
2  Remove the instrument cluster retaining screws (see illustration).
3  Pull out the cluster, disconnect the electrical connector (see illustration) and remove the cluster.
4  Installation is the reverse of removal.

**11.2  To detach the instrument cluster from the instrument panel, remove these four screws**

**11.3  Pull out the cluster, depress this release tab and disconnect the electrical connector from the cluster**

# 12-12 CHASSIS ELECTRICAL SYSTEM

## 12 Radio and speakers - removal and installation

> **WARNING:**
> The models covered by this manual are equipped with a Supplemental Restraint System (SRS), more commonly known as airbags. Always disarm the airbag system before working in the vicinity of any airbag system component to avoid the possibility of accidental deployment of the airbag, which could cause personal injury (see Section 26).

### RADIO

*Refer to illustrations 12.3, 12.4a and 12.4b*

1  Remove the center trim panel (see Chapter 11).
2  Remove the heater and air conditioning control assembly (see Chapter 3).
3  Remove the radio mounting screws (see illustration).
4  Pull the radio out from the dash, then disconnect the electrical connectors and the antenna cable from the backside of the radio (see illustrations).
5  Installation is the reverse of removal.

### SPEAKERS

#### Door speakers

6  Remove the door trim panel (see Chapter 11).
7  Remove the speaker mounting screw.
8  Pull out the speaker and disconnect the electrical connector.
9  When installing the speaker, make sure that the lower mounting flange of the speaker is aligned with and fits inside the lower edge of the speaker mounting hole in the door.
10  Installation is otherwise the reverse of removal.

### Tweeters

*Refer to illustrations 12.12 and 12.13*

11  Remove the A-pillar trim panel (see Chapter 11).
12  Disconnect the tweeter electrical connector (see illustration).
13  To remove the tweeter assembly from the A-pillar trim panel, grasp the tweeter grille firmly and pull off the grille and the tweeter as a single assembly (see illustration).
14  Carefully disengage the three retainers with a small screwdriver and separate the tweeter from the grille.
15  Installation is the reverse of removal.

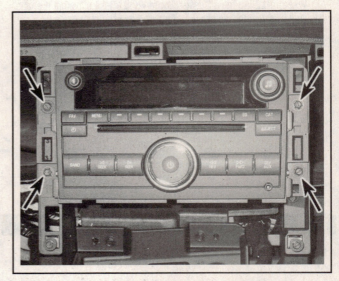

**12.3  To detach the radio unit from the instrument panel, remove these four screws**

**12.4a  Pull out the radio unit, depress these release tabs and disconnect the two electrical connectors**

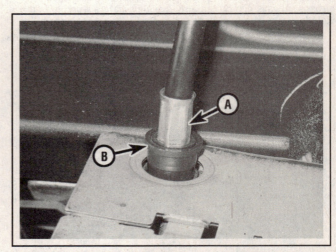

**12.4b  Push the antenna lead (A) toward the radio, pull back the locking ring (B), then pull the cable and locking ring out of the radio**

# CHASSIS ELECTRICAL SYSTEM     12-13

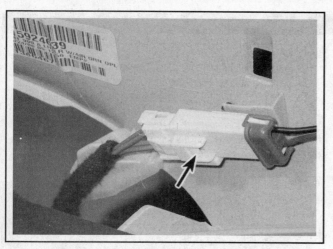

**12.12  To disconnect the tweeter electrical connector, depress this release tab**

**12.13  To detach the tweeter assembly from the A-pillar trim, grasp the grille firmly and pull it off by hand**

## 13  Antenna and cable - removal and installation

### ANTENNA MAST

♦ Refer to illustration 13.1

1   Unscrew the antenna mast (see illustration).
2   Installation is the reverse of removal.

### ANTENNA MAST MOUNTING BASE AND CABLE

♦ Refer to illustrations 13.4 and 13.6

3   Remove the antenna mast. Remove the cowl cover (see Chapter 11).
4   Unbolt the antenna mounting base (see illustration).
5   Remove the top trim panel from the instrument panel (see Chapter 11).

6   Disconnect the antenna mast cable from the extension cable (see illustration).
7   Attach a piece of wire of sufficient length to the old antenna mast cable and carefully pull out the cable through the grommet located below the antenna mounting base.
8   Attach the new antenna mast cable to the wire and pull it back through the grommet below the antenna mast base.
9   Installation is otherwise the reverse of removal.

### ANTENNA EXTENSION CABLE

10   The antenna extension cable, which connects the antenna mounting base cable to the radio, is routed through the instrument panel, which you must remove (see Chapter 11) to replace it.

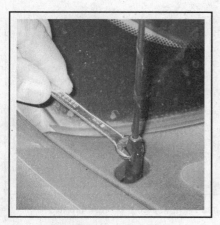

**13.1  To remove the antenna mast, simply unscrew it with a wrench**

**13.4  To detach the antenna mounting base, remove these bolts**

**13.6  Disconnect the antenna mast cable from the extension cable at this coaxial connector**

## 12-14 CHASSIS ELECTRICAL SYSTEM

## 14 Headlight housing - replacement

### TRUCK MODELS

♦ Refer to illustrations 14.4, 14.5 and 14.6

1  On Chevrolet models, remove the radiator grille (see Chapter 11).
2  On GMC models, remove the trim piece that covers the gap between the top of the front bumper and the underside of the grille and headlight housings (see Chapter 11).
3  Loosen the front wheel lug nuts, raise the vehicle, place it securely on jackstands and remove the front wheel. Remove the front part of the fenderwell liner (see Chapter 11).
4  Working through the wheel well opening, disconnect the headlight housing electrical connector (see illustration).
5  Loosen, but don't remove, the lower outer headlight housing mounting bolt (see illustration).
6  Remove the two upper headlight housing bolts (see illustration).
7  Carefully pull out the headlight housing. Note the two locator pins on the outer edge of the housing. When installing the headlight housing, make sure that these locator pins are aligned with their mounting holes in the front fender bracket.
8  If you're replacing either headlight bulb, refer to Section 16. If you're replacing any other bulbs in the headlight housing, refer to Section 17.
9  Installation is otherwise the reverse of removal.
10  Adjust the headlights when you're done (see Section 15).

### AVALANCHE, SUBURBAN AND TAHOE MODELS

♦ Refer to illustrations 14.14 and 14.16

11  Loosen the front wheel lug nuts, raise the vehicle, place it securely on jackstands and remove the front wheel. Remove the front part of the fenderwell liner (see Chapter 11).
12  Remove the two upper headlight housing bolts (see illustration 14.6).
13  Remove the bumper cover (see Chapter 11).
14  Remove the lower headlight housing bolt (see illustration).
15  Grasp the headlight housing firmly at its upper inner and lower outer edges and pull it forward to disengage the locating pins from the fender panel. Pull the outer side of the headlight housing forward until the locating pin disengages from its mounting hole in the radiator support bracket.
16  Pull out the headlight housing and disconnect the electrical connector (see illustration).
17  Installation is the reverse of removal.

14.4  To disconnect the headlight housing electrical connector, depress this release tab and pull off the connector (truck models)

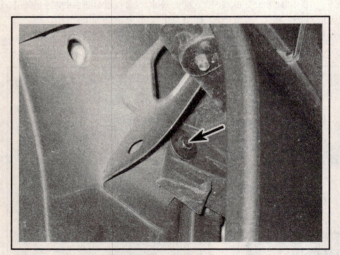

14.5  Loosen (but don't remove) the lower headlight housing mounting bolt (truck models)

14.6  To detach the headlight housing, remove the two upper bolts

# CHASSIS ELECTRICAL SYSTEM

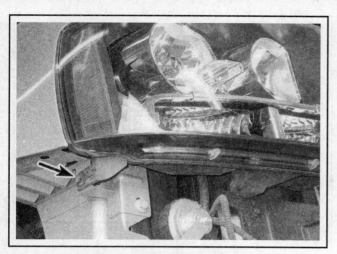

14.14 To detach the lower part of the headlight housing, remove this bolt (Avalanche, Suburban and Tahoe models)

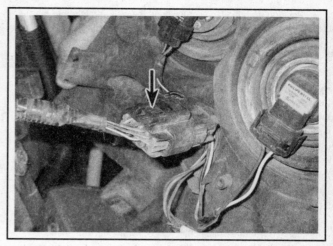

14.16 To disconnect the main headlight housing electrical connector, depress this release tab (Avalanche, Suburban and Tahoe models)

## YUKON MODELS

18  Loosen the front wheel lug nuts, raise the vehicle, place it securely on jackstands and remove the front wheel. Remove the front part of the fenderwell liner (see Chapter 11).

19  Remove the six upper bumper cover bolts (up to the hood latch) and the lower rear bumper cover bolt from the support bracket. Loosen the two bumper cover-to-fender bolts from underneath the bumper cover and pull the outer end of the bumper cover straight out until it disengages from the fender support bracket. Pull the bumper cover forward and down to allow sufficient clearance to remove the headlight housing.

20  Loosen the two lower outer headlight housing mounting bolts.

21  Remove the two upper headlight housing mounting bolts.

22  Grasp the upper inner and lower outer edges of the headlight housing and pull it forward to disengage the locator tab.

23  Pull the outer edge of the headlight housing forward to disengage the other two locator tabs from the radiator support bracket.

24  Pull out the headlight housing and disconnect the electrical connector.

25  Installation is the reverse of removal.

## 15  Headlights - adjustment

♦ Refer to illustrations 15.1 and 15.3

➡Note: *The headlights must be aimed correctly. If adjusted incorrectly they could blind the driver of an oncoming vehicle and cause a serious accident or seriously reduce your ability to see the road. The headlights should be checked for proper aim every 12 months and any time a new headlight is installed or front end bodywork is performed. It should be emphasized that the following procedure is only an interim step that will provide temporary adjustment until a properly equipped shop can adjust the headlights.*

1  The vertical adjustment screws are located behind each headlight housing (see illustration). There are no horizontal adjustment screws.

2  There are several methods for adjusting the headlights. The simplest method requires masking tape, a blank wall and a level floor.

15.1  To turn the headlight housing vertical adjustment screw, insert an Allen wrench into the top

## 12-16 CHASSIS ELECTRICAL SYSTEM

3  Position masking tape vertically on the wall in relation to the vehicle centerline and the centerlines of both headlights (see illustration).

4  Position a horizontal tape line in relation to the centerline of all the headlights.

**Note:** It might be easier to position the tape on the wall with the vehicle parked only a few inches away.

5  Adjustment should be made with the vehicle parked 25 feet from the wall, sitting level, the gas tank half-full and no heavy load in the vehicle.

6  Starting with the low beam adjustment, position the high intensity zone so it is two inches below the horizontal line. Adjustment is made by turning the adjusting screw clockwise to raise the beam and counterclockwise to lower the beam.

7  With the high beams on, the high intensity zone should be vertically centered with the exact center just below the horizontal line.

**Note:** It might not be possible to position the headlight aim exactly for both high and low beams. If a compromise must be made, keep in mind that the low beams are the most used and have the greatest effect on safety.

8  Have the headlights adjusted by a dealer service department or service station at the earliest opportunity.

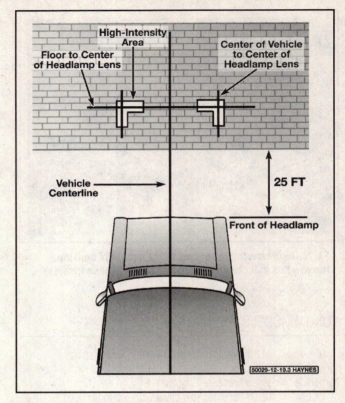

15.3 Headlight adjustment details

## 16 Headlight bulb - replacement

▶ Refer to illustrations 16.3a, 16.3b and 16.3c

**※ WARNING:**

Halogen gas-filled bulbs are under pressure and can shatter if the surface is scratched or the bulb is dropped. Wear eye protection and handle the bulb carefully, grasping it only by the bulb holder. Do not touch the surface of the bulb with your fingers because the oil from your skin could cause it to overheat and fail prematurely. If you do touch the bulb surface, clean it with rubbing alcohol.

1  If you're replacing a right-side bulb on a truck model, remove the air filter housing (see Chapter 4). If you're replacing a left-side bulb on a truck model and the vehicle is equipped with an auxiliary battery, remove the battery (see Chapter 5, Section 1).

2  If you're working on an SUV model, remove the headlight housing (see Section 14).

3  Disconnect the headlight bulb electrical connector, rotate the bulb holder counterclockwise and remove it from the housing (see illustrations).

4  Installation is the reverse of removal.

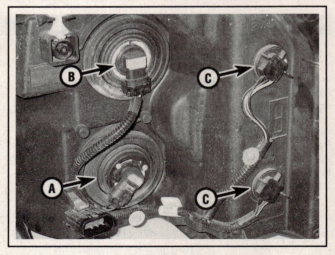

16.3a Truck model headlight housing bulbs (truck model shown, Yukon similar)

A  High beam headlight bulb
B  Low beam headlight bulb
C  Turn signal/sidemarker bulbs

# CHASSIS ELECTRICAL SYSTEM  12-17

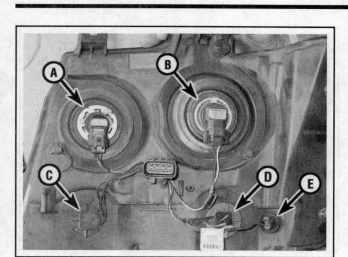

**16.3b  Avalanche, Suburban and Tahoe headlight housing**

- A  Low beam headlight bulb
- B  High beam headlight bulb
- C  Daytime running light bulb
- D  Parking light/turn signal light bulb
- E  Front sidemarker light bulb

**16.3c  Turn the headlight bulb holder counterclockwise and remove it from the housing**

## 17  Bulb replacement

### EXTERIOR LIGHT BULBS

#### Front parking light/turn signal/sidemarker/daytime running light bulbs

◆ Refer to illustration 17.2

1  Remove the headlight housing (see Section 14).
2  Turn the front parking bulb holder counterclockwise (see illustration) and pull it out of the headlight housing.
3  To remove the old bulb from its holder, pull it straight out. To install a new bulb in the bulb holder, push it straight into the socket until it stops.
4  Install the headlight housing.

#### Fog light bulbs

**⚠ WARNING:**

Halogen gas-filled bulbs are under pressure and can shatter if the surface is scratched or the bulb is dropped. Wear eye protection and handle the bulb carefully, grasping it only by the bulb holder. Do not touch the surface of the bulb with your fingers because the oil from your skin could cause it to overheat and fail prematurely. If you do touch the bulb surface, clean it with rubbing alcohol.

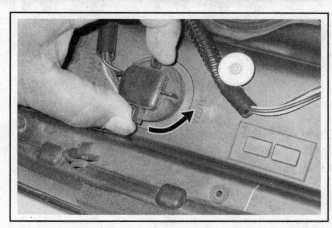

**17.2  To remove a front parking light/turn signal bulb holder from the headlight housing, turn it counterclockwise and pull it out of the housing (Chevy truck unit shown, GMC similar)**

5  Loosen the front wheel lug nuts, raise the vehicle, place it securely on jackstands and remove the front wheel.
6  If you're working on an SUV or Avalanche model, remove the front part of the fenderwell liner (see Chapter 11). On truck models the fog light can be accessed by reaching up from behind the front air deflector.

## 12-18 CHASSIS ELECTRICAL SYSTEM

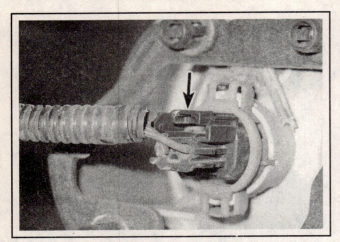

17.7 To disconnect the electrical connector from the fog light bulb holder, depress this release tab

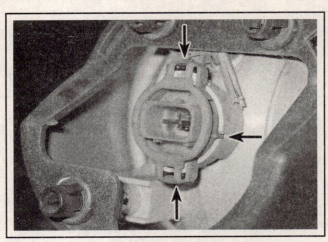

17.8 To remove the fog light bulb holder, squeeze these two tabs and pull out the holder. When installing the new bulb holder, make sure the ridges on the holder are aligned with the slots in the housing

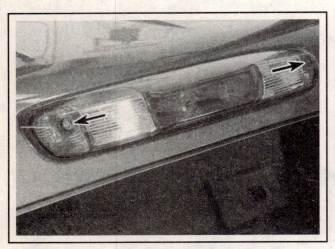

17.15 To detach the center high-mounted brake light housing, remove these two screws (truck models)

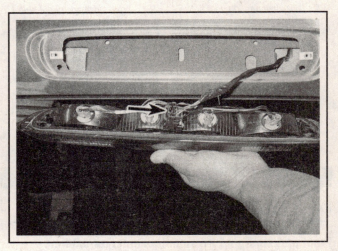

17.16 To disconnect the electrical connector from the center high-mounted brake light housing, depress the release tab and pull off the connector (truck models)

**Chevy and GMC trucks and Chevy Avalanche, Suburban and Tahoe models**

▶ Refer to illustrations 17.7 and 17.8

7 Locate the fog light electrical connector (see illustration) and disconnect it.
8 To remove the bulb holder, squeeze the upper and lower tabs on the bulb holder and pull out the holder (see illustration). The bulb and bulb holder are a single assembly.
9 To install the new bulb, align the ridges on the bulb holder with the slots in the fog light housing and push it into place.
10 Installation is otherwise the reverse of removal.

**GMC Yukon models**

11 Disconnect the electrical connector from the fog light bulb holder.
12 To remove the fog light bulb holder, turn it counterclockwise and pull it out of the fog light housing.
13 To install the new fog light bulb holder, insert it into the mounting hole in the bumper cover and turn it clockwise until it locks into place.
14 Installation is otherwise the reverse of removal.

**Center high-mounted brake light bulbs**

**Truck models**

▶ Refer to illustrations 17.15 and 17.16

15 Remove the two center high-mounted brake light housing retaining screws (see illustration).
16 Remove the center high-mounted brake light housing (see illustration) and disconnect the electrical connector.
17 To remove a bulb holder from the center high-mounted brake light housing, turn it counterclockwise and pull it out of the housing.
18 To remove a bulb from the bulb holder, pull it straight out of the holder. To install a bulb in its holder, push it straight into the holder until it stops.
19 Installation is the reverse of removal.

# CHASSIS ELECTRICAL SYSTEM 12-19

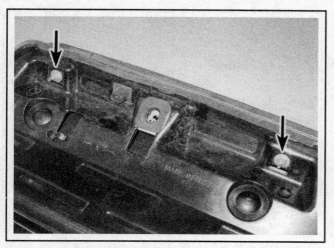

17.21 To detach the center high-mounted brake light housing from the spoiler, remove these two nuts (SUV models)

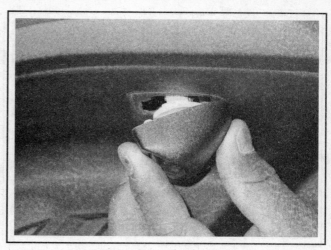

17.29 Carefully pry the license plate light housing out of the bumper

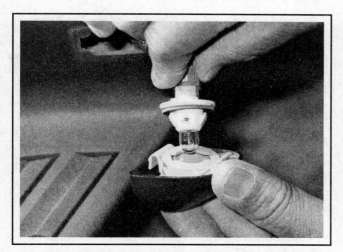

17.30 To remove the bulb holder from the license plate light housing, turn it counterclockwise

17.35 To detach the taillight housing, remove the two housing retaining screws

### SUV models

▶ **Refer to illustration 17.21**

20 Remove the spoiler from the liftgate (see Chapter 11, Section 23).
21 Remove the center high-mounted brake light assembly mounting nuts (see illustration).
22 Remove the center high-mounted brake light assembly from the center high-mounted brake light housing/spoiler.
23 Installation is the reverse of removal.

### Avalanche models

24 Remove the center cargo bridge panel (see Chapter 11).
25 Remove the two center high-mounted brake light clips.
26 Remove the center high-mounted brake light housing retaining screw.
27 Remove the center high-mounted brake light housing from the center cargo bridge panel.
28 Installation is the reverse of removal.

### License plate light bulbs

▶ **Refer to illustrations 17.29 and 17.30**

29 Carefully pry the license plate light housing out of the bumper (see illustration).
30 Remove the bulb holder from the license plate light housing (see illustration).
31 To remove the old bulb from its holder, simply pull it straight out.
32 To install a new license plate light bulb, simply push it straight into the socket until it stops.
33 Installation is the reverse of removal.

### Taillight bulbs

▶ **Refer to illustrations 17.35 and 17.36**

34 Open the tailgate or liftgate.
35 Remove the taillight housing retaining screws and remove the taillight housing (see illustration).

# 12-20  CHASSIS ELECTRICAL SYSTEM

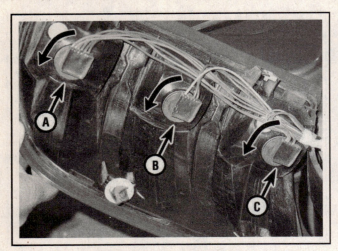

**17.36 Typical taillight bulb arrangement (Chevy Silverado shown; refer to your owner's manual for bulb arrangement on other models)**

- A  Brake light/turn signal/taillight bulb
- B  Back-up light bulb
- C  Brake light/turn signal/taillight bulb

36  Turn the bulb holder counterclockwise to remove it from the housing (see illustration).
37  To remove the bulb from its holder, pull it straight out. Push the new bulb straight into the socket until it stops.
38  When installing the taillight housing, make sure that the locator pins snap into place.

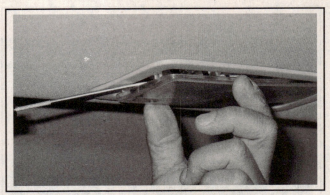

**17.39 Carefully pry loose the dome light lens**

## INTERIOR LIGHTS

### Dome light bulbs

♦ Refer to illustration 17.39

39  Pry off the dome light or map reading light lens (see illustration).
40  Remove the bulb by pulling it straight down.

### ❋❋ WARNING:

**Don't pry on the glass.**

41  To install a new bulb, push it up between the conductors until it snaps into place.
42  When installing the lens, make sure that it snaps into place.

### Map/reading light bulbs

43  Carefully pry off the map/reading light lens.
44  To replace the bulb, simply pull it straight out.
45  Push the new bulb into the socket until it stops.
46  When installing the lens, make sure that the three tabs on the lens are aligned with their slots housing, then push it into place.

## 18  Wiper motors - replacement

### WINDSHIELD WIPER MOTOR

♦ Refer to illustrations 18.1, 18.5, 18.7 and 18.8

1  Pry off the windshield wiper trim caps (see illustration).
2  Remove the windshield wiper retaining nuts and washers. Mark the position of each wiper arm in relation to its shaft, then remove the wiper arms.
3  Remove the cowl cover (see Chapter 11).
4  Disconnect the windshield wiper motor electrical connector.
5  Remove the three mounting bolts that secure the windshield wiper motor and linkage assembly to the cowl (see illustration).
6  Lift the windshield wiper motor and linkage assembly out of the cowl area.
7  Pry the linkage off the wiper motor crank arm (see illustration).
8  Remove the nut that attaches the actuator arm to the motor shaft (see illustration). Mark the relationship of the actuator arm to the motor shaft, then remove the actuator arm from the shaft.
9  Remove the motor mounting screws and remove the motor.

10  Installation is the reverse of removal. Be sure to align the marks you made on the actuator arm and the motor shaft, and on the windshield wiper arms and the wiper arm shafts.

### REAR WIPER MOTOR (SUV MODELS)

♦ Refer to illustrations 18.11 and 18.16

11  Remove the trim cap from the wiper arm (see illustration).
12  Remove the wiper arm retaining nut.
13  Mark the relationship of the wiper arm to the wiper motor shaft, then remove the arm.
14  Remove the liftgate trim panel (see Chapter 11).
15  Disconnect the wiper motor electrical connector.
16  Remove the wiper motor mounting fasteners (see illustration).
17  Installation is the reverse of removal. When installing the wiper arm, be sure to align the marks that you made on the wiper arm and the wiper motor shaft.

## CHASSIS ELECTRICAL SYSTEM 12-21

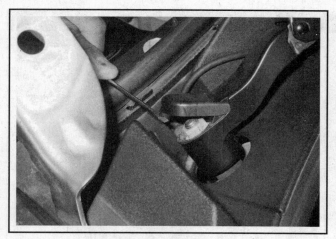

18.1 Pry open each trim cap, remove each wiper arm retaining nut and mark the relationship of each arm to its shaft, then remove the arms

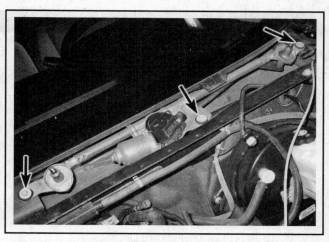

18.5 To detach the windshield wiper motor and linkage assembly from the cowl, remove these three bolts

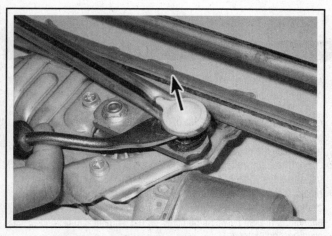

18.7 Carefully pry the linkage arm loose from the crank arm

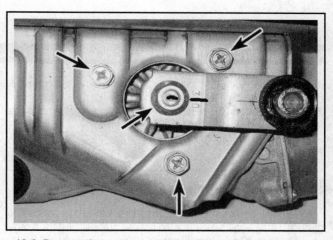

18.8 Remove the crank arm nut, mark the relationship of the crank arm to the motor shaft, remove the crank arm, then remove the three motor mounting bolts

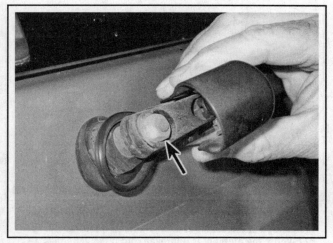

18.11 Remove the trim cap and the wiper arm retaining nut, then mark the relationship of the wiper arm to its shaft before removing the arm

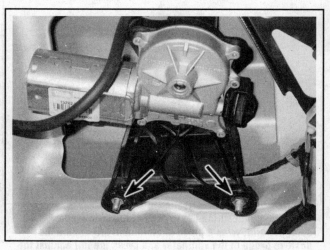

18.16 To detach the rear wiper motor from the liftgate, remove the two fasteners

# 12-22 CHASSIS ELECTRICAL SYSTEM

## 19 Horn - replacement

▶ Refer to illustration 19.2

➡ Note: The two horns are located at the front corners of the vehicle, below each headlight housing.

1  If you're working on an SUV model or a Chevy truck model, remove the headlight housing (see Section 14). If you're working on a GMC truck model, remove the fasteners securing the front of the inner fender liner, then remove the bumper cover trim panel from below the headlight.
2  Disconnect the electrical connector from the horn (see illustration).
3  Remove the horn mounting bracket bolt and remove the horn and bracket.
4  Detach the horn from its mounting bracket.
5  Installation is the reverse of removal.

**19.2  Disconnect the electrical connector from the horn, then remove the horn mounting bolt**

## 20 Rear window defogger - check and repair

1  The rear window defogger consists of a number of horizontal elements baked onto the glass surface.
2  Small breaks in the element can be repaired without removing the rear window.

### CHECK

▶ Refer to illustrations 20.4, 20.5 and 20.7

3  Turn the ignition switch and defogger system switches to the ON position. Using a voltmeter, place the positive probe against the defogger grid positive terminal and the negative probe against the ground terminal. If battery voltage is not indicated, check the fuse, defogger switch and related wiring. If voltage is indicated, but all or part of the defogger doesn't heat, proceed with the following tests.
4  When measuring voltage during the next two tests, wrap a piece of aluminum foil around the tip of the voltmeter positive probe and press the foil against the heating element with your finger (see illustration). Place the negative probe on the defogger grid ground terminal.
5  Check the voltage at the center of each heating element (see illustration). If the voltage is 5 or 6-volts, the element is okay (there is no break). If the voltage is zero, the element is broken between the center of the element and the positive end. If the voltage is 10 to 12-volts the element is broken between the center of the element and ground. Check each heating element.
6  Connect the negative lead to a good body ground. The reading

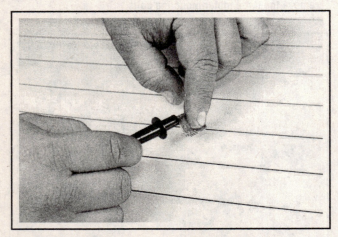

**20.4  When measuring the voltage at the rear window defogger grid, wrap a piece of aluminum foil around the positive probe of the voltmeter and press the foil against the wire with your finger**

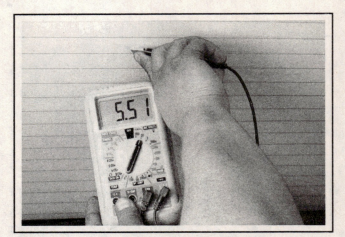

**20.5  To determine if a heating element has broken, check the voltage at the center of each element; if the voltage is 5 or 6-volts, the element is unbroken, but if the voltage is 10 or 12-volts, the element is broken between the center and the ground side. If there is no voltage, the element is broken between the center and the positive side**

# CHASSIS ELECTRICAL SYSTEM   12-23

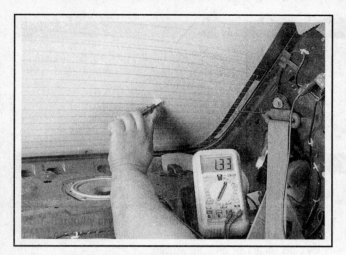

20.7  To find the break, place the voltmeter negative lead against the defogger ground terminal, place the voltmeter positive lead with the foil strip against the heating element at the positive terminal end and slide it toward the negative terminal end. The point at which the voltmeter reading changes abruptly is the point at which the element is broken

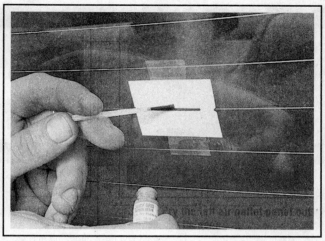

20.13  To use a defogger repair kit, apply masking tape to the inside of the window at the damaged area, then brush on the special conductive coating

should stay the same. If it doesn't, the ground connection is bad.

7   To find the break, place the voltmeter negative probe against the defogger ground terminal. Place the voltmeter positive probe with the foil strip against the heating element at the positive terminal end and slide it toward the negative terminal end. The point at which the voltmeter deflects from several volts to zero is the point at which the heating element is broken (see illustration).

## REPAIR

▶ Refer to illustration 20.13

8   Repair the break in the element using a repair kit specifically recommended for this purpose, available at most auto parts stores. Included in this kit is plastic conductive epoxy.

9   Prior to repairing a break, turn off the system and allow it to cool off for a few minutes.

10   Lightly buff the element area with fine steel wool, then clean it thoroughly with rubbing alcohol.

11   Use masking tape to mask off the area being repaired.

12   Thoroughly mix the epoxy, following the instructions provided with the repair kit.

13   Apply the epoxy material to the slit in the masking tape, overlapping the undamaged area about 3/4-inch on either end (see illustration).

14   Allow the repair to cure for 24 hours before removing the tape and using the system.

## 21  Electric side view mirrors - general information

1   Most electric rear view mirrors use two motors to move the glass; one for up and down adjustments and one for left-right adjustments.

2   The control switch has a selector portion that sends voltage to the left or right side mirror. With the ignition ON but the engine OFF, roll down the windows and operate the mirror control switch through all functions (LEFT-RIGHT and UP-DOWN) for both the left and right side mirrors.

3   Listen carefully for the sound of the electric motors running in the mirrors.

4   If the motors can be heard but the mirror glass doesn't move, there's a problem with the drive mechanism inside the mirror.

5   If the mirrors do not operate and no sound comes from the mirrors, check the fuse (see Section 3).

6   If the fuse is OK, remove the mirror control switch. Have the switch continuity checked by a dealership service department or other qualified automobile repair facility.

7   Make sure the mirror is properly grounded.

8   If the mirror still doesn't work, remove the mirror and check the wires at the mirror for voltage.

9   If there's not voltage in each switch position, check the circuit between the mirror and control switch for opens and shorts.

10   If there's voltage, remove the mirror and test it off the vehicle with jumper wires. Replace the mirror if it fails this test.

## 22 Cruise control system - general information

The Powertrain Control Module (PCM) controls the cruise control system electronically via the electronic throttle control system. If you have problems with the cruise control system, check for the presence of trouble codes stored in the PCM (see Chapter 6). If that doesn't turn up any problems, have it checked by a dealer service department or other qualified repair shop.

## 23 Power window system - general information

1. The power window system operates electric motors, mounted in the doors, which lower and raise the windows. The system consists of the control switches, the motors, regulators, glass mechanisms, the body control module and associated wiring.

2. The power windows can be lowered and raised from the master control switch by the driver or by remote switches located at the individual windows. Each window has a separate motor, which is reversible. The position of the control switch determines the polarity and therefore the direction of operation.

3. The circuit is protected by a fuse and a circuit breaker. Each motor is also equipped with an internal circuit breaker; this prevents one stuck window from disabling the whole system.

4. Most models have a window lockout switch at the master control switch which, when activated, disables the switches at the rear windows and, sometimes, the switch at the passenger's window also. Always check these items before troubleshooting a window problem.

5. These procedures are general in nature, so if you can't find the problem using them, take the vehicle to a dealer service department or other properly equipped repair facility.

6. If the power windows won't operate, always check the fuse and circuit breaker first.

7. If only the rear windows are inoperative, or if the windows only operate from the master control switch, check the rear window lockout switch for continuity in the unlocked position. Replace it if it doesn't have continuity.

8. Check the wiring between the switches and fuse panel for continuity. Repair the wiring, if necessary.

9. If only one window is inoperative from the master control switch, try the other control switch at the window.

➡ **Note: This doesn't apply to the driver's door window.**

10. If the same window works from one switch, but not the other, check the switch for continuity.

11. If the switch tests OK, check for a short or open in the circuit between the affected switch and the window motor.

12. If one window is inoperative from both switches, remove the trim panel from the affected door and check for voltage at the switch and at the motor while the switch is operated.

13. If voltage is reaching the motor, disconnect the glass from the regulator (see Chapter 11). Move the window up and down by hand while checking for binding and damage. Also check for binding and damage to the regulator. If the regulator is not damaged and the window moves up and down smoothly, replace the motor. If there's binding or damage, lubricate, repair or replace parts, as necessary.

14. If voltage isn't reaching the motor, check the wiring in the circuit for continuity between the switches and body control module, and between the body control module and the motors (see the wiring diagrams at the end of this Chapter).

## 24 Power door lock system - general information

1. A power door lock system operates the door lock actuators mounted in each door. The system consists of the switches, actuators, the body control module and associated wiring. Diagnosis can usually be limited to simple checks of the wiring connections and actuators for minor faults that can be easily repaired.

2. Power door lock systems are operated by bi-directional solenoids located in the doors. The lock switches have two operating positions: Lock and Unlock. When activated, the switch sends a signal to the body control module to lock or unlock the doors. Depending on which way the switch is activated, the control module reverses polarity to the solenoids, allowing the two sides of the circuit to be used alternately as the feed (positive) and ground side.

3. Some vehicles may have an anti-theft system incorporated into the power locks. If you are unable to locate the trouble using the following general Steps, consult a dealer service department or other qualified repair shop.

4. Always check the circuit protection first. Some vehicles use a combination of circuit breakers and fuses.

5. Operate the door lock switches in both directions (Lock and Unlock) with the engine off. Listen for the click of the solenoids operating.

6. Test the switches for continuity. Remove the switches and have them checked by a dealer service department or other qualified automobile repair facility.

7. Check the wiring between the switches, body control module and solenoids for continuity. Repair the wiring if there's no continuity.

8. If all but one of the lock solenoids operate, remove the trim panel from the door with the problem (see Chapter 11) and check for voltage at the solenoid while the lock switch is operated. One of the wires should have voltage in the Lock position; the other should have voltage in the Unlock position.

9. If the inoperative solenoid is receiving voltage, replace the solenoid.

10. If the inoperative solenoid isn't receiving voltage, check the for an open circuit condition in the wire between the switch and the body control module, and between lock solenoid and the body control module (see the wiring diagrams at the end of this Chapter).

## 25  Daytime Running Lights (DRL) - general information

The Daytime Running Lights (DRL) system illuminates the headlights whenever the engine is running. The only exception is with the engine running and the parking brake engaged. Once the parking brake is released, the lights will remain on as long as the ignition switch is on, even if the parking brake is later applied.

The DRL system supplies reduced power to the headlights so they won't be too bright for daytime use, while prolonging headlight life.

## 26  Airbag system - general information and precautions

### GENERAL INFORMATION

1  All models are equipped with two front airbags, formally known as the Supplemental Inflatable Restraint (SIR) system. This system is designed to protect the driver and the front seat passenger from serious injury in the event of a frontal collision. It consists of an array of external and internal (inside the SDM) information sensors (decelerometers), the Inflatable Restraint Sensing and Diagnostic Module (SDM), the inflator modules (a driver's airbag in the steering wheel and a passenger airbag in the dash) and the wiring and connectors tying all these components together. An optional pair of side-impact airbags, also known as roof rail or side curtain airbags, is available for protection against side impacts. The side-impact airbags, if equipped, are located along the left and right edges of the headliner, above the doors.

### AIRBAG/INFLATOR MODULES

#### Driver's airbag/inflator module

2  The airbag inflator module in the steering wheel consists of a housing, the cushion (airbag), an initiating device and a canister of gas-generating material. The initiator is part of the inflator module deployment loop. When a collision occurs, the SDM sends current through the deployment loop to the initiator. Current passing through the initiator ignites the material in the canister, producing a rapidly expanding gas, which inflates the airbag almost instantaneously. Seconds after the airbag inflates, it deflates almost as quickly through airbag vent holes and/or the airbag fabric.

3  When the SDM sends current to the initiator, it travels through the airbag circuit to the steering column. From there, a clockspring on the steering wheel delivers the current to the module initiator. This clockspring assembly, which is the final segment of the airbag ignition circuit, functions as the bridge between the end of the airbag circuit on the (fixed) steering column and the beginning of the circuit on the (rotating) steering wheel. It's designed to maintain a closed circuit between the steering column and the steering wheel regardless of the position of the steering wheel. For this reason, removing and installing the clockspring is critical to the performance of the driver's side airbag. For information on how to remove and install the driver's side airbag, refer to Chapter 10, Section 16.

#### Passenger's airbag/inflator module

4  The passenger's airbag/inflator module is mounted above the glove compartment. It's similar in design to the driver's airbag except that it doesn't use a clockspring. When deployed by the SDM, the passenger's airbag bursts through the dashboard above the glove box. Although this area looks like it's simply part of the dashboard, it's actually a trim cover with a perforated seam that allows the cover to separate from the dash when the passenger's airbag inflates.

#### Side impact airbag/inflator (roof rail) modules

5  The (optional) side-impact airbag/inflator (roof rail) modules are mounted along the outer edges of the headliner, right above the door openings. They extend from the A-pillar (front windshield pillar) to the C-pillar (rear window pillar). Each module consists of a housing, an inflatable airbag, an initiator and a canister of gas-generating material. Each roof rail module employs its own side impact sensor (SIS), which contains a sensing device that monitors changes in vehicle acceleration and velocity. This data is sent to the SDM, which compares it with its program. When the data exceeds a certain threshold, the SDM determines that the vehicle has been hit hard enough on one side or the other to warrant deployment of the roof rail on that side. The SDM doesn't deploy the roof rail airbags on both sides, just on the side being hit. Then the SDM sends current to the roof rail initiator to inflate the airbag, ripping open the headliner trim as it deploys to protect the occupant(s) on the left or right side of the vehicle. Side impact airbag/inflator modules are long enough to protect the driver and a left-side rear-seat passenger, or a front seat passenger and right-side rear-seat passenger.

### INFLATABLE RESTRAINT SENSING AND DIAGNOSTIC MODULE (SDM)

6  The SDM is the computer module that controls the airbag system. Besides a microprocessor, the SDM also includes an array of sensors. Some of them are inside the SDM itself. Other external sensors are located throughout the vehicle. All of the sensors, internal and external, send a continuous voltage signal to the SDM, which compares this data to values stored in its memory. When these signals exceed a threshold value (when the SDM determines that the vehicle is decelerating more quickly than the threshold value), the SDM allows current to flow through the circuit to the appropriate airbag module(s), which initiates deployment of the airbag(s).

7  For more information about the airbag system in your vehicle, refer to your owner's manual.

# 12-26 CHASSIS ELECTRICAL SYSTEM

## DISARMING THE SYSTEM AND OTHER PRECAUTIONS

**⁂ WARNING:**

**Failure to follow these precautions could result in accidental deployment of the airbag and personal injury.**

8  Whenever working in the vicinity of the steering wheel, instrument panel or any of the other SIR system components, the system must be disarmed. To disarm the system:
   a) Point the wheels straight ahead and turn the key to the Lock position.
   b) Disconnect the cable from the negative battery terminal. Refer to Chapter 5, Section 1 for the disconnecting procedure.
   c) Wait at least two minutes for the back-up power supply to be depleted.

9  Whenever handling an airbag module, always keep the airbag opening (the trim side) pointed away from your body. Never place the airbag module on a bench or other surface with the airbag opening facing the surface. Always place the airbag module in a safe location with the airbag opening (the upholstered side) facing up.

10  Never measure the resistance of any SIR component or use any electrical test equipment on any of the wiring or components. An ohmmeter has a built-in battery supply that could accidentally deploy the airbag.

11  Never dispose of a live airbag/inflator module. Return it to a dealer service department or other qualified repair shop for safe deployment and disposal.

12  Never use electrical welding equipment in the vicinity of any airbag components. The connectors for the system are easy to spot because they're bright yellow. Do NOT disconnect or tamper with these connectors, or you run the risk of setting a Diagnostic Trouble Code (DTC) in the SDM. Like the PCM, the SDM has a malfunction indicator light, known as the AIR BAG indicator light, on the instrument cluster. When you turn the ignition key to ON, the SDM checks out all of the system components and circuits. If everything is okay, the AIR BAG indicator light goes off, just like the PCM's Malfunction Indicator Light (MIL). But if there's a problem somewhere, the light stays on, and will remain on until the problem is repaired and the DTC(s) cleared from the SDM's memory.

## IMPACT SEAT BELT RETRACTORS

13  All models are equipped with pyrotechnic (explosive) units in the front seat belt retracting mechanisms for both the lap and shoulder belts. During an impact that would trigger the airbag system, the airbag control unit also triggers the seat belt retractors. When the pyrotechnic charges go off, they accelerate the retractors to instantly take up any slack in the seat belt system to more fully prepare the driver and front seat passenger for impact.

14  The airbag system should be disabled any time work is done to or around the seats.

**⁂ WARNING:**

**Never strike the pillars or floorpan with a hammer or use an impact-driver tool in these areas unless the system is disabled.**

## 27  Wiring diagrams - general information

Since it isn't possible to include all wiring diagrams for every year covered by this manual, the following diagrams are those that are typical and most commonly needed.

Prior to troubleshooting any circuits, check the fuse and circuit breakers (if equipped) to make sure they're in good condition. Make sure the battery is properly charged and check the cable connections (see Chapter 1).

When checking a circuit, make sure that all connectors are clean, with no broken or loose terminals. When unplugging a connector, do not pull on the wires; pull only on the connectors.

## CHASSIS ELECTRICAL SYSTEM 12-27

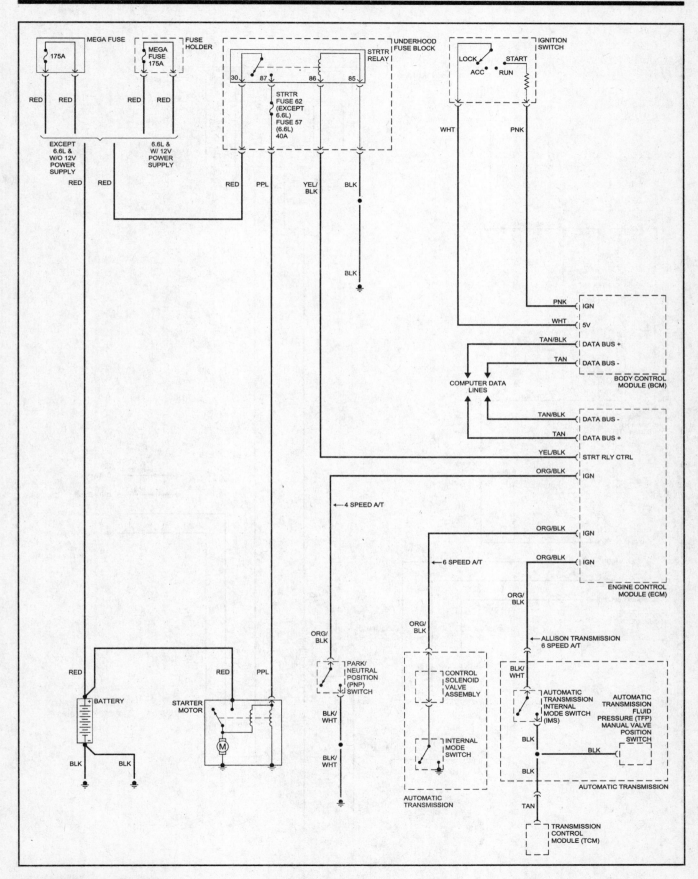

**Starting system - all models**

## 12-28 CHASSIS ELECTRICAL SYSTEM

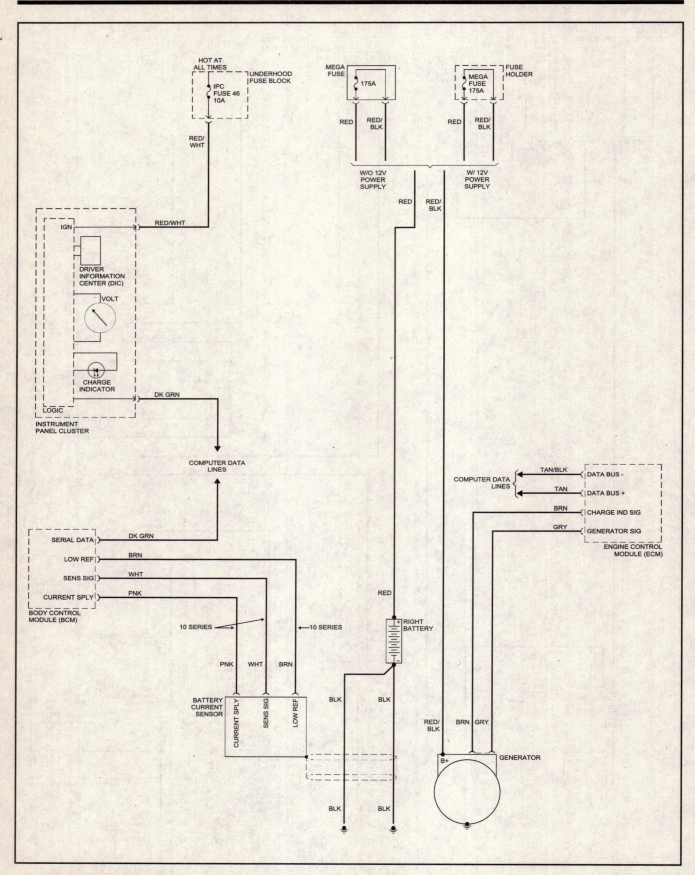

**Charging system - truck models**

## CHASSIS ELECTRICAL SYSTEM  12-29

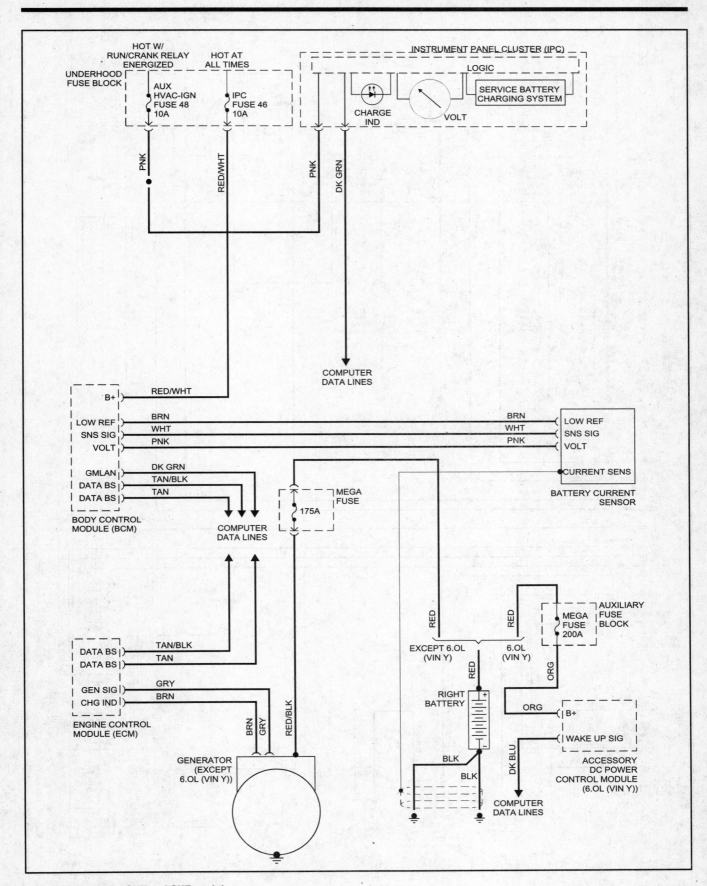

**Charging system - SUV and SUT models**

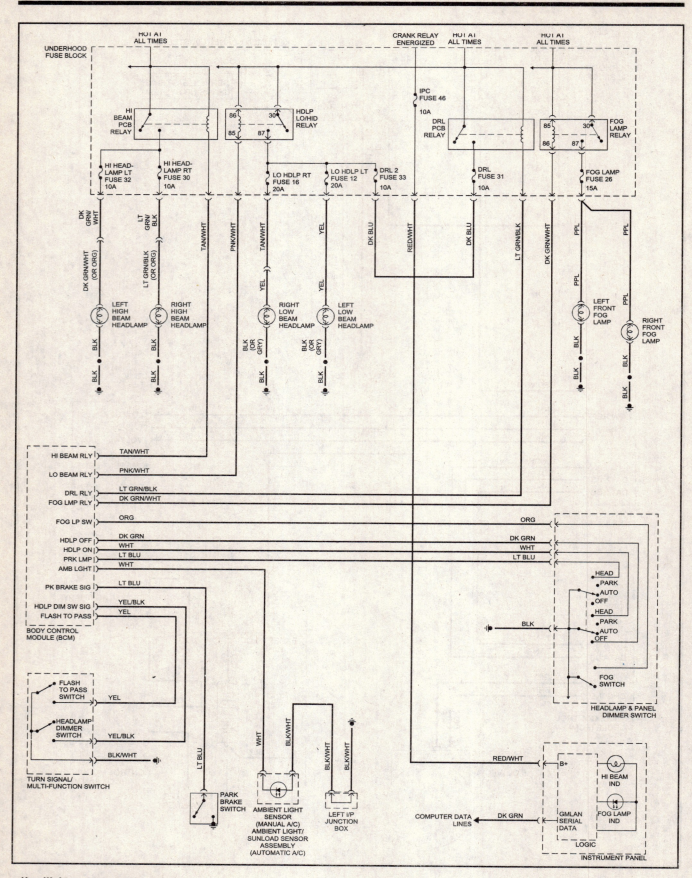

Headlight system - truck models

## CHASSIS ELECTRICAL SYSTEM 12-31

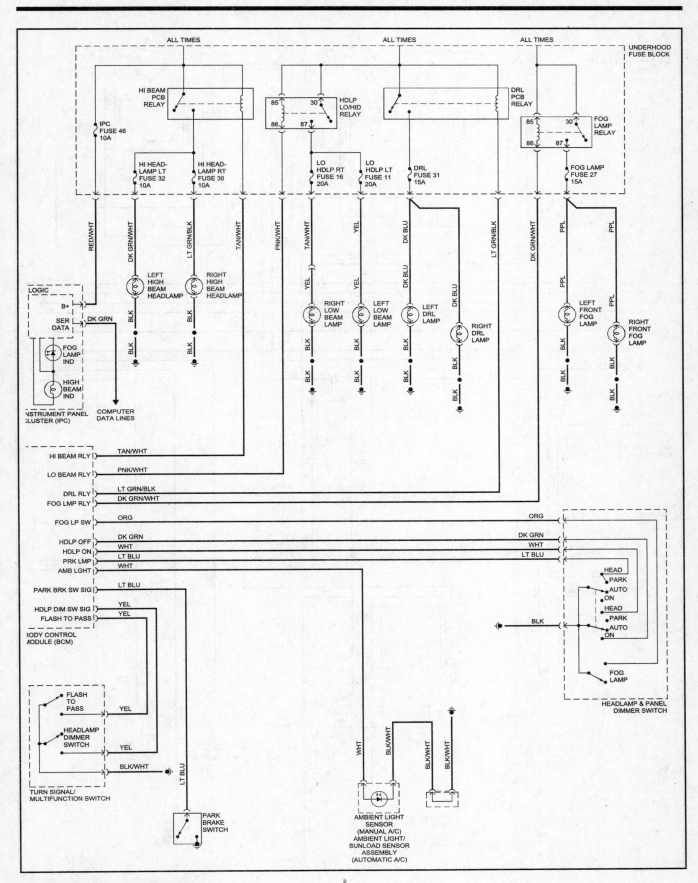

Headlight system - SUV and SUT models

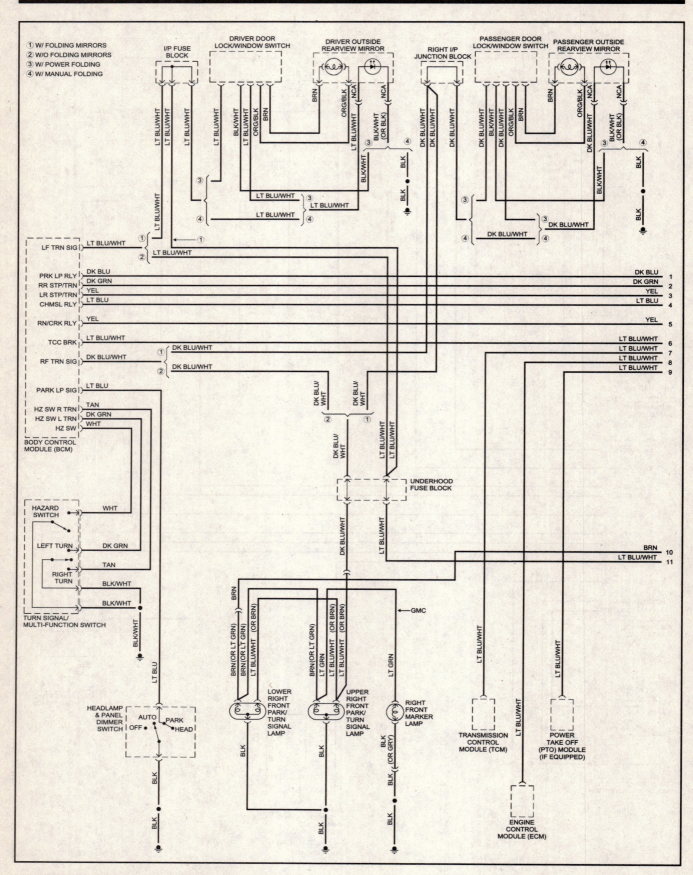

**Exterior lighting system - truck models (1 of 3)**

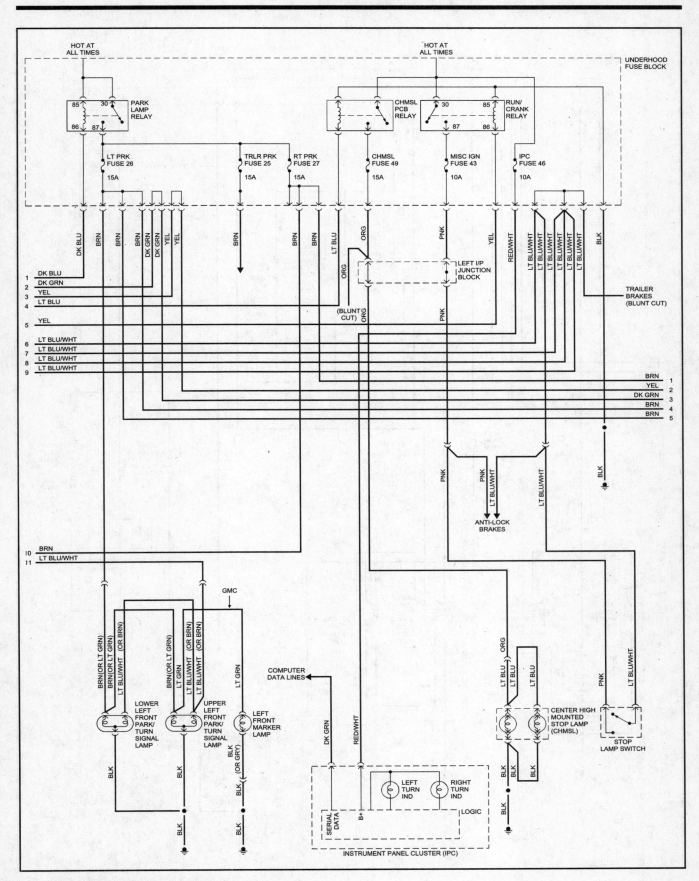

**Exterior lighting system - truck models (2 of 3)**

## 12-34 CHASSIS ELECTRICAL SYSTEM

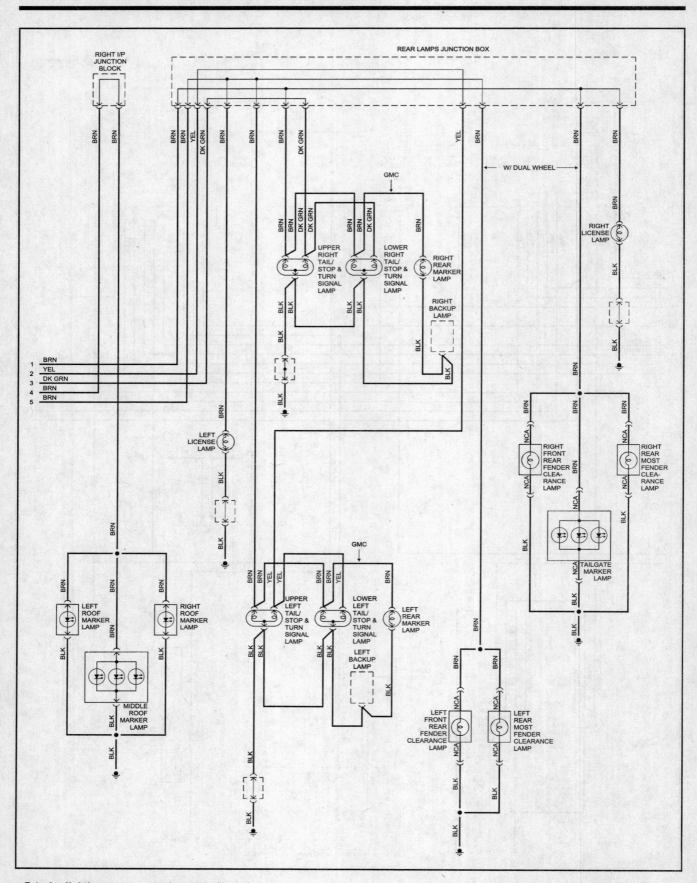

Exterior lighting system - truck models (3 of 3)

## CHASSIS ELECTRICAL SYSTEM 12-35

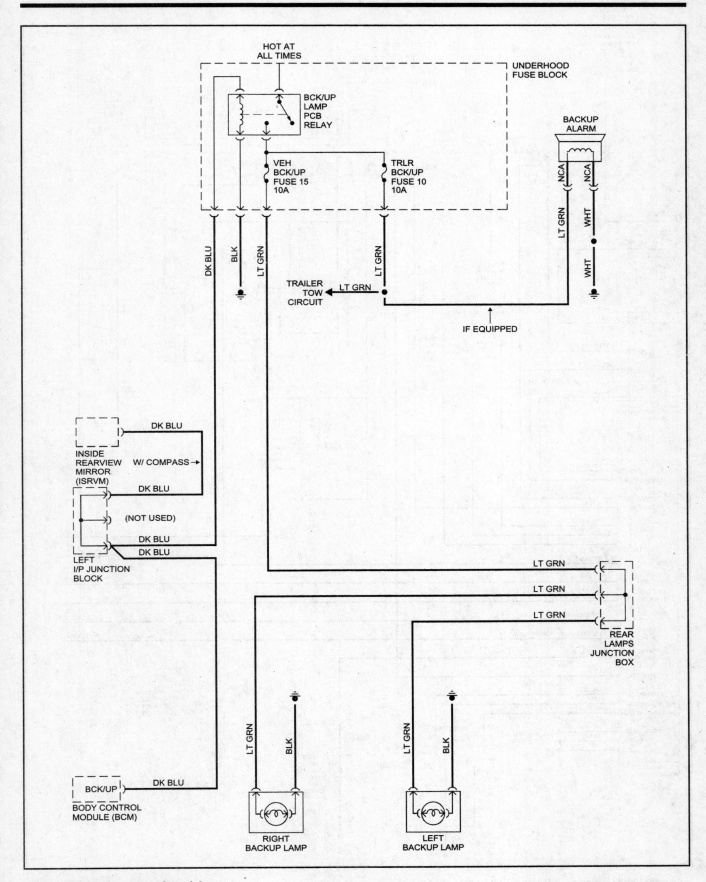

**Back-up light system - truck models**

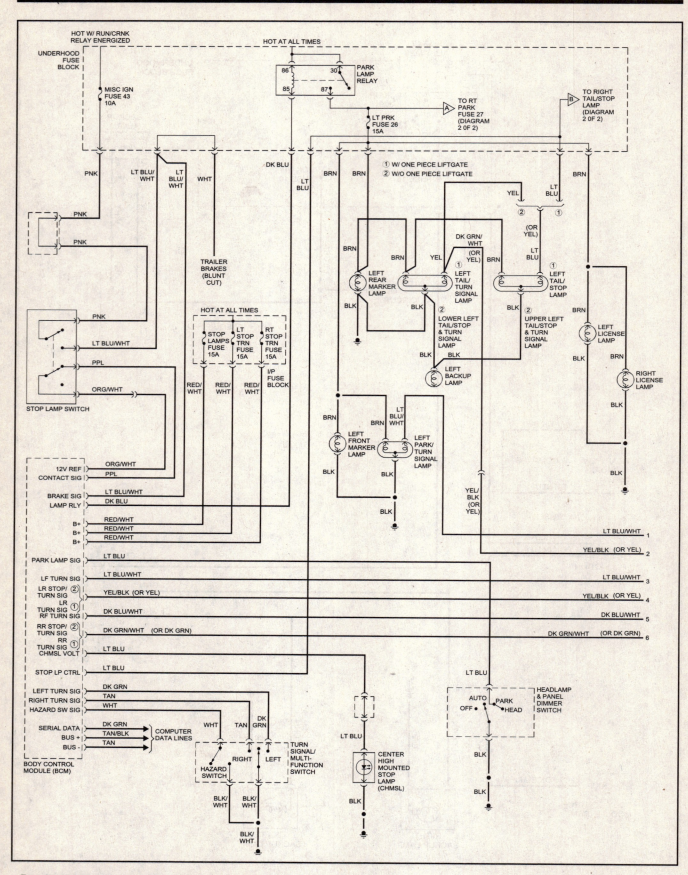

Exterior lighting system - SUV and SUT models (1 of 2)

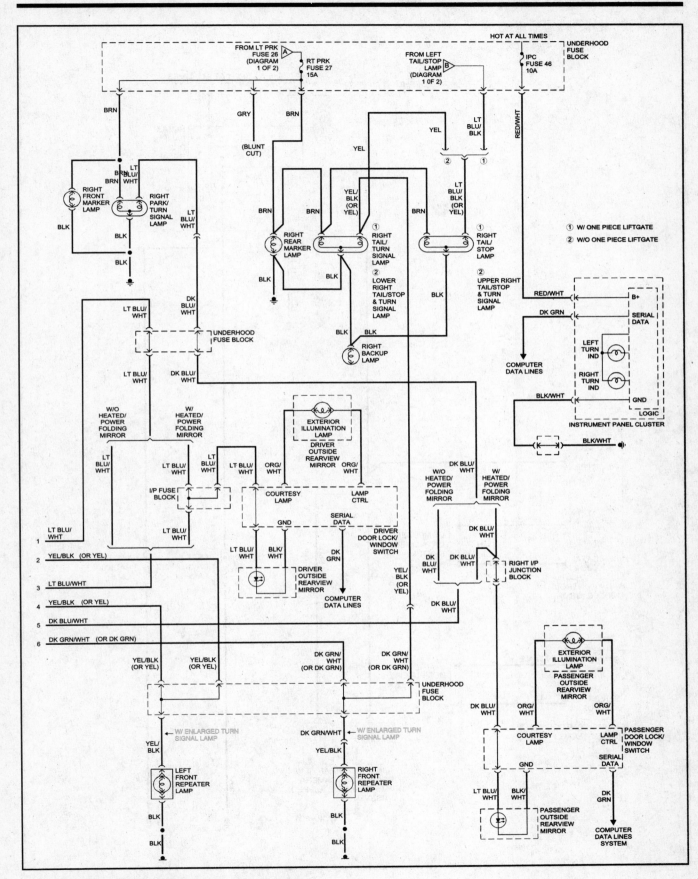

Exterior lighting system - SUV and SUT models (2 of 2)

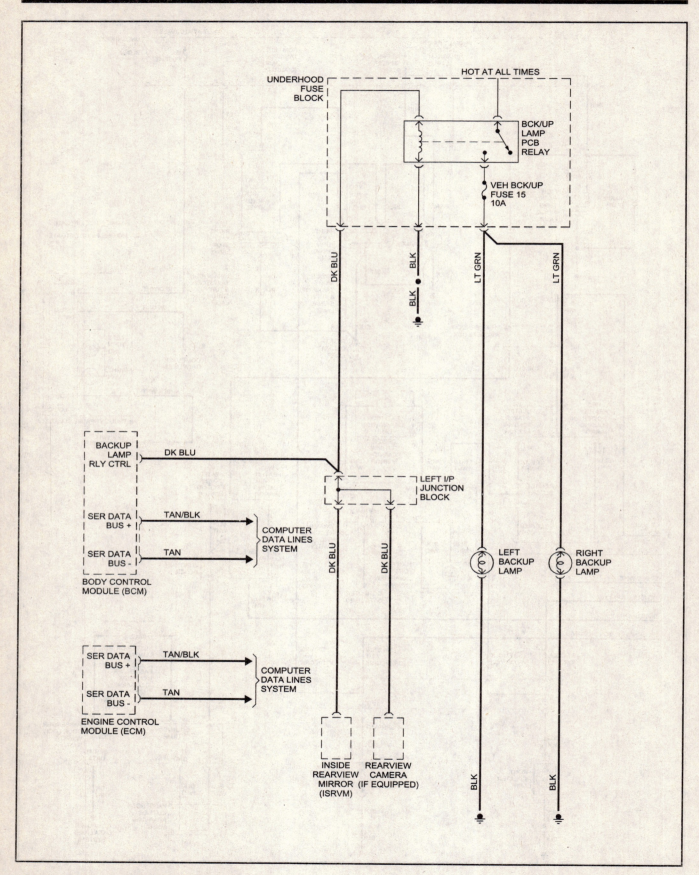

Back-up light system - SUV and SUT models

# CHASSIS ELECTRICAL SYSTEM 12-39

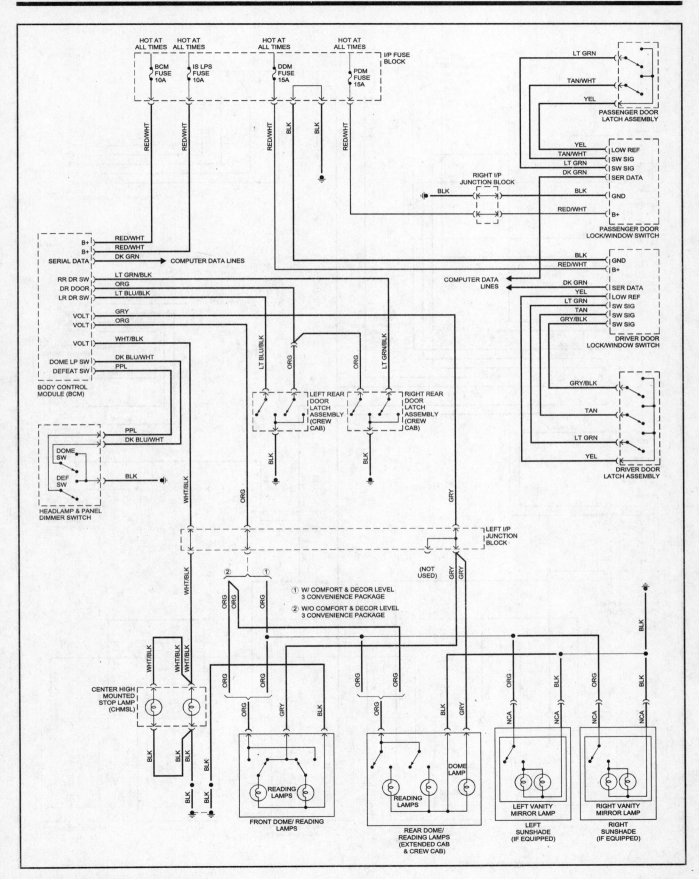

**Courtesy light system - truck models**

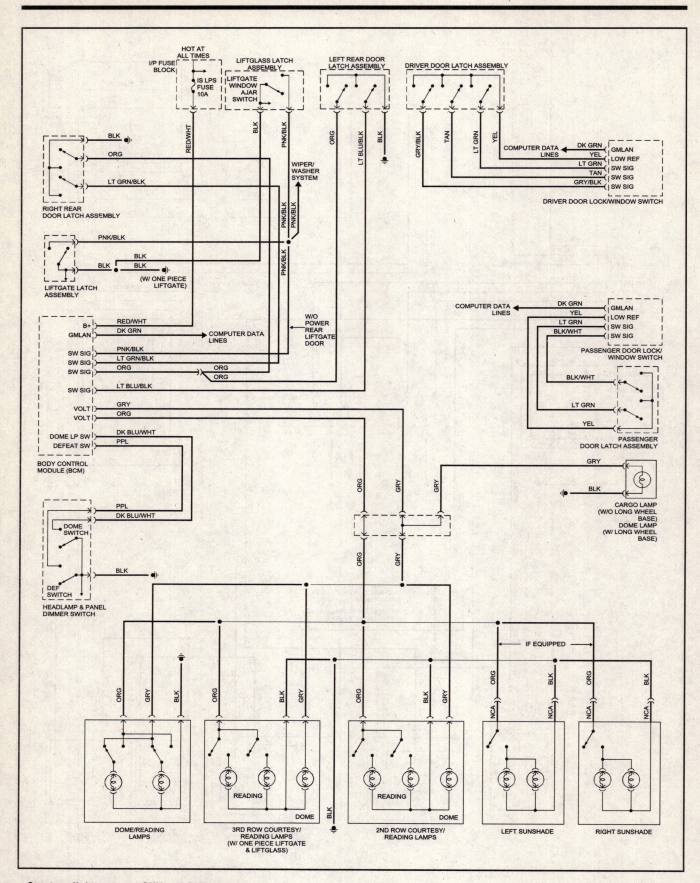

**Courtesy light system - SUV and SUT models**

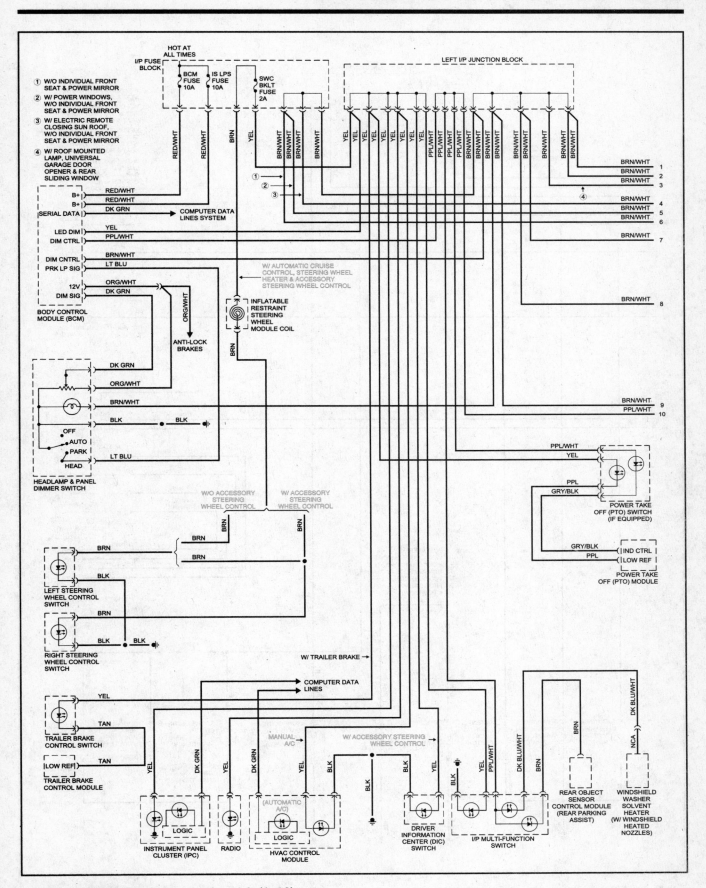

**Instrument illumination system - truck models (1 of 2)**

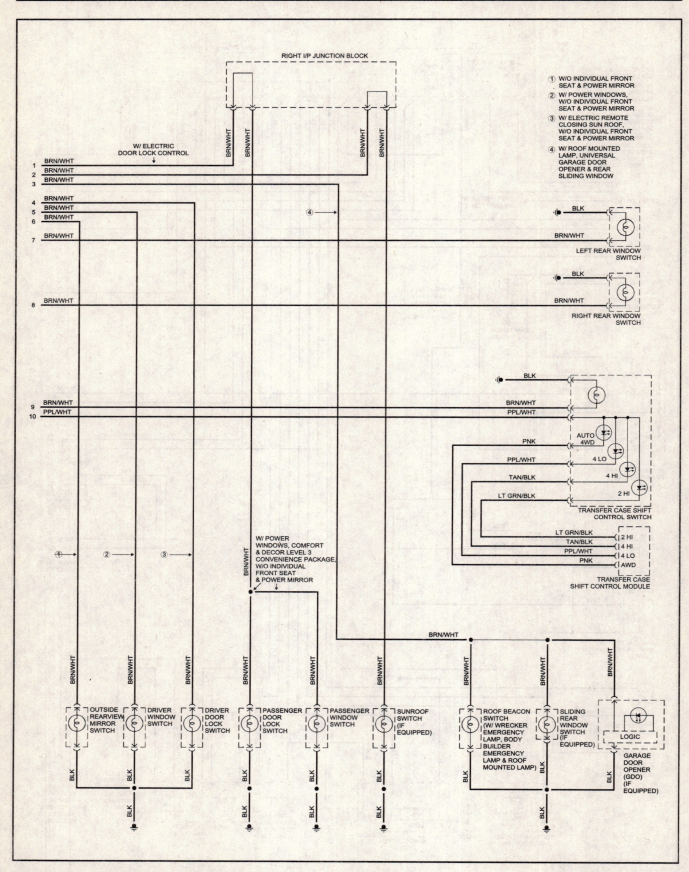

**Instrument illumination system - truck models (2 of 2)**

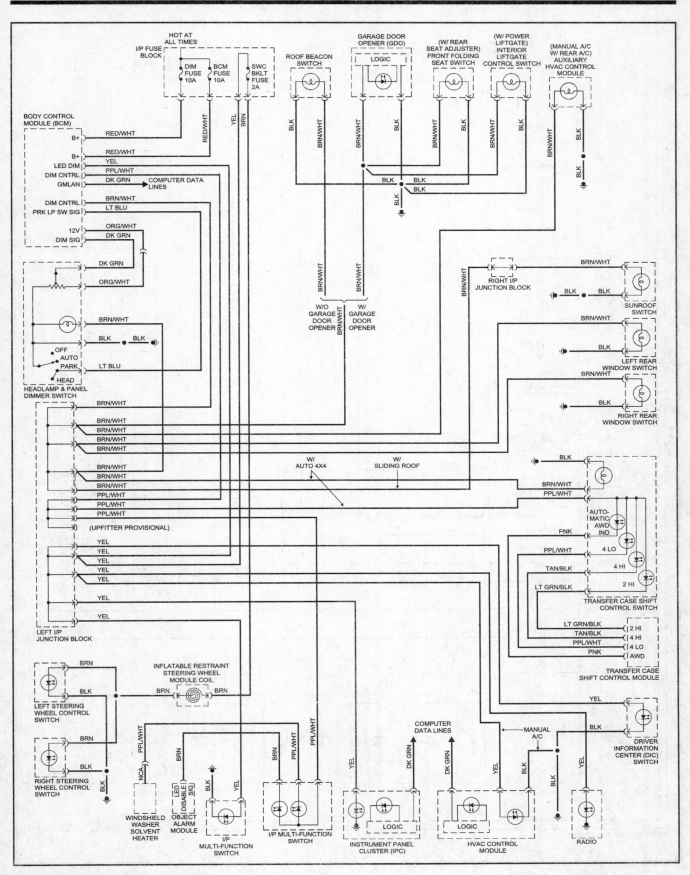

**Instrument illumination system - SUV and SUT models**

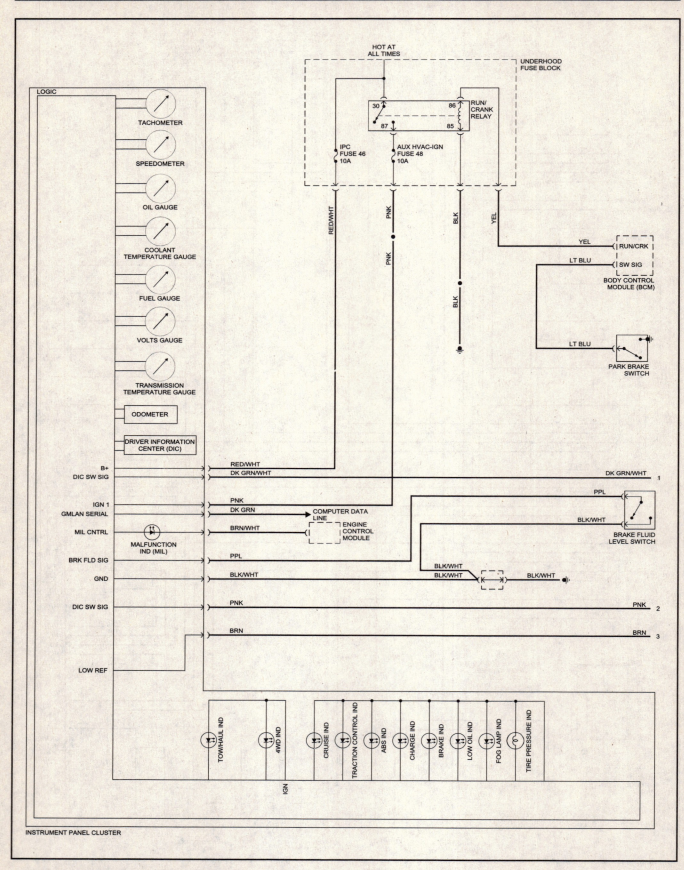

**Gauges and warning lights system - all models (1 of 2)**

## CHASSIS ELECTRICAL SYSTEM 12-45

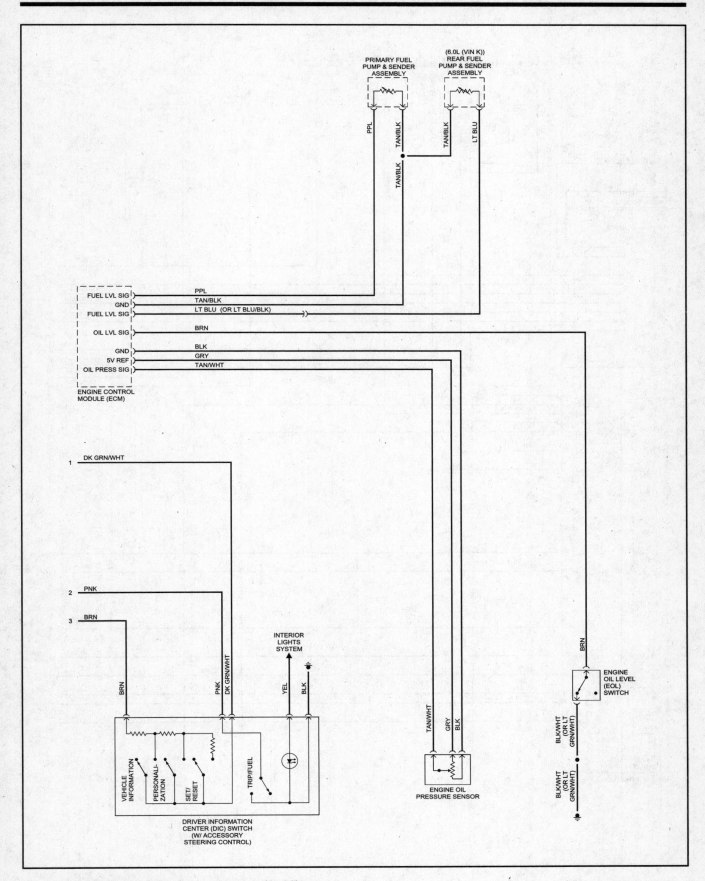

**Gauges and warning lights system - all models (2 of 2)**

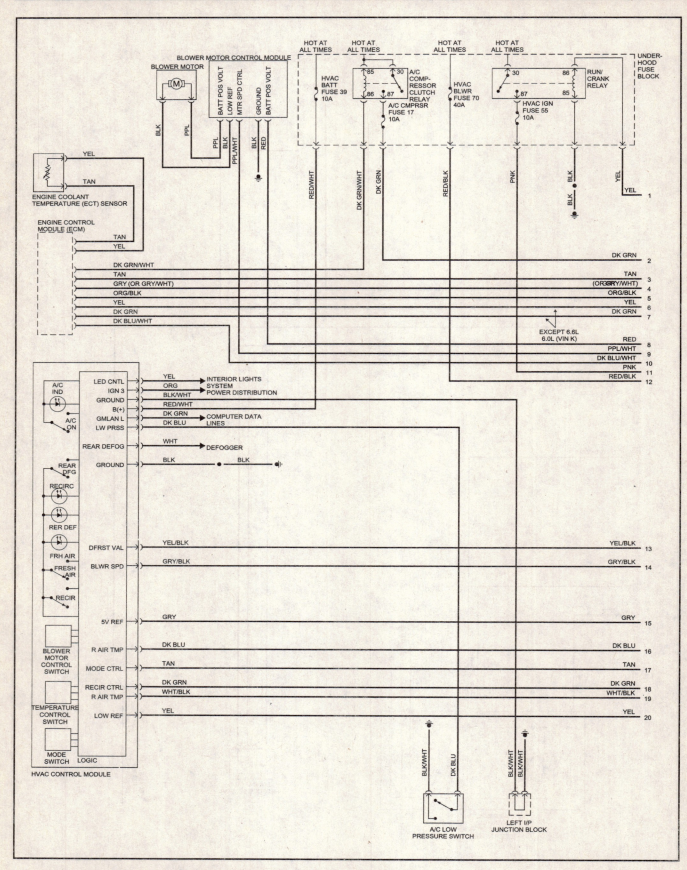

Air conditioning (manual) and engine cooling fans system - truck models (1 of 3)

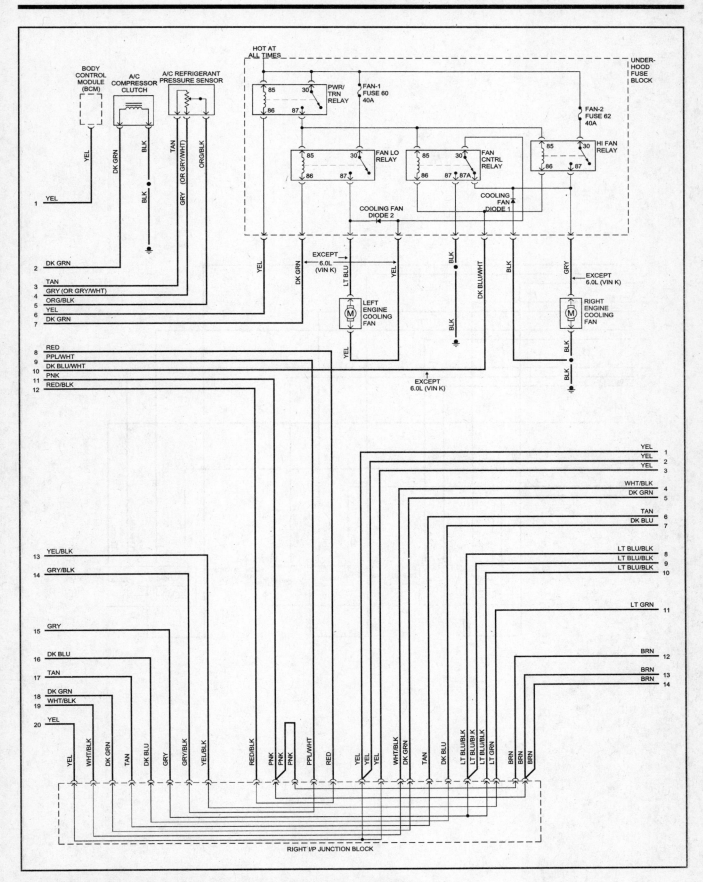

Air conditioning (manual) and engine cooling fans system - truck models (2 of 3)

## 12-48 CHASSIS ELECTRICAL SYSTEM

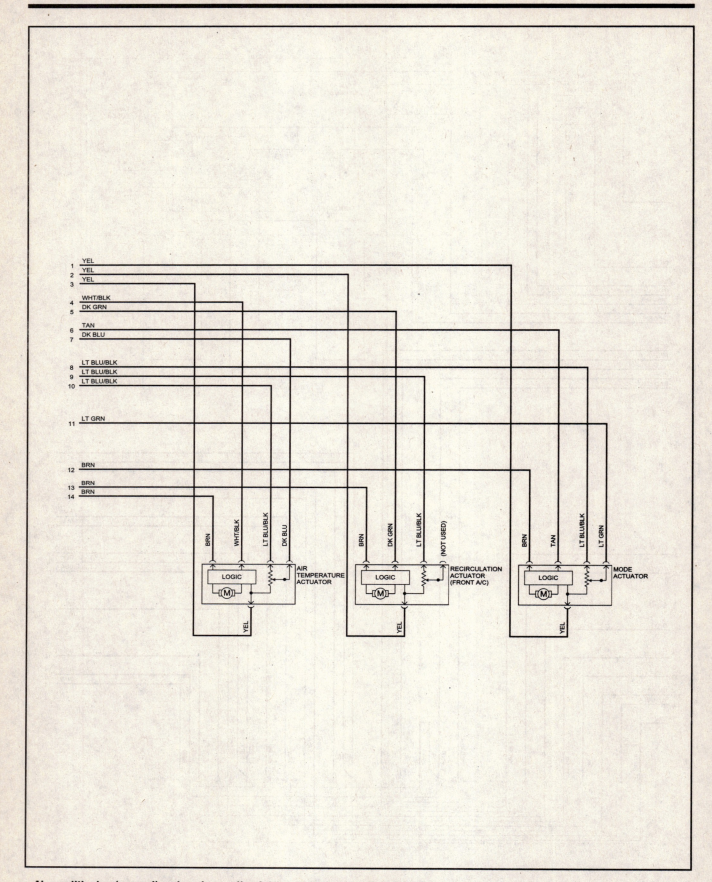

Air conditioning (manual) and engine cooling fans system - truck models (3 of 3)

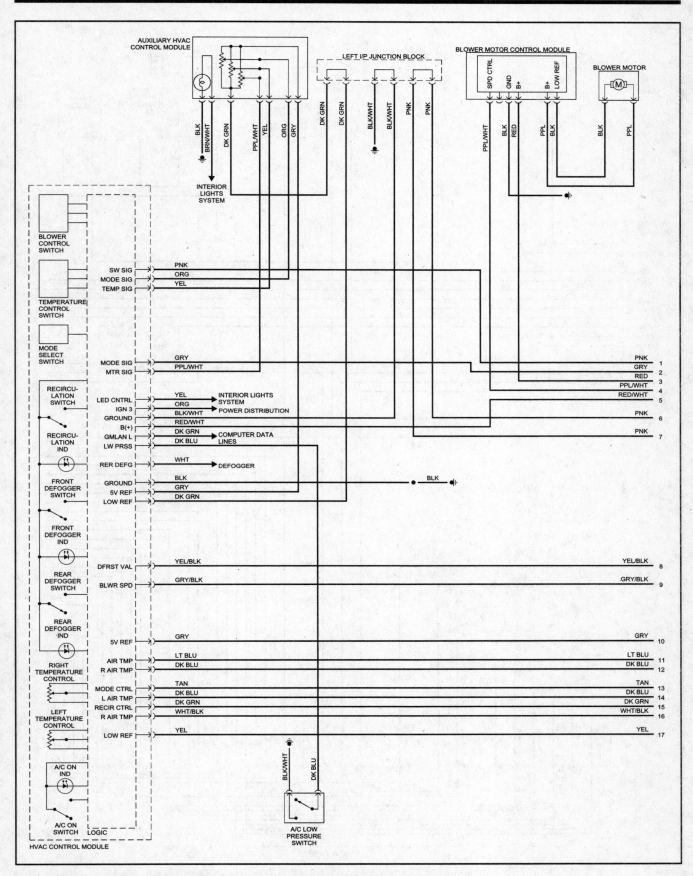

**Air conditioning (manual) and engine cooling fans system - SUV and SUT models (1 of 4)**

## 12-50 CHASSIS ELECTRICAL SYSTEM

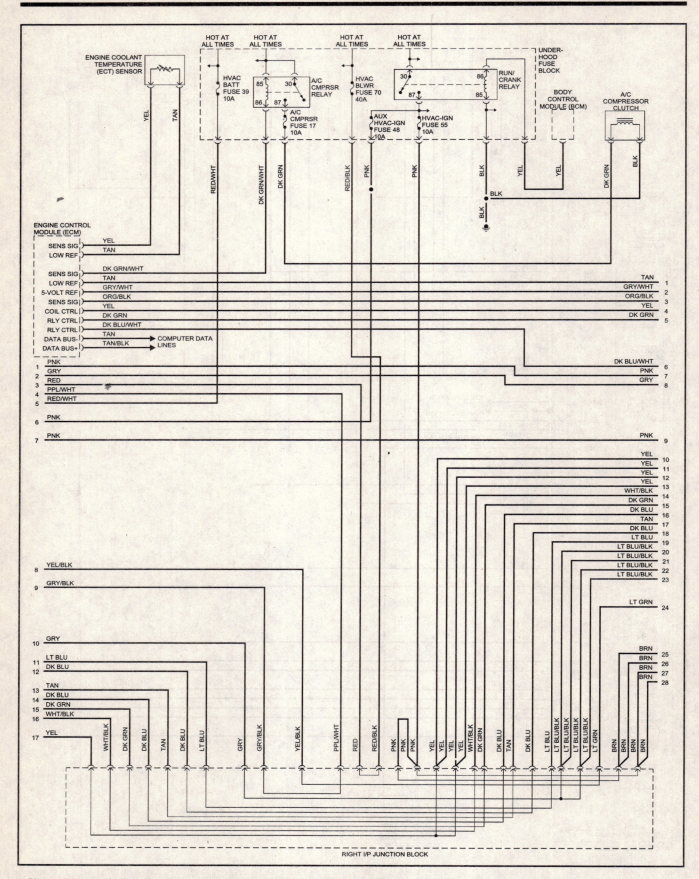

**Air conditioning (manual) and engine cooling fans system - SUV and SUT models (2 of 4)**

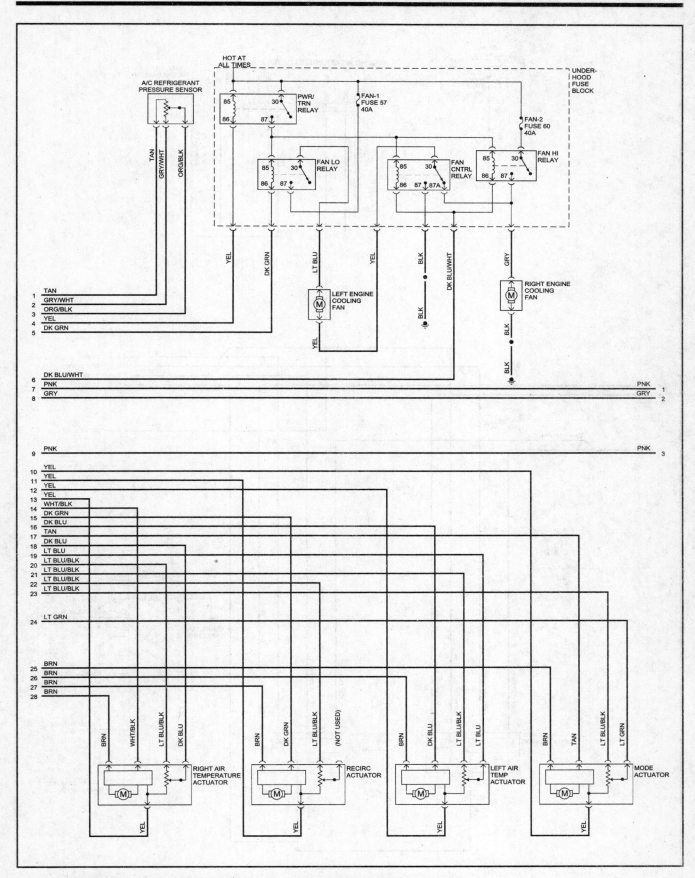

**Air conditioning (manual) and engine cooling fans system - SUV and SUT models (3 of 4)**

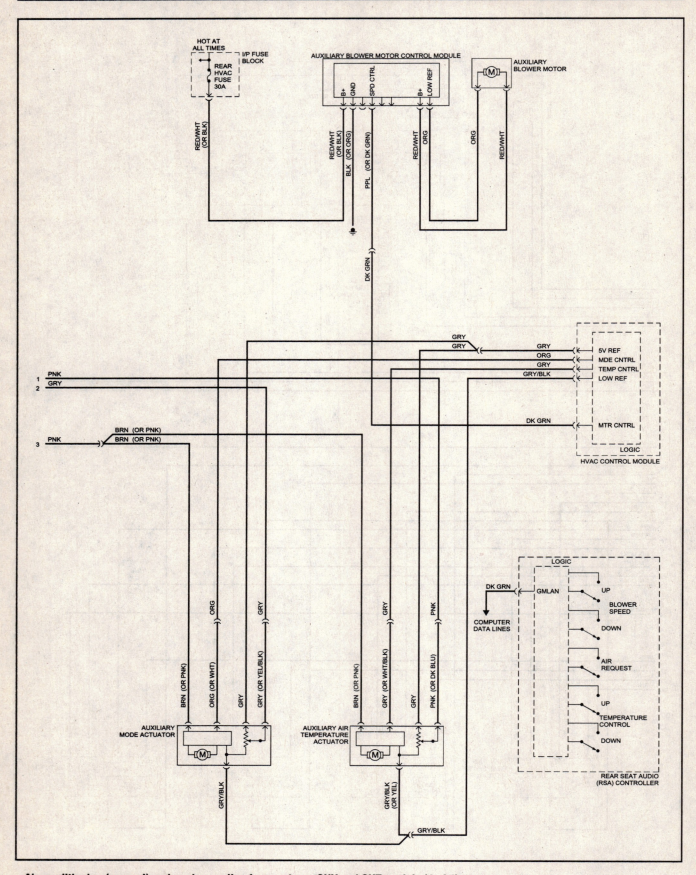

**Air conditioning (manual) and engine cooling fans system - SUV and SUT models (4 of 4)**

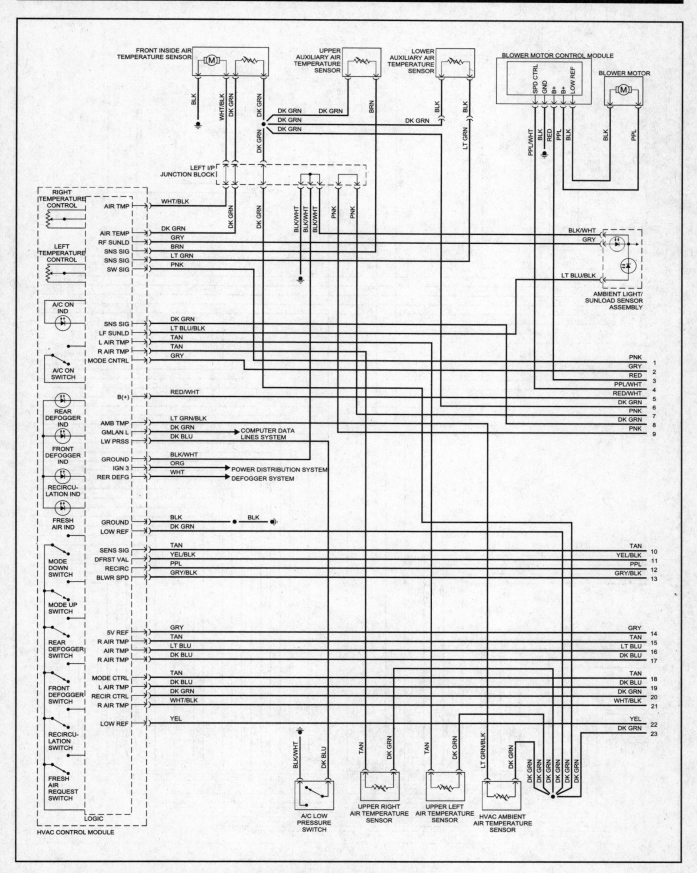

**Air conditioning (automatic) and engine cooling fans system - all models (1 of 4)**

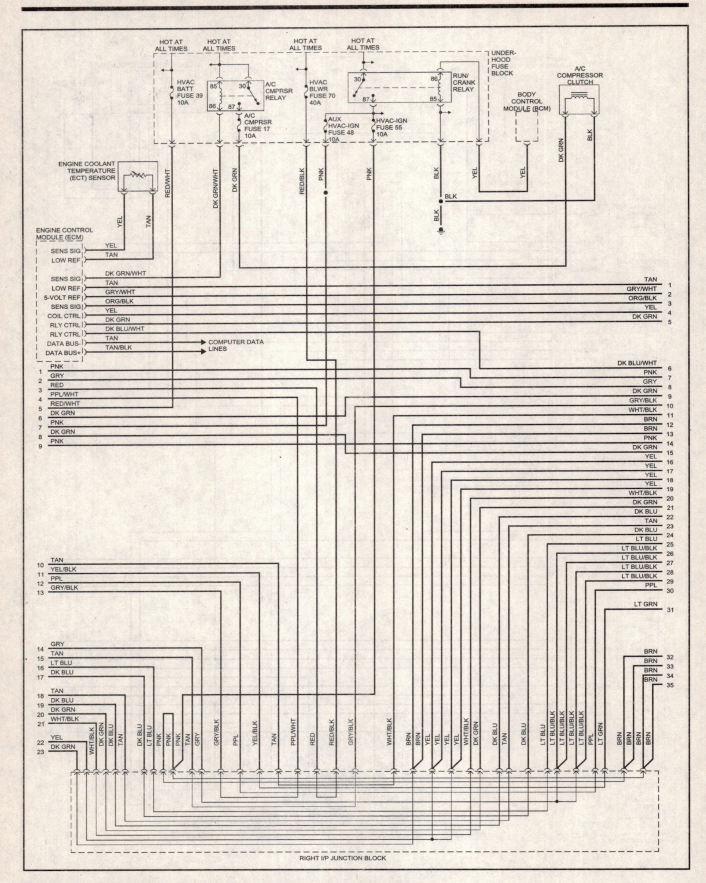

Air conditioning (automatic) and engine cooling fans system - all models (2 of 4)

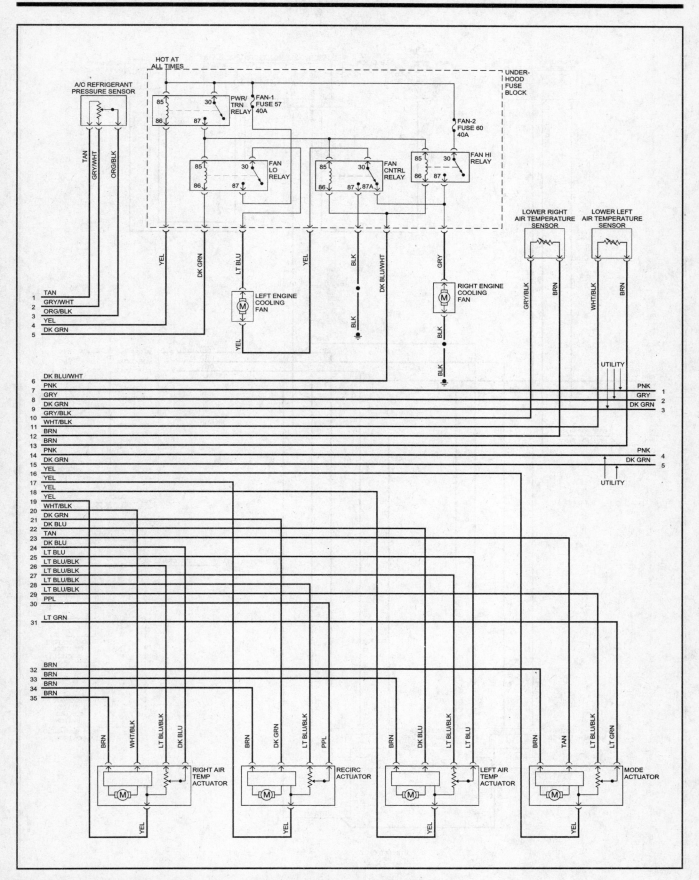

Air conditioning (automatic) and engine cooling fans system - all models (3 of 4)

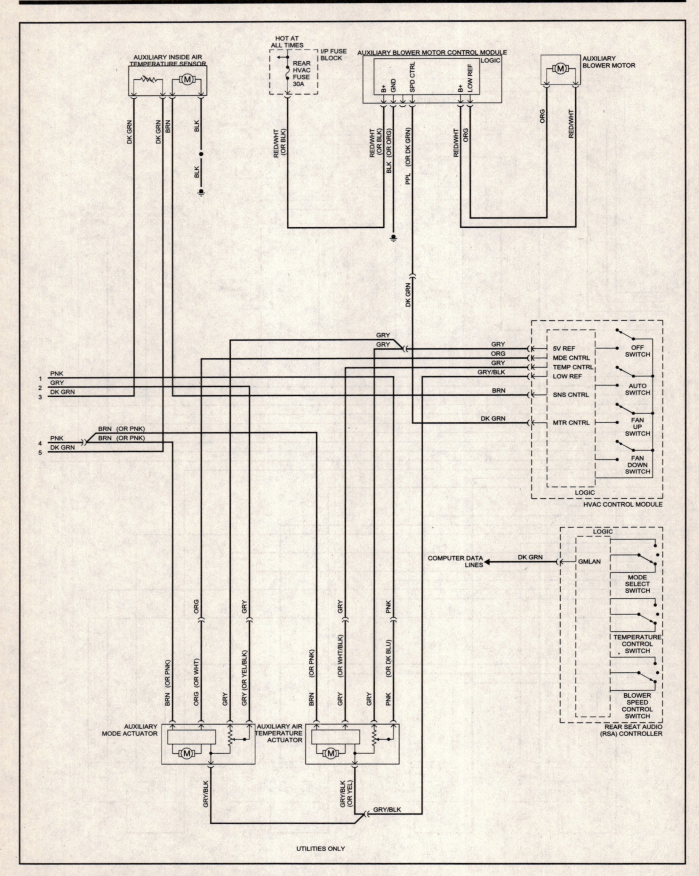

Air conditioning (automatic) and engine cooling fans system - all models (4 of 4)

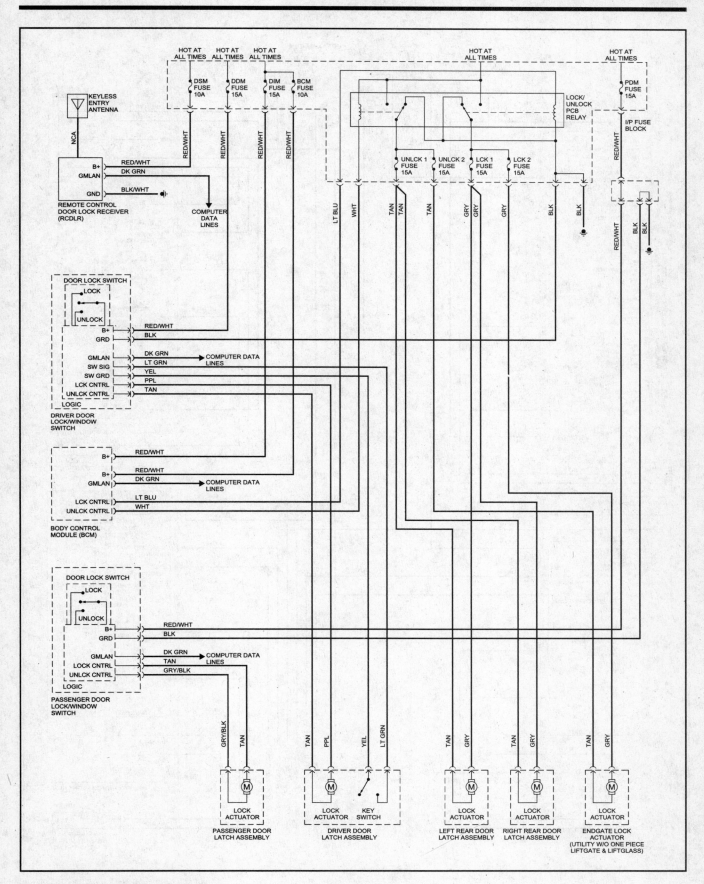

Power door locks system - all models

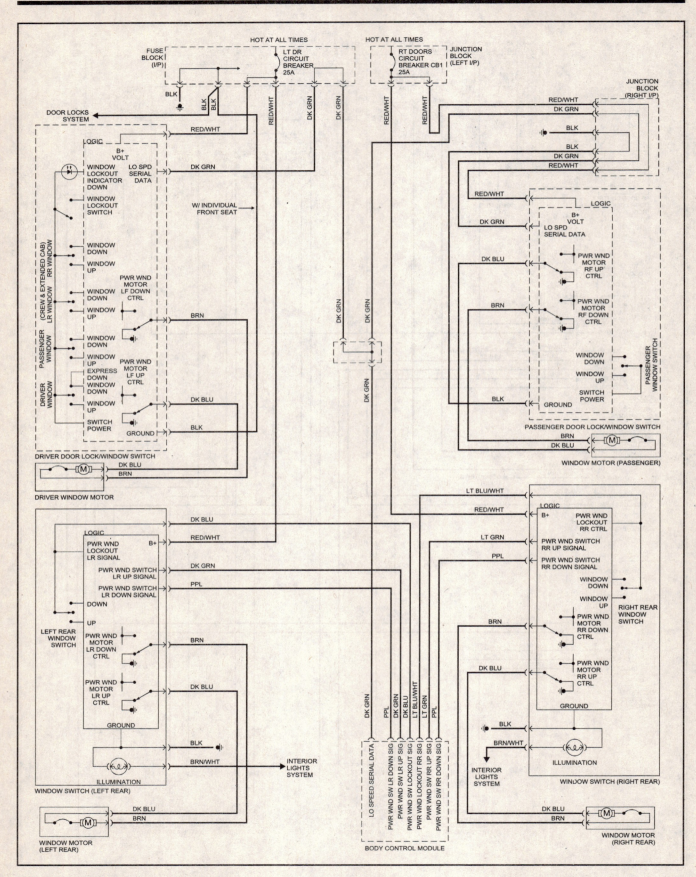

**Power windows system - truck models**

## CHASSIS ELECTRICAL SYSTEM 12-59

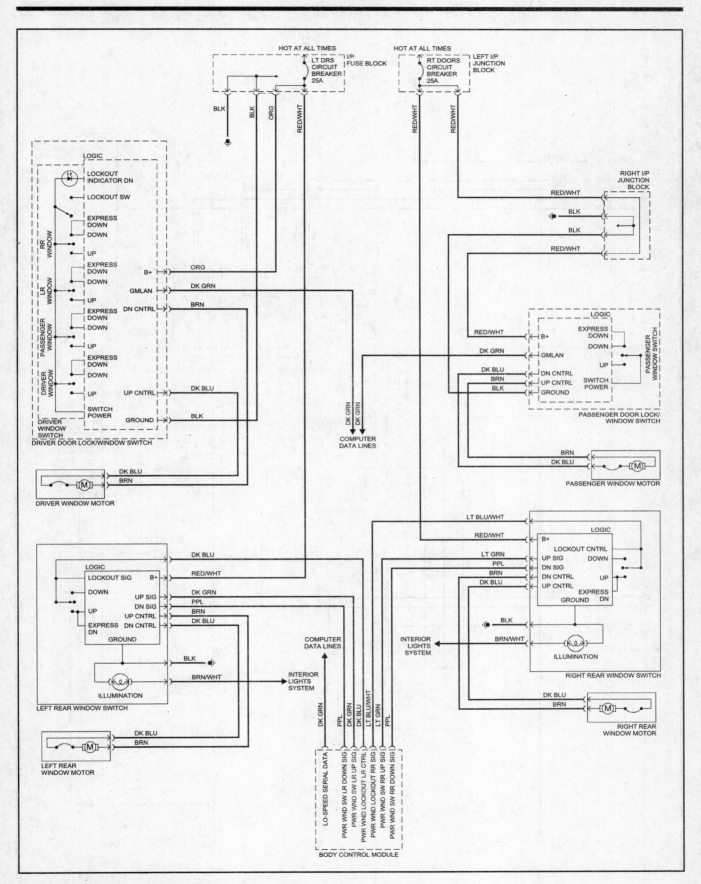

**Power windows system - SUV and SUT models**

## 12-60 CHASSIS ELECTRICAL SYSTEM

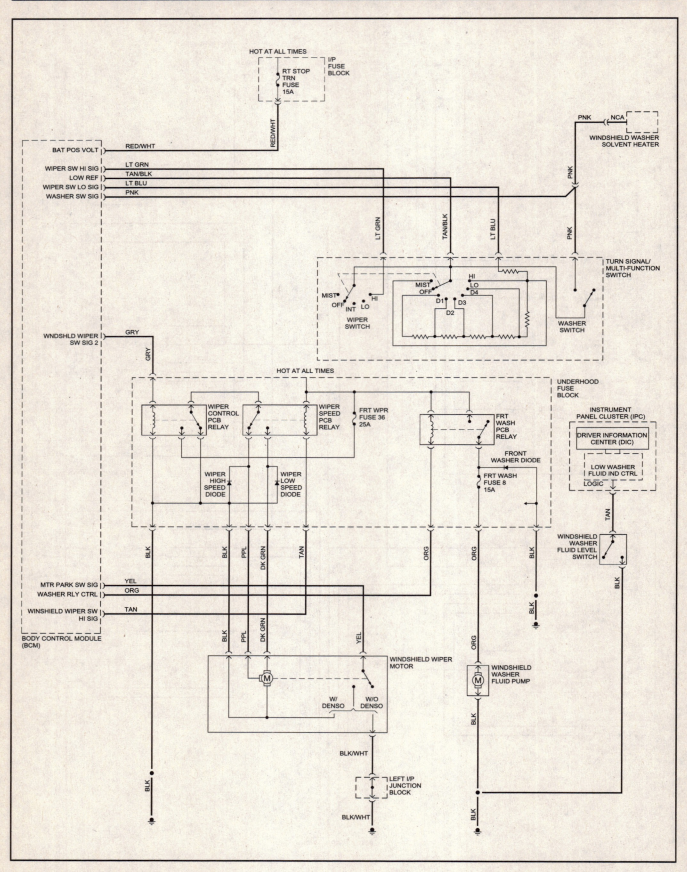

**Wiper/washer system (front) - all models**

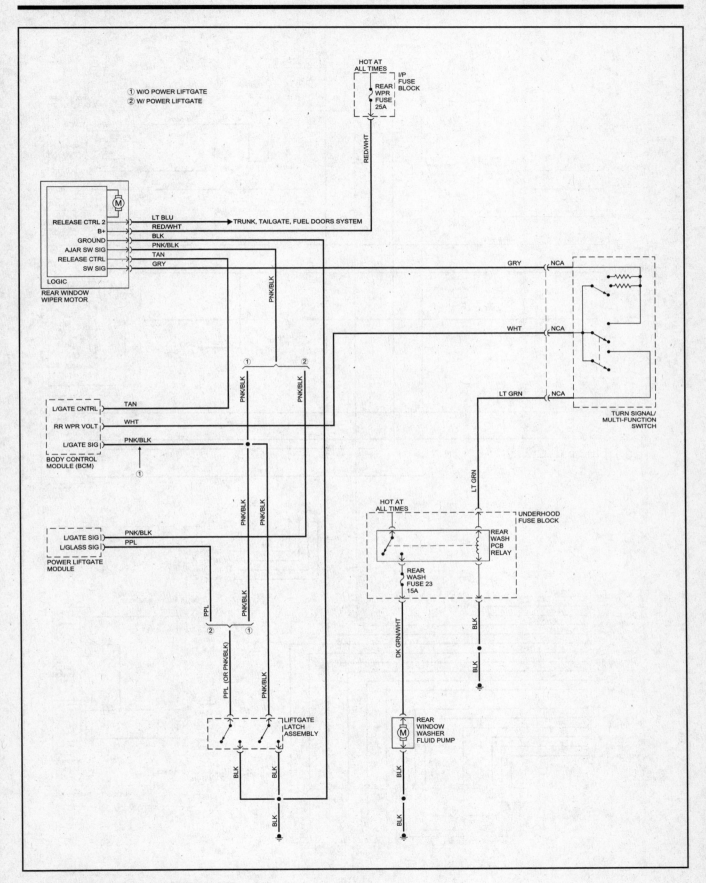

**Rear wiper/washer system - SUV and SUT models**

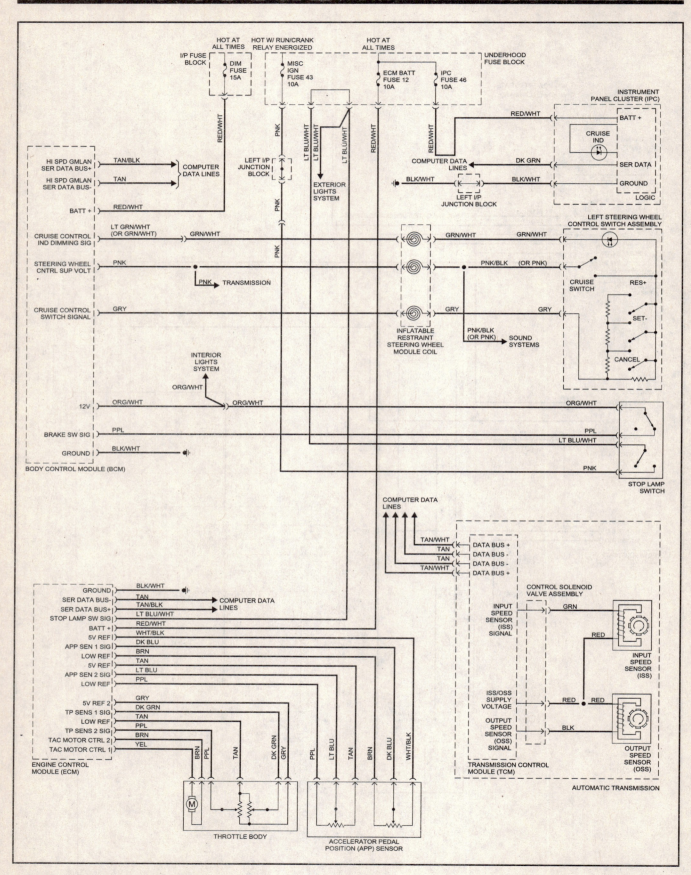

**Cruise control system - 6.0L engine**

# CHASSIS ELECTRICAL SYSTEM 12-63

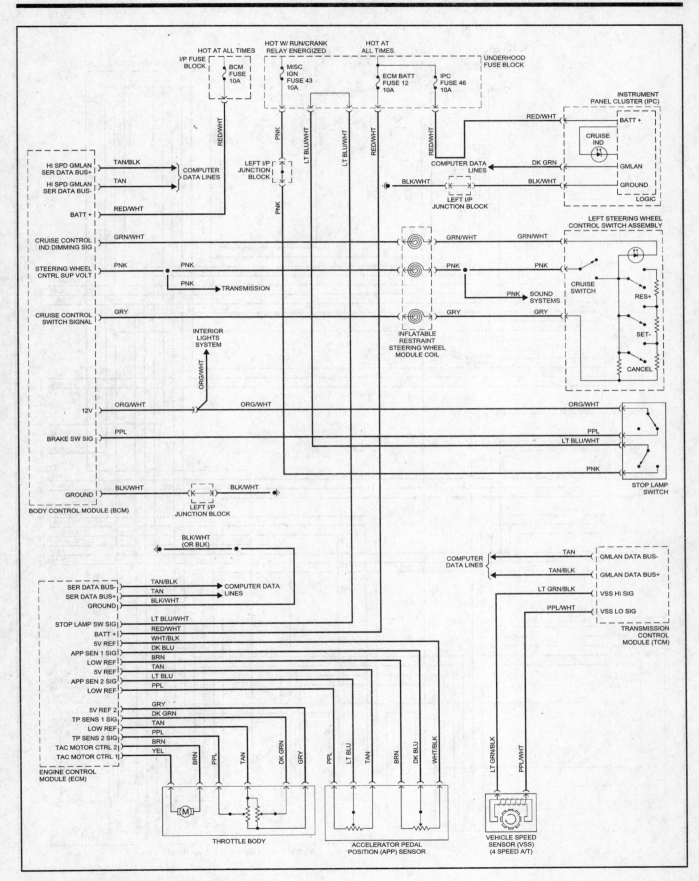

Cruise control system - all except 6.0L engine

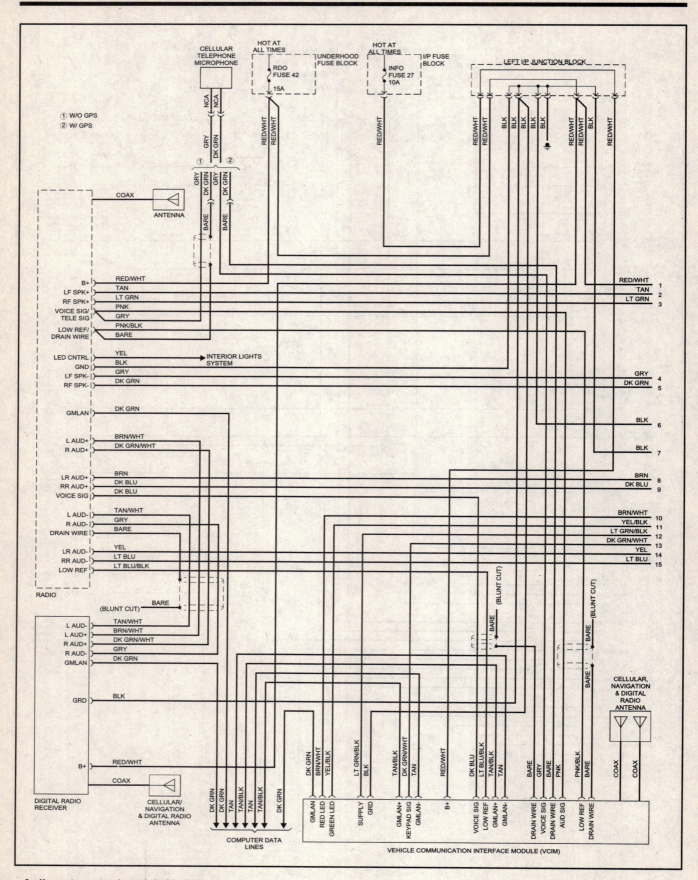

Audio system - truck models (1 of 3)

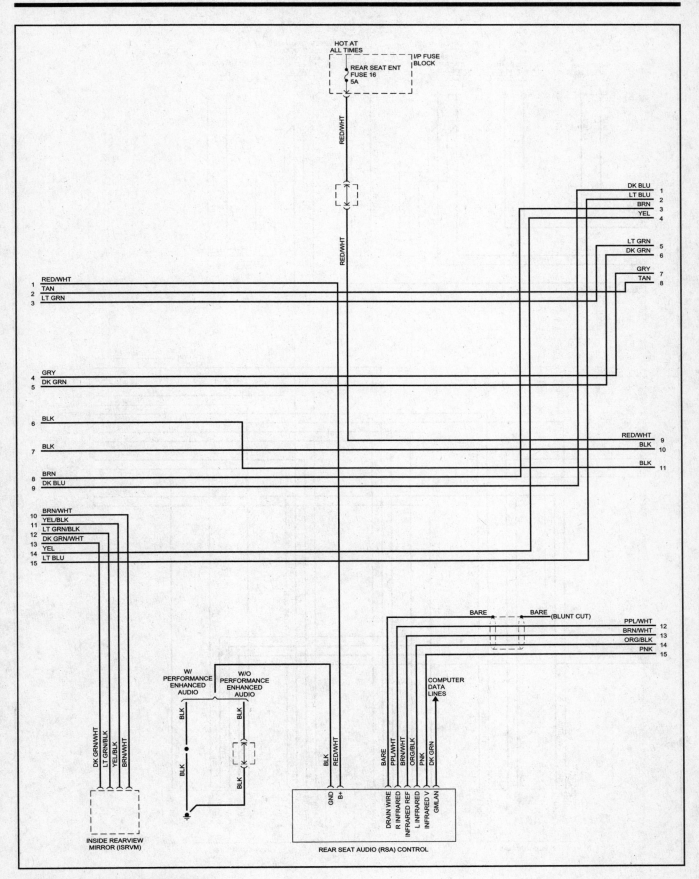
Audio system - truck models (2 of 3)

## 12-66 CHASSIS ELECTRICAL SYSTEM

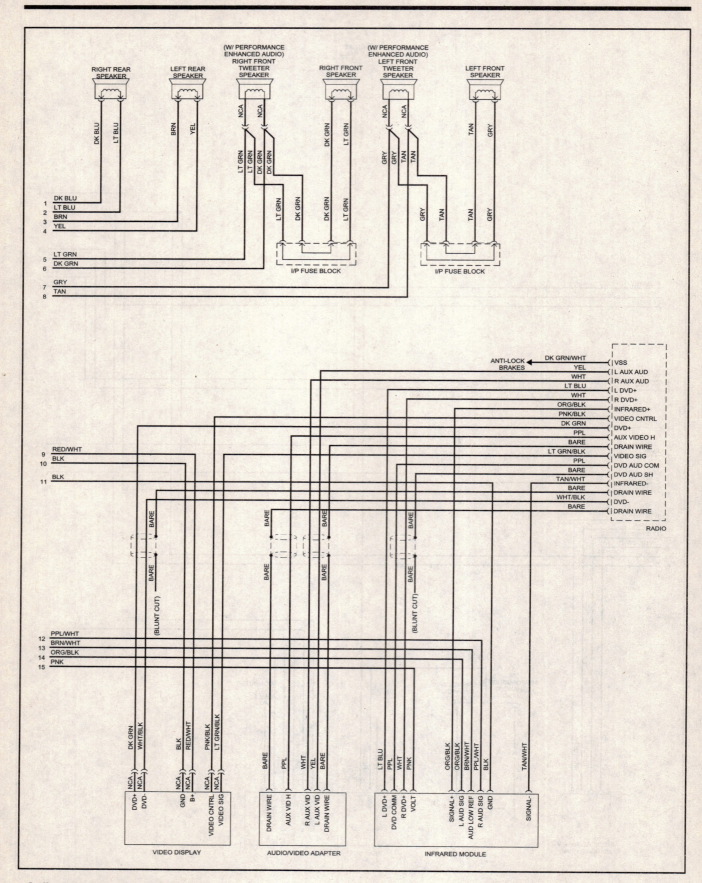

**Audio system - truck models (3 of 3)**

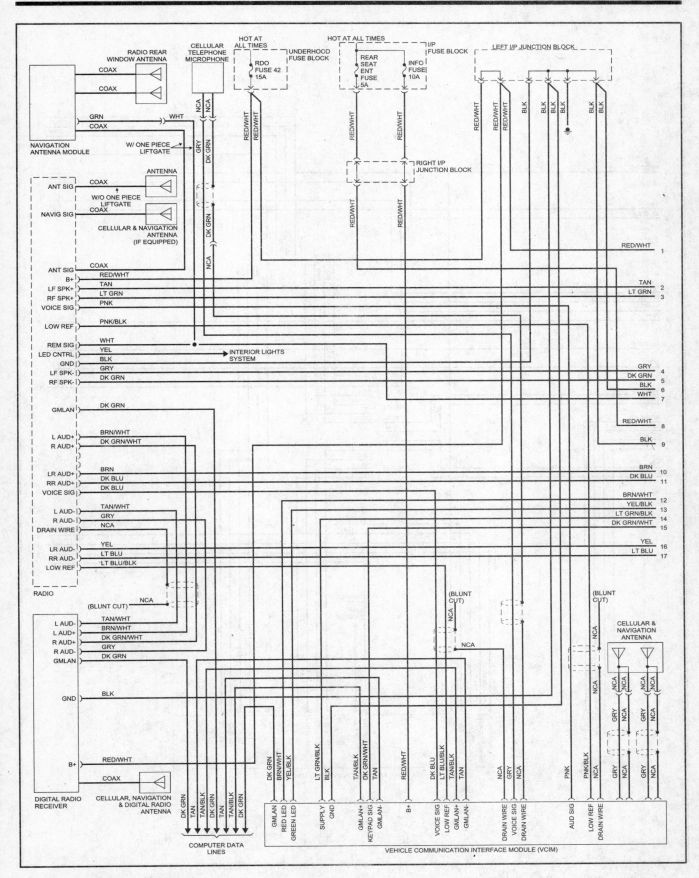

**Audio system - SUV and SUT models (1 of 3)**

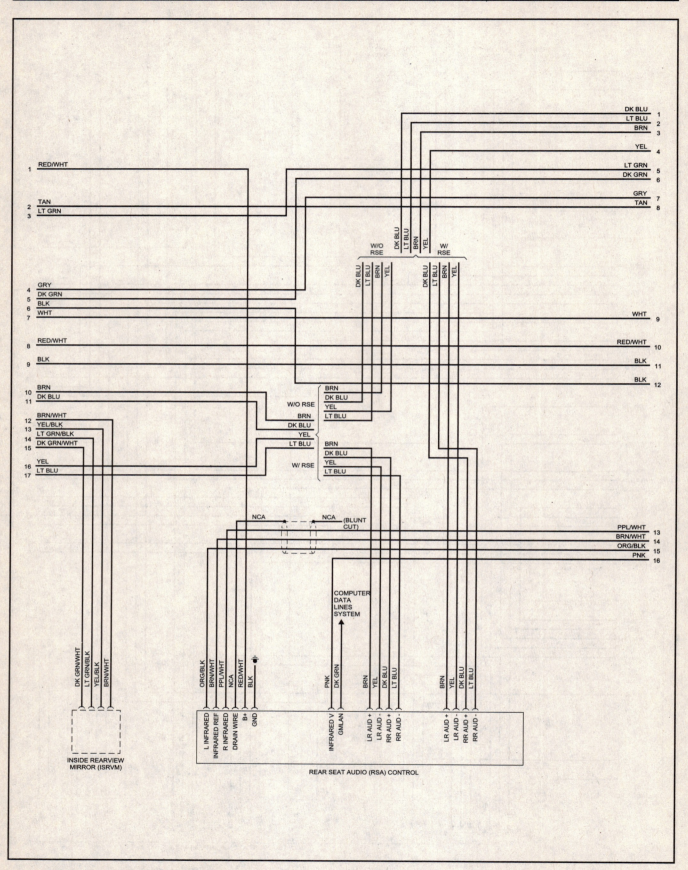

Audio system - SUV and SUT models (2 of 3)

# CHASSIS ELECTRICAL SYSTEM 12-69

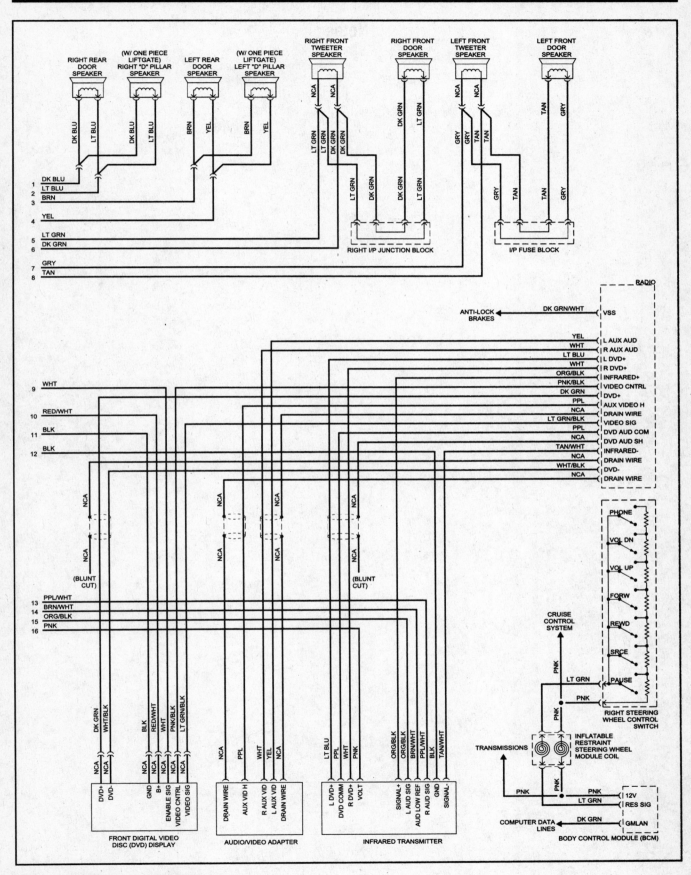

**Audio system - SUV and SUT models (3 of 3)**

## 12-70 CHASSIS ELECTRICAL SYSTEM

**Notes**

# GLOSSARY

**AIR/FUEL RATIO:** The ratio of air-to-gasoline by weight in the fuel mixture drawn into the engine.

**AIR INJECTION:** One method of reducing harmful exhaust emissions by injecting air into each of the exhaust ports of an engine. The fresh air entering the hot exhaust manifold causes any remaining fuel to be burned before it can exit the tailpipe.

**ALTERNATOR:** A device used for converting mechanical energy into electrical energy.

**AMMETER:** An instrument, calibrated in amperes, used to measure the flow of an electrical current in a circuit. Ammeters are always connected in series with the circuit being tested.

**AMPERE:** The rate of flow of electrical current present when one volt of electrical pressure is applied against one ohm of electrical resistance.

**ANALOG COMPUTER:** Any microprocessor that uses similar (analogous) electrical signals to make its calculations.

**ARMATURE:** A laminated, soft iron core wrapped by a wire that converts electrical energy to mechanical energy as in a motor or relay. When rotated in a magnetic field, it changes mechanical energy into electrical energy as in a generator.

**ATMOSPHERIC PRESSURE:** The pressure on the Earth's surface caused by the weight of the air in the atmosphere. At sea level, this pressure is 14.7 psi at 32°F (101 kPa at 0°C).

**ATOMIZATION:** The breaking down of a liquid into a fine mist that can be suspended in air.

**AXIAL PLAY:** Movement parallel to a shaft or bearing bore.

**BACKFIRE:** The sudden combustion of gases in the intake or exhaust system that results in a loud explosion.

**BACKLASH:** The clearance or play between two parts, such as meshed gears.

**BACKPRESSURE:** Restrictions in the exhaust system that slow the exit of exhaust gases from the combustion chamber.

**BAKELITE:** A heat resistant, plastic insulator material commonly used in printed circuit boards and transistorized components.

**BALL BEARING:** A bearing made up of hardened inner and outer races between which hardened steel balls roll.

**BALLAST RESISTOR:** A resistor in the primary ignition circuit that lowers voltage after the engine is started to reduce wear on ignition components.

**BEARING:** A friction reducing, supportive device usually located between a stationary part and a moving part.

**BIMETAL TEMPERATURE SENSOR:** Any sensor or switch made of two dissimilar types of metal that bend when heated or cooled due to the different expansion rates of the alloys. These types of sensors usually function as an on/off switch.

**BLOWBY:** Combustion gases, composed of water vapor and unburned fuel, that leak past the piston rings into the crankcase during normal engine operation. These gases are removed by the PCV system to prevent the buildup of harmful acids in the crankcase.

**BRAKE PAD:** A brake shoe and lining assembly used with disc brakes.

**BRAKE SHOE:** The backing for the brake lining. The term is, however, usually applied to the assembly of the brake backing and lining.

**BUSHING:** A liner, usually removable, for a bearing; an anti-friction liner used in place of a bearing.

**CALIPER:** A hydraulically activated device in a disc brake system, which is mounted straddling the brake rotor (disc). The caliper contains at least one piston and two brake pads. Hydraulic pressure on the piston(s) forces the pads against the rotor.

**CAMSHAFT:** A shaft in the engine on which are the lobes (cams) which operate the valves. The camshaft is driven by the crankshaft, via a belt, chain or gears, at one half the crankshaft speed.

**CAPACITOR:** A device which stores an electrical charge.

**CARBON MONOXIDE (CO):** A colorless, odorless gas given off as a normal byproduct of combustion. It is poisonous and extremely dangerous in confined areas, building up slowly to toxic levels without warning if adequate ventilation is not available.

**CARBURETOR:** A device, usually mounted on the intake manifold of an engine, which mixes the air and fuel in the proper proportion to allow even combustion.

**CATALYTIC CONVERTER:** A device installed in the exhaust system, like a muffler, that converts harmful byproducts of combustion into carbon dioxide and water vapor by means of a heat-producing chemical reaction.

**CENTRIFUGAL ADVANCE:** A mechanical method of advancing the spark timing by using flyweights in the distributor that react to centrifugal force generated by the distributor shaft rotation.

**CHECK VALVE:** Any one-way valve installed to permit the flow of air, fuel or vacuum in one direction only.

**CHOKE:** A device, usually a moveable valve, placed in the intake path of a carburetor to restrict the flow of air.

**CIRCUIT:** Any unbroken path through which an electrical current can flow. Also used to describe fuel flow in some instances.

**CIRCUIT BREAKER:** A switch which protects an electrical circuit from overload by opening the circuit when the current flow exceeds a predetermined level. Some circuit breakers must be reset manually, while most reset automatically.

# GL-2 GLOSSARY

**COIL (IGNITION):** A transformer in the ignition circuit which steps up the voltage provided to the spark plugs.

**COMBINATION MANIFOLD:** An assembly which includes both the intake and exhaust manifolds in one casting.

**COMBINATION VALVE:** A device used in some fuel systems that routes fuel vapors to a charcoal storage canister instead of venting them into the atmosphere. The valve relieves fuel tank pressure and allows fresh air into the tank as the fuel level drops to prevent a vapor lock situation.

**COMPRESSION RATIO:** The comparison of the total volume of the cylinder and combustion chamber with the piston at BDC and the piston at TDC.

**CONDENSER:** 1. An electrical device which acts to store an electrical charge, preventing voltage surges. 2. A radiator-like device in the air conditioning system in which refrigerant gas condenses into a liquid, giving off heat.

**CONDUCTOR:** Any material through which an electrical current can be transmitted easily.

**CONTINUITY:** Continuous or complete circuit. Can be checked with an ohmmeter.

**COUNTERSHAFT:** An intermediate shaft which is rotated by a mainshaft and transmits, in turn, that rotation to a working part.

**CRANKCASE:** The lower part of an engine in which the crankshaft and related parts operate.

**CRANKSHAFT:** The main driving shaft of an engine which receives reciprocating motion from the pistons and converts it to rotary motion.

**CYLINDER:** In an engine, the round hole in the engine block in which the piston(s) ride.

**CYLINDER BLOCK:** The main structural member of an engine in which is found the cylinders, crankshaft and other principal parts.

**CYLINDER HEAD:** The detachable portion of the engine, usually fastened to the top of the cylinder block and containing all or most of the combustion chambers. On overhead valve engines, it contains the valves and their operating parts. On overhead cam engines, it contains the camshaft as well.

**DEAD CENTER:** The extreme top or bottom of the piston stroke.

**DETONATION:** An unwanted explosion of the air/fuel mixture in the combustion chamber caused by excess heat and compression, advanced timing, or an overly lean mixture. Also referred to as "ping".

**DIAPHRAGM:** A thin, flexible wall separating two cavities, such as in a vacuum advance unit.

**DIESELING:** A condition in which hot spots in the combustion chamber cause the engine to run on after the key is turned off.

**DIFFERENTIAL:** A geared assembly which allows the transmission of motion between drive axles, giving one axle the ability to turn faster than the other.

**DIODE:** An electrical device that will allow current to flow in one direction only.

**DISC BRAKE:** A hydraulic braking assembly consisting of a brake disc, or rotor, mounted on an axle, and a caliper assembly containing, usually two brake pads which are activated by hydraulic pressure. The pads are forced against the sides of the disc, creating friction which slows the vehicle.

**DISTRIBUTOR:** A mechanically driven device on an engine which is responsible for electrically firing the spark plug at a predetermined point of the piston stroke.

**DOWEL PIN:** A pin, inserted in mating holes in two different parts allowing those parts to maintain a fixed relationship.

**DRUM BRAKE:** A braking system which consists of two brake shoes and one or two wheel cylinders, mounted on a fixed backing plate, and a brake drum, mounted on an axle, which revolves around the assembly.

**DWELL:** The rate, measured in degrees of shaft rotation, at which an electrical circuit cycles on and off.

**ELECTRONIC CONTROL UNIT (ECU):** Ignition module, module, amplifier or igniter. See Module for definition.

**ELECTRONIC IGNITION:** A system in which the timing and firing of the spark plugs is controlled by an electronic control unit, usually called a module. These systems have no points or condenser.

**END-PLAY:** The measured amount of axial movement in a shaft.

**ENGINE:** A device that converts heat into mechanical energy.

**EXHAUST MANIFOLD:** A set of cast passages or pipes which conduct exhaust gases from the engine.

**FEELER GAUGE:** A blade, usually metal, or precisely predetermined thickness, used to measure the clearance between two parts.

**FIRING ORDER:** The order in which combustion occurs in the cylinders of an engine. Also the order in which spark is distributed to the plugs by the distributor.

**FLOODING:** The presence of too much fuel in the intake manifold and combustion chamber which prevents the air/fuel mixture from firing, thereby causing a no-start situation.

**FLYWHEEL:** A disc shaped part bolted to the rear end of the crankshaft. Around the outer perimeter is affixed the ring gear. The starter drive engages the ring gear, turning the flywheel, which rotates the crankshaft, imparting the initial starting motion to the engine.

**FOOT POUND (ft. lbs. or sometimes, ft.lb.)**: The amount of energy or work needed to raise an item weighing one pound, a distance of one foot.

# GLOSSARY GL-3

**FUSE:** A protective device in a circuit which prevents circuit overload by breaking the circuit when a specific amperage is present. The device is constructed around a strip or wire of a lower amperage rating than the circuit it is designed to protect. When an amperage higher than that stamped on the fuse is present in the circuit, the strip or wire melts, opening the circuit.

**GEAR RATIO:** The ratio between the number of teeth on meshing gears.

**GENERATOR:** A device which converts mechanical energy into electrical energy.

**HEAT RANGE:** The measure of a spark plug's ability to dissipate heat from its firing end. The higher the heat range, the hotter the plug fires.

**HUB:** The center part of a wheel or gear.

**HYDROCARBON (HC):** Any chemical compound made up of hydrogen and carbon. A major pollutant formed by the engine as a byproduct of combustion.

**HYDROMETER:** An instrument used to measure the specific gravity of a solution.

**INCH POUND (inch lbs.; sometimes in.lb. or in. lbs.):** One twelfth of a foot pound.

**INDUCTION:** A means of transferring electrical energy in the form of a magnetic field. Principle used in the ignition coil to increase voltage.

**INJECTOR:** A device which receives metered fuel under relatively low pressure and is activated to inject the fuel into the engine under relatively high pressure at a predetermined time.

**INPUT SHAFT:** The shaft to which torque is applied, usually carrying the driving gear or gears.

**INTAKE MANIFOLD:** A casting of passages or pipes used to conduct air or a fuel/air mixture to the cylinders.

**JOURNAL:** The bearing surface within which a shaft operates.

**KEY:** A small block usually fitted in a notch between a shaft and a hub to prevent slippage of the two parts.

**MANIFOLD:** A casting of passages or set of pipes which connect the cylinders to an inlet or outlet source.

**MANIFOLD VACUUM:** Low pressure in an engine intake manifold formed just below the throttle plates. Manifold vacuum is highest at idle and drops under acceleration.

**MASTER CYLINDER:** The primary fluid pressurizing device in a hydraulic system. In automotive use, it is found in brake and hydraulic clutch systems and is pedal activated, either directly or, in a power brake system, through the power booster.

**MODULE:** Electronic control unit, amplifier or igniter of solid state or integrated design which controls the current flow in the ignition primary circuit based on input from the pick-up coil. When the module opens the primary circuit, high secondary voltage is induced in the coil.

**NEEDLE BEARING:** A bearing which consists of a number (usually a large number) of long, thin rollers.

**OHM:** ($\Omega$) The unit used to measure the resistance of conductor-to-electrical flow. One ohm is the amount of resistance that limits current flow to one ampere in a circuit with one volt of pressure.

**OHMMETER:** An instrument used for measuring the resistance, in ohms, in an electrical circuit.

**OUTPUT SHAFT:** The shaft which transmits torque from a device, such as a transmission.

**OVERDRIVE:** A gear assembly which produces more shaft revolutions than that transmitted to it.

**OVERHEAD CAMSHAFT (OHC):** An engine configuration in which the camshaft is mounted on top of the cylinder head and operates the valve either directly or by means of rocker arms.

**OVERHEAD VALVE (OHV):** An engine configuration in which all of the valves are located in the cylinder head and the camshaft is located in the cylinder block. The camshaft operates the valves via lifters and pushrods.

**OXIDES OF NITROGEN (NOx):** Chemical compounds of nitrogen produced as a byproduct of combustion. They combine with hydrocarbons to produce smog.

**OXYGEN SENSOR:** Use with the feedback system to sense the presence of oxygen in the exhaust gas and signal the computer which can reference the voltage signal to an air/fuel ratio.

**PINION:** The smaller of two meshing gears.

**PISTON RING:** An open-ended ring with fits into a groove on the outer diameter of the piston. Its chief function is to form a seal between the piston and cylinder wall. Most automotive pistons have three rings: two for compression sealing; one for oil sealing.

**PRELOAD:** A predetermined load placed on a bearing during assembly or by adjustment.

**PRIMARY CIRCUIT:** the low voltage side of the ignition system which consists of the ignition switch, ballast resistor or resistance wire, bypass, coil, electronic control unit and pick-up coil as well as the connecting wires and harnesses.

**PRESS FIT:** The mating of two parts under pressure, due to the inner diameter of one being smaller than the outer diameter of the other, or vice versa; an interference fit.

**RACE:** The surface on the inner or outer ring of a bearing on which the balls, needles or rollers move.

**REGULATOR:** A device which maintains the amperage and/or voltage levels of a circuit at predetermined values.

# GL-4 GLOSSARY

**RELAY:** A switch which automatically opens and/or closes a circuit.

**RESISTANCE:** The opposition to the flow of current through a circuit or electrical device, and is measured in ohms. Resistance is equal to the voltage divided by the amperage.

**RESISTOR:** A device, usually made of wire, which offers a preset amount of resistance in an electrical circuit.

**RING GEAR:** The name given to a ring-shaped gear attached to a differential case, or affixed to a flywheel or as part of a planetary gear set.

**ROLLER BEARING:** A bearing made up of hardened inner and outer races between which hardened steel rollers move.

**ROTOR:** 1. The disc-shaped part of a disc brake assembly, upon which the brake pads bear; also called, brake disc. 2. The device mounted atop the distributor shaft, which passes current to the distributor cap tower contacts.

**SECONDARY CIRCUIT:** The high voltage side of the ignition system, usually above 20,000 volts. The secondary includes the ignition coil, coil wire, distributor cap and rotor, spark plug wires and spark plugs.

**SENDING UNIT:** A mechanical, electrical, hydraulic or electromagnetic device which transmits information to a gauge.

**SENSOR:** Any device designed to measure engine operating conditions or ambient pressures and temperatures. Usually electronic in nature and designed to send a voltage signal to an on-board computer, some sensors may operate as a simple on/off switch or they may provide a variable voltage signal (like a potentiometer) as conditions or measured parameters change.

**SHIM:** Spacers of precise, predetermined thickness used between parts to establish a proper working relationship.

**SLAVE CYLINDER:** In automotive use, a device in the hydraulic clutch system which is activated by hydraulic force, disengaging the clutch.

**SOLENOID:** A coil used to produce a magnetic field, the effect of which is to produce work.

**SPARK PLUG:** A device screwed into the combustion chamber of a spark ignition engine. The basic construction is a conductive core inside of a ceramic insulator, mounted in an outer conductive base. An electrical charge from the spark plug wire travels along the conductive core and jumps a preset air gap to a grounding point or points at the end of the conductive base. The resultant spark ignites the fuel/air mixture in the combustion chamber.

**SPLINES:** Ridges machined or cast onto the outer diameter of a shaft or inner diameter of a bore to enable parts to mate without rotation.

**TACHOMETER:** A device used to measure the rotary speed of an engine, shaft, gear, etc., usually in rotations per minute.

**THERMOSTAT:** A valve, located in the cooling system of an engine, which is closed when cold and opens gradually in response to engine heating, controlling the temperature of the coolant and rate of coolant flow.

**TOP DEAD CENTER (TDC):** The point at which the piston reaches the top of its travel on the compression stroke.

**TORQUE:** The twisting force applied to an object.

**TORQUE CONVERTER:** A turbine used to transmit power from a driving member to a driven member via hydraulic action, providing changes in drive ratio and torque. In automotive use, it links the driveplate at the rear of the engine to the automatic transmission.

**TRANSDUCER:** A device used to change a force into an electrical signal.

**TRANSISTOR:** A semi-conductor component which can be actuated by a small voltage to perform an electrical switching function.

**TUNE-UP:** A regular maintenance function, usually associated with the replacement and adjustment of parts and components in the electrical and fuel systems of a vehicle for the purpose of attaining optimum performance.

**TURBOCHARGER:** An exhaust driven pump which compresses intake air and forces it into the combustion chambers at higher than atmospheric pressures. The increased air pressure allows more fuel to be burned and results in increased horsepower being produced.

**VACUUM ADVANCE:** A device which advances the ignition timing in response to increased engine vacuum.

**VACUUM GAUGE:** An instrument used to measure the presence of vacuum in a chamber.

**VALVE:** A device which control the pressure, direction of flow or rate of flow of a liquid or gas.

**VALVE CLEARANCE:** The measured gap between the end of the valve stem and the rocker arm, cam lobe or follower that activates the valve.

**VISCOSITY:** The rating of a liquid's internal resistance to flow.

**VOLTMETER:** An instrument used for measuring electrical force in units called volts. Voltmeters are always connected parallel with the circuit being tested.

**WHEEL CYLINDER:** Found in the automotive drum brake assembly, it is a device, actuated by hydraulic pressure, which, through internal pistons, pushes the brake shoes outward against the drums.

# MASTER INDEX

## A

**ABOUT THIS MANUAL, 0-5**
**ACCELERATOR PEDAL POSITION (APP) SENSOR, REPLACEMENT, 6-15**
**ACCUMULATOR, AIR CONDITIONING, REMOVAL AND INSTALLATION, 3-19**
**ACKNOWLEDGEMENTS, 0-2**
**AIR CONDITIONING**
accumulator, removal and installation, 3-19
and heating system, check and maintenance, 3-16
compressor, removal and installation, 3-20
condenser, removal and installation, 3-21
expansion (orifice) tube, replacement, 3-23
low pressure switch and pressure sensor, removal and installation, 3-23
**AIR FILTER**
element, replacement, 1-27
housing, removal and installation, 4-12
**AIRBAG SYSTEM, GENERAL INFORMATION AND PRECAUTIONS, 12-25**
**ALIGNMENT, FRONT END GENERAL INFORMATION, 10-27**
**ALTERNATOR, REMOVAL AND INSTALLATION, 5-11**
**ANTENNA AND CABLE, REMOVAL AND INSTALLATION, 12-13**
**ANTIFREEZE, GENERAL INFORMATION, 3-3**
**ANTI-LOCK BRAKE SYSTEM (ABS), TRACTION CONTROL (TCS) AND VEHICLE STABILITY CONTROL (STABILITRAK) SYSTEMS, GENERAL INFORMATION, 9-2**
**AUTOMATIC TRANSMISSION, 7A-1**
diagnosis, general, 7A-2
extension housing oil seal (2WD), replacement, 7A-6
fluid
    and filter change, 1-34
    level check, 1-12
    type, 1-37
Park/Lock system, description and component replacement, 7A-5
removal and installation, 7A-8
shift cable, removal, installation and adjustment, 7A-3
transmission mount, check and replacement, 7A-7

# MASTER INDEX

**AUTOMOTIVE CHEMICALS AND LUBRICANTS, 0-21**
**AXLE(S)**
assembly, rear, removal and installation, 8-11
description and check, 8-6
front shift motor (4WD models), replacement, 8-19
**AXLESHAFT**
bearing (rear, semi-floating axle), replacement, 8-9
oil seal (rear, semi-floating axle), replacement, 8-8
oil seals and bearings (front, 4WD models), replacement, 8-21
rear, removal and installation, 8-7
right axleshaft, tube, bearing and shift fork (4WD and AWD models), removal, component replacement and installation, 8-20

# B

**BALLJOINTS, CHECK AND REPLACEMENT, 10-10**
**BATTERY**
cables, check and replacement, 5-5
check and replacement, 5-3
check, maintenance and charging, 1-15
electrolyte level check, 1-9
precautions and disconnection, 5-2
**BLOWER MOTOR AND CONTROL MODULE, REMOVAL AND INSTALLATION, 3-10**
**BODY, 11-1**
**BODY REPAIR**
major damage, 11-8
minor damage, 11-4
**BODY, MAINTENANCE, 11-3**
**BOOSTER BATTERY (JUMP) STARTING, 0-20**
**BRAKES, 9-1**
Anti-lock Brake System (ABS), general information, 9-2
caliper, removal and installation, 9-7
disc, inspection, removal and installation, 9-7
fluid
    change, 1-27
    level check, 1-9
    type, 1-37
general information and precautions, 9-2
hoses and lines, check and replacement, 9-18
hydraulic system, bleeding, 9-19
light switch, check, adjustment and replacement, 9-23
master cylinder, removal, installation and reservoir/O-ring replacement, 9-14
pads, replacement, 9-4
parking brake
    adjustment, 9-21
    shoes, replacement, 9-21
pedal travel, check, 9-21
power brake booster, check, removal and installation, 9-20
shoes, replacement
    2008 and earlier models, 9-9
    2009 and later models, 9-11
system check, 1-24
Traction Control (TCS), general information, 9-2
vehicle stability control (StabiliTrak), general information, 9-2
wheel cylinder, removal and installation, 9-15
**BULB REPLACEMENT, 12-17**
**BUMPERS, REMOVAL AND INSTALLATION, 11-12**
**BUYING PARTS, 0-10**

# C

**CABLE REPLACEMENT**
antenna, 12-13
battery, 5-5
hood release, 11-11
**CALIPER, DISC BRAKE, REMOVAL AND INSTALLATION, 9-7**
**CAMSHAFT AND LIFTERS, REMOVAL, INSPECTION AND INSTALLATION**
V6 engine, 2A-14
V8 engines, 2B-16
**CAMSHAFT POSITION (CMP) ACTUATOR SYSTEM, DESCRIPTION AND COMPONENT REPLACEMENT, 6-28**
**CAMSHAFT POSITION (CMP) SENSOR, REPLACEMENT, 6-15**
**CAPACITIES, FLUIDS AND LUBRICANTS, 1-37**
**CATALYTIC CONVERTERS, DESCRIPTION, CHECK AND REPLACEMENT, 6-24**
**CENTER CONSOLE, REMOVAL AND INSTALLATION, 11-27**
**CENTER SUPPORT BEARING, DRIVESHAFT, REPLACEMENT, 8-4**
**CHARGING SYSTEM**
alternator, removal and installation, 5-11
check, 5-10
general information and precautions, 5-10
**CHASSIS ELECTRICAL SYSTEM, 12-1**
**CHASSIS LUBRICATION, 1-22**
**CHEMICALS AND LUBRICANTS, 0-21**
**CIRCUIT BREAKERS, GENERAL INFORMATION, 12-5**
**COIL PACK, IGNITION, REPLACEMENT, 5-9**
**COIL SPRING, REMOVAL AND INSTALLATION**
and shock absorber, front, 10-5
rear, 10-15
**COMPRESSOR, AIR CONDITIONING, REMOVAL AND INSTALLATION, 3-20**
**CONDENSER, AIR CONDITIONING, REMOVAL AND INSTALLATION, 3-21**
**CONTROL ARM, REMOVAL AND INSTALLATION**
lower, 10-9
upper, 10-8
**CONTROL ASSEMBLY, HEATER AND AIR CONDITIONING, REMOVAL AND INSTALLATION, 3-13**
**CONVERSION FACTORS, 0-22**

# MASTER INDEX  IND-3

**COOLANT**
expansion tank, removal and installation, 3-7
general information, 3-3
level check, 1-8
temperature gauge sending unit, check and replacement, 3-10
type, 1-37
**COOLANT TEMPERATURE (ECT) SENSOR, REPLACEMENT, 6-17**
**COOLING SYSTEM**
check, 1-20
general information, 3-2
servicing (draining, flushing and refilling), 1-31
**COOLING, HEATING AND AIR CONDITIONING SYSTEMS, 3-1**
**COWL COVERS, REMOVAL AND INSTALLATION, 11-15**
**CRANKSHAFT BALANCER, REMOVAL AND INSTALLATION, V8 ENGINES, 2B-12**
**CRANKSHAFT FRONT OIL SEAL, REPLACEMENT**
V6 engine, 2A-12
V8 engines, 2B-12
**CRANKSHAFT POSITION (CKP) SENSOR, REPLACEMENT, 6-16**
**CRANKSHAFT, REMOVAL AND INSTALLATION, 2C-17**
**CRUISE CONTROL SYSTEM, GENERAL INFORMATION, 12-24**
**CYLINDER COMPRESSION CHECK, 2C-4**
**CYLINDER DEACTIVATION SYSTEM, DESCRIPTION AND COMPONENT REPLACEMENT, 6-28**
**CYLINDER HEADS, REMOVAL AND INSTALLATION**
V6 engine, 2A-9
V8 engines, 2B-10

# D

**DASHBOARD SWITCHES, REPLACEMENT, 12-9**
**DASHBOARD TRIM PANELS, REMOVAL AND INSTALLATION, 11-28**
**DAYTIME RUNNING LIGHTS (DRL), GENERAL INFORMATION, 12-25**
**DEFOGGER, REAR WINDOW, CHECK AND REPAIR, 12-22**
**DIAGNOSIS, 0-25**
**DIAGNOSTIC TROUBLE CODES (DTCS), ACCESSING, 6-2**
**DIFFERENTIAL**
carrier (front), removal and installation, 8-21
lubricant
    change, 1-36
    level check, 1-21
    type, 1-37
pinion oil seal, replacement, 8-10
**DISC BRAKE**
caliper, removal and installation, 9-7
disc, inspection, removal and installation, 9-7
pads, replacement, 9-4

**DOOR**
latch, lock cylinder and handles, removal and installation, 11-19
removal, installation and adjustment, 11-18
trim panels, removal and installation, 11-16
window glass regulator, removal and installation, 11-21
window glass, removal and installation, 11-20
**DRIVEAXLE (4WD MODELS)**
boot
    check, 1-32
    replacement, 8-13
general information and inspection, 8-12
removal and installation, 8-12
**DRIVEBELT**
check, adjustment and replacement, 1-17
tensioner, replacement, 1-18
**DRIVELINE, 8-1**
**DRIVEPLATE, REMOVAL AND INSTALLATION**
V6 engine, 2A-18
V8 engines, 2B-20
**DRIVESHAFT**
and universal joints, general information and inspection, 8-2
center support bearing, replacement, 8-4
general information and inspection, 8-2
removal and installation, 8-3
universal joints, replacement, 8-4
**DRUM BRAKE SHOES, REPLACEMENT**
2008 and earlier, 9-9
2009 and later models, 9-11

# E

**ELECTRIC SIDE VIEW MIRRORS, GENERAL INFORMATION, 12-23**
**ELECTRICAL TROUBLESHOOTING, GENERAL INFORMATION, 12-2**
**EMISSIONS AND ENGINE CONTROL SYSTEMS, 6-1**
**ENGINE**
oil life monitor, resetting, 1-14
oil type and viscosity, 1-37
**ENGINE AND EMISSIONS CONTROL SYSTEMS, GENERAL INFORMATION, 6-2**
**ENGINE COOLANT TEMPERATURE (ECT) SENSOR, REPLACEMENT, 6-17**
**ENGINE COOLANT, LEVEL CHECK, 1-8**
**ENGINE COOLING FANS, CHECK AND REPLACEMENT, 3-5**
**ENGINE ELECTRICAL SYSTEMS, 5-1**
**ENGINE MOUNTS, CHECK AND REPLACEMENT**
V6 engine, 2A-19
V8 engines, 2B-21
**ENGINE OIL**
and oil filter change, 1-12
level check, 1-7

# IND-4  MASTER INDEX

**ENGINE OVERHAUL**
disassembly sequence, 2C-10
reassembly sequence, 2C-20
**ENGINE REBUILDING ALTERNATIVES, 2C-7**
**ENGINE, REMOVAL AND INSTALLATION, 2C-8**
**ENGINE REMOVAL, METHODS AND PRECAUTIONS, 2C-7**
**ENGINE, GENERAL OVERHAUL PROCEDURES, 2C-1**
crankshaft, removal and installation, 2C-17
cylinder compression check, 2C-4
engine overhaul
    disassembly sequence, 2C-10
    reassembly sequence, 2C-20
engine rebuilding alternatives, 2C-7
engine removal, methods and precautions, 2C-7
engine, removal and installation, 2C-8
initial start-up and break-in after overhaul, 2C-20
oil pressure check, 2C-4
pistons and connecting rods, removal
    and installation, 2C-11
rear main oil seal retainer installation, 2C-19
vacuum gauge diagnostic checks, 2C-5
**ENGINE, IN-VEHICLE REPAIR PROCEDURES**
V6 engine, 2A-1
    camshaft and lifters, removal, inspection and installation, 2A-14
    crankshaft front oil seal, replacement, 2A-12
    cylinder heads, removal and installation, 2A-9
    driveplate, removal and installation, 2A-18
    engine mounts, check and replacement, 2A-19
    exhaust manifolds, removal and installation, 2A-8
    intake manifold, removal and installation, 2A-6
    oil pan, removal and installation, 2A-16
    oil pump, removal and installation, 2A-17
    rear main oil seal, replacement, 2A-18
    repair operations possible with the engine in
        the vehicle, 2A-2
    rocker arms and pushrods, removal, inspection
        and installation, 2A-4
    timing chain, removal and installation, 2A-13
    Top Dead Center (TDC) for number 1 piston, locating, 2A-2
    torque specifications, 2A-20
    valve covers, removal and installation, 2A-3
    valve springs, retainers and seals,
        replacement, 2A-5
    vibration damper, removal and installation, 2A-11
V8 engines, 2B-1
    camshaft and lifters, removal, inspection and installation, 2B-16
    crankshaft balancer, removal and installation, 2B-12
    crankshaft front oil seal, removal
        and installation, 2B-12
    cylinder heads, removal and installation, 2B-10
    driveplate, removal and installation, 2B-20
    engine mounts, check and replacement, 2B-21
    exhaust manifolds, removal and installation, 2B-9
    intake manifold, removal and installation, 2B-7
    oil pan, removal and installation, 2B-18
    oil pump, removal, inspection and installation, 2B-19
    rear main oil seal, replacement, 2B-21
    repair operations possible with the engine in
        the vehicle, 2B-2
    rocker arms and pushrods, removal, inspection
        and installation, 2B-4
    timing chain, removal, inspection
        and installation, 2B-13
    Top Dead Center (TDC) for number 1 piston, locating, 2B-2
    torque specifications, 2B-23
    valve covers, removal and installation, 2B-3
    valve springs, retainers and seals, replacement, 2B-5
    vibration damper, removal and installation, 2B-12
**EVAPORATIVE EMISSIONS CONTROL (EVAP) SYSTEM, DESCRIPTION AND COMPONENT REPLACEMENT, 6-25**
**EXHAUST MANIFOLDS, REMOVAL AND INSTALLATION**
V6 engine, 2A-8
V8 engines, 2B-9
**EXHAUST SYSTEM**
check, 1-26
servicing, general information, 4-20
**EXPANSION (ORIFICE) TUBE, AIR CONDITIONING, REPLACEMENT, 3-23**
**EXTENSION HOUSING OIL SEAL (2WD), REPLACEMENT, 7A-6**

# F

**FANS, ENGINE COOLING, CHECK AND REPLACEMENT, 3-5**
**FASTENERS, TRIM, REMOVAL, 11-2**
**FAULT FINDING, 0-25**
**FENDER, FRONT, REMOVAL AND INSTALLATION, 11-14**
**FILTER REPLACEMENT**
engine air, 1-27
engine oil, 1-12
**FIRING ORDER, 1-38**
**FLUID LEVEL CHECKS, 1-7**
automatic transmission, 1-12
battery electrolyte, 1-9
brake fluid, 1-9
engine coolant, 1-8
engine oil, 1-7
power steering, 1-11
windshield washer, 1-8
**FLUIDS AND LUBRICANTS**
capacities, 1-37
recommended, 1-37
**FRACTION/DECIMAL/MILLIMETER EQUIVALENTS, 0-23**
**FRONT AXLE SHIFT MOTOR (4WD MODELS), REPLACEMENT, 8-19**
**FRONT DIFFERENTIAL CARRIER, REMOVAL AND INSTALLATION, 8-21**
**FRONT END ALIGNMENT, GENERAL INFORMATION, 10-27**

# MASTER INDEX  IND-5

**FRONT HUB AND BEARING ASSEMBLY, REMOVAL AND INSTALLATION, 10-11**
**FUEL**
lines and fittings, general information, 4-4
meter body and injectors (V6 models), removal and installation, 4-16
pressure relief procedure, 4-2
pressure sensor, replacement, 4-11
pump flow control module, replacement, 4-11
pump/fuel level sending unit
    component replacement, 4-10
    removal and installation, 4-9
pump/fuel pressure, check, 4-2
rail and injectors (V8 models), removal and installation, 4-18
Sequential Fuel Injection (SFI) system
    general check, 4-14
    general information, 4-13
system check, 1-23
tank
    cleaning and repair, general information, 4-9
    removal and installation, 4-7
**FUEL AND EXHAUST SYSTEMS, 4-1**
**FUSES, GENERAL INFORMATION, 12-4**

## G

**GENERAL ENGINE OVERHAUL PROCEDURES, 2C-1**
crankshaft, removal and installation, 2C-17
cylinder compression check, 2C-4
engine overhaul
    disassembly sequence, 2C-10
    reassembly sequence, 2C-20
engine rebuilding alternatives, 2C-7
engine removal, methods and precautions, 2C-7
engine, removal and installation, 2C-8
initial start-up and break-in after overhaul, 2C-20
oil pressure check, 2C-4
pistons and connecting rods, removal and installation, 2C-11
rear main oil seal retainer installation, 2C-19
vacuum gauge diagnostic checks, 2C-5
**GRILLE, RADIATOR, REMOVAL AND INSTALLATION, 11-8**

## H

**HAZARD FLASHERS, GENERAL INFORMATION, 12-7**
**HEADLIGHT**
adjustment, 12-15
bulb, replacement, 12-16
housing, replacement, 12-14
**HEATER AND AIR CONDITIONING CONTROL ASSEMBLY, REMOVAL AND INSTALLATION, 3-13**
**HEATER CORE, REMOVAL AND INSTALLATION, 3-13**
**HEATING AND AIR CONDITIONING SYSTEM, CHECK AND MAINTENANCE, 3-16**
**HINGES AND LOCKS, MAINTENANCE, 11-8**
**HOOD LATCH AND RELEASE CABLE, REMOVAL AND INSTALLATION, 11-11**
**HOOD, REMOVAL, INSTALLATION AND ADJUSTMENT, 11-9**
**HORN, REPLACEMENT, 12-22**
**HUB AND BEARING ASSEMBLY, FRONT, REMOVAL AND INSTALLATION, 10-11**

## I

**IGNITION SWITCH AND KEY LOCK CYLINDER, REPLACEMENT, 12-8**
**IGNITION SYSTEM**
check, 5-9
coil pack, replacement, 5-9
general information, 5-8
**INITIAL START-UP AND BREAK-IN AFTER OVERHAUL, 2C-20**
**INJECTORS, FUEL, REMOVAL AND INSTALLATION**
V6 models, 4-16
V8 models, 4-18
**INSTRUMENT CLUSTER, REMOVAL AND INSTALLATION, 12-11**
**INSTRUMENT PANEL, SERVICE POSITIONING, 11-33**
**INTAKE AIR TEMPERATURE (IAT) SENSOR, REPLACEMENT, 6-18**
**INTAKE MANIFOLD, REMOVAL AND INSTALLATION**
V6 engine, 2A-6
V8 engines, 2B-7
**INTERMEDIATE SHAFT AND COUPLER, REMOVAL AND INSTALLATION, 10-19**
**INTRODUCTION, 0-5**

## J

**JACKING AND TOWING, 0-19**
**JUMP STARTING, 0-20**

## K

**KEY LOCK CYLINDER**
and ignition switch, replacement, 12-8
and latch, door, removal and installation, 11-19
**KEYLESS ENTRY REMOTE TRANSMITTER, BATTERY REPLACEMENT AND PROGRAMMING, 12-6**
**KNEE BOLSTER, REMOVAL AND INSTALLATION, 11-28**
**KNOCK SENSOR, REPLACEMENT, 6-18**

## L

**LEAF SPRING, REMOVAL AND INSTALLATION, 10-14**
**LIFTGATE**
glass, removal and installation, 11-24
removal and installation, 11-24
trim panels, latch and support struts, removal
 and installation, 11-25
**LOW PRESSURE SWITCH AND PRESSURE SENSOR, AIR CONDITIONING, REMOVAL AND INSTALLATION, 3-23**
**LOWER CONTROL ARM, REMOVAL AND INSTALLATION, 10-9**
**LUBRICANTS AND CHEMICALS, 0-21**
**LUBRICANTS AND FLUIDS**
capacities, 1-37
recommended, 1-37

## M

**MAINTENANCE**
routine, 1-1
schedule, 1-2
techniques, tools and working facilities, 0-11
**MANIFOLD ABSOLUTE PRESSURE (MAP) SENSOR, REPLACEMENT, 6-19**
**MASS AIR FLOW (MAF) SENSOR, REPLACEMENT, 6-20**
**MASTER CYLINDER, REMOVAL, INSTALLATION AND RESERVOIR/O-RING REPLACEMENT, 9-16**
**MIRRORS**
electric side view, general information, 12-23
removal and installation, 11-22
**MONITOR, ENGINE OIL LIFE, RESETTING, 1-14**
**MULTI-FUNCTION SWITCH, REPLACEMENT, 12-7**

## O

**OIL LIFE MONITOR, RESETTING, 1-14**
**OIL PAN, REMOVAL AND INSTALLATION**
V6 engine, 2A-16
V8 engines, 2B-18
**OIL PRESSURE CHECK, 2C-4**
**OIL PUMP, REMOVAL AND INSTALLATION**
V6 engine, 2A-17
V8 engines, 2B-19
**OIL, ENGINE**
level check, 1-7
type and viscosity, 1-37
**ON-BOARD DIAGNOSTIC (OBD) SYSTEM AND DIAGNOSTIC TROUBLE CODES (DTCS), 6-2**
**ORIFICE (EXPANSION) TUBE, AIR CONDITIONING, REPLACEMENT, 3-23**
**OXYGEN SENSORS, GENERAL INFORMATION AND REPLACEMENT, 6-20**

## P

**PADS, DISC BRAKE, REPLACEMENT, 9-4**
**PARK/LOCK SYSTEM, DESCRIPTION AND COMPONENT REPLACEMENT, 7A-5**
**PARKING BRAKE**
adjustment, 9-21
shoes, replacement, 9-21
**PARTS, REPLACEMENT, BUYING, 0-11**
**PINION OIL SEAL, REPLACEMENT, 8-10**
**PISTONS AND CONNECTING RODS, REMOVAL AND INSTALLATION, 2C-11**
**POSITIVE CRANKCASE VENTILATION (PCV) SYSTEM**
general description and check, 6-28
valve replacement (V6 engine), 1-36
**POWER BRAKE BOOSTER, CHECK, REMOVAL AND INSTALLATION, 9-20**
**POWER DOOR LOCK SYSTEM, GENERAL INFORMATION, 12-24**
**POWER STEERING**
fluid
 level check, 1-11
 type, 1-37
pump, removal and installation, 10-24
system, bleeding, 10-26
**POWER WINDOW SYSTEM, GENERAL INFORMATION, 12-24**
**POWERTRAIN CONTROL MODULE (PCM), REMOVAL AND INSTALLATION, 6-23**

## R

**RADIATOR**
and coolant expansion tank, removal and installation, 3-7
grille, removal and installation, 11-8
**RADIO AND SPEAKERS, REMOVAL AND INSTALLATION, 12-12**
**REAR HUB, WHEEL BEARING AND SEAL (FULL-FLOATING AXLE), REMOVAL, BEARING/SEAL REPLACEMENT AND INSTALLATION, 8-9**
**REAR MAIN OIL SEAL, REPLACEMENT**
V6 engine, 2A-18
V8 engines, 2B-21
**REAR MAIN OIL SEAL RETAINER INSTALLATION, 2C-19**
**REAR WINDOW DEFOGGER, CHECK AND REPAIR, 12-22**
**RECALL INFORMATION, 0-7**
**RECOMMENDED LUBRICANTS AND FLUIDS, 1-37**

**RELAYS, GENERAL INFORMATION AND TESTING, 12-5**
**REMOTE TRANSMITTER, KEYLESS ENTRY, BATTERY REPLACEMENT AND PROGRAMMING, 12-6**
**REPAIR OPERATIONS POSSIBLE WITH THE ENGINE IN THE VEHICLE**
V6 engine, 2A-2
V8 engines, 2B-2
**REPLACEMENT PARTS, BUYING, 0-10**
**RIGHT AXLESHAFT, TUBE, BEARING AND SHIFT FORK (4WD AND AWD MODELS), REMOVAL, COMPONENT REPLACEMENT AND INSTALLATION, 8-20**
**ROCKER ARMS AND PUSHRODS, REMOVAL, INSPECTION AND INSTALLATION**
V6 engine, 2A-4
V8 engines, 2B-4
**ROTATING THE TIRES, 1-21**
**ROTOR, BRAKE, INSPECTION, REMOVAL AND INSTALLATION, 9-7**
**ROUTINE MAINTENANCE, 1-1**
**ROUTINE MAINTENANCE SCHEDULE, 1-2**

# S

**SAFETY FIRST!, 0-24**
**SAFETY RECALL INFORMATION, 0-7**
**SCHEDULED MAINTENANCE, 1-1**
**SEAT BELT CHECK, 1-14**
**SEATS, REMOVAL AND INSTALLATION, 11-35**
**SEQUENTIAL FUEL INJECTION (SFI) SYSTEM**
general check, 4-14
general information, 4-13
**SHIFT**
cable, removal, installation and adjustment, 7A-3
lever, transfer case (manual-shift models), removal and installation, 7B-2
linkage, transfer case (manual-shift models), adjustment, 7B-2
motor, front axle (4WD models), replacement, 8-19
motor/actuator, transfer case (electric-shift models), replacement, 7B-3
**SHOCK ABSORBER, REMOVAL AND INSTALLATION**
front, 10-5
rear, 10-13
**SHOES, DRUM BRAKE, REPLACEMENT**
2008 and earlier models, 9-9
2009 and later models, 9-11
**SPARE TIRE, INSTALLING, 0-19**
**SPARK PLUG**
replacement, 1-28
torque, 1-39
type and gap, 1-38
wire check and replacement, 1-36

**SPEAKERS, REMOVAL AND INSTALLATION, 12-12**
**STABILITRAK, GENERAL INFORMATION, 9-2**
**STABILIZER BAR AND BUSHINGS, REMOVAL AND INSTALLATION**
front, 10-6
rear, 10-13
**STARTER MOTOR**
and circuit, check, 5-12
removal and installation, 5-13
**STARTING SYSTEM, GENERAL INFORMATION AND PRECAUTIONS, 5-11**
**STEERING**
column
covers, removal and installation, 11-33
switches, replacement, 12-5
removal and installation, 10-18
gear boots, replacement, 10-21
gear, removal and installation, 10-23
intermediate shaft and coupler, removal and installation, 10-19
knuckle, removal and installation, 10-12
linkage, inspection, removal and installation, 10-22
wheel, removal and installation, 10-16
**STOP LIGHT SWITCH, CHECK, ADJUSTMENT AND REPLACEMENT, 9-23**
**SUSPENSION AND STEERING SYSTEMS, GENERAL INFORMATION, 10-2**
**SUSPENSION ARMS, REAR, REMOVAL AND INSTALLATION, 10-15**
**SUSPENSION, STEERING AND DRIVEAXLE BOOT CHECK, 1-32**

# T

**TAILGATE**
latch and handle, removal and installation, 11-23
removal and installation, 11-22
**TENSIONER, DRIVEBELT, REPLACEMENT, 1-18**
**THERMOSTAT, CHECK AND REPLACEMENT, 3-3**
**THROTTLE BODY, INSPECTION, REMOVAL AND INSTALLATION, 4-15**
**TIE-ROD ENDS, REMOVAL AND INSTALLATION, 10-20**
**TIMING CHAIN, REMOVAL AND INSTALLATION**
V6 engine, 2A-13
V8 engines, 2B-13
**TIRE AND TIRE PRESSURE CHECKS, 1-9**
**TIRE ROTATION, 1-21**
**TIRE, SPARE, INSTALLING, 0-19**
**TOOLS AND WORKING FACILITIES, 0-11**
**TOP DEAD CENTER (TDC) FOR NUMBER 1 PISTON, LOCATING**
V6 engine, 2A-2
V8 engines, 2B-2

# IND-8 MASTER INDEX

**TORQUE SPECIFICATIONS**
cylinder head bolts
    V6 engine, 2A-20
    V8 engines, 2B-24
spark plugs, 1-39
thermostat housing bolts, 3-24
water pump bolts, 3-24
wheel lug nuts, 1-39
*Other torque specifications can be found in the Chapter that deals with the component being serviced*

**TORSION BAR, REMOVAL AND INSTALLATION, 10-6**
**TOWING, 0-19**
**TRACK BAR, REMOVAL AND INSTALLATION, 10-16**
**TRACTION CONTROL (TCS), GENERAL INFORMATION, 9-2**
**TRANSFER CASE (4WD MODELS), 7B-1**
control module (electric-shift models), replacement, 7B-3
control switch (electric-shift models), replacement, 7B-2
lubricant
    change, 1-35
    level check, 1-27
    type, 1-37
oil seal, replacement, 7B-4
overhaul, general information, 7B-5
removal and installation, 7B-4
shift
    lever (manual-shift models), removal and installation, 7B-2
    linkage (manual-shift models), adjustment, 7B-2
    motor/actuator (electric-shift models), replacement, 7B-3
speed sensors (electric-shift models), general information, 7B-3

**TRANSMISSION, AUTOMATIC, 7A-1**
diagnosis, general, 7A-2
extension housing oil seal (2WD), replacement, 7A-6
fluid
    and filter change, 1-34
    level check, 1-12
    type, 1-37
Park/Lock system, description and component replacement, 7A-5
removal and installation, 7A-8
shift cable, removal, installation and adjustment, 7A-3
transmission mount, check and replacement, 7A-7

**TRANSMISSION RANGE (TR) SENSOR, REPLACEMENT AND ADJUSTMENT, 6-22**
**TRIM FASTENERS, REMOVAL, 11-2**
**TRIM PANELS, REMOVAL AND INSTALLATION**
door, 11-16
liftgate, 11-25

**TROUBLE CODES, ACCESSING, 6-2**
**TROUBLESHOOTING, 0-25**
**TUNE-UP**
and routine maintenance, 1-1
general information, 1-6

**TURN SIGNAL/HAZARD FLASHER/MULTI-FUNCTION SWITCH, REPLACEMENT, 12-7**
**TURN SIGNALS AND HAZARD FLASHERS, GENERAL INFORMATION, 12-7**

## U

**UNDERHOOD HOSE CHECK AND REPLACEMENT, 1-19**
**UNIVERSAL JOINTS, REPLACEMENT, 8-4**
**UPHOLSTERY AND CARPETS, MAINTENANCE, 11-4**
**UPPER CONTROL ARM, REMOVAL AND INSTALLATION, 10-8**

## V

**V6 ENGINE, 2A-1**
camshaft and lifters, removal, inspection and installation, 2A-14
crankshaft front oil seal, replacement, 2A-12
cylinder heads, removal and installation, 2A-9
driveplate, removal and installation, 2A-18
engine mounts, check and replacement, 2A-19
exhaust manifolds, removal and installation, 2A-8
intake manifold, removal and installation, 2A-6
oil pan, removal and installation, 2A-16
oil pump, removal and installation, 2A-17
rear main oil seal, replacement, 2A-18
repair operations possible with the engine in the vehicle, 2A-2
rocker arms and pushrods, removal, inspection and installation, 2A-4
timing chain, removal and installation, 2A-13
Top Dead Center (TDC) for number 1 piston, locating, 2A-2
torque specifications, 2A-20
valve covers, removal and installation, 2A-3
valve springs, retainers and seals, replacement, 2A-5
vibration damper, removal and installation, 2A-11

**V8 ENGINES, 2B-1**
camshafts and lifters, removal, inspection and installation, 2B-16
crankshaft balancer, removal and installation, 2B-12
crankshaft front oil seal, removal and installation, 2B-12
cylinder heads, removal and installation, 2B-10
driveplate, removal and installation, 2B-20
engine mounts, check and replacement, 2B-21
exhaust manifolds, removal and installation, 2B-9
intake manifold, removal and installation, 2B-7
oil pan, removal and installation, 2B-18
oil pump, removal, inspection and installation, 2B-19
rear main oil seal, replacement, 2B-21
repair operations possible with the engine in the vehicle, 2B-2
rocker arms and pushrods, removal, inspection and installation, 2B-4
timing chain, removal, inspection and installation, 2B-13
Top Dead Center (TDC) for number 1 piston, locating, 2B-2

torque specifications, 2B-23
valve covers, removal and installation, 2B-3
valve springs, retainers and seals, replacement, 2B-5
vibration damper, removal and installation, 2B-12
**VACUUM GAUGE DIAGNOSTIC CHECKS, 2C-5**
**VALVE COVERS, REMOVAL AND INSTALLATION**
V6 engine, 2A-3
V8 engines, 2B-3
**VALVE SPRINGS, RETAINERS AND SEALS, REPLACEMENT**
V6 engine, 2A-5
V8 engines, 2B-5
**VEHICLE IDENTIFICATION NUMBERS, 0-6**
**VEHICLE SPEED SENSOR (VSS), REPLACEMENT, 6-23**
**VEHICLE STABILITY CONTROL (STABILITRAK), GENERAL INFORMATION, 9-2**
**VIBRATION DAMPER, REMOVAL AND INSTALLATION**
V6 engine, 2A-11
V8 engines, 2B-12
**VINYL TRIM, MAINTENANCE, 11-4**

# W

**WATER PUMP, CHECK AND REPLACEMENT, 3-8**
**WHEEL ALIGNMENT, GENERAL INFORMATION, 10-27**
**WHEEL BEARING**
front, removal and installation, 10-11
rear (full-floating axle), removal, bearing/seal replacement and installation, 8-9
**WHEEL CYLINDER, REMOVAL AND INSTALLATION, 9-15**
**WHEELS AND TIRES, GENERAL INFORMATION, 10-26**
**WHEELS STUDS, REPLACEMENT, 10-12**
**WINDOW GLASS, DOOR, REMOVAL AND INSTALLATION, 11-20**
**WINDOW GLASS REGULATOR, REMOVAL AND INSTALLATION, 11-21**
**WINDSHIELD AND FIXED GLASS, REPLACEMENT, 11-8**
**WINDSHIELD WASHER FLUID, LEVEL CHECK, 1-8**
**WINDSHIELD WIPER BLADE INSPECTION AND REPLACEMENT, 1-15**
**WIPER MOTORS, REPLACEMENT, 12-20**
**WIRING DIAGRAMS, GENERAL INFORMATION, 12-26**
**WORKING FACILITIES, 0-11**

# NOTES